# 9th International Conference on Photonics and Imaging in Biology and Medicine 2010



# (PIBM 2010)

**Journal of Physics: Conference Series Volume 277**

## Wuhan, China
## 2 - 5 November 2010

Editors:

**Qingming Luo**
**Valery V. Tuchin**
**Lihong V. Wang**

**ISBN: 978-1-61782-661-0**
**ISSN: 1742-6588**

**Printed from e-media with permission by:**

Curran Associates, Inc.
57 Morehouse Lane
Red Hook, NY 12571

**Some format issues inherent in the e-media version may also appear in this print version.**



Printed by Curran Associates, Inc. (2011)

For permission requests, please contact the Institute of Physics
at the address below.

Institute of Physics
Dirac House, Temple Back
Bristol BS1 6BE UK

Phone: 44 1 17 929 7481
Fax: 44 1 17 920 0979

techtracking@iop.org

**Additional copies of this publication are available from:**

Curran Associates, Inc.
57 Morehouse Lane
Red Hook, NY 12571 USA
Phone: 845-758-0400
Fax: 845-758-2634
Email: curran@proceedings.com
Web: www.proceedings.com

# TABLE OF CONTENTS

## PLENARY AND INVITED PAPERS:



## REGULAR PAPERS:

## PREFACE

The 9th International Conference on Photonics and Imaging in Biology and Medicine (PIBM 2010), combined with the 3rd Photonics and Optoelectronics Meetings (POEM 2010), was held from November 2–5, 2010, at Wuhan Science & Technology Convention & Exhibition Center, Wuhan, PR China. The present volume contains papers from a selection from the invited, oral, and poster presentations.

PIBM is the largest international biomedical photonics conference series in Asia. It was initially held at HUST bi-yearly from 1999. After being held three times in Wuhan (1999, 2001 and 2003), it was hosted once in Tianjin (2005), before returning to Wuhan every year since 2006. PIBM is designed to bring together scientists, engineers and clinical researchers from a variety of disciplines engaged in applying optical science, photonics and imaging technologies to problems in biology and medicine. The scope of this conference ranges from basic research to instrumentation engineering, and biological and clinical studies. It is recognized as one of the largest and most comprehensive international conferences in China, and represents the highest level of worldwide research in this field. In the past ten years, 7 volumes of proceedings with a total of 672 papers were published by SPIE (International Society for Optical Engineering), and a volume with 75 papers was published by World Scientific Publishing Co. in 2007. Proceedings of PIBM 1999, 2001, 2003, 2005, 2006, 2008 and 2009 were indexed by EI Compendex, while proceedings of PIBM 1999, 2001, 2003, 2005, 2006, and 2007 were indexed by SCI. Some excellent papers were recommended for publication in the peer-reviewed *Journal of Innovative Optical Health Sciences* (JIOHS). An increasing number of young researchers present and exchange their innovative ideas on this friendly and professional platform, which has made PIBM an unforgettable annual meeting in Wuhan.

This year PIBM attracted distinguished scholars in the field of biomedical photonics and imaging from all over the world, including the United States, Russia, Australia, Canada, Israel, France, Ireland, Japan, Korea and China. The major topics covered at the conference and presented in this volume include: *Photonic Therapeutics, Diagnostics and Instrumentations*; *Tissue Optics and Laser Tissue Interaction*; *Biomedical Spectroscopy and Microscopy*; *Multimodal and Hybrid Biomedical Imaging*; and *Optical Molecular Imaging*. The conference voted for the three best student papers; awards were presented to the participant students whose posters were recognized as excellent and who took part in the oral presentation competition. The conference received 133 submitted abstracts, and this volume of the *Journal of Physics: Conference Series* includes a selection of 53 excellent submissions.

The Conference Secretariat and Local Organizing Committee deserve recognition for planning a smoothly run and productive conference with comprehensive, instructive lectures and innovative work displayed in poster presentations. The faculties and students from Britton Chance Center for Biomedical Photonics were dedicated to their work in reception and service during the conference. It is a pleasure to thank all of them for their efficient and hard work. We are also grateful for the financial support from 111 Project

(B07038), and the assistance in organization and coordination from Wuhan National Laboratory for Optoelectronics and Huazhong University of Science and Technology.

Finally, we would like to thank all the authors for their contributions to PIBM 2010 and all the members of the Committees for their cooperation and time spent reviewing submissions. Special thanks are due to the Advisory Committee members Shu Chien, Aaron Ciechanover, Steve Dahms, Da Hsuan Feng, Steven R Goodman, Brian Salzberg, Fujia Yang, Jianquan Yao, Baoyong Zheng and Olivia Ho Cheng for their participation on-site, and their significant contributions to the conference.

Wuhan, PR China
December, 2010

*Qingming Luo*
*Lihong V Wang*
*Valery V Tuchin*
Conference Chairs

**9th International Conference on Photonics and Imaging in Biology and Medicine (PIBM 2010)**
2–5 November 2010
Wuhan, China

**Editors**

*Qingming Luo*, Huazhong University of Science and Technology (China)
*Lihong V Wang* , Washington University in St. Louis (USA)
*Valery V Tuchin*, Saratov State University (Russia)

**Sponsored and Organized by**

Huazhong University of Science and Technology (China)
Wuhan National Laboratory for Optoelectronics (China)
Britton Chance Center for Biomedical Photonics (China)

**Technical Co-sponsored by**

IBOS-International Biomedical Optics Society
The Chinese Optical Society
The Biophysical Society of China

**Co-organized by**

Key Laboratory of Biomedical Photonics, Ministry of Education (China)
Virtual Research Center of Biomedical Photonics, Ministry of Education (China)
Hubei Bioinformatics and Molecular Imaging Key Laboratory (China)

## CONFERENCE COMMITTEES

### Honorary Chairs

*Britton Chance*, University of Pennsylvania (USA)
*Bingkun Zhou*, Tsinghua University (China)

### Conference Chairs

*Qingming Luo*, Huazhong University of Science and Technology (China)
*Lihong V Wang* , Washington University in St. Louis (USA)
*Valery V Tuchin*, Saratov State University (Russia)

### Advisory Committee

*Sydney Brenner*, The Salk Institute in La Jolla, California (USA)
*Howard Chen*, K&L Gates (USA)
*Jing Cheng*, Tsinghua University (China)
*Shu Chien*, University of California, San Diego (USA)
*Paul Ching-Wu Chu*, University of Houston (USA)
*Aaron Ciechanover*, Technion-Israel Institute of Technology, Haifa (Israel)
*A Stephen Dahms*, Alfred E Mann Foundation for Biomedical Engineering (USA)
*Da Hsuan Feng*, National Cheng Kung University (Taiwan, China)
*Steven R Goodman*, SUNY Upstate Medical University (USA)
*Barry Halliwell*, National University of Singapore (Singapore)
*John Hart*, The University of Texas at Dallas (USA)
*George Radda*, Agency for Science, Technology and Research (A*STAR) (Singapore)
*Zihe Rao*, Nankai University (China)
*Brian M Salzberg*, University of Pennsylvania (USA)
*Ruey-Jen Sung*, Stanford University (USA)
*A Dean Sherry*, The University of Texas at Dallas (USA)
*Bruce Tromberg*, University of California/Irvine (USA)
*Fujia Yang*, Nottingham University (UK)
*Jianquan Yao*, Tianjin University (China)
*Yixin Zeng*, Sun Yat-sen University Cancer Center (China)
*Baoyong Zheng*, Hua Wei Technologies Corporation, Inc (China)

### Program Committee

*Wei R Chen*, University of Central Oklahoma (USA)
*Zhongping Chen*, University of California/Irvine (USA)
*Arthur Chiou*, National Yang-Ming University (Taiwan, China)
*Frank Y S Chuang*, University of California, Davis (USA)
*Zhihua Ding*, Zhejiang University (China)
*Congwu Du*, Brookhaven National Laboratory (USA)
*Stefan Haacke*, Strasbourg University – IPCMS-DON (France)
*Weiping Han*, Agency for Science, Technology and Research (A*STAR) (Singapore)
*Zheng Huang*, University of Colorado Health Sciences Center (USA)
*Zhiwei Huang*, National University of Singapore (Singapore)
*Steven L Jacques*, Oregon Health & Science University (USA)

*Fu-Jen Kao*, National Yang-Ming University (Taiwan, China)
*Hideaki Koizumi*, Hitachi, Ltd (Japan)
*Xingde Li*, Johns Hopkins University (USA)
*Yong-qing Li*, East Carolina University (USA)
*Chengyi Liu*, South China Normal University (China)
*Hong Liu*, University of Oklahoma (USA)
*Zuhong Lu*, Southeast University (China)
*Dennis L Matthews*, University of California/Davis (USA)
*Avraham Mayevsky*, Bar Ilan University (Israel)
*Stephen P Morgan*, University of Nottingham (UK)
*Shoko Nioka*, University of Pennsylvania (USA)
*Yingtian Pan*, State University of New York at Stony Brook (USA)
*Alexander V Priezzhev*, MV Lomonosov Moscow State University (Russia)
*Jianan Y Qu*, The Hongkong University of Science and Technology (Hong Kong, China)
*Colin J R Sheppard*, National University of Singapore (Singapore)
*Mamoru Tamura*, Tsinghua University (China)
*Sergey Ulyanov*, Saratov State University (Russia)
*Ruikang K Wang*, Oregon Health & Science University (USA)
*Xunbin Wei*, Fudan University (China)
*Da Xing*, South China Normal University (China)
*Haishan Zeng*, BC Cancer Research Centre (Canada)
*Gang Zheng*, University of Toronto (Canada)
*Dongping Zhong*, The Ohio State University (USA)

**Organizing Committee**
*Ling Fu* (Chair), Huazhong University of Science and Technology (China)
*Yuandi Zhao* (Chair), Huazhong University of Science and Technology (China)
*Hui Gong*, Huazhong University of Science and Technology (China)
*Pengcheng Li*, Huazhong University of Science and Technology (China)
*Bifeng Liu*, Huazhong University of Science and Technology (China)
*Qian Liu*, Huazhong University of Science and Technology (China)
*Shaoqun Zeng*, Huazhong University of Science and Technology (China)
*Zhihong Zhang*, Huazhong University of Science and Technology (China)
*Dan Zhu*, Huazhong University of Science and Technology (China)

**Local Secretariat**
*Hua Shi*, Huazhong University of Science and Technology (China)

# Co-publication with *Journal of Physics: Conference Series* volume 276

This is a co-publication with *Journal of Physics: Conference Series* volume 276, the 3rd International Photonics & OptoElectronics Meetings (POEM 2010). These conferences were held together from November 2–5, 2010, at Wuhan Science & Technology Convention & Exhibition Center, Wuhan, PR China.

**PEER REVIEW**

All papers published in this volume of *Journal of Physics: Conference Series* have been peer reviewed through processes administered by the proceedings Editors. Reviews were conducted by expert referees to the professional and scientific standards expected of a proceedings journal published by IOP Publishing.

# Mechanotransduction in Endothelial Cells Studied with Fluorescence Imaging

Shu Chien
Departments of Bioengineering and Medicine and Institute of Engineering in Medicine, University of California, San Diego, La Jolla, California 92093-0427

## ABSTRACT

Mechanotransduction involves the conversion of mechanical stimuli to intracellular signaling to modulate gene and protein expressions and hence cellular functions in endothelial cells, thus playing importance roles in the regulation of homeostasis in health and disease. The aim of this paper is to investigate the dynamics of mechanotransduction in endothelial cells by the use of fluorescent resonance energy transfer (FRET) to study the temporal and spatial activation of Src kinase and focal adhesion kinase, both of which play critical roles in many cellular processes. The results have contributed to the elucidation of the roles of these two important signaling molecules and their interactions in mediating mechanotransduction.

## INTRODUCTION

Mechanical stresses resulting from blood flow and pressure can modulate cellular functions by activating sequentially mechano-sensors, signaling pathways, and gene and protein expressions (1). Most studies on these signaling processes have been conducted by using biochemical or molecular procedures to study a large number of cells subjected to experimental conditions. With the increasing recognition of spatial and temporal variations in mechanotransduction, it is necessary to have technologies that would enable investigations on single cells and subcellular components in 3D space and with time-dependent dynamics. To this end, fluorescent resonance energy transfer (FRET) has been used to study the temporal and spatial activation of Src kinase and focal adhesion kinase (FAK), which are known to play critical roles in many cellular processes, including motility/migration, mechanotransduction, and disease states such as cancer. We developed genetically encoded single-molecule Src biosensor (2) and FAK biosensor (3), which enable the imaging and quantification of temporal-spatial activation of Src and FAK activities in live cells in response to mechanical and chemical stimuli. These biosensors have enabled monitoring key signaling transduction cascades and mechanobiology in live cells with temporal-spatial characterization. We have also developed FRET technology to monitor FAK and Src activities at lipid rafts and non-raft regions of the plasma membrane in live cells (4), thus elucidating the dynamics and interplay of these molecules in mediating mechanotransduction.

## THE SRC AND FAK FRET BIOSENSORS

### *Design of the Src and FAK FRET Biosensors*

We designed FRET-based biosensors for Src (2) and FAK (3). The Src biosensor is composed of a Src SH2 domain, a flexible linker peptide, and a Src substrate peptide encompassing Tyr 662 and 664, which were originally concatenated between CFP and YFP, but in its current version between s ECFP and YPet to improve the sensitivity and range of the biosensor (4). The FAK biosensor specifically detects the autophosphorylation of Tyr397 and its subsequent binding to the Src SH2 domain, which are the crucial molecular events in the FAK activation and represent the FAK activity. The FAK biosensor contains the SH2 domain derived from c-Src, a flexible linker peptide, and a specific substrate sequence from FAK encompassing Tyr397, which are concatenated between ECFP and YPet. The active FAK can phosphorylate the Tyr397 in the substrate peptide of the biosensor (*trans*-activation), and the phosphorylated substrate can then bind to the intramolecular SH2 domain to cause an increase in the ECFP/YPet emission ratio, which represents an enhanced FAK activity.

Incubation of the purified FAK biosensor with active FAK led to an increase of the ECFP/YPet emission ratio, reflecting the phosphorylation of the biosensor by FAK. Src kinase, which caused significant FRET changes of the Src biosensor, had no effect on FRET change of the FAK biosensor, indicating the specificity of the FAK biosensor.

*Characterization of the FAK Biosensor in Mammalian Cells*

We examined this ECFP/YPet FAK biosensor in mouth embryonic fibroblasts (MEF). The ECFP/YPet emission ratios of the FAK biosensor were significantly higher in MEF cells adhered on fibronectin than those in suspension (Fig. 2A), consistent with the knowledge that the integrin-mediated cell adhesion process can induce FAK activation. In FAK knockout (FAK-/-) MEFs, transfection with the constitutively active FAK (FAK Δ1-375) and wild-type FAK (FAK WT) caused significantly higher ECFP/YPet emission ratios than the negative mutant of FAK (FAK NT), the kinase-dead FAK (FAK KD), or an empty vector. The expression of wild-type Src (Src WT) did not cause a significant FRET change of the FAK biosensor in FAK-/- cells, indicating that the FAK biosensor can detect FAK activity with a high specificity in mammalian cells.

The FRET response in FAK-/- cells expressing the active FAK (FAK Δ1-375) was eliminated after the mutation of Tyr397 in the substrate peptide to Phe or the mutation of Arg175 in the SH2 domain to Val, indicating that the FRET response of the FAK biosensor upon FAK activation is attributable to the intramolecular interaction between the phosphorylated substrate (Tyr397) and the SH2 domain (Arg175).

## FAK ACTIVITIES AT DIFFERENT MICRODOMAINS OF THE PLASMA MEMBRANE

FAK plays important roles in regulating the subcellular localization of lipid rafts and the associated signaling events. A Lyn-FAK biosensor was constructed by fusing a raft-targeting motif derived from the Lyn kinase at the N-terminus of the cytosolic FAK (Cyto-FAK) biosensor to monitor the FAK activity at membrane rafts. As a control, a non-raft targeting sequence from KRas was fused at the C-terminus of the

Cyto-FAK biosensor for the generation of KRas-FAK biosensor.

## FAK ACTIVATION BY PDGF

*The PDGF-induced FAK Activation at Lipid rafts*

PDGF caused significant FRET change of the Lyn-FAK biosensor, but not the KRas-FAK biosensor, indicating that the phosphorylation at Tyr 397 is responsible for the PDGF-induced FRET change of the Lyn-FAK biosensor, which targets the rafts. The FAK inhibitor PF228 reversed the PDGF-induced FRET response of Lyn-FAK biosensor. The PDGF-induced FRET signal of the Lyn-FAK biosensor was suppressed after cholesterol extraction and raft disruption by methyl-β-cyclodextrin (MβCD), verifying the localization and functionality of the biosensor at rafts.

*The PDGF-induced FAK Activation Is Dependent on Src*

The Src kinase inhibitor PP1 completely blocked the PDGF-induced FRET response of the Lyn-FAK biosensor, indicating that the FAK activation at rafts upon PDGF stimulation is dependent on Src.

To test whether FAK also regulates Src activation in response to PDGF, we examined the effect of the FAK inhibitor PF228 on FRET response of the ECFP/YPet Lyn-Src biosensor targeted at rafts. PF228 partially inhibited, but did not prevent, the PDGF-induced FRET response of the Lyn-Src biosensor, suggesting that FAK activity is not essential for the PDGF-induced Src activation.

## FAK ACTIVATION BY CELL ADHESION

*FAK Activity Is Essential for the Src Activation during Cell Adhesion*

In contrast to the PDGF-induced FAK activation, the adhesion-induced FAK activation was not inhibited by blockage of Src activity with PP1. The inhibition of FAK kinase activity by PF228 completely blocked the adhesion-induced Src activation. Therefore, in contrast to the PDGF-induced signaling, the adhesion-mediated FAK activation acts upstream to Src.

## SUMMARY AND CONCLUSIONS

The FAK biosensors that can target the lipid rafts and non-raft regions of the plasma membrane have allowed the elucidation of the subcellular regulation of FAK activities. The FAK activity at rafts was found to be significantly higher than that at non-raft regions in response to either PDGF stimulation or integrin-mediated adhesion. These newly developed biosensors also allowed the deciphering of the hierarchical interactions between FAK and Src in response to different modes of stimulation. In response to PDGF, Src activation is upstream to FAK. In contrast, in response to adhesion, FAK activation is upstream to Src.

## ACKNOWLEDGMENTS

The author is indebted to Dr. Yingxiao Wang and his colleagues at the University of Illinois, Urbana-Champaign (Jihye Seong, Mingxing Ouyang, Po-Chao Wen, Shaoying Lu, and Nicholas M. Llewellyn) for their work on the FAK biosensors targeting the rafts and non-raft regions.

**REFERENCES**

1. Chien, S.: Mechanotransduction and endothelial cell homeostasis: the wisdom of the cell. Am. J. Physiology: Heart and Circulatory Physiology, 292:H1209-1224, 2007.
2. Wang, Y.; Botvinick, E.L.; Zhao, Y.; Berns, M.; Usami, S.; Tsien, R.Y., and Chien, S.: Visualizing the mechanical activation of Src. Nature, 434:1040-1045, 2005.
3. Seong, J.; Ouyang, M.; Wen, P.-C.; Lu, S.; Llewellyn, N.M.: Schlaepfer, D.D.; Guan, J.L.; Chien, S., and Wang, Y.: FRET Detection of FAK Signaling at Lipid Rafts. To be published.
4. Ouyang, M.; Sun, J.; Chien, S., and Wang, Y.:_Determination of hierarchical relationship of Src and Rac at subcellular locations with FRET biosensors. Proc Natl Acad Sci U S A. 105(38):14353-14358, 2008.

# A novel instrument aimed at measuring hypertrophic scar formation

**J. C. Ramella-Roman[1] T. A. Nguyen[1], A.R. Pavlovich[2], P. Lemaillet[1], M.H. Jordan[2], J. W. Shupp[2]**

[1]Department of Biomedical Engineering, The Catholic University of America, USA
[2]The Burn Center, Department of Surgery, Washington Hospital Center-MedStar Health Research Institute, DC

Production Editor, *Journal of Physics: Conference Series*,
IOP Publishing, Dirac House, Temple Back, Bristol BS1 6BE, UK

E-mail: ramella@cua.edu

Hypertrophic scaring is a condition that can occur after thermal trauma and can lead to loss of function and psyco-social complications. Assessment of the development of a hypertrophic scar currently relies on visual inspection and grading. In this paper we propose a system based on polarized illumination and collection capable of capturing images of scars and quantify their roughness. Models of rough surface roughness are ultimately used to quantify the level of roughness of scarred skin compared to normal skin.

## 1. Introduction

Hypertrophic scaring is still a poorly understood and devastating morbidity following injury, particularly thermal trauma with prevalence reported as high as 67% [1]. Scar contractures can lead to significant reduction in function and inhibit patients from returning to work, participating in leisure activities and even render them unable to provide care for themselves. In addition these patients suffer great psycho-social setbacks due to their disfigured appearance [2]. There is a continued need for better assessment techniques to diagnose and treat patients with scar. This is underscored by the fact that greater than 30% of the respondents to an American Burn Association Rehabilitation committee's survey on research needs in burn care indicate that scar management should be a priority [3].

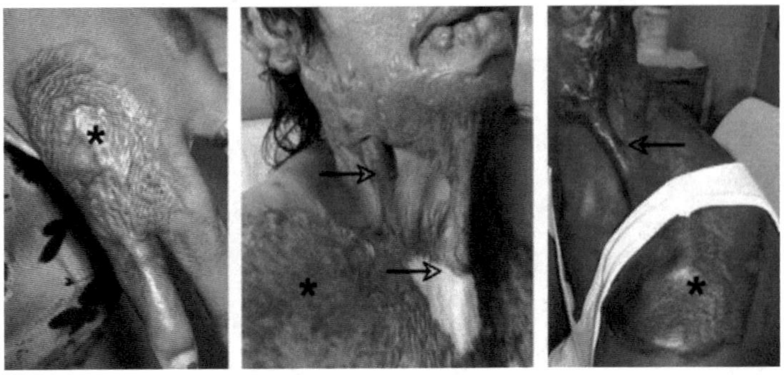

Fig.1 Examples of patients with hypertrophic scar from burn injury. Arrows denote large high-tension scar bands that are inhibiting neck movement. Asterisks indicate areas with raised scar.

The assessment of scar is limited by the lack of accurate objective measurement tools. The Vancouver Scar Scale (VSS), currently the accepted clinical scar assessment tool, was introduced [4, 5] in the nineties and relies on the physician subjective evaluation of skin pliability, height, vascularity, and pigmentation, in a rating of 0 to 4. Recently several groups have proposed improvements to the scale, both with more pertinent metrics [6, 7], as well as the introduction of novel instrumentation. [8,9, 10,11,12].

Further work is needed in a controlled model to assess the natural history of scar. The research effort described here proposes a new instrument aimed at obtaining quantitative parameters of hypertrophic scar formation based on skin roughness and optical properties.

**2. Theory**

Polarimetric techniques are extremely sensitive to roughness [13]. Assessment of skin roughness including wrinkles has been successfully performed on skin using spectro-polarimetric techniques [14,15]. Jacques *et al.* used polarized light imaging to determine the margins of certain skin cancers relying on the contrast provided by a cancer induced disruption of the underlying collagen matrix [16,17]. Jacques *et al.* also showed examples of scar tissue exhibiting lower degree of polarization compared to normal tissue possibly due to randomized collagen restructuring [18].

Studies on rough surface scattering [14] on non-biological samples have shown that polarized light can be used to discriminate different scattering mechanisms. The rough-surface-scattering component can be highlighted by changing the direction of the light source, as well as the polarization of the illuminator and collector. Figure 1 shows the geometry for out-of-plane measurements, where $\theta_i$ is the incident polar angle, $\theta_s$ is the scattering polar angle, and $\phi_s$ is the azimuth, or out-of-plane, angle.

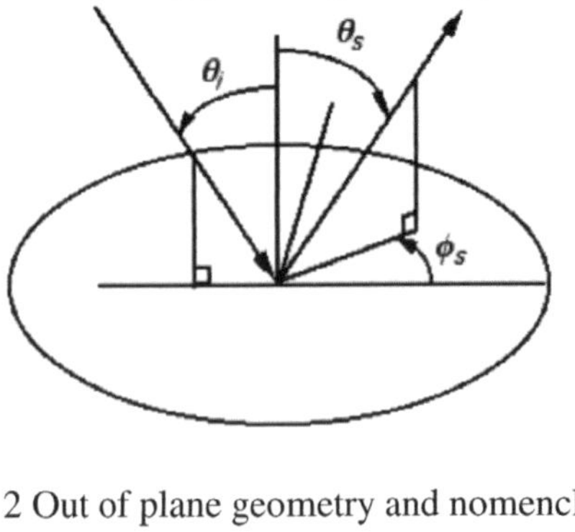


Fig. 2 Out of plane geometry and nomenclature

For a given scattering geometry, we measure the Stokes vector intensity ***S*** of the back-reflected light, modulating the detector polarizing elements. The Stokes vector can be divided into its polarized $I_{pol}$ and unpolarized components $I_{unpol}$

$$\mathbf{S} = [S_o \quad S_1 \quad S_2 \quad S_3]^{\mathrm{T}} = I_{\mathrm{unpol}} [1 \quad 0 \quad 0 \quad 0]^{\mathrm{T}} + I_{\mathrm{pol}} \left[1 \quad S_1/S_{\mathrm{pol}} \quad S_2/S_{\mathrm{pol}} \quad S_3/S_{\mathrm{pol}}\right]^{\mathrm{T}} \qquad 1$$

$$S_{\mathrm{pol}} = \left(S_1^2 + S_2^2 + S_3^2\right)^{1/2} \qquad 2$$

We also characterize the polarization in terms of the principal angle of the polarization ellipse, *η=arctangent(S2/S1)/2*. All these parameters are exquisitely sensitive to rough surface scattering as shown below.

We simulate the effect of different surface roughness on the polarized portion of the remitted beam for different incident angles using the MIST program [19]. Shown in Fig. 3 is the result as the root mean square slope (RMS) of a skin sample is allowed to change between 0 and 1 in steps of 0.2 (RMS slope for skin has been found to lie between 0.2 and 0.4 [20]. The deviation from the normal (Fig.3 on the right) will have a strong polarized signature.

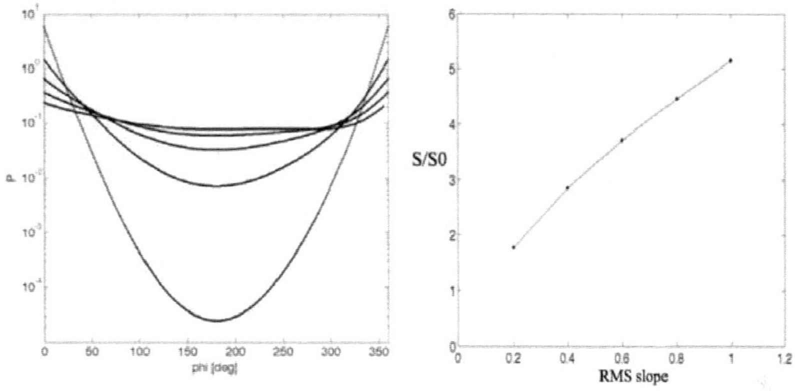


Fig. 3 The polarized portion of the remitted beam *P* (Eq.2) as the RMS slope varies between 0 and 1. Clear changes in polarized signature are visible. On the right hand side the sum of the polarized response at every angle (S) divided by the initial the polarized response (S0).

## 3. Material and Methods

In previous studies we developed a polarimetric system with no moving parts that acquires Stokes vector images as functions of illumination direction, hence allowing for the calculation of the previously described parameters [14]. The system consists of sixteen illumination tubes directed towards a sample position. Each illumination tube, has a tri-color light emitting diode (LED), a polarizer, and a lens and provides collimated and linearly polarized light centered at wavelengths of 472 nm, 525 nm, and 633 nm. Each polarizer is aligned so that the incident illumination is linearly polarized with its electric field 45° from the plane of incidence. One illuminator has a polar angle of $\theta_i = 0°$, six have a polar angle $\theta_i = 24°$, and nine have a polar angle $\theta_i = 49°$. The system was modified to include a 12-bit digital charge-coupled device (CCD) capable of up to 60 frames per second acquisition. The camera views the sample with a polar angle of $\theta_s = 49°$. Two liquid crystal variable retarders (LC1 and LC2) followed by a polarizer (POL) modulate the scattered light before reaching the camera. The rotation of the retarders and the polarizer remain fixed, while the retardances are chosen to span the Poincaré sphere and enable Stokes vector measurements.

The system was calibrated with a well-characterized rough gold standard; a facet model was able to describe the modulation of $\eta$ with an error << 1%.

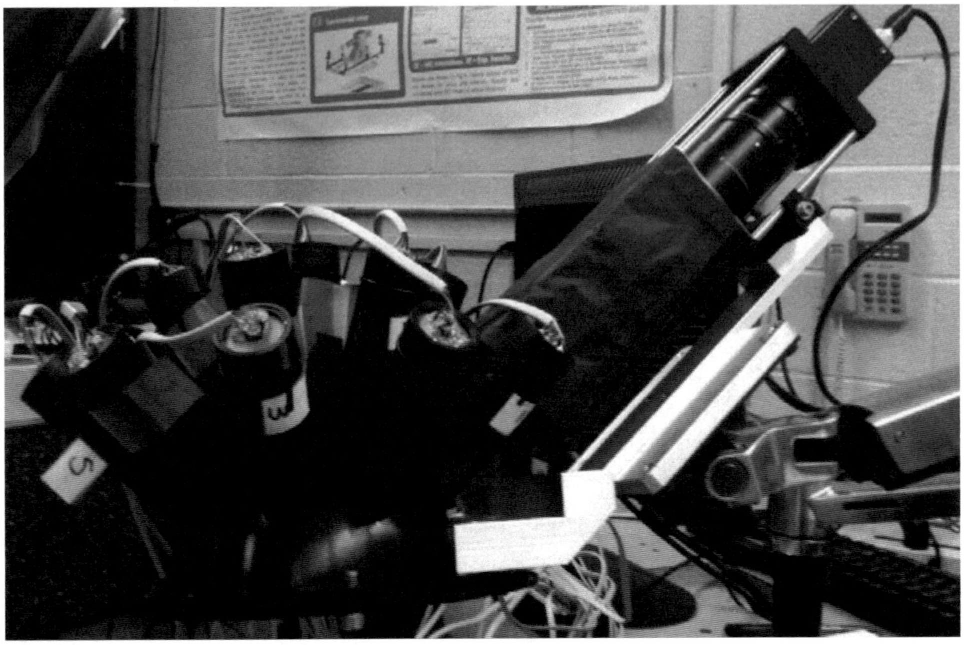

Fig. 4 Experimental layout for imaging Stokes vector polarimeter

**4. Results**

Images of a foot thermal scar of a 26 years old volunteer where obtained with the current system all illumination angles where used and the Stokes vector calculated for each position. From the Stokes vector we calculated the degree of linear polarization (PLin) the degree of circular polarization (PCirc) the polarized and unpolarized portion of the remitted light (S0Pol and S0unpol respectively) and the principle angle of polarization $\eta$ for each pixel in the images. Figures 5 and 6 show two such sets for different incident angles. As shown below the roughness of the scar stands out compared to the normal skin background, and the polarized response is strongly influenced by the illumination angle.

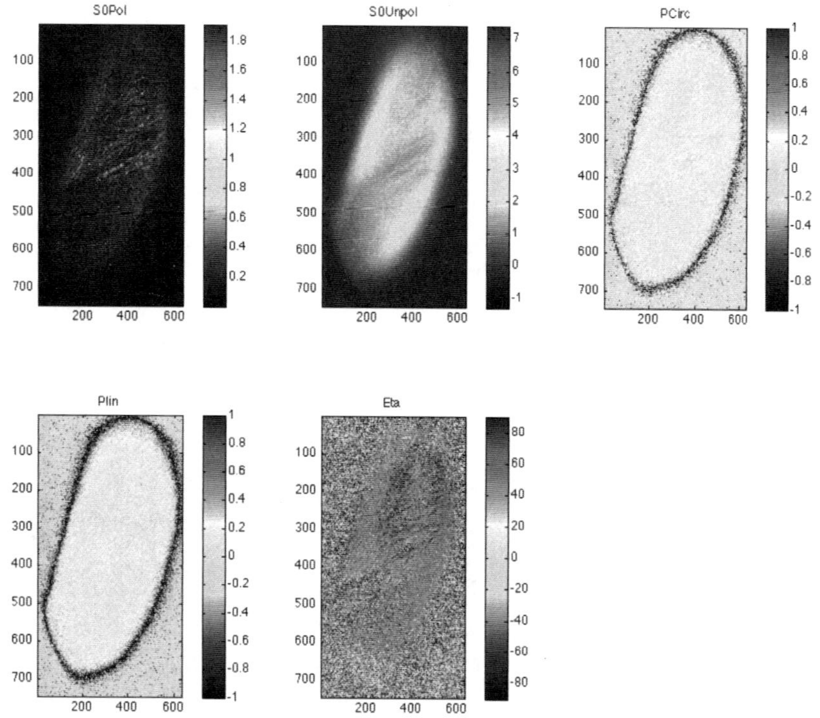


Fig. 5 Images obtained for a azimuth angle $\phi$=180 degrees

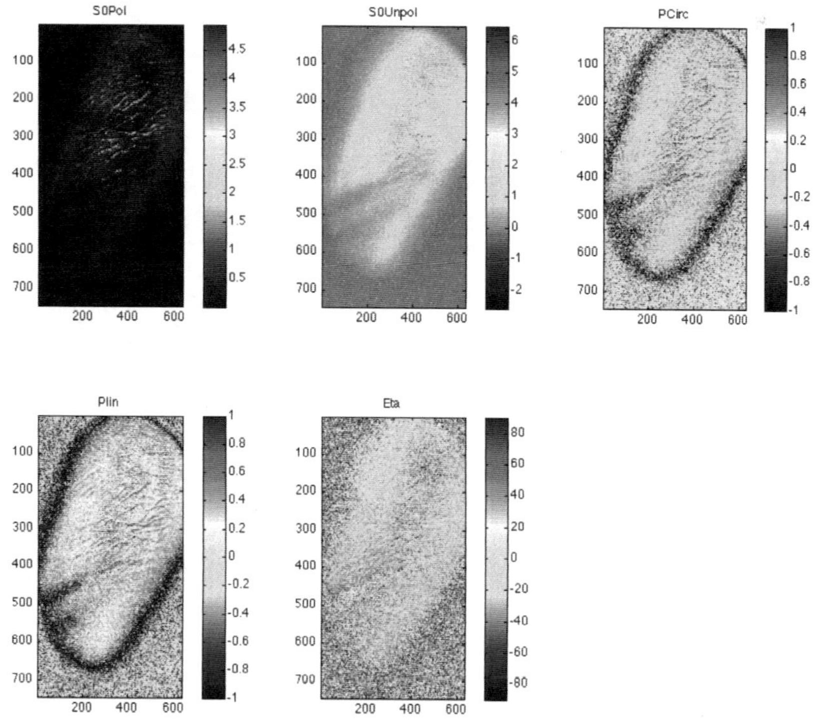


Fig. 6 Images obtained for a azimuth angle $\phi$=36 degrees

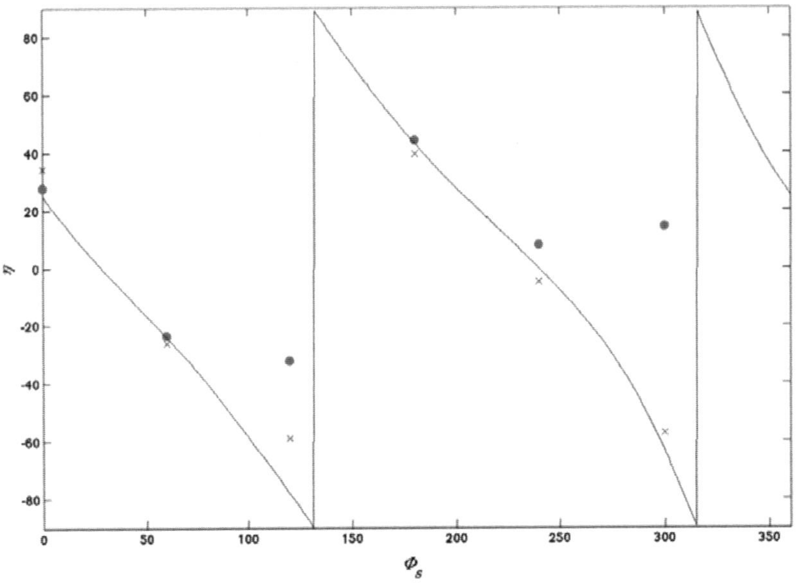


Fig. 7 Principle angle of polarization for different incident illumination azimuthal angles, $\theta$= 24 degrees.

Finally a region of interest of (100 pixels x 100 pixels) was selected on the visible portion of the scar, the resulting principle angle of polarization was plotted versus the incident angle, Fig. 7 (black crosses) symbols. A facet model was used to simulate the polarized behaviour of the scar (solid line). The same process was repeated for a portion of the intact skin (red filled symbols). A clear difference in polarized remittance can be seen.

**Conclusions**

The characterization of hypertrophic scars is currently conducted with visual inspection. New tools mostly based on spectral remittance are becoming more common. We have introduced an imaging system aimed at characterizing scars using their roughness. The advantage of this technique is in its sensitivity, so that not only rough scarred skin can be measured but also normal skin, (less rough). The technique has also the advantage of producing quantifiable metrics hence allowing for more effective studies on scar treatment and reduction.

**References**


1. Bombaro, K. M., Engrav, L. H., Carrougher, G. J., Wiechman, S. A., Faucher, L., Costa, B. A., Heimbach, D. M., Rivara, F. P., and Honari, S. What is the prevalence of hypertrophic scarring following burns? Burns 29: 299-302, 2003.
2. Robert, R., Meyer, W., Bishop, S., Rosenberg, L., Murphy, L., and Blakeney, P. Disfiguring burn scars and adolescent self-esteem Burns 25: 581-585, 1999.
3. Richard, R., Baryza, M. J., Carr, J. A., Dewey, W. S., Dougherty, M. E., Forbes-Duchart, L., Franzen, B. J., Healey, T., Lester, M. E., Li, S. K., Moore, M., Nakamura, D., Nedelec, B., Niszczak, J., Parry, I. S., Quick, C. D., Serghiou, M., Ward, R. S., Ware, L., and Young, A. Burn rehabilitation and research: proceedings of a consensus summit J Burn Care Res 30: 543-573, 2009.
4. Sullivan, T., Smith, J., Kermode, J., McIver, E., and Courtemanche, D. J. Rating the burn scar J Burn Care Rehabil 11: 256–60, 1990.

5. Powers, P. S., Sarkar, S., Goldgof, D. B., Cruse, C. W., and Tsap, L. V. Scar assessment: current problems and future solutions J Burn Care Rehabil 20: 54–60, 1999.
6. Yeong, E. K., Mann, R., and Engrav, L. H. Improved burn scar assessment with use of a new scar-rating scale J Burn Care Rehabil 18: 353–355, 1997.
7. Nedelec, B., Shankowsky, H. A., Tredget, E. E. Rating the resolving hypertrophic scar: comparison of the Vancouver Scar Scale and scar volume J Burn Care Rehabil 21: 205–212, 2000.
8. Boyce, S. T., Supp, A. P., Wickett, R. R., and Warden, G. D. Biophysical instrumentation for non-invasive assessment of skin wounds J Burn Care Rehabil 18: S198, 1997.
9. Boyce, S. T., Supp, A. P., Wickett, R. R., Hoath, S. B., and Warden, G. D. Assessment with the dermal torque meter of skin pliability after treatment of burns with cultured skin substitutes J Burn Care Rehabil 21: 55–63, 2000.
10. Barel, A. O., Courage, W., and Clarys, P. Suction method for measurement of skin mechanical properties: the cutometer. In: Serup J, Jemec GBE, editors Handbook of noninvasive methods and the skin, Boca Raton (FL): CRC Press Inc. 329–34,1995.
11. Fong, S. S. L., Hung, L. K., and Cheng, J. C. Y. The cutometer and ultrasonography in the assessment of postburn hypertrophic scar: a preliminary study Burns 23: S12–8,1997.
12. Spann, K., Mileski, W. J., Atiles, L., Purdue, G., and Hunt, J. The 1996 Clinical Research Award. Use of a pneumatonometer in burn scar assessment J Burn Care Rehabil 17: 515–7, 1996.
13. Germer, T. A. Polarized light scattering by microroughness and small defects in dielectric layers J. Opt. Soc. Am. A 18: 1279–1288, 2001.
14. Ramella-Roman, J. C., Boulbry, B., and Germer, T. A. A novel hemispherical spectro-polarimetric scattering instrument for skin lesions imaging Proceeding of the SPIE, Photonic Therapeutics and Diagnostics II, N. Kollias, Eds., 6078, 2006.
15. Bargo, P. R., and Kollias, N. Measurement of skin texture through polarization imaging British Journal of Dermatology 724-731, 2010
16. Jacques, S. L., Ramella-Roman, J. C., and Lee, K. Imaging skin pathology with polarized light J. of Biomedical Optics 7: 329-340, 2002.
17. Ramella-Roman, J. C., Lee, K., Prahl, S. A., and Jacques, S. L. Design, testing and clinical studies of a hand-held polarized light camera J. of Biomedical Optics 9: 1305-1310, 2004.
18. Jacques, S. L., Ramella-Roman, J. C., and Lee, K. Imaging Superficial Tissues with Polarized Light Lasers Surg. Med 26: 119-129, 2000.
19. Germer, T. A. Modeled Integrated Scatter Tool (MIST) http://physics.nist.gov/scatmech
20. Moller, A. Rendering Techniques, Tomas Akenine-Moller (Editor), 2006.

**Acknowledgments**
The authors wish to acknowledge the help of Thomas Germer of NIST for clarifications on the use of MIST program.

# Stimulated bioluminescence by fluid shear stress associated with pipe flow

**CAO Jing[1,2], WANG Jiang-an [1], WU Ronghua[1]**

1. Col. of Electronic Eng.,Naval University of Engineering, Wuhan 430033, China
2. Dept. of Weaponry Eng., Naval University of Engineering, Wuhan 430033, China

E-mail: caojing981@126.com

**Abstract:** Dinoflagellate can be stimulated bioluminescence by hydrodynamic agitation. Two typical dinoflagellate (*Lingulodinium polyedrum* and *Pyrocystis noctiluca* ) was choosed to research stimulated bioluminescence. The bioluminescence intensity and shear stress intensity were measured using fully developed pipe flow. There is shear stress threshold to agitate organism bioluminescence. From these experiment, the response thresholds of the stimulated bioluminscence always occurred in laminar flows at a shear stress level of 0.6-3 dyn/cm$^2$. At the same time, the spectral characteristc of dinoflagellate was recorded, the wavelength of them is about 470nm, and the full width at half maximum is approximate 30nm.

## 1. Introduction

Bioluminescence is a ubiquitous and conspicuous nighttime phenomena in the sea waters. Bioluminescent plankton can be stimulated by flow fields associated with ship wakes[1], breaking waves[2], swimming dolphins[3], small fish, divers, and swells moving past submerged objects. Marine bioluminescent organisma occur at all accessible depths and domains, oftern in immense numbers [4]. And the typical organisms responsible for this displays are primarily dinoflagellates, unicellular marine plankton which are the most abundant sources of bioluminescence in near-surface, coastal waters. Because they can be cultured, dinoflagellates are suitable for laboratory study using well-characterized flow fields. It has been suggested that concentrations of dinoflagellates as low as 100 cells per liter are sufficient for bioluminescence at night to highlight moving objects[5]. Therefore, the research of stimulated bioluminescence can have many application, such as, flow visualization technology, object detection and identification in water at night in the future.

It is a common experience that luminous marine organisms can be mechanically stimulated to produce bioluminescence. Ship's wakes and breaking waves are often common source of observable bioluminescence. However, in terms of quantifiable hydrodynamic parameters, the conditions related to bioluminescent excitation have been poorly characterized. It was thought that investigating the response of bioluminescence in a well described flow field would provide some quantitative measure of the attendant state of mechanical agitation. Pipe flow was chosen as it is one of the most studied and best hydrodynamically characterized over a wide range of flow conditions[1]. A series of experiments were performed in the laboratory where the bioluminescence of seawater flowing through a glass pipe was measured at different flow rate. To detection stimulated bioluminescence of ocean organisms in the future, the spectral properties of 2 species of the dinoflagellates were also researched.

## 2. Pipe flow field stimulated system and theory

In order to compare the response of bioluminescence in different quantitative flow field, and to determine the threshold values of wall shear stress which are necessary to stimulated bioluminescences, the study of bioluminescence in pipe flow was employed. Pipe flow was chosen to stimulate bioluminescence because (1) it is well characterized by hydrodynamic method, (2) wall shear stress $\tau$ can be easily calculated by the pressure drop, (3) it offers a wide range of laminar and turbulent stimula, (4) and new organisms are constantly entering the flow field. In fully developed laminar flow there is a parabolic velocity profile which results in a linear gradient of shear. The highest magnitude of shear is at the wall of the pipe; shear decreases to zero at the centerline.

### 2.1. The theory of fully developed pipe flow

Pipe flow is considered fully developed when its velocity profile is no longer evolving downstream. With no mean acceleration, the pressure drop, dP/dx, across the face of any concentric, cylindrical control volume of radius r must be balanced by the shear stress acting on its perimeter. This relationship can be expressed as

$$\tau(r) = -\frac{dp}{dx}\frac{r}{2} \tag{1}$$

with at the pipe wall, R, becoms

$$\tau_{wall} = -\frac{dp}{dx}\frac{R}{2} \tag{2}$$

then the relatioship of the shear stress in flow field and the wall shear stress can be reduced to $\tau(r) = \tau_{wall}\frac{r}{R}$. The formula expresses a general law of the linear stress distribution, regardless of whether the flow is laminar or tubulent. In order to determine if the pipe flowwas fully developed, laminar, turbulent, or transitional, the Daecy friction factor[1]:

$$\lambda = -\frac{dp}{dx}\frac{D}{\frac{1}{2}\rho U_{avg}^{2}} \tag{3}$$

and the Reynolds number is defined as,

$$\mathrm{Re} = \rho\frac{U_{avg}D}{\mu} \tag{4}$$

where $U_{avg}$ is the average flow speed, $\rho$ is the fluid density, $\mu$ is the dynamical viscosity, $D$ is the pipe diameter.

When the pipe flow is in the laminar flow, the sole source of shear stress in laminar flow is due to viscosity. The viscous stress developed in laminar flows of most common fluids are found to be equal to the product of the kinematic viscosity and the local velocity gradient. Fluids adhering to the linearship between shear stress and velocity gradient are referred to as Newtonian. In fully developed laminar pipe flow of a Newtonian fluid the following relationship holds:

$$\tau(r)_{la\min ar} = \mu\frac{du}{dr} \tag{5}$$

where $u(r)$ is the velocity along the pipe axis, measured at radius $r$ from the pipe centerline. In fully developed laminar pipe flow, the theoretical relation[6] between $\lambda$ and Re is

$$\lambda=64/\mathrm{Re} \tag{6}$$

which can also be expressed, via equations (2), (3), and (6), as

$$\tau_{wall} = \frac{8\rho U_{avg}^{2}}{\mathrm{Re}} \tag{7}$$

According to equation (5), equation (7) can be described as

$$\tau_{wall} = \frac{8\mu}{D} U_{avg} \propto U_{avg} \tag{8}$$

When the pipe flow is in the turbulent flow, the empirical relation by Blasius[5] is

$$\lambda=0.3164/Re^{1/4}$$

The same way as before, the wall shear stress in tubulent can be expressed as

$$\tau_{wall}=0.04\rho U_{avg}^{2}/Re^{1/4}=0.04(\mu\rho^{3}/D)^{1/4}U_{avg}^{7/4}$$

From these equations, it is important to realize that $\tau_{wall}$ increase with the flow speed in both laminar and turbulent flow. Therefore, in the experiment, the wall shear stress can be changed with the pipe flow speed.

## 2.2. Bioluminescence experimental apparatus

### *2.2.1. Flow-field measurments*

Cultures were exposed to fully developed pipe flow as showed in Figuture 1. The experimental apparatus consisted of a 50 liter transparent-plastic aquarium, arranged to discharge seawater vertically downward into a gently contraction, then through a 6.5mm internal diameter clear polycarbonate glass pipe 1 meter in length. The contraction was carefully designed using a cubic equation to give the smallest boundary layer thickness at the entrance to the test pipe. Volume flow rate, measured at the discharge with a rising-ball flow meter, was divided by the pipe area to obtain the average water velocity. Flow field through the pipe was manually controlled by a valve. The mean velocity ($U_{avg}$) of each steady flow was determined by weighing the amount of water collected over a measured time and dividing by the pipe area.The pipe was equipped with pressure taps, the pressure drop along the pipe was measured with a variable reluctance differential transducer The relationship between pressure drop and flow rate was used to characterize if the flow was fully developed and laminar, turbulent, or transitional.

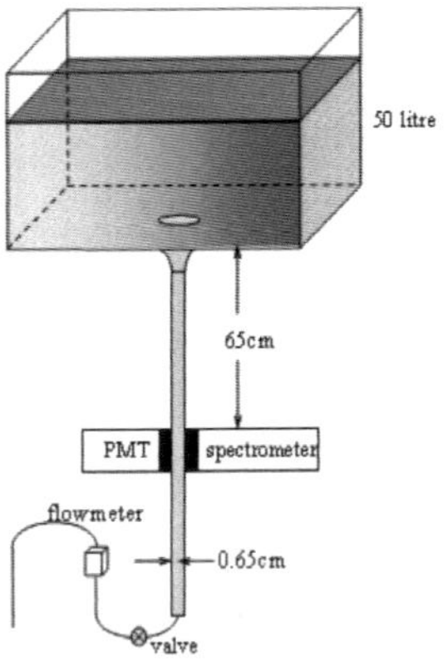


**Figure 1.** Schematic diagram of Pipe flow.

a. *Lingulodinium polyedrum*

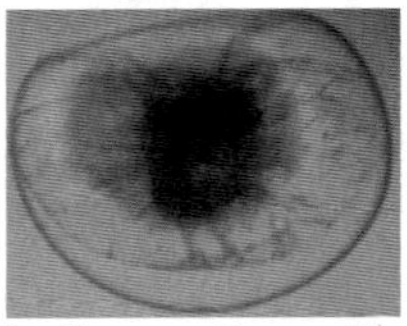

b. *Pyrocystis noctiluca*

**Figure 2.** Dinoflagellate culture in microscope.

### *2.2.2. Experimental organism*

*Lingulodinium polyedrum* and *Pyrocystis noctiluca* are the common oceanic species of dinoflagellate. They were grown in f/2 medium minus silicate and maintained in a culture chamber at 20±2°C on 12:12 light:dark cycle. Day illumination was provided by cool-white fluorescent tubes at approximatly 10w/m$^2$. Subjective day always stared at 17:00 and subjective night at 05:00. Experiments were always conducted between 07:30 and 11:30, so the cells were in the middle of the dark phase. Cell abundance was determined by counting samples under an optic microscope. The images of live cells of the bioluminescent dinoflagellates are displayed in Figture 2. Cultures used for the experiments were inoculated every 4 weeks. And the concentrations which was determined by counting samples under a microscope used for the experiments were 28 -500 cell/ml or as otherwise stated. At the beginning of the darkphase, the pipe flow apparatus fulled with organisms was covered with opaque sheeting. When the organism bioluminescence is maximal, the test commenced in the middle dark phase.

*2.2.3. The bioluminescence measurements*

Bioluminescence was measured by photomultiplier and its property was analysised by spectrometer. A photomultiplier tube located 65cm (100 pipe diameters) from the pipe inlet was used in photon count mode to measure the light flux emitted by the dinoflagellates. The photon number was integrated over a chosen period of 100ms. The sensitive area diameter of PMT is 25mm, and the dark current is equal to less than 100 counts/s at 25°C. The PMT is sensitive enough to detect signle flashes. The exported pulses of PMT was counted by a counter controlled by a computer through software. At the same time, the optical properties of bioluminescence were recorded by spectrometer.

## 3. The experimental data and analysis

3.1. The relationship between bioluminescence and wall shear stress

According to hydrodynamics the relation between Darcy friction factor and Reynolds number indicates whether flow is laminar, turbulent, fully developed, or transitional. Calculations of mean velocity and wall shear stress confirmed that the cells in fully developed flow when the measurements of bioluminescence were made (Figuture 3). As showed in figuture, the empirical data is consistent with the theoretical value. So the measure method of flow field is feasible.

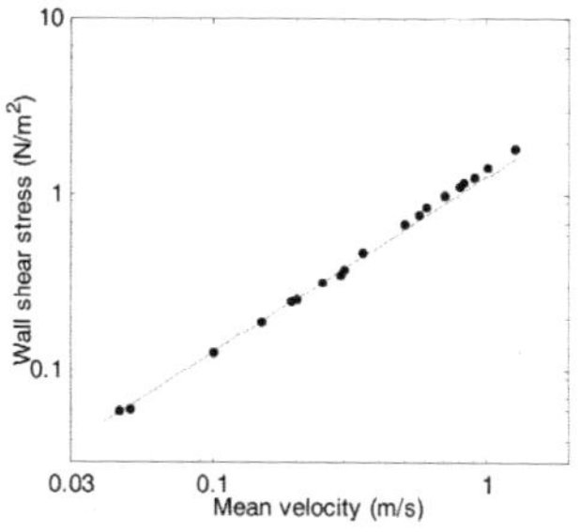


**Figure 3.** Relationship between wall shear stress and mean velocity for pipe flow.The solid line is the theoretical relationship for laminar flow, symbols represent empirical values.

When cultures of dinoflagellate were subjected to wall shear stress of pipe flow, the response pattern was characterized by an excitation threshold as showed in figure 4. Culture response varied with flow conditions. Flows with a wall shear stress below or near the threshold value, luminescence was similar to background levels. As the flow rate and thus the wall sshear stress increased, so did the

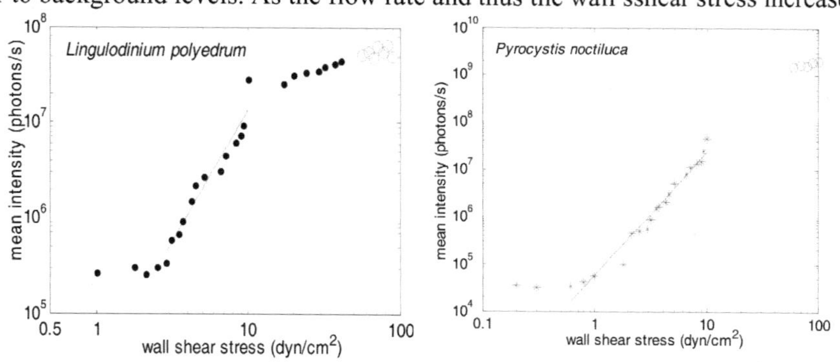


**Figure 4.** Mean bioluminescence intensity as a functon of wall shear stress.

rate of flashes and bioluminescence intensity of individual flashes. From the figture, the threshold of the *L. polyedrum* is presetented about 3 dyn/cm$^2$, and more than that of the *P. noctiluca* about 0.6 dyn/cm2. The line respresents the least-square power law regression between wall shear stress values and mean tintensity. For L. polyedrum, the stresses within the range of 1-10dyn/cm2 light emission was proportional to the2.8 power of shear stress(R2=0.94). And to P. noctiluca the slope of line was about 2.6 ($R^2$=0.97).

3.2. The spectral properties of bioluminescence

Measurement of bioluminescence spectral properties are complicated by the frequently dim and transient nature of marine organisms luminescence. The dinoflagellate spectra are confinde to the range of 460-490nm as showned in figture 5. In this experiment, the maximum wavelength of *L. polyedrum* is approximate 474nm, full width at half maximum (FWHM) is about 30nm; and for *P. noctiluca*, $\lambda_{max}$=471nm, FWHM=32nm.

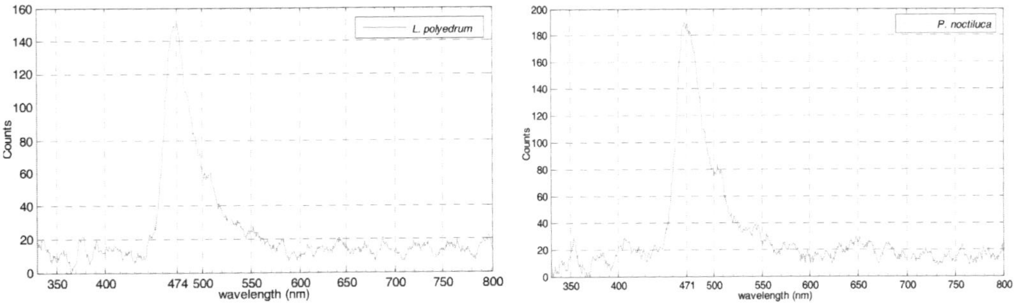


**Figure 5.** Tthe spectral characteristics of dinoflagellate. The cell concentration of *L. polyedrum* is 463 cells $ml^{-1}$; that of the *P. noctiluca* is about 28 cells $ml^{-1}$.

## 4. Conclusions

According to previous reporter, bioluminescence intensity will depend on the speicies abundance and concentration of luminescent organisms present, as well as their agitation, light and nutrient prehistory[3].The present experiment shows that pipe flow as the source to stimulate organism bioluminescence is feasible. And the study also present that dinoflagellate such as *L. polyedrum*, *P. noctiluca* was stimulated bioluminescence when they responsed to well-characterized hydrodynamic forces. The response of dinoflagellate to hydrodynamic stimulation was characterized by wall shear stress, and at the same values of wall shear stress, the response was similar for laminar and turbulent flows.

The thresholds of wall shear stress that stimulated bioluminescence occurred depending on dinoflagellate species. To some extent, the stronger of the wall shear stress became, the more mean bioluminescence intensity produced by dinoflagellate. On the basis of these properties, the study of ship wake bioluminescence and flow visualization technology can be developed, furthermore these technology can have many applications, for examples, controlling and guiding torpedoes.

## References

[1] Rohr J, Schoonmaker J, Losee J, Latz M I and Hyman M 1999 Flow visualization in the ocean-implications of laboratory bioluminescence experiments *OCEANS '99 MTS/IEEE. Riding the Crest into the 21st Century* **1** 145

[2] Stokes M D, Deane G B, Latz M I, Rohr J. 2004 Bioluminescence imaging of wave-induced turbulence *J.l Geophys. Res.* **109** C01004

[3] Rohr J, Latz M I, Fallon S, Nauen J C and Hendricks E 1998 Experimental approaches towards interpreting dolphin-stimulated bioluminscence *J. Exp. Bio.* **201** 1447

[4] Herring P J 2002 *The biology of the deep ocean* Oxford University Press

[5] MORIN J G 1983 Coastal bioluminescence: patterns and functions. *Bull. mar. Sci.* **33** 787

[6] Schlichting H 1979 *Boundary-layer theory* McGraw-Hill

# Optimal discrimination threshold for the detection of singlet oxygen luminescence

**Defu Chen, Huiyun Lin, Yi Shen, Buhong Li*, Shusen Xie**

Key Laboratory of OptoElectronic Science and Technology for Medicine of Ministry of Education, Fujian Provincial Key Laboratory for Photonics Technology, Fujian Normal University, Fujian 350007, China

Corresponding author: bhli@fjnu.edu.cn

**Abstract**. Direct detection of singlet oxygen ($^1O_2$) luminescence around 1270 nm is the golden standard of $^1O_2$ identification. In this study, the effect of the discrimination threshold on the detection of $^1O_2$ luminescence that generated from the photoirradiation of Rose Bengal (RB) was evaluated by using a self-developed photon-counting detection system. The obtained results show that the discrimination threshold for photon counting has a significant impact on the intensity and shape of the measured $^1O_2$ luminescence, which resulted in the variation of $^1O_2$ lifetimes. The optimal discrimination threshold is determined to be about -0.0412 V, and the corresponding $^1O_2$ lifetime in air-saturated distilled water is 4.26±0.06 μs.

## 1. Introduction

Photodynamic therapy (PDT) is a rapidly developing treatment modality that has been successfully used in the treatment of oncologic and non-oncological diseases [1,2]. Robust individualized dosimetry for PDT is vital for achieving a precise and effective PDT treatment [3,4]. Singlet oxygen ($^1O_2$) is widely known to be the major mediator for the destruction of target tissue during PDT [2-4]. Therefore, the quantitative measurement of $^1O_2$ during PDT treatment is clinically meaningful for PDT dosimetry. Most recently, with the advances in photon counting techniques and high sensitive photomultiplier tube (PMT) for near-infrared luminescence detection, the direct detection of $^1O_2$ near-infrared luminescence around 1270 nm has attracted growing attention for PDT dosimetry [5,6]. The advantage of this photophysical technique for PDT dosimetry is that it can circumvent the complicated interactions between the photosensitizer, light and molecular oxygen during PDT treatment.

The most challenging work for $^1O_2$ luminescence detection is to achieve a satisfied signal to noise ratio (SNR) and to obtain the convincing data. In particular, the measured $^1O_2$ lifetimes in aqueous environments have been previously reported in the range from 3.0 to 4.4 μs [7,8], and there seems still to be so much uncertainty among these values. As we know, the measurement of $^1O_2$ luminescence was mainly affected by the excitation light source, the optical collected unit, the photon counting technique, and the data processing approach. In this study, the effect of the discrimination threshold for photon counter on the detection of $^1O_2$ luminescence that generated from the photoirradiation of Rose Bengal (RB) was evaluated by using our self-developed photon-counting detection system [9].

## 2. Materials and methods

2.1. Theoretical model

Based on the well-establish photophysical model for determination of the $^1O_2$ rise and decay times from its luminescence, a simplified four-parameter photo physical model was proposed for fitting the measured $^1O_2$ luminescence data [9]:

$$I(t) = A(\frac{\tau_D}{\tau_T - \tau_D})[\exp(\frac{-t}{\tau_T}) - \exp(\frac{-t}{\tau_D})] + B \quad (1)$$

, where A and B are the intensities of $^1O_2$ luminescence and background signal of the detection system, respectively. $\tau_T$ and $\tau_D$ are the lifetimes of the triplet-state photosensitizer and $^1O_2$.

2.2. Chemicals

Rose Bengal (RB) (Sigma-Aldrich., St. Louis, USA) was selected as a model photosensitizer for $^1O_2$ generation. A 100 μM stock solution of RB was made up in distilled water and stored at 4 °C in the dark. The fresh air-saturated samples were prepared from this stock solution prior to the measurements. Preparations were done in near dark conditions.

2.3. Experiment setup

The detection system used to measure time-resolved $^1O_2$ luminescence is illustrated schematically in Figure 1. The present system has been described in detail elsewhere [9], with the exception that the 6-position motorized filter wheel has been replaced by a narrow-band filters centered at 1270 nm (OD3 blocking, 20-nm full-width at half-maximum (FWHM); Omega Optical, Brattleboro, USA) and the optical collection geometry has been re-optimized, which has a much larger numerical aperture for luminescence collection.

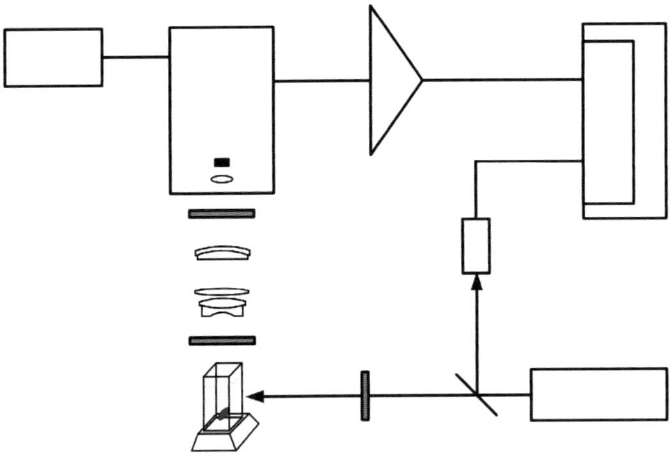

Figure 1. Schematic of the $^1O_2$ luminescence detection system

Briefly, a diode-pumped, Q-switched, frequency-doubled Nd:YLF laser (QG-523-500; Crystalaser Inc., Reno, NV, USA) with a emission wavelength of 523 nm was used for excitation. The pulse repetition rate is set to 12 kHz while average output power of pulse was adjusted to 20 mW by using a neutral-density attenuator. The 2.5 mL samples were held in a standard 10 mm pathlength quartz cuvette mounted on a hotplate-stirrer unit. This allowed the samples to be continuously stirred with a magnetic stir bar and maintained the source-sample-detector geometry constant between experiments. The cuvette was open so that the solutions were exposed to room air at the top. The exciting luminescence was collected through a 1000 nm long-pass filter (Omega Optical, Brattleboro, USA), the optical collected unit and a narrow-band filter that placed in front of the PMT (H10330-45, Hamamatsu Corp., Japan). The working voltage of the PMT was set to -900 V, which is provided by a high voltage module controller. The luminescence counts were recorded by a fast photon counter

((MSA-300, Becker & Hickl GmbH, Berlin, Germany). In order to achieve a sufficient SNR, the $^1O_2$ luminescence counts were summed over 285000 laser pulses for the following measurements.

2.4. Discrimination threshold for $^1O_2$ measurements
The MSA-300 integrated a discriminator, which has an adjustable pulse-height discrimination threshold from -0.5000 to 0.5000 V. In our detection system, the pulses from the pre-amplifier were negative pulses. The discriminator compares the input pulses with the preset discrimination threshold to divide them into two groups: one group is lower and the other is higher than the preset discrimination threshold. The higher pulses are shaped to have a constant level to be counted, while the lower pulses will be eliminated during counting [10].

2.5. Data processing
All the measured data were processed and graphed using OriginPro8.0 software (OriginLab, Northampton, MA, USA). Data are means plus-minus the standard deviation (SD) of three independent experiments each performed in duplicate.

**3. Results and discussion**

3.1. Time-resolved $^1O_2$ luminescence spectra
The representative time-resolved $^1O_2$ luminescence spectra of RB were shown in Figure 2, which were measured with the discrimination threshold ($V_{th}$) of -0.1800 and -0.2000 V, respectively. As indicated in Figure 2, the discrimination threshold has significant impact on both the intensity and the shape of $^1O_2$ luminescence.

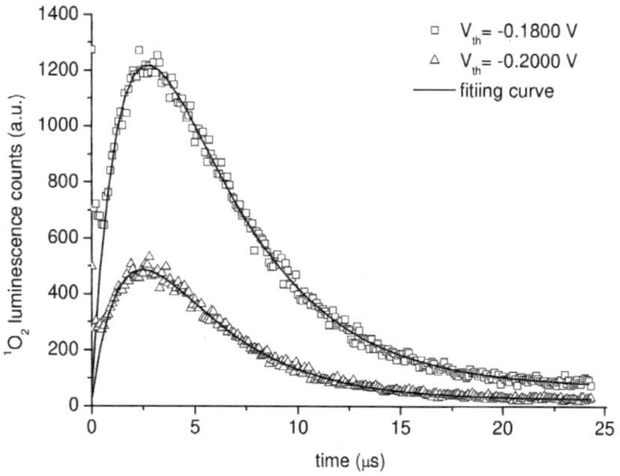


Figure 2. Time-resolved $^1O_2$ luminescence spectra of 6 μM RB.

When the discrimination threshold was set to -0.2510 V, the intensity of the $^1O_2$ luminescence was extremely low and this value just allowing counter to count the $^1O_2$ luminescence. In the following measurements, the time-resolved $^1O_2$ luminescence spectra of 6 μM RB were measured, in which the discrimination threshold was continually increased from -0.2510 to -0.0059 V in *ca.* 0.010 V step. The $^1O_2$ lifetimes can be derived from the fitting curves by using the equation (1), as shown in Figure 3. There has a turning point for the $^1O_2$ lifetime with the discrimination threshold value of -0.1800 V, as marked by the red arrow. It can be clearly seen that $^1O_2$ lifetimes have great variability when the discrimination threshold is lower than -0.1800 V, while the $^1O_2$ lifetimes are most same once the discrimination threshold is higher than -0.1800 V.

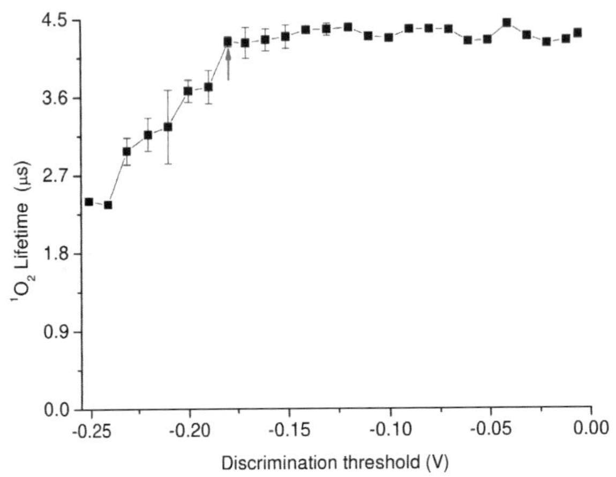


Figure 3. The $^1O_2$ lifetimes versus the discrimination threshold for $^1O_2$ luminescence measurements.

3.2. Optimization of the discrimination threshold

In order to determine the optimal discrimination threshold, the time-resolved $^1O_2$ luminescence spectra of 6 μM RB were measured with the discrimination thresholds that was initially set from -0.1800 V and then continually increased from -0.1800 V to -0.0059 V in *ca.* 0.01 V step. According to the time-resolved characteristics of the $^1O_2$ luminescence spectra in Figure 2, the background counts can be clearly found in the early time points (*ca.* 0~1 μs) and beyond *ca.* 20 μs. Therefore, we can define the integration signal counts from 1 to 20 μs in the time-resolved spectra with and without irradiation as "signal and background" and "background", respectively. Figure 4 shows the integration counts of the "signal and background" and "background" with different discrimination thresholds for each measurement. The optimal discrimination threshold can be determined to be about -0.0412 V from the gentle slope portion, which is indicated by red arrow in Figure 4. In this case, the unwanted background counts can be maximally suppressed [10].

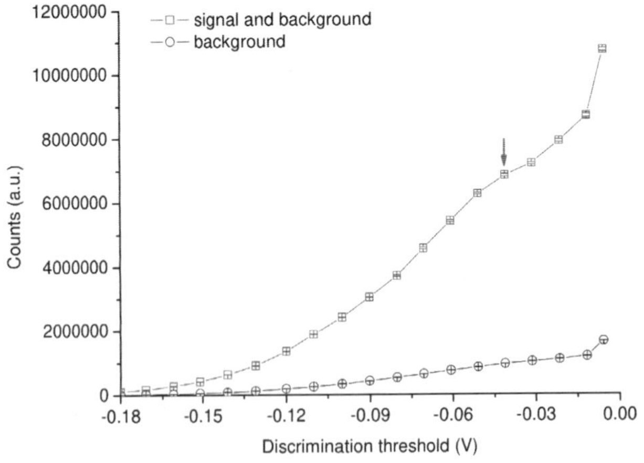


Figure 4. The integration counts of signal and background, and background versus the discrimination threshold for $^1O_2$ luminescence measurements.

3.3. Determination of the $^1O_2$ lifetime with an optimal discrimination threshold

When the optimal discrimination threshold was set to -0.0412 V, the time-resolved $^1O_2$ luminescence spectra of RB were measured with different concentrations. After each measurement, the time-resolved $^1O_2$ luminescence spectrum was mathematically fitted by using the equation (1) to yield the $^1O_2$ lifetime. As shown in Figure 5, there is no significant difference among the obtained $^1O_2$ lifetimes, as expected. The $^1O_2$ lifetimes were determined to be about 4.26±0.06 μs, which is in good agreement with the previously published values [8]. However, further comparative experiments for different photosensitizers or solvent environments should be performed.

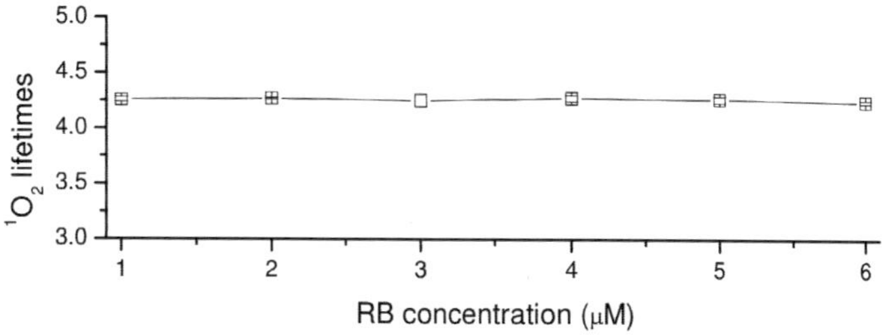


Figure 5. The $^1O_2$ lifetimes against the RB concentrations from 1 to 6 μM for $^1O_2$ luminescence measurements.

## 4. Conclusions

The effect of the discrimination threshold on the detection of $^1O_2$ luminescence that generated from the photoirradiation of photosensitizer RB was investigated by using our self-developed photon-counting detection system. Our preliminary results show that the discrimination threshold for photon counting impacts both the intensity and the shape of $^1O_2$ luminescence, which resulted in the variation of $^1O_2$ lifetimes. In order to develop the $^1O_2$ luminescence dosimetry for PDT treatment, the discrimination threshold is a crucial parameter for the $^1O_2$ luminescence detection system that needs to be optimized.

**Acknowledgements**

This work was supported by the National Natural Science Foundation of China (60978070), the program for New Century Excellent Talents in University of China (NCET-10-0012), the Key Grant of Chinese Ministry of Education (209063) and the Research Program of Fujian Provincial Educational Department (JA08031).

**References**

[1] Ogilby P R 2010 *Chem. Soc. Rev.* **39** 3181
[2] Schweitzer C and Schmidt R 2003 *Chem. Rev.* **103** 1685
[3] Wilson B C, Patterson M S and Lilge L 1997 *Lasers Med. Sci.* **12** 182
[4] Jarvi M T, Niedre M J, Patterson M S and Wilson B C 2006 *Photochem. Photobiol.* **82** 1198
[5] Schlothauer J, Hackbarth S and Röder B 2009 *Laser Phys. Lett.* **6** 216
[6] Jiménez-Banzo A, Ragàs X, Kapusta P and Nonell S 2008 *Photochem. Photobiol. Sci.* **7** 1003
[7] Egorov S Y, Kamalov V F, Koroteev N I, Krasnovsky A A, Toleutaev B N and Zinukov S V 1989 *Chem. Phys. Lett.* **163** 421
[8] Lee S, Zhu L, Minhaj A M, Hinds M F, Vu D H, Rosen D I, Davis S J and Hasan T 2008 *J. Biomed. Opt.* **13** 034010
[9] Li B H, Lin H Y, Chen D F, Wang M, and Xie S S 2010 *Chin. Opt. Lett.* **8** 86
[10] Photon counting using photomultiplier tubes, 2001 Hamamatsu Corp., Hamamatsu, Japan

# Iterative Regularization Denoising Method Based on OSV Model for BioMedical Image Denoising

**Chen Guan-nan[1], Chen Rong, Huang Zu-fang, Lin Ju-qiang, Feng Shang-yuan, Li Yong-zeng, Teng Zhong-jian**

Key Laboratory of OptoElectronic Science and Technology for Medicine, Ministry of Education, Fujian Normal University, Fuzhou 350007, China

E-mail: edado@fjnu.edu.cn

**Abstract** Biomedical image denoising algorithm based on gradient dependent energy functional often compromised the biomedical image features like textures or certain details. This paper proposes an iterative regularization denoising method based on OSV model for biomedical image denoising. By using iterative regularization, the oscillating patterns of texture and detail are added back to fit and compute the original OSV model,and the iterative behavior avoids overfull smoothing while denoising the features of textures and details to a certain extent. In addition, the iterative procedure is proposed in this paper, and the proposed algorithm also be proved the convergence property. Experimental results show that the proposed method can achieve a batter result in preserving not only the features of textures for biomedical image denoising but also the details for biomedical image.

## 1. Introduction

Image denoising methods aim at recovering the original image from a noisy measurement. Such a research has significant theoretic values in the general fields of image processing[1]. In the past decade, the representative research achievements obtained at edges and details preserving smoothing based on nonlinear image filter method, including minimizer of energy functional method, regularization method, anisotropic diffusion method and nonlinear digital filter method. Compared the other denoising methods, Image denoising algorithm based on gradient dependent energy functional preserved more edges of the biomedical image than other denoising algorithm, but it often compromised the biomedical image features like textures or certain details[2].

At Present, the details of the texture and details of biomedical image denoising technique consists mainly of two types: one is to adjust parameters of the adaptive variational method [3], at different scales of space down noise, thereby maintaining the texture; another is to medicine image space by the BV space up to the G space [4-7] or other space in order to maintain the texture and detail, in the instance of the application of both methods achieve a certain effect, but all these method will lose the detail and texture in some extent[9]. This paper would propose an denoising method based on OSV model for biomedical image denoising, iterative regularization is added to modified the noises image, in order to achieve better results.

## 2. Method

[1] To whom any correspondence should be addressed.

In this paper, the oscillate decomposition model of image Characteristic Waveforms with textures or details Based on Meyer is introduced. Meyer's model devides the image into two parts: f=u+v, where u is the part of structure feature information of decomposition image, and v is the part of textures details or noise information. The model expatiate the oscillate decomposition from the theoretical perspective, and establish the decomposition and denoising model based on minimizer of energy functional method. Through using of dual space(G space), the model improve and expand the application of BV space. The G space is defined as:

$$G=\{v \mid v=\partial_x g_1(x,y)+\partial_y g_2(x,y). g_1, g_2 \in L^{\infty}(\Omega)\} \tag{1}$$

The norm of the space is:

$$\|v\|_G = \inf_{g=(g_1,g_2)} \{\left\|\sqrt{g_1^2+g_2^2}\right\|_{L^{\infty}} \mid v=\partial_x g_1+\partial_y g_2\} \tag{2}$$

If image $f \in L_2(\Omega), \Omega \subset R^2$, the Meye's model is

$$\inf_u \{E(u)=\int_{\Omega} |\nabla u| + \lambda \|v\|_G, f=u+v\} \tag{3}$$

Where the oscillate degree of textures information v could be computed with $\|v\|_G$. Because the value is difficult to be computed in G space, there is no general method to the minimize the Euler-Lagrange equation. Some approximation models for Meye's model are proposed, and the model with better effect is OSV model which is improved from VO model by Osher, Sole and Vese:

$$\inf_u \{E(u)=\int_{\Omega} |\nabla u| + \lambda \int_{\Omega} |\nabla \Delta^{-1}(f-u)|^2, f=u+v\}$$
$$=\|u\|_{BV(\Omega)} + \lambda \|f-u\|^2_{H^{-1}(\Omega)}, f=u+v \tag{4}$$

Through using OSV model and other different norm of the model can preserved overmuch smoothing the image texture details when denosing with the variational models of ROF in a certain extent, but it still has many improvement of space.

## 3. Numerical analyses

In order to preserve more features of textures and the details for biomedical image, we fixed the original OSV model, added an iterative regularization. The iterative behavior avoids overfull smoothing while denoising the features of textures and details to a certain extent.

Step 1: We fix the original OSV model as:

$$u_1 = \arg\min\{\int |\nabla u| + \lambda \int_{\Omega} |\nabla \Delta^{-1}(f-u)|^2\} \tag{5}$$

And define $n_1 = \dfrac{\nabla u_1}{|\nabla u_1|}$

Step 2:Through further revision, the fixed OSV model can express as:

$$u_1 = \arg\min\{\int (|\nabla u| - \Delta^{-1} n_1 \nabla u) + \lambda \int_{\Omega} |\nabla \Delta^{-1}(f-u)|^2\} \tag{6}$$

Apparently , $-\int n_1 \nabla u = \int u \nabla \cdot n_1 = \int u \nabla (\nabla \cdot \dfrac{\nabla u_1}{|\nabla u_1|})$

At the same time, $f = u_1 + v_1$, so $\nabla \cdot \dfrac{\nabla u_1}{|\nabla u_1|} = -2\lambda (f-u_1) = -2\lambda v_1$

And $-\int n_1 \nabla u = -2 \int \lambda u v_1$, we substitut it into the result of step 2, then continue to calculate the equation:

$$u_2 = \arg\min\{\int(|\nabla u|) + \lambda\int_\Omega |\nabla\Delta^{-1}(f-u)^2 - 2uv_1|\}$$
$$= \arg\min\{\int(|\nabla u|) + \lambda\int_\Omega |\nabla\Delta^{-1}(f+v_1-u)|^2 - \lambda\int(v_1^2+2v_1 f)\} \quad (7)$$

Where the last item $\lambda\int(v_1^2+2v_1 f)$ isn't restricted by u. Therefore, we only need to calculate in front of two items:

$$u_2 = \arg\min\{\int(|\nabla u|) + \lambda\int_\Omega |\nabla\Delta^{-1}(f-u+v_1)|^2\} \quad (8)$$

The iterative regularization model:

In our method, the noised item v1 computed from the previous calculations is added to the result f of original image, then the original equation is constantly updated and modified to get more accurate results.

The proposed algorithm as follows:

- initialize u0=0,v0=0;
- let k=0,1,2,……using OSV model to calculate the results uk+1:

$$u_{k+1} = \arg\min\{\int(|\nabla u|) + \lambda\int_\Omega |\nabla\Delta^{-1}(f-u_k+v)|^2\} \quad (9)$$

- Update vk+1:

$$v_{k+1} = f + v_k - u_{k+1} \quad (10)$$

In the iterative process of solving, we need to consider the variational problem of step 2, The corresponding Euler-Lagrange equation as follow:

$$2\lambda\Delta^{-1}(f+v-u) = div(\frac{\nabla u}{|\nabla u|}) \quad (11)$$

The iterative regularization model is:

$$u_t = -\frac{1}{2\lambda}\Delta[div(\frac{\nabla u}{|\nabla u|})] - (u-v-f) \quad (12)$$

## 4. Results

In order to validate the result of image enhancement and denoising worked with the scheme in this paper, we choose heart image which add standard deviations std =15 of the white Gaussian noise, compare the effects with different models.

Figure1 shows the results of image denoising with original OSV model, our models with parameters: iterations=3, and our models with parameters: iterations=6. Contrasting Fig. 1(d) to Fig. 1(e), we can notice that the original OSV model could only preserve the great vessels of the heart in Fig. 1(c), while the details of the capillary vessels are smoothed in excess. the details in Fig. 1(d,e) are preserved great better than the same parts in Fig. 1(c), the capillary vessels are reserved in more distal. And PSNR in Fig. 1(d.e) are 18.8391 and 19.4425, are also bigger than PSNR in Fig. 1(c) 18.0885.

Experimental results show that this iterative regularization method obtain good results in both subjective and objective, the proposed method can achieve a batter result in preserving not only the features of textures for biomedical image denoising but also the details for biomedical image.

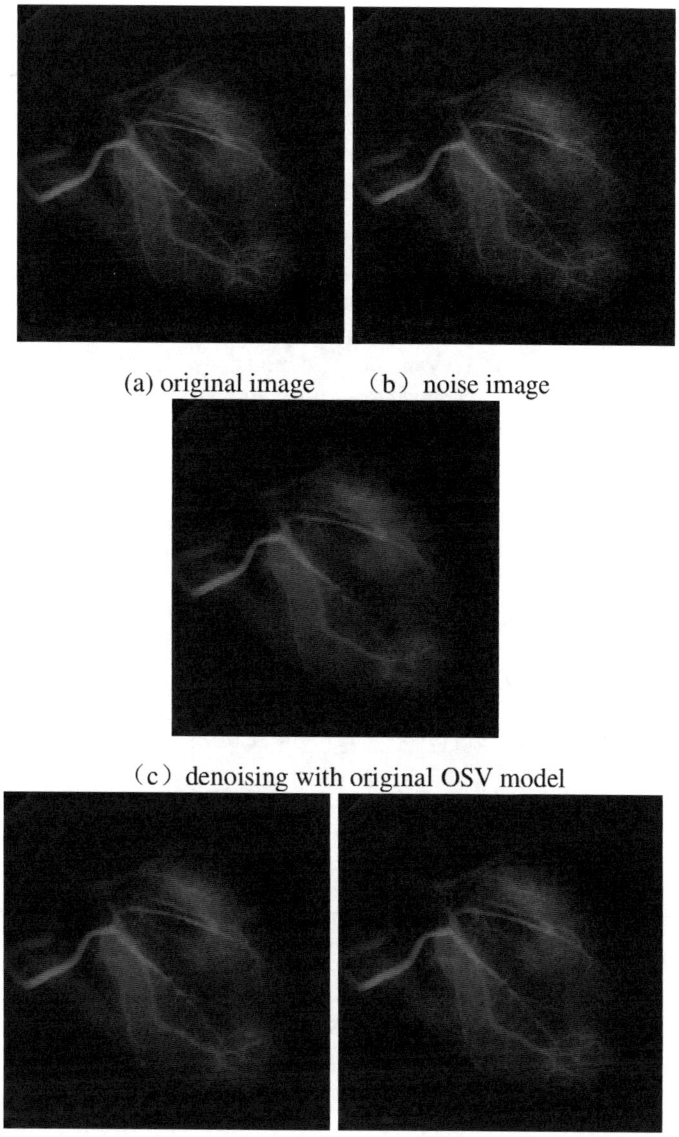

(a) original image (b) noise image

(c) denoising with original OSV model

(d) denoising with our model(ite=3) (e) denoising with our model(ite=6)

Fig. 1(a-e). Filtering a heart image.

## 5. Conclusions

Some processes based on the original OSV model filters are presented, moreover, the principles, and the problem of them are analyzed. A new iterative regularization denoising Method Based on OSV Model is presented for image smoothing in this work. The noised item computed from the previous calculations is added to fix the result of original image. Then, the new Iterative Regularization is constantly updated and modified to get more accurate results. The new denoising method based on OSV Model can not only well restrain noise but also keep much more textures details of a biomedical image,. Experimental results also show that the effectiveness for biomedical images denoising with our method is better than the original method's.

**6. Acknowledgements**

This work was supported by the National Natural Science Foundation of China (Nos. 60778046 and 60910106016), the Science and Technology Project of Fujian Province (Nos. 2010J05130, 2009J01276 and 2008J0016), the Project of Science Foundation of Ministry of Health and United Fujian Provincial Health and Education Project for Tackling the Key Research (No.WKJ2008-2-046).

**References**

[1] CHAN T F, SHEN J, VESE L. Variational PDE models in image processing[J]. Notices Amer Math Soc, 2003, 50:14-26.

[2] STRONG D, CHAN T F. Edge preserving and scale dependent properties of total variation regularization[C]. Inv Probl, 2003, 19: 165-187.

[3] Gilboa G, Sochen N, Zeevi Y Y. Variational denoising of partly-textured images by spatially varying constraints[J ]. IEEE Transactions on Image Processing. 2006, 15(8): 2281-2289.

[4] Meyer Y. Oscillating Patterns in Image Processing and Nonlinear Evolution Equations[M]. University Lecture Scries. Boston, USA: American Mathematical Society, 2001.

[5] Vese L A, Osher S J. Modeling Textures with Total Variation Minimization and Oscillating Patterns in Image Processing[J]. Joumal of Scientific Computing, 2003, 19(11): 553-572.

[6] Osher S, Sole A, Vese L. Image Decomposition and Restoration Using Total Variation Minimization and the H-Norm[J]. Journal of Multiscale Modeling and Simulation, 2003, 1(3):349-370.

[7] Aujol J F, Chambolle A. Dual Norms and Image Decomposition Models. International Journal of Computer Vision, 2005, 63(1):85-104.

[8] Chen Yunmei, Levine S, Rat M. Variable Exponent, Linear Growth Functionals in Image Processing[J]. SIAM Journal on Applied Mathematics, 2006, 66(4): 1383-1406.

# Comparison of Two Strategies for the Synthesis of Upconverting Nanoparticles as Biological labels

Sisi Cui, Haiyan Chen, Yueqing Gu*
Department of Biomedical Engineering, School of Life Science and Technology, China Pharmaceutical University, Nanjing 210009, People's Republic of China

**Abstracts**

Oleic acid-coated upconverting nanoparticles were prepared by hydrothermal and solvothermal synthesis. The properties of as-prepared $NaYF_4$:Yb,Er nanoparticles were characterized, including transmission electron microscope, X-ray diffraction pattern and photoluminescence spectrum. The effects of reaction conditions (reaction time and temperature) and the concentration of rare earth dopants have been investigated in each strategy. Optical properties of upconverting nanoparticles synthesized by two strategies were compared as well. For biological application, the surface ligands were exchanged from oleic acid to hexanedioic acid with carboxyl-functionalized groups. The results indicated upconverting nanospheres ($45 \pm 2$ nm) and nanoplates ($150 \pm 5$ nm) prepared by solvothermal synthesis had higher photoluminescence intensity with uniform size distribution, whereas hydrothermal synthesis produced small upconverting nanocubes (8-20 nm). Upconverting nanoparticles in hydrothermal synthesis showed better optical properties while increasing the reaction time and decreasing the molar ratio of water. The ratio of green and red emissions of $NaYF_4$:Yb,Er nanoparticles could be adjusted through changing the concentration of rare earth dopants. Furthermore, hexanedioic acid mediated ligand-exchange process transferred $NaYF_4$ nanoparticles from hydrophobic to hydrophilic with great potential as biological labels.

**Key Words:** upconverting nanoparticles; hydrothermal synthesis; solvothermal synthesis; ligand-exchange; comparison

## 1. Introduction

Fluorescence-based detection is widely used in biological labeling and imaging. Conventional fluorescent labeling materials such as organic dyes and semiconductor quantum dots (QDs) show visible and near-infrared (NIR) fluorescence under excitation of high-energy light. Organic dyes are easily accessible and versatile to use but they do have considerable disadvantages. Commonly, organic dyes are susceptible to photobleaching or degradation in certain conditions. In recent years nanoparticle technology has become a research focus as its fundamental and practical importance is well known, especially in the case of fluorescent materials. For example, QDs with small diameter show brighter and more stable

* Corresponding author. Fax: +86-25-83271046
*E-mail address:* cpuyueqing@163.com(Y.Q. Gu).

fluorescence than common organic dyes. Additionally, their emission wavelength can be tuned from visible to NIR light by variation of their diameter [1,2]. However, the application of QDs still has many limitations, especially the toxicity of their heavy metal components while *in vivo* applications. On the other hand, UV excitation of down-conversion materials will cause severe damage to biological tissues and strong background fluorescence as *in vivo* biolabels. Compared to the above down-conversion fluorescent materials, upconverting nanoparticles (UCNs), which emit higher-energy photos after absorbing lower-energy NIR photos, have many advantages including narrow emission, deep tissue penetration, high chemical and optical stability, less photodamage to living organisms, weak autofluorescence from cells and tissues, low background fluorescence and high sensitivity for detection.

In the past decade, upconversion in lanthanide-doped materials have been proposed to be more suitable for ideal biolabels. The upconversion process is mainly located in the lattice of the host materials in which the lanthanide ion is embedded in. To the best of our knowledge, among the reported upconversion materials, $NaYF_4$ is one of the most efficient host materials for visible upconversion fluorescence. There are two phases of the crystal structure of $NaYF_4$ cubic phase and hexagonal phase. Upconversion efficiency of the pure hexagonal phase $NaYF_4$ is much higher than cubic phase $NaYF_4$. The emission wavelength of the UCNs is not size-dependent as it is for QDs and multicolor emission can easily be accompalished by varying host crystal and rare earth dopants. Many studies have been performed to obtain $NaYF_4$ nanoparticles (nanocubics, nanospheres, nanoplates, nanorods, etc) via hydrothermal synthesis [3,4], solvothermal synthesis[5], thermal decomposition[6,7], sol-gel and homogenous precipitation. Among these routes, sol-gel and homogenous precipitation are low temperature methods. However, the UCNs synthesized using these two methods always have low crystallinity and require post-treatment or annealing at high temperature up to 1000 °C. The powder products with large size cannot be easily redispersed in the solvent which make them unsuitable for biological application. Thermal decomposition of mixed trifluoroacetates is an effective method to synthesize high quality cubic phase and hexagonal phase $NaYF_4$, but pyrolysis of trifloroacetates produces very toxic fluorinated and oxyfluorinated carbon species. Thus, the experimental procedure must be carried out carefully in the ventilation. Moreover, a liquid-solid solvothermal reaction has been reported to produce oil-dispersed hexagonal phase $NaYF_4$, whereas the excessive NaF reactants are difficult to be removed. Very recently, Zhang et al. [5] proposed a efficient and user-friendly solothermal method for the synthesis of oleic acid (OA)-coated UCNs with controllable shape and upconversion fluorescence. Researchers also produced hydrophobic and hydrophilic UCNs through hydrothermal synthesis with proper photoluminescence intensity (PL) for biological application. Both solvothermal and hydrothermal synthesis can obtain UCNs with small size (<50 nm), high lumiscence and good redispersibility.

Although recent advances in the synthetic methods for UCNs, the UCNs with hydrophobic ligands on their surface still cannot be used directly in biological applications due to low solubility in water and unfavorable surface properties. Therefore, it is prerequisite to prepare water-soluble UCNs bearing functional groups (such as –COOH, $-NH_2$ or –SH) on their surface to conjugate with biomolecules, which will promote the development of UCN-based biolables. Three strategies have been developed for transferring hydrophobic OA-coated UCNs to hydrophilic UCNs with functionalized groups on the surface for bio-conjugation. One strategy is ligand-exchange processes which replace the original organic ligands with hydrophilic ones, such as hexanedioic acid (HDA) [8], citrate [9], thioglycollic acid (TGA) [10] and so on. The procedure has usually been performed at high reaction temperature, but it has less adverse effect on the quality of the UCNs. The second strategy is based on the encapsulation of hydrophobic UCNs with amphiphilic copolymer (such as polyvinylpyrrolidone, poly(acrylic acid), polyethylene glycol and polyethylenimine) or silica[11-13]. However, the processes are complicated and not easy to control the reaction precisely. The third strategy involves converting surface OA ligands through certain chemical reactions like Lemieux-von Rudloff oxidation which selectively oxidize a carbon-carbon double bond to give two carboxylic acids [14]. Hydrophilic UCNs can be stable in aqueous solution for more than one month.

Hydrophilic UCNs with surface functionalized groups have great potential for biological applications, including bioimaging for cancer cells *in vitro* and real-time targeted imaging of tumor bearing mouse *in vivo*, photodynamic therapy, immunoassay for antibody, bioassay for RNA and DNA based on fluorescence resonant energy transfer (FRET), etc [14-16]. UCNs absorb NIR light with deep tissues penetration and less autofluorescence and convert it to visible light or even NIR light (about 800 nm). Meanwhile, UCNs have outstanding luminescent stability under long-term excitation in comparison with down-conversion fluorescence materials.

In this study, UCNs with OA ligands were prepared by hydrothermal and solvothermal synthesis. As-prepared hydrophobic OA-coated UCNs were converted into hydrophilic via a ligand-exchange process with HDA. The transmission electron microscope (TEM), X-ray diffraction (XRD) pattern and upconverting emission spectrum of OA-coated UCNs were characterized in each strategy. The effects of reaction time and temperature for the synthesis of UCNs were investigated. Moreover, the ratio of green and red emission was adjusted by changing the concentration of rare earth dopants. Optical property of UCNs before and after ligand-exchange process was measured as well. The differences of as-prepared UCNs prepared by hydrothermal and solvothermal synthesis were compared for further biological applications.

## 2. Experimental section

### 2.1 Materials

$Y(NO_3)_3 \cdot 6H_2O$, $Yb(NO_3)_3 \cdot 5H_2O$ and $Er(NO_3)_3 \cdot 5H_2O$ (purity >99.9%) were purchased from Aladdin. Rare earth chlorides ($LnCl_3$, Ln: Y, Yb and Er) were prepared by dissolving the corresponding oxides ($Y_2O_3$, $Yb_2O_3$ and $Er_2O_3$ from Aladdin and China National Medicine Corporation Ltd.) in 10% hydrochloric solution and then evaporating the water completely. Diethylene glycol (DEG; 98%) were purchased from Aladdin. Hexanedioic acid, NaOH, $NH_4F$ and NaF were purchased from China National Medicine Corporation Ltd. 1-octadecene (90%) and oleic acid (90%) were all purchased from Sigma-Aldrich and used as starting materials without further purification.

### 2.2 Synthesis of oleic acid-coated upconverting nanoparticles

2.2.1 Hydrothermal synthesis of $NaYF_4$:Yb,Er nanoparticles

NaYF4:Yb,Er nanoparticles were prepared by a modified hydrothermal process [4]. In a typical synthesis, NaOH (1.2 g, 30 mmol), ethanol (10 mL), and oleic acid (20 mL) and different amount of water were mixed under agitation to form a homogeneous solution. Subsequently, 0.6 mmol (total amounts) of rare-earth nitrate (1.2 mL, 0.5 mol/L $LnNO_3$) aqueous solution was added under magnetic stirring. Then 1.0 M aqueous NaF (4 mL) solution was added dropwise to the above solution. After agitating for another 10 min, the mixture was transferred to a 50 mL autoclave, sealed, and hydrothermally treated at 160 °C for 12 h. The autoclave was cooled to room-temperature naturally, and the products were collected at the bottom of the container by cyclohexane, washed with cyclohexane/ethanol (1:6 v/v) for three times and centrifuged to obtained power products.

2.2.2 Solvothermal synthesis of $NaYF_4$:Yb,Er nanoparticles

$NaYF_4$:Yb,Er nanoparticles were synthesized following this protocol: Rare earth chlorides solution (0.5 mol/L) were prepared by dissolving the corresponding oxides in 10% hydrochloric solution. Different molar ratio of $RECl_3$ (Y: Yb: Er) were added to a 100 ml flask containing 6 ml oleic acid (OA) and 15 ml 1-octadecene (ODE). The solution was heated to 160 °C for 30 min and then cooled down to room temperature. Thereafter, 10 mL methanol solution of $NH_4F$ (0.148 g) and NaOH (0.100 g) was added dropwise under stirring for 30 min. After methanol evaporated, the solution was heated to 300 °C under nitrogen for 1 h and cooled down to room temperature. The resulting nanoparticles were precipitated by the addition of ethanol, collected by centrifugation. The products were washed with cyclohexane/ethanol (1:6 v/v) for three times.

### 2.3 Ligand-exchange of oleic acid-coated $NaYF_4$:Yb,Er nanoparticles with HDA

Ligand-exchange of UCNs was prepared following Zhang's report [8]. Ligand exchange was carried out as follows: The DEG solution (10 mL) containing HDA (500 mg) was heated up to 110 °C for 30 min with vigorous stirring under nitrogen. A chloroform solution of UCNs (20 mg) was injected into the above solution, then heated to 240 °C and kept at this temperature for about 1.5 h until the solution became clear. After the solution was cooled to room temperature, excess aqueous solution was added. The UCNs were isolated by centrifugation and decantation. Finally the sample was washed three times with pure water and redispersed in deionized water for analysis.

### 2.4 Characterization

Powder XRD measurement was carried out using a Philips X'Pert PRO X-ray diffractometer. Sizes and morphologies of UCNs were determined at 200 kV using a JEOL JEM-2100 high-resolution transmission electron microscope (HR-TEM). Up-conversion luminescence spectra were measured with S2000 eight-channel optical fiber spectrographotometer (Ocean Optics corporation, America) and a 0-2000 mW adjustable laser (980 nm, Shaanxi Scitower Photoelectricity Equipment co., LTD., China) was used as the external excitation source. All optical measurements were performed at room temperature.

## 3. Results and discussion

### 3.1 Size and morphology of $NaYF_4$ nanoparticles

TEM images of UCNs with different synthesis routes and conditions were shown in figure 1. Nanocubes were generated by hydrothermal synthesis with sizes of 8-12 nm, as the volume ratio of OA, ethanol and water is 20:10:9 (Figure 1A). While reducing the volume of water to 4 mL, the size distribution enlarged to 15 ± 5 nm (Figure 1B). A few nanorods with the length of 900 nm and the width of 20 nm could be observed in this sample (data were not shown). The results indicate the amount of water influence the growth of UCNs, since OA and produced UCNs are not dissolved in water. OA coated UCNs can easily grow with less molar ratio of water and become larger in size.

For UCNs prepared by solvothermal synthesis, nanospheres and nanoplates were obtained. The size distribution was 45 ± 2 nm in the reaction system of 6 mL OA and 15 mL ODE. While increasing the reaction time and temperature without changing the ratio of reagents, the shape of as-prepared UCNs was almost changed to plate-like crystals with larger size (150 ± 5 nm). A few cubic impurities could also be observed in figure 1D.

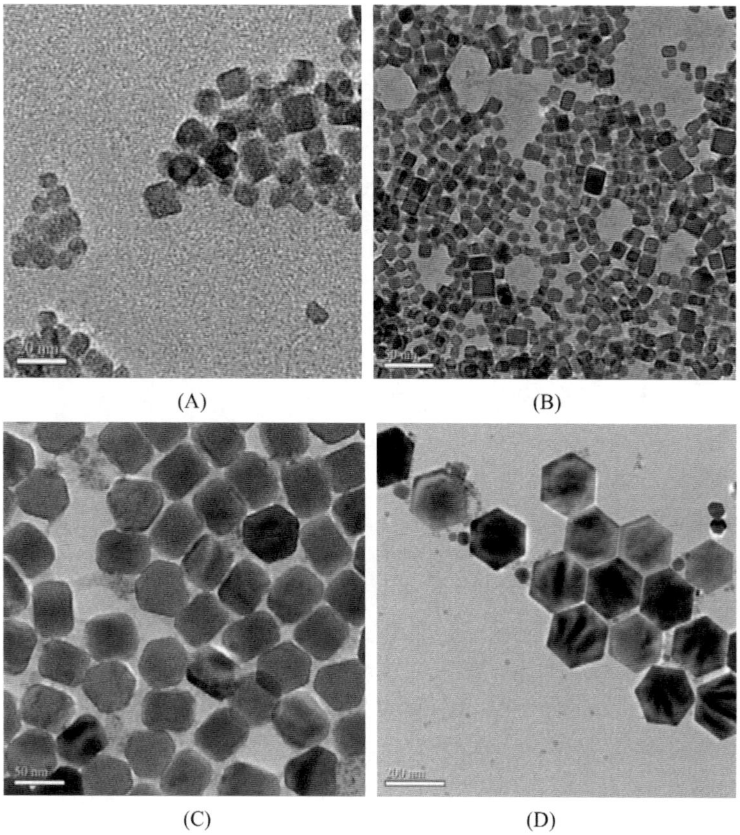


Figure.1 TEM images of $NaYF_4$:Yb,Er nanoparticles with different synthesis methods and conditions: hydrothermal synthesis (A) OA/ethanol/water=20:10:9 (v/v/v) (ruler=20 nm) and (B) OA/ethanol/water=10:5:2 (v/v/v) (ruler=50 nm); solvothermal synthesis at different reaction temperatures (C) 300 °C for 1 h (ruler=50 nm) and (D) 310 °C for 1 h (ruler=200 nm).

In comparison of these two synthesis strategies, hydrothermal synthesis with more amount of water can generate small upconverting nanocubes (less than 20 nm). However, solvothermal synthesis always produced nanoplates and nanospheres with more uniform size distribution than hydrothermal synthesis, the shape of which can be adjusted through adding different amount of OA and ODE. According to Li's report [3], hydrothermal synthesis also produced nanorods with large size when increasing hydrothermal reaction time and temperature. All the as-prepared UCNs were monodisperse suggesting that the long-chain OA ligands on their surface prevent aggregation.

**3.2 Structure of $NaYF_4$ nanoparticles**

The X-ray diffraction (XRD) pattern result of the $NaYF_4$ nanoparticles prepared by solvothermal synthesis was shown in figure 2. The diffraction peaks positions and intensities confirmed the existence of hexagonal (JCPDS card 16-0334) and cubic (JCPDs card 77-2042) phase $NaYF_4$ nanoparticles (standard cards were not shown). Most of OA-coated UCNs synthesizing via solvothermal route were hexagonal phase. The concentration of OA ligands affected the growth rates of hexagonal-phase $NaYF_4$ nanoparticles in different direction.

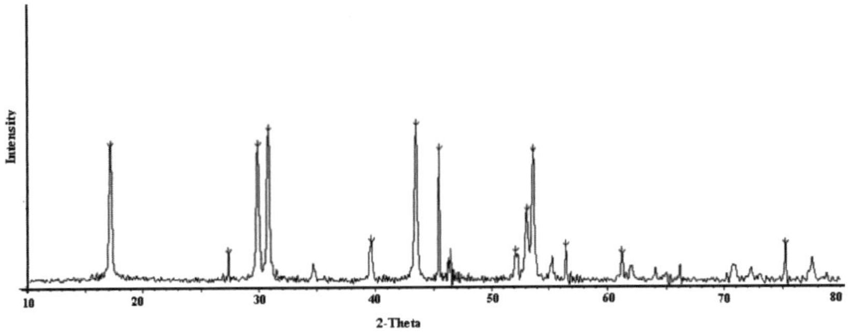


Figure.2 XRD pattern of oleic acid coated-$NaYF_4$:Yb,Er nanoparticles prepared by solvothermal synthesis

### 3.3 Optical characterization of $NaYF_4$ nanoparticles

Photoluminescence spectrum of prepared $NaYF_4$ nanoparticles powder were measured at room temperature and given in figure 3. The typical emission peaks of the $NaYF_4$:Yb,Er nanoparticles were located at 407, 521, 539 and 651 nm, which due to the transitions between energy levels $^4H_{9/2}$, $^4H_{11/2}$, $^4S_{3/2}$ and $^4F_{9/2}$ to $^4I_{15/2}$ of $Er^{3+}$.

Figure 3A presented upconverting emission of $NaYF_4$:Yb,Er nanoparticles synthesizing through hydrothermal method. PL intensities in different reacting condition demonstrated the amount of water affected the emission intensity of UCNs. UCNs with larger size had higher PL intensity. For hydrothermal synthesis, decreasing the amount of water and increasing the reaction time and temperature can both enhance upconverting emission of UCNs (Figure 4). Moreover, for solvothermal synthesis, reducing the amount of OA and increasing reaction temperature (below the boiling point of ODE) and time will increase the PL emission.

Figure.3 Upconversion spectrum of $NaYF_4$:Yb,Er powders with different synthesis stratergies and rare earth ion doped ratio: (A) $NaYF_4$:20 mol%Yb,2 mol%Er nanoparticles prepared by hydrothermal synthesis; (B) $NaYF_4$:18 mol%Yb,2 mol%Er nanoparticles prepared by solvothermal synthesis.

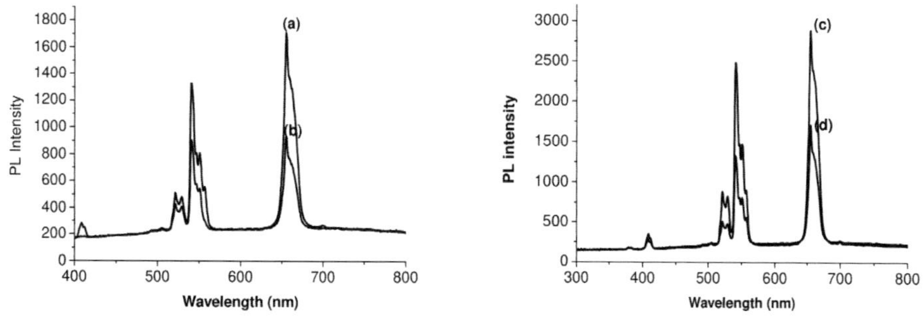


(A) (B)

Figure 4 The effects of reaction temperature for PL intensity of UCNs: (A) hydrothermal synthesis160 °C for 12 h (a) and180 °C for 12 h (b); (B) solvothermal synthesis 300 °C (c) and 310°C (d).

Photostability of OA-coated UCNs were investigated in 90 minutes (Figure 5). UCNs have higher optical and chemical stability than conventional organic dyes and QDs, which is significant for bioimaging in cells or in animal models. Nanoplates prepared by solvothermal route showed stronger upconverting emission compared with nanospheres (solvothermal synthesis) and nanocubes (hydrothermal synthesis). Previous reports revealed that UCNs with hexagonal phase showed 4.4 times stronger PL intensity than cubic phase powders [5], and the different is even bigger when the size of UCN powder is of nanoscale order. $NaYF_4$:Yb,Er nanoparticles prepared by hydrothermal synthesis with 4 mL water showed approximate upconverting emission comparing with solvothermal synthesized NaYF4:Yb,Er nanospheres (6 mL OA, 300 °C for 1 h).

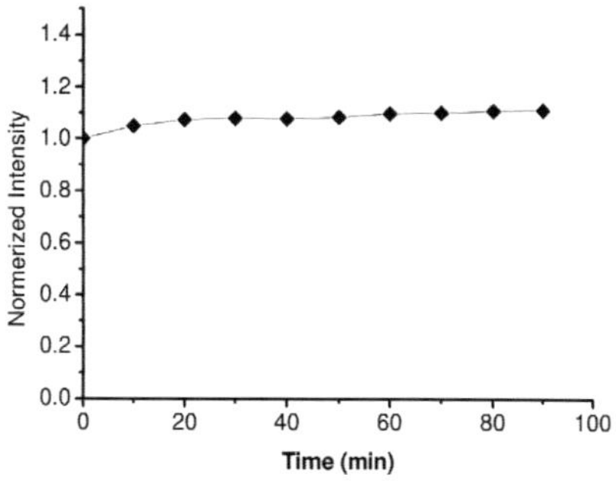


Figure 5 Photostability of OA-coated UCNs.

### 3.3 Ligand-exchange of $NaYF_4$ nanoparticles

$NaYF_4$ nanoparticles were synthesized by hydrothermal and solvothermal route in the presence of OA molecules. In this reaction process, OA molecules were coated onto the outer face of the UCNs through the interaction between their carboxyl groups and rare earth ions. Since the alkyl chains of OA were left outside, the as-prepared UCNs were hydrophobic and easily dispersed in nonpolar solvents, such as hexane, cyclohexane, chloroform, etc. After the addition of polar solvents like ethanol and acetone, OA coated UCNs were deposited and centrifuged to obtain power products.

To convert hydrophobic OA coated UCNs into carboxyl functionalized derivatives, an efficient ligand-exchange process were carried out to obtain hydrophilic UCNs. HDA molecules were used to replace the original OA ligands in DEG solvent at high reaction temperature without obvious adverse effects on the optical property of the UCNs (Figure 6). The HDA molecules with six carbons have proper length to chelate to the surface of UCNs. High reaction temperature can enhance the efficiency of the ligand-exchange process. The HDA coated UCNs showed strong upconverting emission and high water solubility. Most importantly, HDA-coated UCNs with carboxyl-functionalized groups become a promising candidate for biological labeling.

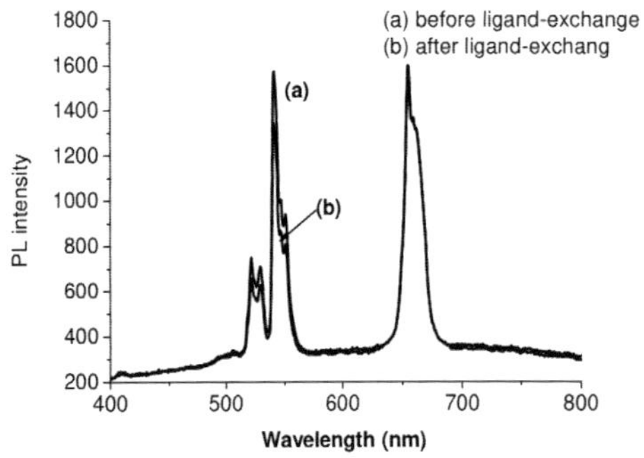


Figure 6 Photoluminescence spectrum of UCNs before (A) and after (B) HDA ligand-exchange.

### 3.4 The effect of rare earth doped concentration on the PL intensity ratio of green and red emissions

Upconveting emission peaks of $NaYF_4$:Yb,Er nanoparticles were located at 407, 521, 539 and 651 nm, which could show green and yellow color observed by naked eyes. Red color can be observed through suitable optical filters. The intensity of green (540 nm) and red (650 nm) emission peaks can be adjusted by changing the ratio of rare earth dopants. Table 1 presented the effects of rare earth doped concentration on the PL intensity ratio of green and red emissions. When the concentration of $Er^{3+}$ is constant, increasing the concentration of doped $Yb^{3+}$ ions could obviously enhance red emission, but had little effects on the green emission of $NaYF_4$:Yb,Er nanoparticles (Figure 7). Adjustable green and red emissions are useful for multicolor biological labeling.

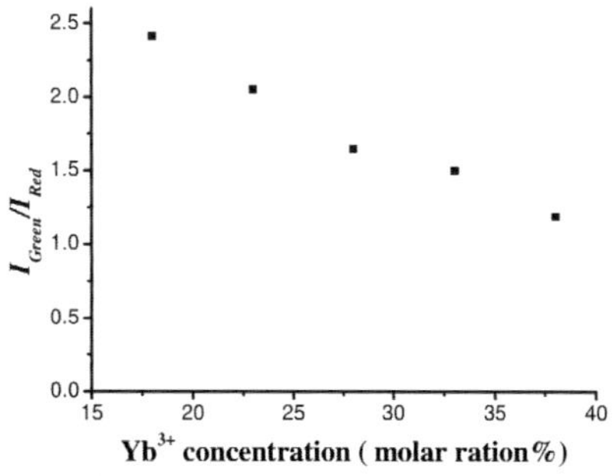


Figure 7 PL intensity ratio of green and red light with different doped $Yb^{3+}$ ions

| Y (molar%) | Yb (molar%) | Er (molar%) | PL intensity ration of green and red light |
|---|---|---|---|
| 80 | 18 | 2 | 2.414 |
| 75 | 23 | 2 | 2.054 |
| 70 | 28 | 2 | 1.644 |
| 65 | 33 | 2 | 1.499 |
| 60 | 38 | 2 | 1.188 |

Table 1 PL intensity ratio of green and red light with different molar ratio of rare earth dopants

## 4. Conclusion

In this study, we have synthesized $NaYF_4$:Yb,Er nanoparticles via hydrothermal and solvothermal routes and transferred hydrophobic OA-coated UCNs to hydrophilic HDA-coated UCNs with higher water solubility. TEM results indicated the as-prepared UCNs were monodisperse with uniform size distribution. Nanocubes, nanospheres and nanoplates were obtained by changing the molar ratio of reagents. UCNs prepared with less amount of water in hydrothermal synthesis and synthesized by solvothermal route showed strong upconversion fluorescence. Moreover, HDA ligand-exchange hardly had any adverse effects on luminescence properties of the UCNs. The carboxyl-functionalized groups of HDA can further conjugate with tumor targeting ligands, antibodies, DNA, proteins, etc. High luminescent UCNs have great potential for biological labeling, bioimaging *in vitro* and *in vivo*. Future work will involve targeted localization of these UCNs tumors *in vivo* in small animals for targeted bioimaging and photodynamic therapy.

## Reference

[1] Chen H. Y., Wang Y. Q., Xu J., Ji J. Z., Zhang J., Hu Y. Z., Gu Y. Q., "Non-invasive near infrared fluorescence imaging of CdHgTe quantum dots in mouse model," J. Fluoresc. 18(5), 801-811 (2008).

[2] Cao J., Xue B., Li H., Deng D. W., Gu Y. Q., "Facile synthesis of high-quality water-soluble N-acetyl-L-cysteine-capped $Zn_{1-x}CdxSe/ZnS$ core/shell quantum dots emitting in the violet-green spectral range," J. Colloid Interface Sci., 348(2), 369-376 (2010).

[3]Wang L. Y., Li Y. D., "$Na(Y_{1.5}Na_{0.5})F_6$ single-crystal nanorods as mutilcolor luminescent materials," Nano. Lett., 6(8),1645-1649 (2006)

[4] Wang L. Y., Li Y. D., "Controlled synthesis and luminescence of lanthanide doped $NaYF_4$ nanocrystals," Chem. Mater., 19(4), 727-734 (2007).

[5] Li Z. Q., Zhang Y., "An efficient and user-friendly method for the synthesis of hexagonal-phase NaYF4:Yb,Er/Tm nanocrystals with controllable shape and upconversion fluorescence," Nanotechnology, 19(34), 345606 (2008).

[6] Boyer J. C., Vetrone F., Cuccia L. A., Capobianco J. A., Lin V. S. Y. "Synthesis of Colloidal Upconverting NaYF4 Nanocrystals Doped with $Er^{3+}$, $Yb^{3+}$ and $Tm^{3+}$, $Yb^{3+}$ via Thermal Decomposition of Lanthanide Trifluoroacetate Precursors." J. Am. Chem. Soc., 128(23), 7444-7445(2006).

[7] Boyer J. C., Cuccia L. A., Capobianco J. A., "Synthesis of Colloidal Upconverting $NaYF_4$: $Er^{3+}/Yb^{3+}$ and $Tm^{3+}/Yb^{3+}$ Monodisperse Nanocrystals," Nano. Lett., 7(3), 847-852 (2007).

[8] Zhang Q., Song K., Zhao J., Kong X., Sun Y., Liu X., Zhang Y., Zeng Q., Zhang H., "Hexanedioic acid mediated surface-ligand-exchange process for transferring $NaYF_4$:Yb/Er (or Yb/Tm) up-converting nanoparticles from hydrophobic to hydrophilic, " J. Colloid. Interface Sci,. 336(1), 171-175 (2009).

[9] Cao T. J., Yang T. S., Gao Y., Hu H., Li F. Y., "Water-soluble NaYF4:Yb/Erupconversion nanophosphors: Synthesis, characteristics and application in bioimaging," Inorganic Chemistry Communications, 13(3), 392-394 (2010).

[10] Li D., Dong B., Bai X., Wang Y., Song H. W.," Influence of the TGA modification on upconversion luminescence of hexagonal-phase NaYF4:$Yb^{3+}$, $Er^{3+}$ nanoparticles," J. Phys. Chem. C, 114(18), 8219–8226 (2010).

[11] Li Z. Q., Zhang Y., "Monodisperse silica-coated polyvinylpyrrolidone/$NaYF_4$ nanocrystals with multicolor Upconversion Fluorescence Emission," Angew. Chem., 118(46), 7896-7899 (2006).

[12] Wang F., Chatterjee D. K., Li Z. Q., Zhang Y., Fan X. P., Wang M. Q., "Synthesis of polyethylenimine/$NaYF_4$ nanoparticles with upconversion fluorescence," Nanotechnology, 17(23), 5786-5791 (2006).

[13] Chen Z., Chen H., Hu H., Yu M., Li F., Zhang Q., Zhou Z., Yi T., Huang C., "Versatile synthesis strategy for carboxylic acid-functionalized upconverting nanophosphors as biological labels," J. Am. Chem. Soc., 130(10), 3023-3029 (2008).

[14] Qian H. S., Guo H. C., Ho P. C., Mahendran R., Zhang Y., "Mesoporous-silica-coated up-conversion fluorescent nanoparticles for photodynamic therapy, " Small, 5(20), 2285-2290 (2009).

[15] Wang L., Yan R., Huo Z., Wang L., Zeng J., Bao J., Wang X., Peng Q., Li Y., "Fluorescence resonant energy transfer biosensor based on upconversion-luminescent nanoparticles," Angew. Chem. Int. Ed. Engl., 44(37), 6054-6057 (2005).

[16] Chatterjee D. K., Rufaihah A. J., Zhang Y., "Upconversion fluorescence imaging of cells and small animals using lanthanide doped nanocrystals," Biomaterials, 29(7), 937-943 (2008).

# Improvement of skin optical clearing efficacy by topical treatment of glycerol at different temperatures

**Zijian Deng, Caihua Liu, Wei Tao and Dan Zhu***
Britton Chance Center for Biomedical Photonics, Wuhan National Laboratory for Optoelectronics, Huazhong University of Science and Technology, Wuhan, China

E-mail: dawnzh@mail.hust.edu.cn

**Abstract**: In the past decades, laser has been widely used in clinical diagnosis and cosmetic therapy. However, there is limitation for further usage in deeper tissue for high scattering property. Skin optical clearing technique, by introducing optical clearing agents (OCAs) into tissue, will have a potential impact on optical diagnosis and therapy. In this work, anhydrous glycerol at different temperatures of 4, 25, 32 and 45°C were applied respectively to in vitro porcine skin, and reflectance and transmittance spectra were then measured dynamically using a spectrometry combined with integrating sphere system. Further, reduced scattering coefficient and penetration depth were obtained. Results showed that, glycerol at different temperatures could induce the reduced scattering coefficient of in vitro skin to decrease and the penetration depth to increase. 4 and 25°C glycerol had similar effect, decreasing the scattering by 48.2% and 49.7%, and increasing penetration depth by 37.9% and 39.5%, respectively. However, 32 and 45°C glycerol treatment could decrease scattering by 61.6% and 76.6%, and increase penetration depth by 53.3% and 84.1%, respectively. In conclusion, glycerol at higher temperature can induce greater and faster skin optical clearing efficacy.

## 1. Introduction

High scattering property of biological tissue for VIS-NIR light limits the penetration depth of light deep into tissue, which has become the obstacle of optical diagnosis and therapy [1]. However, the tissue scattering can be reduced and the penetration depth can be improved by injecting agents with characteristics of high permeability, high refractive index and biocompatibility into tissues, which is called tissue optical clearing technique and was proposed by Tuchin VV [2-4].

Skin is located on the out surface of organisms and has an important role for clinical optical diagnosis and therapy. But it also functions as a natural barrier. The outmost layer of stratum corneum (SC) and the active epidermis limit the permeation of OCAs into tissue to induce optical clearing efficacy. Therefore, skin optical clearing technique has attracted concerns of many researchers. Currently, the technique mainly focused on two aspects. Firstly, the OCAs were directly applied to dermis, including immersion method for *in vitro* skin [5] or hypodermic injection for *in vivo* skin [6, 7]. But the direct contact of dermis with OCAs at high concentration can disrupt the normal skin structure and function, yet OCAs at low concentration cannot induce good optical clearing efficacy [8]; secondly, the OCAs were applied to skin epidermis to permeate into dermis. And the skin surface had been disrupted by some physical or chemical methods such as sandpaper grinding [9], micro-needle rolling [10], ultrasonic delivery [11], photo irradiation [12], chemical enhancement [13, 14] etc.. However, the above methods need some equipment to implement, and what's more, some even need special technical training. Considering all the above studies, we can conclude that all the existing researches have mainly focused on improving the structure of skin by avoiding or reducing its barrier function. In contrast, investigations about the effect of physical properties of agents on skin optical clearing efficacy have not been reported.

Actually, increasing temperature of drugs has been a common method to improve its permeability in transdermal drug delivery system [15-17]. Literatures have reported that penetrability of aspirin was increased by 15 times when its temperature was increased from 10℃ to 40℃. And the absorption of salicylic acid on the abdominal skin of guinea pig can be increased by 5 times when its temperature is increased from 20℃ to 30℃ [18]. So, do the optical clearing agents have the similar characteristic to skin?

The aim of this study was to investigate the optical clearing efficacy of OCAs at different temperatures. A common used agent——glycerol was chosen in this study. Glycerol at different temperatures (4℃, 25℃, 32℃ and 45℃) was applied onto in vitro porcine skin. The dynamic change of skin optical properties was monitored by spectrometry combined with integrating sphere.

## 2. Materials and methods

### *2.1. Animal model and agent*

In vitro porcine skin was purchased from farmer's market, and its stratum corneum has been wiped off. After removing the subcutaneous fat tissue, the skin with similar thickness and color were selected and cut into 5x5 cm2 pieces as samples ready for use.

Glycerol was an analytical reagent (AR) and was produced by Guangcheng Chemical Reagent Company, Ltd. Tianjin. Before each experiment, glycerol contained in a plastic tube was placed into 4℃ refrigerator or 37℃ water bath to maintain for at least 2 hours. During the experiment, the tube was also kept in the same place.

### *2.2. Detection System*

In this work, a commercially available spectrophotometer (Lambda 950, PerkinElmer, USA) with a 150 mm integrating sphere setup was utilized to measure the reflectance and transmittance of in vitro skin samples. The sphere has an entrance-port and an exit-port of 1 inch in diameter. The scanning wavelength range was 400-1400 nm with 10 nm interval. Particularly, the data at 630 nm have been attained to quantitatively analysis.

### *2.3. Experimental Methods*

The standby skin samples were divided into four groups and there was 5~7 samples in each group. All the experiments were conducted under room temperature at 23±2℃. Skin samples in each group were treated respectively by glycerol at different temperatures of 4, 25, 32 and 45℃.

Before experiments, each skin sample was sandwiched between two glass slides with dimension of 50x50 $mm^2$. Then the thickness of sandwich was measured 4 times by a micrometer from 4 directions to calculate an average as its whole thickness. After that, the native transmittance and reflectance spectra were measured using the spectrophotometer with integrating sphere system. And then all skin samples were put on thermostatic plate with temperature of 32℃ and applied glycerol on the epidermal side in the central region. After 5 min, the glycerol was wiped off and the sandwich thickness, reflectance spectra and transmittance spectra were repeatedly measured from glycerol treating time of 5 min to 30 min, with 5 min interval. Each time after measurement, glycerol was applied to the epidermal side. At last, the thickness of skin sample was determined by subtracting thickness of two glass slides from the whole thickness.

### 2.4. *Data Analysis*

Based on values of reflectance and transmittance as well as the thickness of sample, the reduced scattering coefficient and absorption coefficient was calculated using an inverse adding-doubling (IAD, Scott Prahl, 2007) program. And further, the Penetration depth of photons in skin can be calculated by the following formula [19]:

$$\delta = \frac{1}{\sqrt{3\mu_a(\mu_a + \mu_s^{'})}} \qquad (1)$$

Here $\mu_a$ and $\mu_s'$ are absorption coefficient and reduced scattering coefficient at specific wavelength, respectively.

In the study of organism tissue optical clearing technique, the changes of scattering property and penetration depth are usually used to quantitatively evaluate the effect of optical clearing. In order to avoid influence of tissue absorption, near infrared and infrared light with characteristics of high scattering and low absorption was considered in this study. After glycerol treatment, the relative change of reduced scattering coefficient and penetration depth at 630nm were calculated respectively as follows:

$$\mu_{s-rel}^{'} = \frac{\mu_{st}^{'} - \mu_{s0}^{'}}{\mu_{s0}^{'}} \qquad (2)$$

$$\delta_{rel} = \frac{\delta_t - \delta_0}{\delta_0} \qquad (3)$$

Here $\mu_{s0}^{'}$ and $\delta_0$ represent the original reduced scattering coefficient and original penetration depth of skin, respectively. $\mu_{st}^{'}$ and $\delta_t$ represent the reduced scattering coefficient and penetration depth after glycerol treatment for t min, respectively. $\mu_{s-rel}^{'}$ and $\delta_{rel}$ represent the relative change of reduced scattering coefficient and penetration depth of skin after glycerol treatment.

Finally, in order to compare the significance level ($P<0.05$) of optical clearing efficiency of glycerol at different temperature, SPSS 13.0 software was utilized for statistically analyzing the relative change of reduced scattering coefficient and penetration depth in different groups.

## 3. Results and Analysis

### 3.1. *Spectral characteristics changes of in vitro skin before and after glycerol application*

In the study, the reflection and transmission spectrum of skin before and after treatment by glycerol at different temperature for different time were measured. However, due to the difference of skin thickness, it makes no sense to compare the absolute values of reflection and transmission. Instead, the reduced scattering coefficient can represent skin optical clearing efficacy more precisely. Because there is little impact for optical clearing on skin absorption in the visible and near infrared wavelength bands, here, Figure1 shows the typical reflection spectra, transmission spectra and reduced scattering coefficient spectra between 400-1400 nm before and after treatment for 10min, 20min, 30min with glycerol at 4℃, 25℃, 32℃ and 45℃.

From Figure1, it can be seen that treatment of glycerol at different temperature decreases the reflection spectrum and the reduced scattering coefficient spectrum, and increased the transmittance spectrum. The longer treatment by glycerol, the more decreasing of reflection spectrum and the reduced scattering spectrum, and the more increasing of transmission spectrum. As can be seen from the right column, after treated the same time, the greater change of reduced scattering spectrum was induced by glycerol at higher temperature.

### *3.2. The influence of glycerol at different temperatures on reduced scattering coefficient of in vitro skin*

In order to compare the influence of glycerol treatment at different temperatures on reduced scattering coefficient of in vitro skin, Figure2 shows the relative changes of reduced scattering coefficient at 630nm before and after glycerol treatment for 5-30min. After glycerol application, no matter what the temperature is, the reduced scattering coefficients are all decreased. Glycerol treatment at lower temperature of 4℃ and 25℃ induces the similar decreasing degree and changing tendency. However, glycerol at higher temperature can decrease the reduced scattering coefficient very quickly. For example, during the first 10 min, the 32℃ glycerol treatment decreases the parameter to the same degree as glycerol at lower temperature does, but as the treated time prolongs, the 32℃ glycerol treatment decreases the parameter more quickly. Apparently, after the first 5 min, the 45℃ glycerol treatment can decrease the parameter significantly, and after 15 min, it can decrease the parameter to the same degree as 4℃ or 25℃ glycerol does for 30 min.

After 30 min for glycerol treatment, 4 ℃ and 25 ℃ glycerol decrease the reduced scattering coefficient by 48.2% and 49.7%, respectively, and there is no significant difference ($P$=0.786). The 32℃ glycerol treatment decreased the coefficient by 61.6% and has significant difference compared with other temperatures ($P$<0.05). The 45℃ glycerol decreased the coefficient by 76.6% and has extreme significant difference ($P$<0.01).

### *3.3. The effect of glycerol with different temperature on light penetration depth of in vitro skin*

Figure 3 shows the relative change of penetration depth at 630 nm of in vitro skin treated 30 minutes by glycerol with 4, 25, 32 and 45℃. The relative changes induced by 4 and 25℃ glycerol are almost the same, increased by 37.9% and 39.5%, respectively, without significant difference ($P$=0.625). With the increasing of temperature, the penetration depth relatively changed more significantly. Compared with other three temperatures, the penetration depth induced by treatment of 32 and 45℃ glycerol increases by 53.3% and 84.1%, respectively,

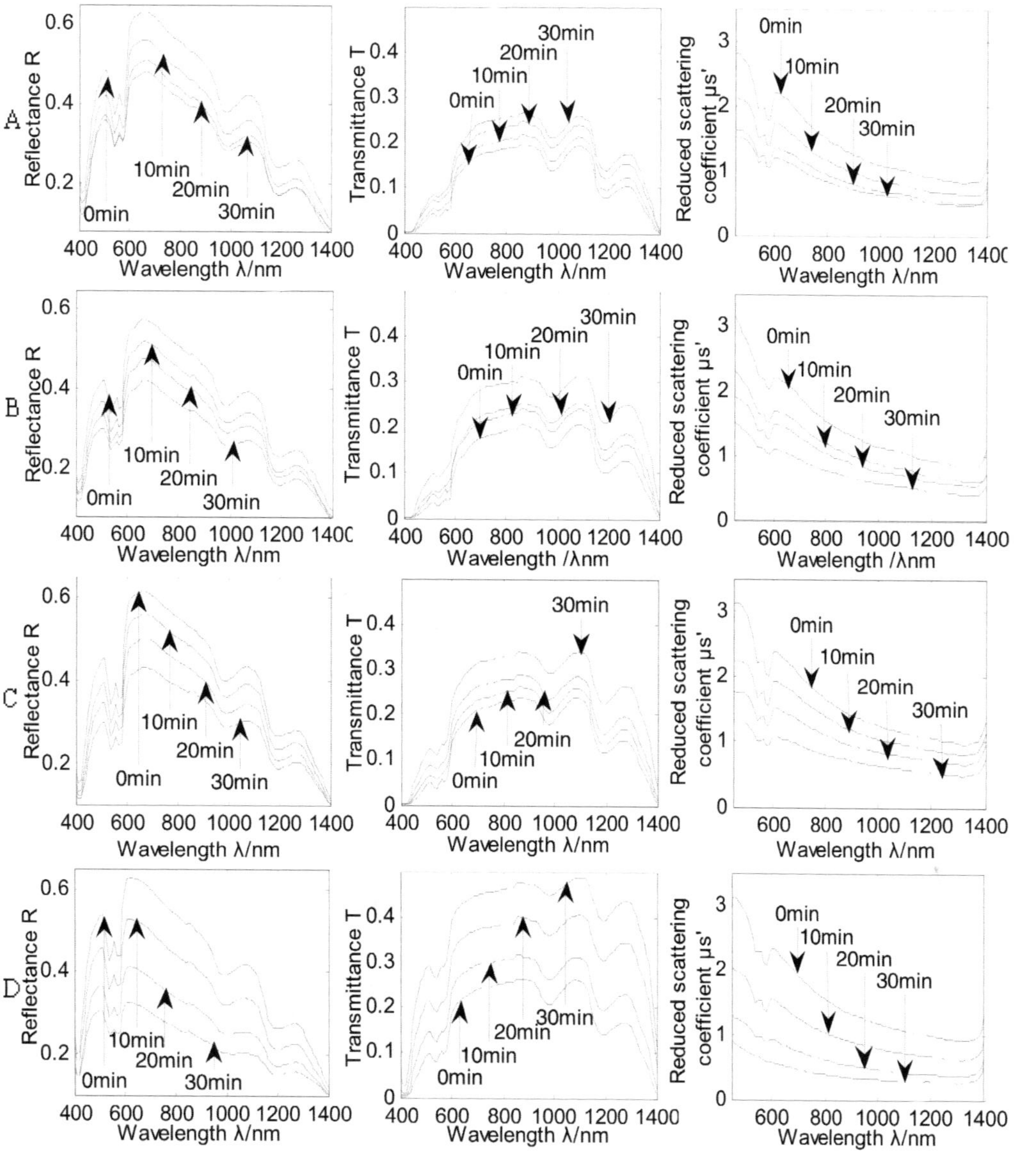


**Figure 1.** Typical changes of skin optical properties induced by anhydrous glycerol. Glycerol temperature: 4, 25, 32 and 45 ℃ for group A, B, C and D; the left, middle and right column represents reflectance, transmittance and reduced scattering coefficient respectively.

and have significant difference ($P<0.05$) and extreme significant difference ($P<0.01$).

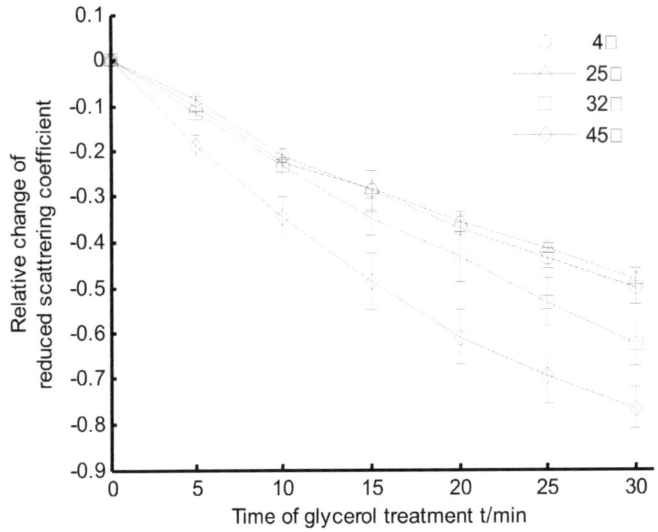


**Figure 2.** Relative reduction of reduced scattering coefficient of *in vitro* skin treated by glycerol at different temperature

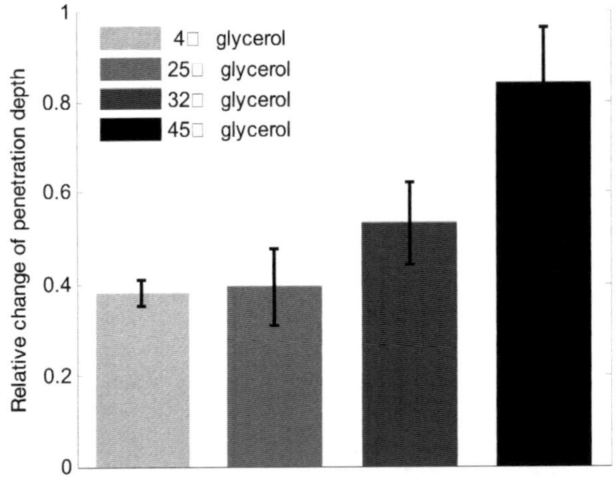


**Figure 3.** Relative reduction of penetration depth on *in vitro* skin induced by glycerol with different temperature

## 4. Discussion

In this work, the influence of OCA's temperature on optical clearing efficacy of in vitro porcine skin was explored. Glycerol was of four different temperature, including refrigeration temperature of 4℃, room temperature of 25℃, normal skin temperature of 32℃ and skin long tolerance temperature of 45℃.

Temperature of OCAs has always been ignored by the previous studies focusing on tissue optical clearing. Actually, OCAs used in the previous studies were either obtained from refrigerator or kept under room temperature which changed with the environment temperature. The different storage temperatures of OCAs may cause different optical efficiency. So it is very important and necessary to invest the influence of glycerol at different temperatures on skin optical clearing efficacy. Though from this study there was no significant difference between the optical efficacy of glycerol at 4 and 25℃, glycerol at higher temperature could induce skin optical clearing much faster and much better. In this study, skin samples were put on a homothermal plate set at temperature of 32℃. The 4℃ Glycerol could gain heat from the skin sample after application and then its temperature may increase, causing the same optical clearing efficacy as 25℃ glycerol did. The 32℃ glycerol had the same temperature as normal skin had, and the force making it penetrate from epidermis deep into dermis was mainly the concentration gradient and molecule diffusion. As the temperature of glycerol increased to 45℃, the thermal motion of glycerol molecule was more intense and the diffusion rate was increased, making it much easier to diffuse deep into dermis. On the other side, increasing of glycerol temperature would decrease glycerol's viscosity. As has been reported in the literature [18] about relations between glycerol's viscosity and its temperature, we could calculate quantitatively that when the temperature increased from 25℃ to 32℃ or even 45℃, its viscosity would decrease from 0.91 Pa.s to 0.51 Pa.s or even 0.2 Pa.s. The decreasing in glycerol's viscosity would facilitate its fluidity in the skin, thus increasing the diffusion rate of glycerol molecule.

From both molecule-kinetic theory and glycerol's viscosity, we can conclude that increasing of glycerol's temperature would increase its perfusion rate and facilitate glycerol perfusion into dermis, which make more glycerol molecule penetrate into dermis to induce optical clearing. The high permeability of glycerol would induce tissue dehydration, that is the tissue fluid with lower index of refraction would be replaced by glycerol with higher index, and make the refractory index of different components in the skin matched, which could decrease light scattering and achieve optical clearing efficacy. Thus, increasing the temperature of glycerol could enhance skin optical clearing significantly. However, there is limitation for skin to endure higher temperature for long time, glycerol with temperature higher than 45℃ should not be applied to skin for long time. For this consideration, it may also effective to enhancing skin optical clearing by increasing skin's temperature or exposing skin to a high-temperature in super-short time [20]. In addition, investigations on the combination of this work and other physical or chemical methods to enhance skin optical clearing are also deserve further study.

## 5. Conclusion

From this study, it can be concluded that glycerol at different temperatures has different effect on skin optical clearing. Compared with normal skin temperature, glycerol at lower temperature has almost the same effect on enhancing optical clearing, and glycerol at higher temperature could induce skin optical skin clearing much faster and more significant, especially, the higher the temperature, the better and faster the enhancing optical clearing efficacy. This study provides an easy and effective way to enhance skin optical clearing.

**Acknowledgement**

This study was supported by the Undergraduate Innovation Technology Foundation in 2010 of Huazhong University of Science and Technology, and the National Nature Science Foundation of China (Grants No. 30770052, 30911120074). The authors also would like to thank VV Tuchin from Saratov State University for his suggesting and help.

**Reference**


[1] Tuchin V.V., Maksimova IL, Zimnyakov DA, *et al.* 1997 Light propagation in tissues with controlled optical properties *J. Biomed. Opt.* **2** 401-17

[2] Tuchin V.V. 2007 *Tissue optics: Light scattering methods and instruments for medical diagnosis* SPIE Tutorial Texts in Optical Engineering TT38: Bellingham, WA. 2000; 2nd edition, (Bellingham: SPIE Press)

[3] Vargas G. 2001 Reduction of light scattering in biological tissue: implications for optical diagnostics and therapeutics *Austin: the University of Texas*

[4] Khan M.H., Choi B., Chess S., *et al.* 2004 Can topically applied optical clearing agents increase the epidermal damage threshold and enhance therapeutic efficacy *Lasers Surg. Med.***35** 93-95

[5] Tuchin V.V. 2005 Optical clearing of tissues and blood using the immersion method *J. Phys. D.:Appl. Phys.* **38** 2497-518

[6] Vargas G., Chan E.K., Barton J.K., *et al.* 1999 Use of an agent to reduce scattering in skin *Lasers Surg. Med.* **24** 133-41

[7] Galanzha E.I., Tuchin V.V., Solovieva A.V., *et al.* 2003 Skin back reflectance and microvascular system functioning at the action of osmotic agents *J. Phys. D.:Appl. Phys.***36** 1739-46.

[8] Mao Z., Han Z., Wen X., Luo Q. and Zhu D. 2008 Influence of glycerol with different concentration on skin optical clearing and morphological changes in vivo *Proc SPIE* **7278** 72781T

[9] Stumpp O., Chen B. and Welch A. J. 2006 Using sandpaper for noninvasive transepidermal optical skin clearing agent delivery *J. Biomed. Opt.* **11** 041118

[10] Yoon J., Son T., Choi E-h., Choi B., *et al.* 2008 Enhancement of optical skin clearing efficacy using a microneedle roller *J. Biomed. Opt.* **13** 021103

[11] Xu X. and Zhu Q. 2008 Sonophoretic delivery for contrast and depth improvement in skin optical coherence tomography *IEEE J Sel Topics Quantum Electron* **14** 56–61.

[12] Liu C., Zhi Z., Tuchin V. V., Luo Q. and Zhu D. 2010 Enhancement of skin optical clearing efficacy using photo-irradiation *Lasers Surg. Med.* **42** 132-140

[13] Mao Z., Han Z. and Zhu D. 2009 Experimental study on *in vivo* rat skin optical clearing at the action of Thiazone *Acta Optica Sinica* **29(7)** 1921-1925

[14] Xu X. and Zhu Q. 2007 Evaluation of skin optical clearing enhancement with Azone as a penetration enhancer *Opt. Commun.***279** 223-228

[15] Jain A.K. and Panchagnula R. 2003 Effect of temperature on imipramine hydrochloride permeation: role of lipid bilayer arrangement and chemical composition of rat skin *Int. J. Pharm.* **250(1)** 287-93

[16] Ogiso T., Hirota T., Iwaki M., *et al.* 1998 Effect of temperature on percutaneous absorption of terodiline, and relationship between penetration and fluidity of the stratum corneum lipids *Int. J. Pharm.* **176(1)** 63-72

[17] Ping Qineng 1998 *Modern Pharmaceutics* (Beijing: Chinese Medicine Science and Technology Press) p 550-558

[18] Cheng N 2008 Formula for the viscosity of a glycerol-water mixture *Ind. Eng. Chem. Res.* **47(9)** 3285-88

[19] Ritz J.P., Roggan A., Isbert C., *et al.* 2001 Optical properties of native and coagulated porcine liver tissue between 400 and 2400 nm *Lasers Surg. Med.* **29** 205–12

[20] Park J.H., Lee J.W., Kim Y.C., *et al.* 2008 The effect of heat on skin permeability *Int. J. Pharm.* **359( 1-2)** 94-103

# A Novel Approach of Cardiac Segmentation In CT Image Based On Spline Interpolation

**Yuan Gao**[1], **Pengcheng Ma**[2]

College of Optoelectronic Science and Engineering, Huazhong University of Science and Technology, Wuhan 430074, China

E-mail: gyolive@gmail.com

**Abstract.** Organ segmentation in CT images is the basis of organ model reconstruction, thus precisely detecting and extracting the organ boundary are keys for reconstruction. In CT image the cardiac are often adjacent to the surrounding tissues and gray gradient between them is too slight, which cause the difficulty of applying classical segmentation method. We proposed a novel algorithm for cardiac segmentation in CT images in this paper, which combines the gray gradient methods and the B-spline interpolation. This algorithm can perfectly detect the boundaries of cardiac, at the same time it could well keep the timeliness because of the automatic processing.

## 1. Introduction

The x-ray computerized tomography (CT) provides a 2D map of details of the organs which are the linear attenuation coefficient mirroring morphological . It is a high spatial resolution and wide dynamic range imaging modality. The X-ray absorption levels of organs and tissues in CT images are represented by different gray. Thus, as the black and white areas shown in X-ray images shadows, the black shadow reflect the low absorption area, representing to low-density areas, such as lungs with gas; white shadow reflect the high absorption areas, representing to high-density areas, such as the skeleton.

In CT image the cardiac are often adjacent to some tissue, such as the parietal, and their gray-scales are within proximity to each other, meaning that there are not obvious gray gradient in the boundary. Because the classical method such as histogram distribution method, Genetic Algorithm, Seed algorithm, ASM (Active Shape Model) ,and ACM (Active Contour Model) are all based on the clear gray gradient, it may not deal with this problem effectively. At the same time, there are many false contours and boundaries in the image causing by the object's movements and random noise resulting from the external field interference. What is more, with analyzing the Series images, besides above features, we found another characteristic that the contours of cardiac and parietal change gently.

Based on these image characteristics, we proposed a new algorithm to achieve heart segmentation. In this algorithm, it could select samples on both sides of the parietal automatically, and then utilize these samples to fit the contour of upper heart, which are adjacent to the parietal, using the spline interpolation theory. We chose the B-spline interpolation to fit those samples. B-spline interpolation is a kind of numerical analysis by nature. The idea of this method is predicting the unknown contours, with selecting samples on the existing or readily available contours. It matches with those three features definitely.

### 2.1 B-spline Interpolation

B-spline interpolation is one significant technology in the interpolation & fitting theory, it belongs to the numerical analysis area apparently. The main idea of interpolation is using the samples to predict the unknown, but relative data. The method concludes the interpolation polynomial method, difference quotient method and spline curve method. B-spline curve is a common technology about spline curve method, it can be always utilized to construct a smooth curve, and widely applied in the computer graphics and CAD (computer-aided design) area.

To ensure the accuracy and save the cost of computation, it always chooses the cubic B-spline curve. In this paper, we also chose the cubic method: assume that sample points are Pi=(xi, yi), i=0, 1, ······, n, the Cubic B-spline can be described as follow:

$$B_i(u) = \sum_{k=-1}^{2} b_k P_{i+k} \tag{1}$$

In the formula, $P_i$ represents the ($x_i$ , $y_i$) point, $b_k$ is the vector base, it doesn't vary regardless of the changes of sample, it is also the weight values. With the weight factor evolving from 0 to 1, B spline interpolation can be obtained using weight sum.

In the cubic B-spline interpolation, the values of $b_k$ could be expressed with the matrix as fellow :

$$B_i(u) = \frac{1}{6}(u^3, u^2, u, 1)\begin{pmatrix} -1 & 3 & -3 & 1 \\ 3 & -6 & 3 & 0 \\ -3 & 0 & 3 & 0 \\ 1 & 4 & 1 & 0 \end{pmatrix}\begin{pmatrix} P_{i-1} \\ P_i \\ P_{i+1} \\ P_{i+2} \end{pmatrix} = \frac{u^T M_b F}{6} \tag{2}$$

If you don't wish to use the Word template provided, please set the margins of your Word document as follows.

### 2.2 Boundary detection algorithm based on LoG operator

LoG(Laplace & Gauss) algorithm utilizes zero cross points of the second derivative to get the object boundary , considering the local maximum of first derivative correspond to the zero cross points. Assume that the original image is g(x,y), we can get output of the LoG operator with the expression as

$$h(x, y) = \nabla^2[G(x, y) * g(x, y)] \tag{3}$$

Apply the Convolution theorem to the expression, it can be written as

$$h(x, y) = [\nabla^2 G(x, y)] * g(x, y) \tag{4}$$

$\nabla^2$ is the Laplace operator.

$$\nabla^2 = \frac{\partial^2}{\partial x^2} + \frac{\partial^2}{\partial y^2} \tag{5}$$

$$G(x, y) = \frac{1}{2\pi\delta^2} e^{-\frac{x^2+y^2}{2\delta^2}} \tag{6}$$

Then, $\nabla^2 G(x,y)$ can be given by

$$\nabla^2 G(x,y) = \frac{\partial^2 G(x,y)}{\partial x^2} + \frac{\partial^2 G(xy)}{\partial y^2} = \frac{1}{\pi\delta^4}\left(\frac{x^2+y^2}{2\delta^2} - 1\right) e^{-\frac{x^2+y^2}{2\delta^2}} \tag{7}$$

For those discrete digital images, LoG operator can often be approximated by a discrete LoG template m (x, y). Parameter δ is related to the template width ω. Their relationship could be described as bellow: if ω is smaller compared to δ, the boundary location can be more accurate, meaning that more slight changes can be detected, on the contrary, if ω is bigger compared to δ, the boundary detected will deviate from the original boundary location, and some significant details could be lost。 Fortunately, there are empirical formula to determine the value selection of ω and δ . Thus, we could calculate the LoG template m(x,y) based.

For the discrete digital image, we can approximate the output of the LoG operator as follow:

$$h(x, y) = m(x,y) * g(x,y) \tag{8}$$

With applying the LoG operator, the output h(x,y) of low gray value will be greater than zero, the output h(x,y) of high gray value will be less than zero conversely. Then the boundary of original image comes out definitely as zero track of the output h(x,y).

### 2.3 Determine the threshold and binarization

At first we could count the all the pixels of boundary points, and then calculated the average value Th, and set the Th as threshold.

Then we can binarize the original image with the following formula:

$$bw(x, y) = \begin{cases} 1 & g(x, y) \text{ is in the area} \\ 0 & g(x, y) \text{ is out of the area} \end{cases} \tag{9}$$

bw(x,y) is the binarization of the image

## 3 Specific steps and results:

Because the gray values of parietal and pulmonary have high contrast, we can use those classical segmentation methods to detect the boundary. In this algorithm, we chose the Log algorithm (Laplace and Gauss) to detect the boundary and extract those eight neighborhood connecting points, and finally extracted the coordinate of each point.

Our aim is to select some uniform sample points on the Parietal, so the choice of samples is based on the spatial characteristics of the lung boundary, if necessary, we should take one or two more points in the upper of heart to amend the direction of curve, this means adding the prior experience of human, which can contribute the accuracy to the segmentation result definitely. Specific Procedure is bellow:

(1)Eliminate some of higher contrast regions, use method of histogram equalization to enhance the contrast of image; (2) Apply the threshold segmentation and binarization methods to obtain contour of lungs, which is part of the parietal; (3)Select uniform samples on the parietal based on Spatial feature, and take one or two samples on the upper of heart interactively; (4)Store coordinates of these sample point, then apply B-spline interpolation method to obtain the upper boundary of the cardiac; (5)Utilize the classical algorithms on the image to obtain the lower boundary of the cardiac; (6)Combine these two part boundaries of the cardiac into one integral boundary.

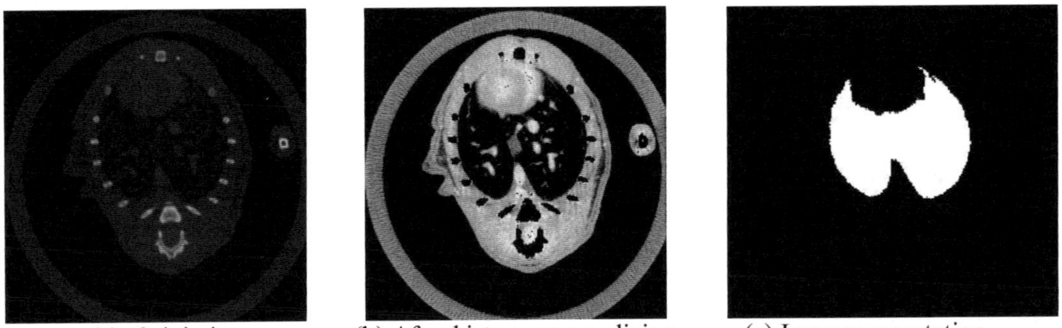

(a) Origin image (b) After histogram equalizing (c) Lung segmentation with LoG

**Figure 1**

Figure 1(a) is the origin image, its size is 250×250; Figure 1(b) is the image obtained by eliminating bones and equalizing the histogram; Figure 1(c) is the lung binarization image, which contributes to part of the parietal contour;

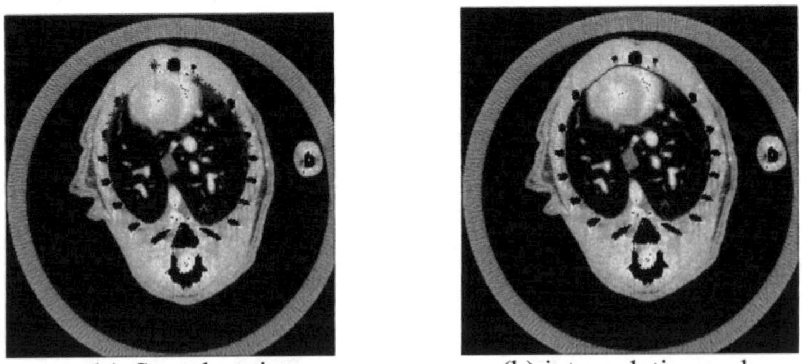

(a) Sample points (b) interpolation and fitting curve

**Figure 2**

Figure 2(a) is the enhanced image with selected points; Figure 2(b) is the upper boundary of cardiac with applying B-spline interpolation and fitting;

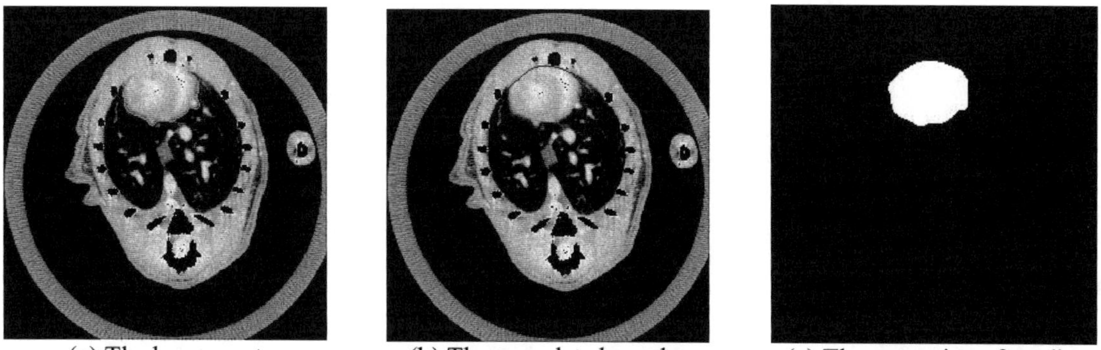

(a) The lower part boundary of cardiac (b) The complete boundary (c) The exatration of cardiac

**Figure 3**

Figure 3(a) is the lower part boundary of the cardiac; Figure 3(b) is combination of those two boundaries, which is definitely the boundary of cardiac; Figure 3(c) is the heart binarization image, it is definitely the foundation of 3D reconstruction

## Conclusion

In this article, we introduced a noble cardiac segmentation algorithm in CT images. This algorithm is mainly based on the gray gradient methods and the B-spline interpolation method. In the algorithm, some sample points was chosen based on the spatial Information of heart, most of them are chosen automatically, one or two sample points are taken interactively to amend the trend of spline if necessary. From the results given above, it can extract heart contour from CT image effectively and premise the timeliness well. The next step of our work is to adjust the parameters to achieve all the segmentation from the series of CT images, the data obtained will be used for the heart 3-D reconstruction.

## Reference


[1] Curtis F.Gerald, Patrick O.Wheatley, "Applied Numerical Analysis", 7th edition, Pearson Education, Inc, ISBN 0-321-13304 -8

[2] Richard L.Burden, J. Douglas Faires, "Numerical Analysis", 7th edition, Thomson Learning, ISBN 0 -543-38216- 9

[3] FU Zengliang, CHEN Xiaojun, YE Ming, LI Feng, WANG Chengtao, "Method for Heart CT Image Segmentation", Computer Engineering, Vol. 35, No. 12, PP. 189 – 191, 2009

[4] Li Kaiyu, "Image Edge Detection Based on B - Spline Interpolation", Journal of Nanjing University of Aeronautics & Astronautics, Vol. 39, No. 2, PP. 198 – 203, 2009

[5] T. F. Cootes, C. J. Taylor, D. H. Cooper, J. Graham, "Active Shape Models - Their Training and Application", Computer Vision and Image Understanding, Vol. 61, NO. 1, January,PP. 38-59, 1995

[6] Michael Kass, Andrew Witkin, Demetri Terzopoulos, "Snakes: Active Contour Models", International Journal of Computer Vision, PP. 321 – 331, 1988

[7] O Ecabert, et al, "Towards Automatic Full Heart Segmentation in Computed-Tomography Images", Computers in Cardiology, IEEE, PP. 223−226, 2005

[8] Bruno Golosio , et al, "A novel multi–threshold method for nodule detection in lung CT", Medical Physics, Vol. 36, No. 8, 2009

# The determination of captopril in Solution by Raman spectroscopy

**Junxiang Gao, Huaimin Gu*, Xiao Dong, fangfang liu**

MOE Key Laboratory of Laser Life Science, College of Biophotonics, South China Normal University, 510631, Guangzhou, China

**Abstract.** Captopril, 1-[(2S)-3-mercapto-2-methyl propionyl]-Lproline, is an angiotensin converting enzyme (ACE) inhibitor, which reduces peripheral resistance and lowers blood pressure. It is widely used in the hypertensive ailments and incongestive heart failure treatment. Due to such crucial pharmacological importance, development of simple and accurate methods for the determination of captopril is desired. In this work, the normal Raman spectra of the captopril in different concentrations were studied, and the relationship between the Raman intensity and the concentrations of the captopril was quantificationally analysed. By selecting appropriate characteristic Raman bands of the cptopril, the solution of some captopril purchased in a local pharmacy was quantificationally determined. A quantificational linear relationship between the Raman intensity and the concentrations of captopril was obtained, and it is little affected by other compounds in the solution of captopril. This study provides an effective technique for the quantificational determination of captopril in solutions, and it has a potential application in the analysis of medicament.

**Keywords:** Raman spectroscopy; Captopril; Quantitative determination

## 1. introduction

Captopril, 1-[(2S)-3-mercapto-2-methyl propionyl]-Lproline, is an angiotensin converting enzyme (ACE) inhibitor. This medication work to block an enzyme system, which causes artery walls to relax, reducing blood pressure. It is widely used in the treatment of hypertension. Other uses of captopril have also been reported, such as decreasing high blood pressure caused by blood vessels in the kidneys, decreasing symptoms of cystinuria, reducing rheumatoid arthritis symptoms, treating Raynaud's phenomena and progression of kidney disease in peoples with diabetes [1,2]. Due to such crucial pharmacological importance, development of simple and accurate methods for the

* Corresponding author. *Tel.:* +86-20-85216972; *E-mail address:* guhm@scnu.edu.cn

determination of captopril is desired. Determination of captopril has previously been reported by capillary electrophoresis, flow injection analysis[3], high performance liquid chromatography, kinetic spectrophotometry[4], FT-Raman spectroscopy[5], voltammetry[6], gas chromatography–mass spectrometry etc[7]. But these methods require many special sample preparation.

Raman spectroscopy, as a scattering technique, is well known for the minimum of sample handling and preparation that is required [8]. Compared with many other common analytical methods, this technique does not require any special sample preparation, so analysed drugs are usually in their unmodified forms. In spite of these advantages, application of Raman spectroscopy to captopril quantification is not widespread. In this paper, we have studied the normal Raman spectra of the captopril in different concentrations from 0.67% to 3.33%. the solution of some captopril purchased in a local pharmacy was quantificationally determined. A quantificational linear relationship between the Raman intensity and the concentrations of captopril was obtained.

## 2. experimental

### *2.1. sample*

Captopril, with a mass fraction purity of 0.99, was purchased from Sigma Aldrich. The samples were used without further purifications. The solutions were prepared by dissolving the sample into doubly distilled water. The Concrete steps were as follows: first, we accurately weighed 5g of the captopril powder, and the powder was dissolved in 150 ml doubly distilled water. The volumes of doubly distilled water, 10 ml, 20 ml, 40 ml, 80 ml, were prepared. Then the solution of the captopril (40 ml) were added to these doubly distilled water, respectively. The final concentrations of captopril were 3.33%, 1.67%, 1.11%, 0.83% and 0.67% respectively. The captopril tablets (12.5 mg/tablet) were purchased from the local pharmacy. The solution of the captopril tablets were prepared by dissolving eight tablets into the 9ml doubly distilled water, and the concentration is 1.11%.

### *2.2. Instruments*

A 514.5 nm argon ion laser was used as Raman excitation. The spectral line width of laser was further spectrally purified with an interference type laser line band-pass filter, and a notch filter (HSPF-514.5-1.0, Kaiser Optical Systems Inc., USA) was used to reject the excitation light while allow Raman signal entering the spectrometer system.

The Raman system was equipped with a BX-41 Olympus microscope and 45× objectives. Spectrometer system (Acton spectro@2300i, Princeton Acton, USA) was used with a 1200 g/mm holographic grating. An air cooled CCD detector (Pixis 256, Princeton Acton, USA) was used for measuring the Raman signal by an integration method. Exposure time of CCD was 50 s and 1 scan. The laser power at the samples was 8 mW.

## 3. Results and discussion

### *3.1. Raman spectrum of captopril in different concentrations*

In Fig. 1 Raman spectra of the captopril in the concentration 3.33% are presented. The range is from

3560 to 230 $cm^{-1}$. The Raman peaks at 2997, 2947, 2895, 2583, 1606 and 1458 $cm^{-1}$ are very obvious. In the quantitative determination of the captopril, in order to ensure accuracy, the Raman bands with high intensity should be selected as the quantitative band. Thus, we will focus on analysis of these bandss and the vibration mode of these bands. The peaks at 2997 and 2947 $cm^{-1}$ should be assigned to the asymmetric CH3 and CH2 stretching vibration, and the peak at 2895 $cm^{-1}$ should be due to the symmetric CH3 stretching mode. The peak at 2583 $cm^{-1}$ corresponds to the SH stretching vibration. The peaks at 1606 $cm^{-1}$ should be assigned to the amide band. The peaks at 1482 and 1458 $cm^{-1}$ are due to the asymmetric and symmetric CH3 bending vibrations, respectively, and the peak at 1340 $cm^{-1}$ should be assigned to the OH bending vibration[9,10]. In addition, the wide band in the range 3100–3500 $cm^{-1}$ is assigned to the liquid water, and the peaks are at 3273 and 3421 $cm^{-1}$. The Raman peak of the liquid water will be chosen as the internal standard peak. It not only simplify the sample preparation process, but also avoid effect of other substances on the determination of captopril.

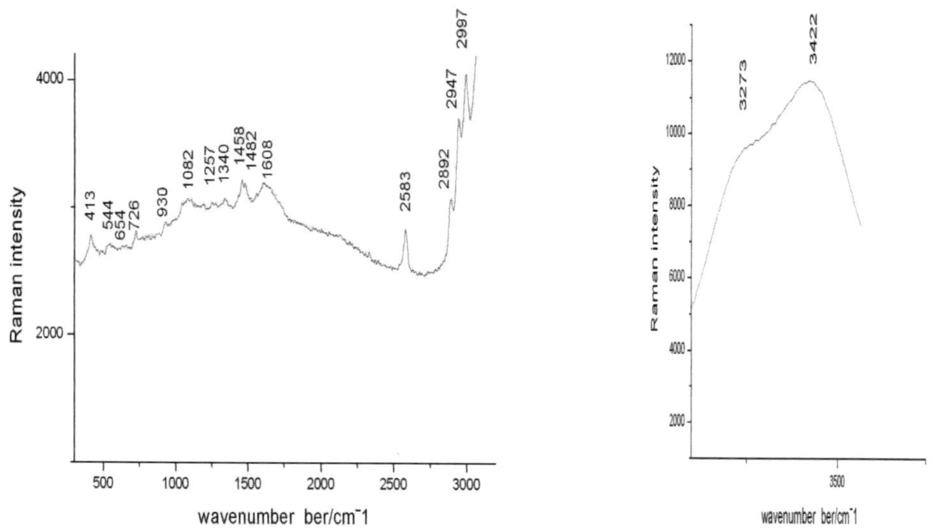


**Figure 1.**Raman spectrum of 3.33% captopril

*3.2. the selection of the internal standard substances and the quantitative peak*

In the direct quantitative determination of the captopril, some uncontrollable factors, such as the different concentration of samples causing changes in refractive index, and the background noise will bring the Raman spectra the unpredictable impact. Therefore, it's difficult to determinate the concentration of captopril by directly comparing the intensity of Raman spectra of different samples, and it requires adding the appropriate internal standard to eliminate the impact. In this paper, the Raman peak of the liquid water at 3421 $cm^{-1}$ was chosen as the internal standard peak[11]. it not only simplify the sample preparation process, but also solve the quantitative problems. It is a convenient internal standard in quantitative analysis of aqueous samples.

In the experiment, the acquisition time of each Raman spectrum is 50s.Because the Raman spectroscopy divided into two section scan, we should choose the Raman bands closing to the Raman peaks of water as the quantitative peak. After testing, as the concentration of the captopril decreases, the intensity of the several peaks closing to the peaks of water decrease, the SH stretching vibration is characteristic Raman bands of the captopril. In this study, we chose the Raman peak of the captopril at 2854 $cm^{-1}$ closing the internal standard as the the quantitative peak of the captopril, and chose the relative peak intensity ratio between 2854 $cm^{-1}$ and 3422 $cm^{-1}$ as the response of the determination of captopril.

*3.3. the standard curve and the linear range*

In Fig. 2 Raman spectra of the captopril in different concentrations are presented. As can be seen from Fig.2, as the concentration of the captopril increases, the relative intensity of Raman spectra of captopril in solution increases, and the peak intensity of the water remains basically unchanged. The radio of the intensity of the characteristic peak of the captopril(Ic) at 2584 $cm^{-1}$ and the water(Iw) at 3422 $cm^{-1}$ constitute the relative intensity ratio(R=Ic/Iw). The measured results can be seen in Table 1. The intensity data of the characteristic peak can be directly read out by the software origin 7.5. In Fig. 2, a quantificational linear relationship between the relative intensity ratio R and the concentrations of captopril is shown. The linear working range is from 0.67% to 3.33%. The linear regression equation is R=0.13085+0.03504C. The linear correlation coefficient is r=0.99359 . The result is very satisfactory for the quantitative analysis and the linear regression equation is very useful. The relative intensity ratio(0.67~3.33%) can be predicted.

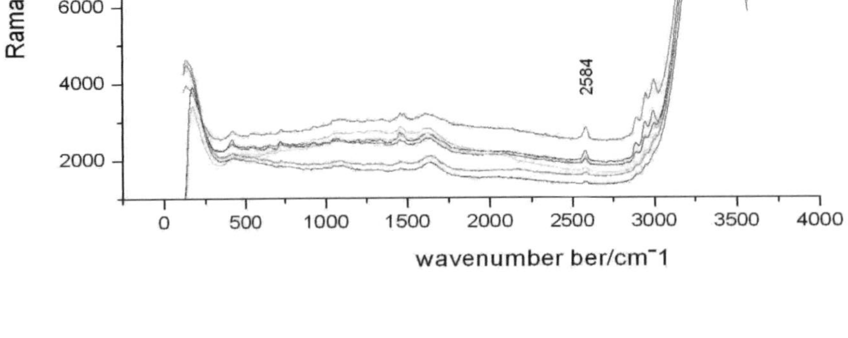


**Figure 2**.Raman spectra in different concentrations of captopril

| Table 1.Intensity and ratio of characteristic peak | | | |
|---|---|---|---|
| Captopril (%) | Ic (2584 $cm^{-1}$) | Iw (3422 $cm^{-1}$) | R |
| 3.33% | 2836 | 11462 | 0.247426 |
| 2.67% | 2208 | 9765 | 0.226114 |
| 2.22% | 2069 | 9961 | 0.20771 |
| 1.67% | 1987 | 10760 | 0.184665 |
| 1.11% | 1783 | 10192 | 0.174941 |
| 0.83% | 1677 | 10234 | 0.163866 |
| 0.67% | 1419 | 9513 | 0.149164 |

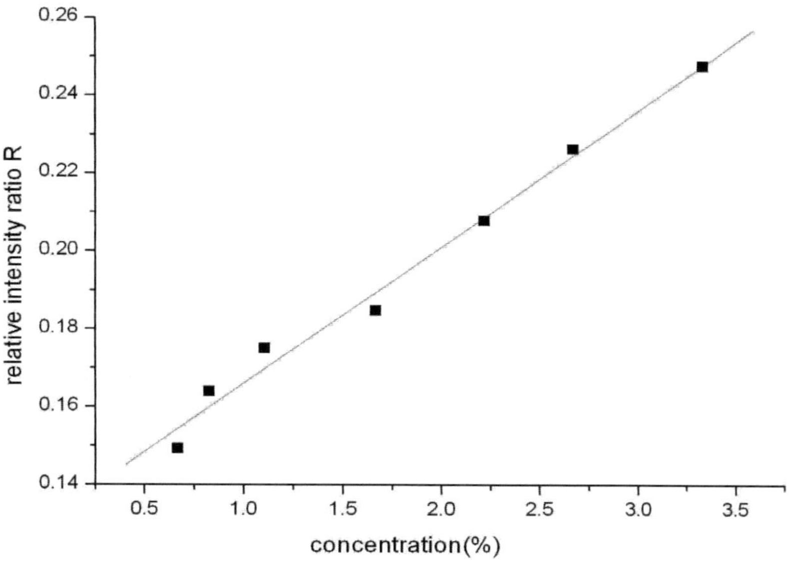


**Figure 3.** Calibration curves errors for captopril in different concentrations

*3.4. the determination of captopril tablets in Solution*

The captopril tablets (12.5mg/tablet) were purchased from the local pharmacy. The solution of the captopril tablets were prepared by dissolving eight tablets into the 9ml doubly distilled water. First, we crushed eight captopril tablets and dissolved the powder into the 9ml doubly distilled water, gently shaked, then the solution was placed for an hour. When the captopril fully dissolved in water we can found a small amount of material sinking to the bottom. A droplet (50μl) of supernatant was dropped on a clean glass slide and was measured on Raman Spectrometer. In Fig. 4 Raman spectra of solution of the captopril tablets in the concentration 1.11% are presented. The relative intensity ratio Rt= 0.171217. From the linear regression equation, we can conclude that the relative intensity ratio of the captopril in the concentration 1.11% is R = 0.174941. The two are very close, and it indicate that the experiment result is satisfactory. This shows that the fitted quantificational linear relationship can be applied to the determination of captopril tablets. By selecting appropriate characteristic Raman bands of the cptopril, the impurities contained in the tablets will not affect the determination of the captopril tablets in the liquid water.

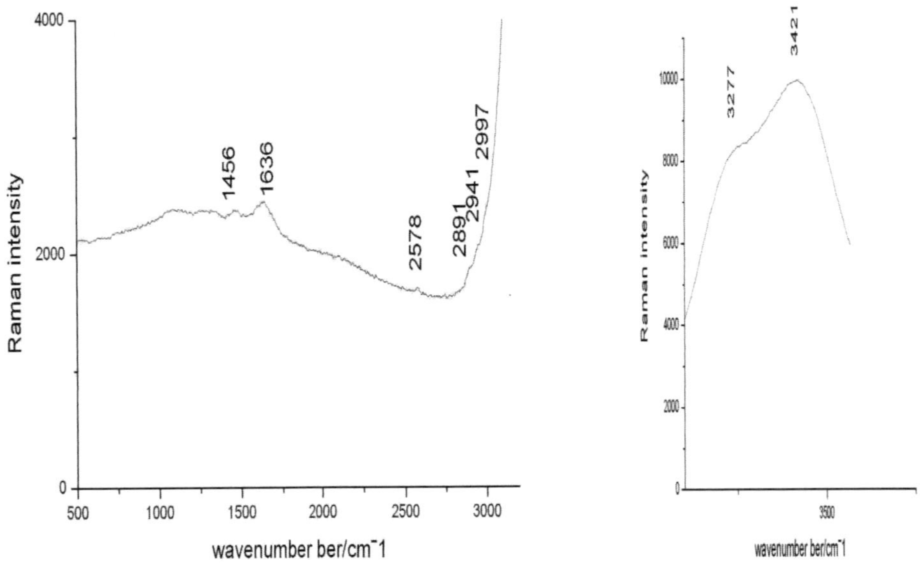


**Figure 4.**Raman spectrum of 1.11% captopril tablets

## 4. Conclusion

This study confirms the Raman spectroscopy can be used to determine the captopril in solution, and a quantificational linear relationship between the Raman intensity and the concentrations of captopril was obtained. The proposed method has the advantages of simplicity and rapidity for the determination of the captopril. The determination of captopril tablets purchased from local market in solution has verified this result and shows the possibility of further application in the business.


## ACKNOWLEDGMENTS

This research is supported by the National Natural Science Foundation of China (60678050).


## References


[1] S. Shahrokhian et al 2005 *Sensors and Actuators* B 109 278–284

[2] S. Mazurek, R. Szostak 2006 *Journal of Pharmaceutical and Biomedical Analysis* 40 1225–1230

[3] F. Tache, A. Farca, A. Medvedovici, V. David 2002 *J. Pharm. Biomed. Anal.* 28 549.

[4] J.A.V. Prior, J.L.M. Santos, J.L.F.C. Lima 2007 *Anal. Chim. Acta* 600 183.

[5] S. Mazurek, R. Szostak 2006 *J. Pharm. Biomed. Anal.* 40 1225.

[6] H. Parham 2005 *B. Zargar. Talanta.* 65 776.

[7] M.E. Franklin, R.S. Addison, P.V. Baker, W.D. Hooper 1998 *J. Chromatogr. B* 705 47.

[8] W.E. Smith and G. Dent 2005 *Modern Raman Spectroscopy – A Practical Approach* (England: John Wiley & Sons Ltd) pp11-12

[9] Jankovics, H., C. Pettinari, F. Marchetti, E. Kamu, L. Nagy, S. Troyanov, L.Pellerito 2003 *J. Inorg.Biochem* 97, 370–376.

[10] Smith, B 1999 *Infrared Spectral Interpretation: A Systematic Approach* (Boca Raton, Florida:CRC Press)

[11] L iuWen-Han , YangWei , Wu Xiao-Qiong , L in Zhen-Xing 2007 *Chinese J. Anal.Chem* 35 (2) : 416～418

# Normal Raman and SERS spectroscopy of the vitamin E

**Tiantian Cai, Huaimin Gu[1], Xiaojuan Yuan, Fangfang Liu**
MOE Key Laboratory of Laser Life Science, College of Biophotonics, South China Normal University, 510631, Guangzhou, China

**Abstract.** In this study, surface-enhanced Raman scattering(SERS)spectra of vitamin E were obtained on colloidal silver(Ag). Alpha-( -) tocopherol which is the only form that is recognized to meet human requirements was selected to study. The analytes (±)- -tocopherol were dissolved in chloroform ($CHCl_3$) and the silver colloid was poured into the compound. Silver colloid was reduced by hydroxylamine hydrochloride. The analytes were the supernatant after standing the mixture for the reason that chloroform have no signals in surface-enhanced Raman scattering in the Ag colloid, and it would not affect the determination of the (±)- - tocopherol. The Normal Raman and SERS spectrum of Vitamin E were contrastively studied to realize how the vitamin E stuck to the silver nanoparticles. The results show the fat-soluble substances can be analysed by SERS. The spectra indicate that the molecules are adsorbed on the surface through the COO- groups by the simultaneous involvement of a and -type coordination. These results suggest some important criteria for consideration in SERS measurements and also provide important insights into the problem of predicting SERS activities for different fat-soluble substances.

**Keywords:** Surface-enhanced Raman spectroscopy; Raman spectroscopy; Vitamin E; Chloroform ($CHCl_3$); Ag colloid

## 1. Introduction

The fundamental knowledge on the structure of atoms or molecules we have today is a result of laser spectroscopic researches. In particular, Raman spectroscopy allows one to derive detailed specific information on a molecular level that other laser spectroscopic methods can provide only to a limited extent. So, a Raman spectrum provides a structural 'fingerprint' of a molecule.[1~3] The advantages of Raman spectroscopy are its unprecedentedly high specificity and its versatility. However, a great disadvantage in any application of Raman spectroscopy results from the extremely small cross section of the Raman process, which is 12–14 orders of magnitude below fluorescence cross sections.

Therefore, in the 1970s, a discovery which showed unexpectedly high Raman signals from pyridine on a rough silver electrode attracted considerable attention[4~6], particularly after experiments in different laboratories gave evidence that the enormously strong Raman signal must be caused by a true enhancement of the Raman scattering (RS) efficiency itself and not by more scattering molecules[7, 8]. Within the next few years, strongly enhanced Raman signals were verified for many different molecules [9] which had been attached to various 'rough' metal surfaces, and the effect was named 'surface-enhanced Raman scattering (SERS)'. The discovery showed promise to overcome the traditionally low sensitivity problem in Raman spectroscopy.

Currently, most SERS techniques use metal aqueous colloids as substrate[10~12], which require the objects studied must be water-soluble compounds, restricting the wild application of SERS technique.

[1] **Corresponding author.** ***Tel.:*** **+86-20-85216972;** ***E-mail address:*** **guhm@scnu.edu.cn**

However, considerable part of the analyte is water-insoluble compound. It's hard to produce the solution of the fixed concentration. It is unfavorable to use Ag colloid directly to measure their Surface Enhanced Raman Spectroscopy (SERS). There is a method to overcome this embarrassment. First, water-insoluble compounds is dissolved in organic solvents; then, the solution which contain the compound is drip onto the metal substrate surface with the surface-enhanced Raman activity; or the metal substrate is immersed in the sample solution and is got out (taken out) after a certain time; and then the metal substrate with compound is measured by the Raman spectrograph after the solvents dry. The main shortcomings of this method is that the substrate or the compound may occur the carbonation reaction in the laser's irradiation, which result in strong spectral background and baseline, increasing the noise, weakening the signal and decreasing spectral resolution.

"Vitamin E" is the collective name for a group of water-insoluble compounds with distinctive antioxidant activities. Naturally occurring vitamin E exists in eight chemical forms (alpha-, beta-, gamma-, and delta-tocopherol and alpha-, beta-, gamma-, and delta-tocotrienol) that have varying levels of biological activity. In this investigation, we selected Alpha- ( -) tocopherol which is the only form that is recognized to meet human requirements to study. According to the dissolved characteristic of Vitamin E, a new method was used to determine the SERS spectra of water-insoluble compounds. This method not only can measure the SERS spectra of water-insoluble compounds directly by using Ag colloid, but also overcome the disadvantages of the aforementioned method.

## 2. Experimental

### *2.1. Instrumentation*

For all measurement, the spectra reported in this study were obtained by use of the 514.5-nm line of an argon ion laser as the excitation source. The Raman system was equipped with a BX-41 Olympus microscope and 45×objective. Spectrometer system (Acton spectro@2300i, Princeton Acton, USA), was used with a 1200 gr/mm holographic grating. An air cooled CCD detector (Pixis 256, Princeton Acton, USA) was used for measuring the Raman signal by an integration method. The intensity of the 514.5-nm laser was 8mW at the sample. The Raman spectrum and absorption spectrum were obtained at room temperature. The absorption spectra were detected by spectrophotometer (Lamda 35, PE company, USA). The data processing was operated by using Origin 7.5 software.

### *2.2. Chemical reagents and Procedure*

Colloidal silver nanoparticles were prepared by using hydroxylamine hydrochloride as reducing agent according to Nicolae Leopold and Bernhard Lendl's method[13]. This method has the advantage of a fast, simple, and is a procedure characterized by a high preparation success rate. What's more, these nanoparticles have the characteristic of a more uniform distribution of size and shape together with the absence of interferences from remainder oxidation product. Some Notes: The cleanliness of the glassware and purity of the materials used are important factors that determine the reproducibility and stability of the colloids. For this reason, all the glassware used for the sols was previously soaked in a 1:1 $HNO_3$:$H_2O$ solution and was rinsed several times with deionized water. All materials used were analytical reagent grade or equivalent. Aqueous stock solutions of the compounds were prepared in nanopure water.

The analytes (±)- -tocopherol (approx 95% HPLC) were purchased from Sigma. They are used without further purification. The (±)- -tocopherol was dissolved in the pure chloroform of the HPLC level chromatograph, stored hermetically in a volumetric flask. The sample was prepared by adding a certain amount of the mixed solution to the equivalent volumes of the Ag colloid, shaking the mixture three minutes. After standing some time, the upper colloids were dripped onto a glass slide, and then the glass slide was placed to the Raman spectroscopy to determine the spectrum.

## 3. Results and discussion

### *3.1. Raman and SERS spectra*

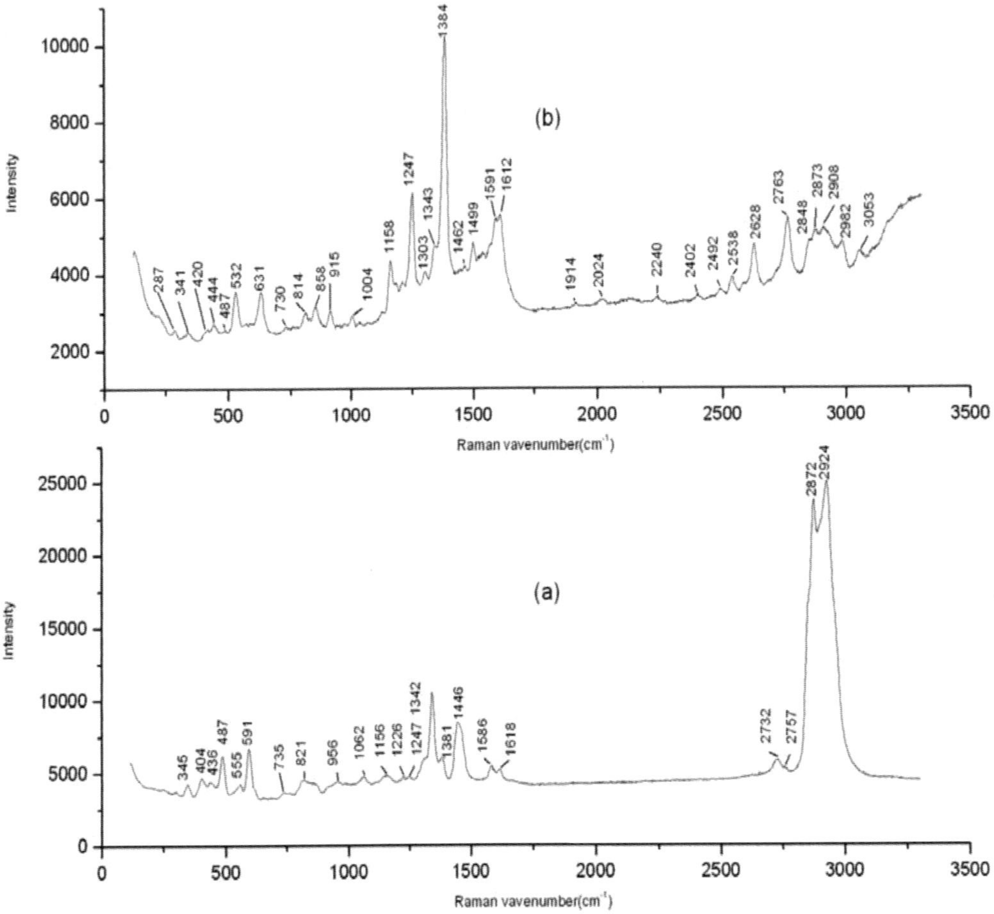


**Figure 1. Raman spectrum of (±)- -tocopherol.**

(a). Raman spectrum of (±)- -tocopherol.
(b). The SERS spectrum of Chloroform solution containing (±)- -tocopherol.

The Raman spectrum of the (±)- -tocopherol in the viscous oily liquid state is shown in Fig.1(a). It can be seen that two strong bands at 2872 $cm^{-1}$ and 2942 $cm^{-1}$ and four weak bands at 487 $cm^{-1}$, 591 $cm^{-1}$, 1342 $cm^{-1}$and 1446 $cm^{-1}$ are in good agreement with the calculated frequencies, though there are some differences of the relative intensity. The Raman spectrum shows a group of characteristic hydrocarbon aliphatic stretching bands at 2732 $cm^{-1}$, 2757 $cm^{-1}$, 2872 $cm^{-1}$ and 2942 $cm^{-1}$. Bands at 1591 $cm^{-1}$ and 1612 $cm^{-1}$ can be ascribed to an asymmetric stretching of the COO- group.

Vitamin E is a drug which is widely prescribed for the antioxidant effect. Traditional methods for its identification are the time-consuming and complicated tests. Therefore, the development of a method that can allow a rapid identification of the drug is desirable. Fig. 1(b) shows the SERS spectrum of the mixture containing (±)- -tocopherol and chloroform adsorbed on Ag colloids. The frequencies at which the bands appear are included in the spectrum to facilitate future reference. In order to study the adsorption process of Vitamin E on Ag colloids, the SERS spectrum of (±)- -tocopherol was compared with the normal Raman spectrum of (±)- -tocopherol. The SERS spectral profile corresponds very well with the normal Raman spectrum and the enhanced signals clearly indicate the achievement of SERS activity for Ag colloids. From these spectra, some notable phenomena should be taken into account. The relative intensities of Raman signals have significant differences between the normal Raman spectra and the SERS spectra though the experimental processes were the same, and some of the intensities of Raman signals are obviously enhanced in the SERS spectra. C-C stretching vibration appearing at 1342/1381/1446 $cm^{-1}$ in the normal Raman spectrum of pure liquid (Fig.1(a)) were observed. In the SERS spectra, at 1343/1384/1462 $cm^{-1}$ spectrum, the C-C stretching vibration appearing were observed. The $\upsilon(C-O)$ band appeared at 1247 $cm^{-1}$ both in the normal Raman spectrum and in SERS. From these results we propose that (±)- -tocopherol is adsorbed on a silver electrode through C-C bonds. Both electromagnetic enhancement and charge transfer enhancement contribute to the intensity of the spectra. Charge transfer enhancement is indicated both by the perturbation from the normal Raman spectrum of (±)- -tocopherol and by the high sensitivity obtainable. There are, however, wave number shifts in some of the major bands and significant differences in the intensities of the weaker bands. First, only $0.1\times10^{-6}$ $mol{\cdot}L^{-1}$ (±)- -tocopherol can obtain a good SERS spectrum, which indicates that the surface-enhanced Raman scattering is very strong. Second, compared with the normal Raman spectra, this SERS spectra appear some new peaks, increasing the amount of information. As we know, the SERS effect is mainly independent of some variables, such as dielectric constants and polarizability. However, due to its high sensitivity, it may be affected easily by some other factors, such as the adsorption behaviour and the distance from the metal surface to the samples, which can bring on changes of surface selection rules and the interaction between molecules and substrates. Due to the absorption function, the vibrations of some molecules are more obvious, and then some new peaks appear in the spectrum.

### 3.2. *Influences from background substrates*

(±)- -tocopherol is insoluble in water, so it can't directly adsorb on Ag colloids. To solve this problem, the pure chloroform was used as the intermediate. In order to learn whether the chloroform would influence experimental results, the SERS spectrum of the chloroform was studied. The chloroform was mixed with Ag colloids at the volume ration of 1:1, and the mixture was shaken for three minutes. After standing layer, the upper colloids were dripped onto a glass slide, and then the glass slide was placed to the Raman spectroscopy to determine the spectrum.

Fig.2 displays the Raman spectrum of the mixture containing the chloroform and the Ag colloids (fig.2(b))and the Raman spectrum of the Ag colloids (fig.2(a)). As can be seen from the figure, both of the Raman signs of the SERS spectrum of the chloroform and the Raman spectrum of the Ag colloids are quite weak. Because the pure chloroform have the very weak SERS signs, the addition of the chloroform have little influence to the Raman spectra of the analyte, and have no affect to the accuracy of identification of analyte. There are no any obvious peaks in the SERS spectrum of the pure chloroform, indicating that the pure chloroform have no SERS effect and does not interfere with the determination of vitamin E. What's more, the SERS of the Ag colloid substrates rarely have any peak. Therefore, the differences between the Normal Raman spectrum and the SERS spectrum mainly comes from the SERS effect which is attributed to the analytes adsorbed on the Ag colloids.

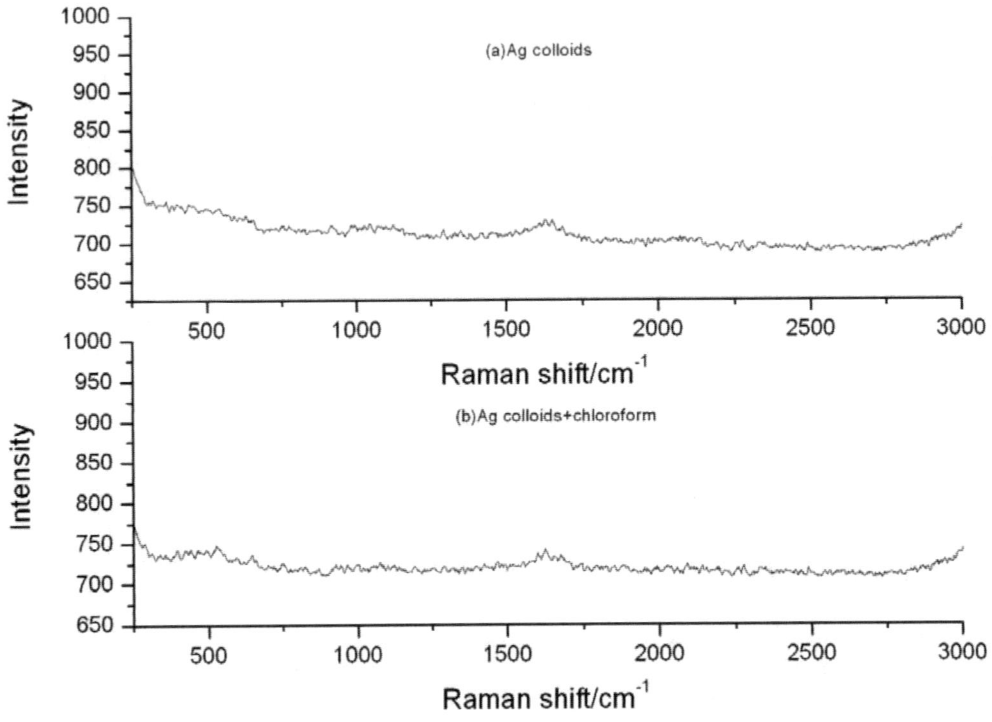


**Figure 2.** (a). Raman spectrum of the Ag colloids.
(b). the SERS spectrum of the pure chloroform solution.

## 4. Conclusion

The experimental results clearly demonstrate that an analytical method based on SERS provides high spectral selectivity and identification power as well as high analytical sensitivity and identification power for the study of the (±)- -tocopherol. The method proposed in this investigation has the extremely important significance to determine the SERS spectrum of the water-insoluble compounds. The increasing kinds of compounds can be determined by using the SERS spectrum in the future. This provides some advantages in SERS applications. In this experiment, the (±)- -tocopherol was used as an example water-insolution analtye to verify this method is feasible. From the results, the SERS spectra of the (±)- -tocopherol is quite a bit different from the normal Raman spectra of the (±)- -tocopherol. The attribution of the peaks was illustrated by comparing SERS spectra with NR(Normal Raman) spectra , and the mechanism and orientation of adsorption were also discussed according to the peaks which was absent in NR spectrum but observed in SERS spectrum. To verify the feasibility of the method for the quantitative analysis, SERE spectra at the same condition of the (±)- -tocopherol were collected. In the SERS spectra, at 1343/1384/1462 $cm^{-1}$ spectrum, the C-C stretching vibration appearing were observed. The intensity of this peak was enhanced compared with the normal spectra.

**Acknowledgments**

This research is supported by the National Natural Science Foundation of China (60678050).

**References**

[1] Katrin Kneipp, Harald Kneipp, Irving Itzkan, Ramachandra R Dasari and Michael S Feld 2002 *J. Phys.:Matter* **14** R597-R624
[2] Schmitt M and Popp J 2006 *J. Raman Spectrosc.* **37** 20-28
[3] AC Albrecht 1961 *J. Chem. Phys.*
[4] Fleishman M Hendra P J and McQuilan A J 1974 *J. Chem. Phys. Lett.* **26** 163-166
[5] S Nie and SR Emory-Science 1974 *J. Science.*
[6] PC Lee, D Meisel 1989 *J. Materials Reserch*
[7] Janmaire D L and Van Duyne R P 1977 *J. Electroanal. Chem. Interfacial Electrochem* **84** 1-20
[8] Albrecht M G and Creighton JA 2001 *J. Amt. Chem. Soc.* **99915** 5215-5217
[9] Seki H J 1986 *J.Electron Spectrosc. Relat. Phenom.* **39** 289-310
[10] BH Loo, YG Lee, EJ Liang and W Kiefer 1998 *J. Chem. Phys. Lett.* **297**
[11] J Wang, D W Li, H W Xin, S Xu and F C Liu 1987 *J. Spectrochim. Acta Part A.* **43** 375
[12] Katrin Kneipp, Ramachandra R. Dasari and Yang Wang 1994 *J.Applied Spectroscopy* **48** 951-955
[13] Nicolae Leopold and Bernhard Lendl 2003 *J. Phy. Chem. B* **107** 5723-5727

# NIR area array CCD-based singlet oxygen luminescence imaging for photodynamic therapy

**Bolin Hu, Yonghong He*, Zhiyi Liu**
Laboratory of Optical Imaging and Sensing, Graduate School at Shenzhen, Tsinghua University, Shenzhen, 518055, China

***Corresponding author email:** heyh@sz.tsinghua.edu.cn

**Abstract:** In this work, a near-infrared CCD-based singlet oxygen luminescence two-dimensional imaging method is proposed to detect singlet oxygen by its 1270nm luminescence. Two-dimensional singlet oxygen images with its near-infrared luminescence during photosensitization could be obtained with a CCD integration time of 1s, without scanning. The data presented shows a linear relationship between the singlet oxygen luminescence intensity and sample concentration. This method provides a detection sensitivity of 0.00189mg/ml (Hematoporphyrin monomethyl Ether dissolved in ethanol) and a spatial resolution better than 100μm. We applied this method *in vivo* to demonstrate its potential in monitoring photodynamic therapy.

## 1. Introduction

Photodynamic therapy (PDT) is an emerging therapy applied in area ranging from cancer treatment to some nonmalignant conditions [1-3]. In PDT, photosensitizer (PS) is incorporated into abnormal tissues and then irradiated with visible light [2,3]. Photoexcited sensitizer transfers energy to the ground-state oxygen via the type-Π photochemical pathway to produce singlet oxygen ($^1O_2$) [4]. Singlet oxygen is an important intermediate in many chemical and biological processes, and is believed to play significant roles in destructing biological cells [4-6]. In the past years, singlet oxygen has been directly detected by its weak 1270 nm luminescence ($a^1\Delta_g \rightarrow X^3\Sigma_g^-$) and most of these works have been done on bulk samples. In these measurements, significant results were acquired though the singlet oxygen luminescence signal was the average value of total sample [7,8]. Considering the inhomogeneities in PS distribution, oxygen concentration and/or the space intensity of exciting light, which can affect the production and behavior of singlet oxygen, methods have been proposed to detect singlet oxygen luminescence with spatial resolution [9-12].

In these methods, single photomultiplier tubes or linear detector arrays were used to obtain singlet oxygen luminescence images from the 1270 nm weak emission. In 2002, Lars Klembt et al. reported a singlet oxygen image with a spatial resolution of 2.5um detected by an

InGaAs linear array [9]. In 2004, I. Zebger et al. used a linear InGaAs device to create a singlet oxygen image of a single cell [10]. In 2005, Mark J Niedre et al. used a laser-scanning system to image singlet oxygen luminescence in an intradermal tumor model in mice [11]. In 2009, Thomas Breitenbach et al. reported singlet-oxygen-based images of a cell obtained by raster-scanned method [12]. These scanning-based researches have greatly prompted the singlet oxygen luminescence imaging, and were of significant advance in basic photobiological research and PDT dosimetry.

In the scanning-based singlet oxygen luminescence imaging methods, there is a high demand for the precision of motion of the moving stage required to load samples, as well as the edge overlap restructuring algorithm used to remove large position errors and join traces in image reconstruction, resulting in more complexity of the system and adverse effect to the measurement accuracy [13]. Moreover because the long integration time in each scan is need for the weak singlet oxygen luminescence signal and image data need to be collected sequentially, scanning image methods need a relatively long time to acquire an image. This may bring adverse influence to the final results because of the changes of parameters mentioned above that can affect the production and behavior of singlet oxygen during the data collection, as well as the time-critical *in vivo* detection of singlet oxygen luminescence signal [14-16]. Thus, a fast-imaging technique with relatively large imaging area and reasonable spatial resolution suitable for clinic application is necessary [17]. In this paper, we developed a two-dimensional singlet oxygen luminescence imaging method without scanning by use of a high quantum efficiency, high uniformity and low dark current infrared area CCD detector. Images of photodynamically generated singlet oxygen luminescence both *in vitro* and *in vivo* have been acquired with an integration time of 1s by this method. PS samples with different concentrations were used to test the relationship between luminescence intensity and sample concentration. The detection sensitivity of the system has been investigated and spatial resolution of the system has been tested by exciting light spots with dark lines of various widths. In *in vivo* experiments, we used the method to acquire singlet oxygen luminescence images from mice injected with PS.

## 2. Methods and materials

### *2.1 optical imaging system design*

The schematic diagram of our experimental system is shown in Fig.1. The whole system includes three parts: 1. the semiconductor laser source for excitation, 2. optical filters/ NIR area CCD detection system, 3. a standard frame grabber carder NI-1427 as the data acquisition system. A semiconductor laser (Changchun New Industries Optoelectronics Technology Co., Ltd.) with a wavelength of 470 nm and a power of 12 mw, was used as light source. The laser light was collimated and fell on a beam expander. The expanded beam was reflected and then an aperture was used to select a spot with a suitable size for both the sample and the detect range of the system. This selected light was used to photo-excite the region of interest in the sample to produce singlet oxygen luminescence signal. An achromatic lens was used to collect the emission from the sample, which included the singlet oxygen luminescence signal, PS fluorescence, back-scattered laser light and autofluorescence from tissues *in vivo*. A longpass filter (LP1150, Thor labs) and a NIR bandpass filter (1270BP20, Omega optical)centered at 1270 nm with a full-width-at-half-maximum (FWHM) of

20 nm and an attenuation OD better than OD5 were used to get rid of PS fluorescence, back-scattered laser light and autofluorescence from tissues *in vivo*, and obtain NIR luminescence signal of the singlet oxygen centered in 1270 nm $\pm$10nm. The transmitted emission from the filters was coupled by a NIR lens onto an InGaAs area detector (XEVA-CL-FPA-1.7-320-TE1-60Hz, Xenics), whose spectra response ranged from 900 nm to 1700 nm with a uniform quantum efficiency of 84%. The CCD surface was cooled at -20 $^0$C to reduce thermal noise. We used a standard frame grabber carder NI-1427 for the image data acquisition.

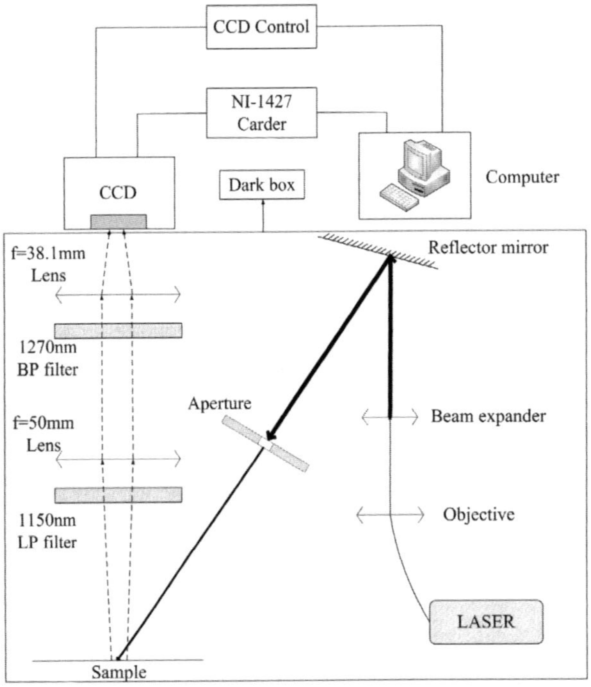


Figure 1.Schematic layout of the NIR area CCD-based singlet oxygen luminescence imaging system.

*2.2 Experimental sections*

In the study, Hematoporphyrin monomethyl Ether (HMME, Shanghai Xianhui Phaunaccutical Co.Ltd) was dissolved in ethanol in different concentration to be illuminated with above-mentioned light source to produce singlet oxygen. All samples were held in a 10 mm ×10 mm ×5 mm quartz cuvette mounted on a sample stage, which was able to maintain the source-sample-detector geometry constant during experiments. The space intensity of the exciting light was 60mw/cm$^2$ and images were captured at an exposure time of 1s in these experiments.

In the study *in vivo*, HMME was dissolved as the introduction from producer to make PS solution with a concentration of 2mg/ml.Two female Kunming mice (Experimental Animal Centre of San Yat-sen University of Medical Science) were used. All the mice were anesthetized with a 1.4mg/g intra-peritoneal dose of Urethane, and then depilated on the zone of dorsal with 8% $Na_2S$ (Guangzhou Chemical Reagent Factory). For the first mouse, HMME solution was subcutaneously injected into the hairless area with a 0.05ml dose for photosensitization. After a waiting period of 30min that allows for the diffusion of PS in the tissues, light exposure was performed with a density of 60mw/cm$^2$. The second one was not injected with HMME solution and irradiated by the same exciting light, as the control group.

**3. Results and Discussions**

3.1 *validation of singlet oxygen luminescence signal*

To demonstrate that the signal detected by our system was indeed derived from singlet oxygen, experiments were performed with three samples of HMME dissolved in alcohol at the concentration of 0.08mg/ml. The first sample was exposed to air, the second was added with 1,4-Diazabicyclo[2,2,2]octane (DABCO), an acknowledged quencher of singlet oxygen, and the last was bubbled with nitrogen gas to displace the dissolved oxygen. Fig. 2 shows the images of samples irradiated by the exciting light of the same fluence rate. The singlet oxygen luminescence signal could be detected and the average signal intensity calculated by integration method is 79.98 counts, compared with the background of 30.98 counts, besides the shapes and details of the singlet oxygen luminescence image are observed in the first sample (Fig. 2(a)). As expected for singlet oxygen luminescence, the near-IR signal disappeared in the sample to which the singlet oxygen quencher DABCO had been added (Fig. 2(b)), as well as the third case that the sample equilibrated with nitrogen (Fig. 2(c)). When nitrogen bubbling stopped and the sample equilibrated with nitrogen previously was exposed to air, the signal could appear again due to the diffusion of air back into the liquid sample (Fig. 2(d)).

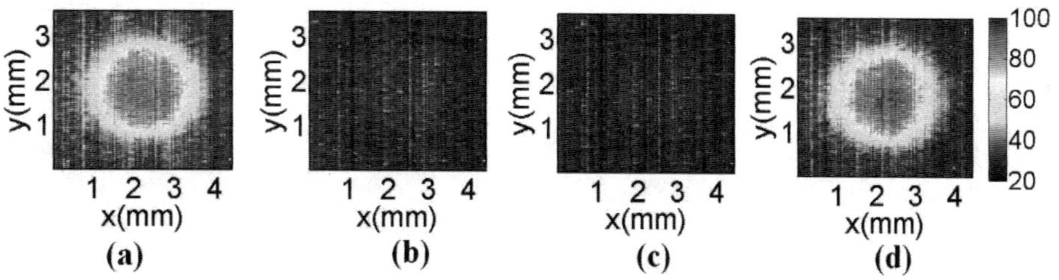


Figure 2. Images of samples in different situations. (a) Sample exposed to air, (b) sample with addition of singlet oxygen quencher DABCO, (c) sample bubbled with nitrogen, (d) sample in (c) exposed to air again.

3.2 *Linear sensing range and detection sensitivity*

HMME dissolved in ethanol with concentrations of 0.01, 0.015, 0.02, 0.04, 0.08, and 0.16mg/ml were prepared for the test of linear sensing range of the system. A background image was recorded upon the irradiation of ethanol without PS dissolved. The singlet oxygen luminescence images were taken immediately after the start of the irradiation. In this series each pixel represents the background-subtracted signal at 1270 nm .Fig. 3(a) shows the series of singlet oxygen luminescence images of samples with different concentrations. It is clear that the singlet oxygen luminescence signal increases with the increasing of sample density for the same fluence rate. From the data in Fig. 3(a) the average singlet oxygen luminescence intensities calculated by integration method are 10.3, 14.2, 18.87, 30.60, 52.70 and 114.40 counts, as plotted in Fig. 3(b). The relationship between the average singlet oxygen luminescence intensity and the concentration of the HMME was shown in Fig. 3(b). It can be found that the average singlet oxygen luminescence intensity shows a linear relationship with the concentration of HMME from 0.01mg/ml to 0.16mg/ml.

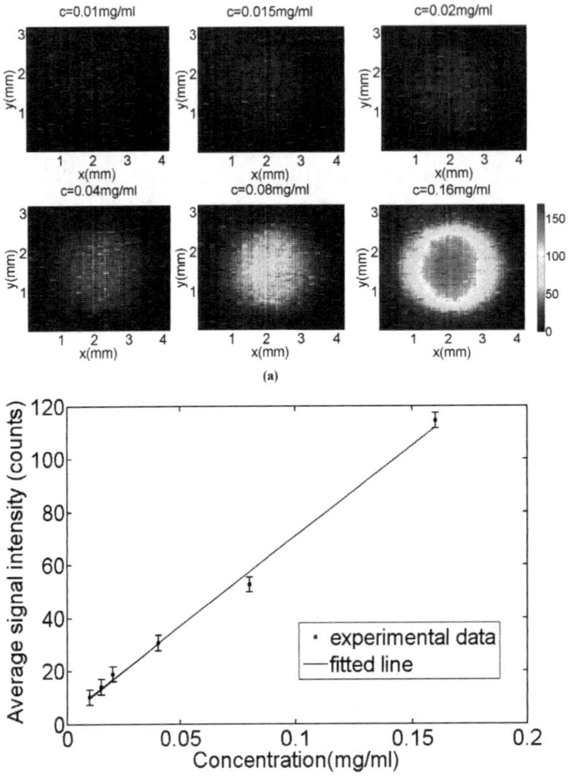


Figure 3. Detection of HMME samples with different concentrations. (a) Images of the HMME dissolved in alcohol with the concentration of 0.01, 0.015, 0.2, 0.4, 0.8, and 0.16mg/ml from the top left to right bottom. (b) The dependence of average singlet oxygen signal intensity on the concentration of HMME samples derived from (a), the error bars represent the standard deviation of three samples with the same concentration in each concentration.

We could calculate the detection sensitivity of the system as follow:

$$S_d = D_s / SNR \qquad (1)$$

Where $S_d$ is the detection sensitivity, $D_s$ the sample density and $SNR$ the signal-to-noise ratio. In this study, we used the 0.08mg/ml HMME solution to test the detection sensitivity and singlet oxygen luminescence image was acquired immediately after the start of the irradiation. Fig. 4(a) is the original image acquired by our system and the intensity graph of marked row A. In Fig. 4(b) and (c), the value of every pixel is replaced by the average of the intensity levels in the 3x3 and 5x5 neighborhood, as a smoothing, linear spatial filter to further enhance the signal-to-noise ratio. The definition of SNR is the ratio of signal intensity to the standard deviation of background. In Fig. 4, the singlet oxygen average intensity is approximately 103.428 counts and the standard deviations of the background are 6.1339, 3.2792 and 2.4442, respectively. Thus the SNR achieved from Fig. 4 are 16.8617, 31.5406 and 42.3156 respectively. According to Eq. (1), as the sample density $D_s$ is 0.08mg/ml, the sensitivity of our system can achieve 0.00189mg/ml with a SNR of 42.3156 as a result of 5×5 smoothing, linear spatial filter. The cost of smoothing, linear spatial filter is the reduction of spatial resolution.

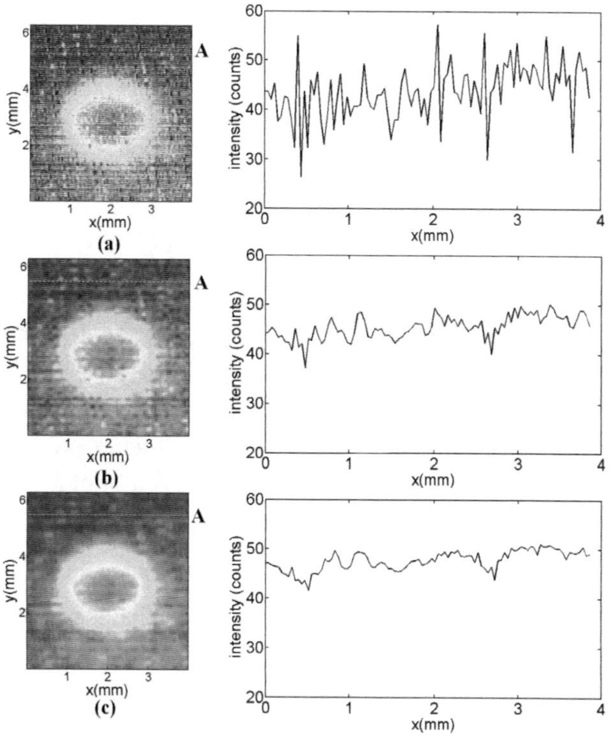


Figure 4. The results of smoothing, linear spatial filter for reducing noise level. (a) oringinal singlet oxygen luminescence image. (b) 3x3 pixels,(c) 5x5 pixels neighborhood are averaged to replace each pixel. Intensity graphs of dashed row A, across the background, are plotted on the right side of these images. The SNRs achieved from (a), (b) and (c) are 16.8617, 31.5406 and 42.3156 respectively.

### 3.3 *Resolving power*

The pixel size of the area CCD is 30μm×30μm and the zoom of this system is 0.762. The spatial resolution of 39.4μm can be calculated in theory. To investigate the spatial resolution of the system, a series of filaments with diameters (D) of 200μm, 150μm and 100μm were put in the center of the cross section of the collimated laser beam sequentially. Thus the exciting light was divided to form a light spot with a dark line in center. Fig. 5 shows one dimensional signal intensities across the centers of the singlet oxygen luminescence images obtained upon irradiation of HMME dissolved in ethanol. As the concavity in the last image can be distinguished, corresponding to the 100μm case, thus the spatial resolution of the system is better than 100μm.

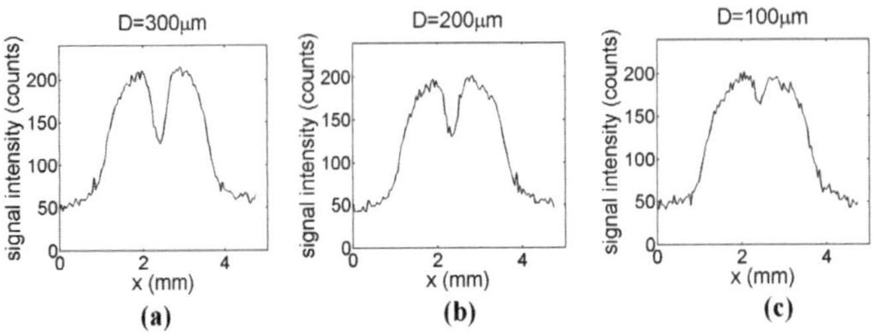


**Figure 5.** The spatial resolution of the system: one-dimensional signal intensities across the centers of singlet oxygen luminescence images with dark lines of width (a) 300μm, (b) 200μm, (c) 100μm.

*3.4 Singlet oxygen luminescence imaging in vivo*

We also used this method on mice to acquire singlet oxygen luminescence imaging *in vivo*. One mouse with 0.05 ml HMME and another mouse without HMME injected were used for this experiment. Fig. 6 shows the results of the *in vivo* experiments. Fig. 6(a) shows a white-light image of an animal taken by commercial digital camera (Olympus, E300), and the area irradiated which was marked by a circle. Fig. 6(b) is one of the images obtained from the mouse without HMME during irradiation. Fig. 6(c) shows a singlet oxygen luminescence image series acquired from the mouse injected with HMME during the light exposure time. Compared with the weak light acquired in the mouse without PS, shown in Fig. 6(b), the singlet oxygen luminescence images could be obtained clearly from the mouse injected with PS *in vivo*, as shown in Fig. 6(c). The image time series of HMME case shows a decreasing singlet oxygen luminescence signal during the long exposure time, compared with the weak light intensity and no intensity depletion in the case without HMME. We deduce that the data without PS mainly represents the scattered background light. The causes of the decrease of singlet oxygen signal are likely to be PS photobleaching, the photochemical depletion of molecular oxygen in tissues [18,19].

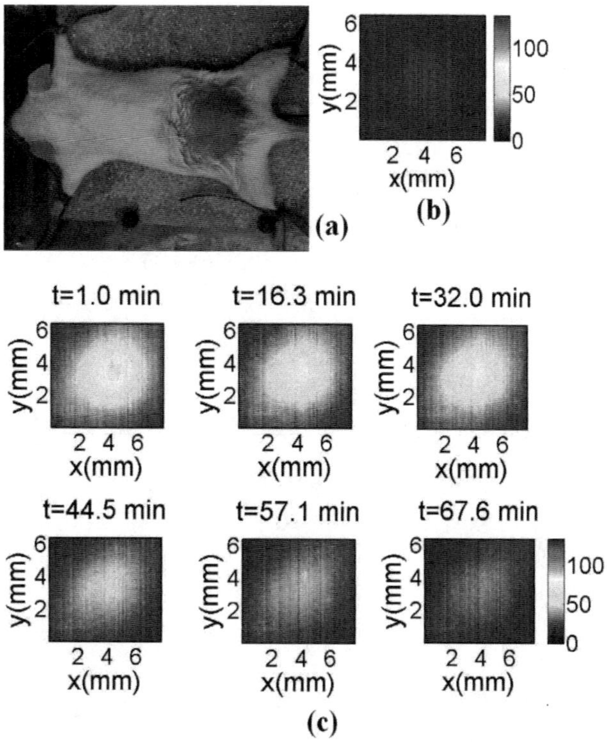


**Figure 6.** (a) white-light image of a Kunming mouse, the area marked by a circle is for irradiation. (b) Example of an image from the mouse without PS during irradiation. (c) Corresponding singlet oxygen luminescence time series-image obtained from the mouse with PS.

**4. Conclusions**

We have proposed a newly developed NIR area CCD-based singlet oxygen luminescence two-dimensional imaging method for photosensitization. Using this method, we demonstrate that singlet oxygen luminescence image could be acquired with an integration time of 1s, without scanning. The experimental system has a detection sensitivity of approximately 0.00189mg/ml (HMME dissolved in ethanol) and a resolving power better than 100μm. We also obtained photodynamically generated singlet oxygen luminescence images *in vivo*. We believe that the proposed NIR area CCD-based singlet oxygen luminescence two-dimensional imaging method is promising for *in vivo* monitoring of singlet oxygen produced in photosensitization.

**Acknowledgements**

This research was made possible with the financial support from the 863 project, China (grant 2006AA06Z402), NSFC China (30770592).

**References**


[1] Dougherty T J., Gomer C J, Henderson B W, Jori G, Kessel D, Korbelik M, Moan J and Peng Q 1998 *J. Natl Cancer Inst.* **90** 889-905

[2] Wilson B C and Patterson M S 2008 *Phys. Med. Biol.* **53** R61-R109

[3] Panjehpour M, Overholt B F, Phan M N, Haydek J M 2005 *Gastrointest. Endosc.* **61** 13-18

[4] Schweitzer C and Schmidt R 2003 *Chem. Rev.* **103** 1685-1757

[5] Li B H, Jarvi M T, Moriyama E H, Wilson B C 2007 *Proc. SPIE.* **6427** 01-07

[6] Kessel D, Luo Y, 1998 *J.photochem. photobiol.* B. **42** 89-95

[7] Krasnovsky A A 1998 *Membr Cell Biol.* **12** 665-90

[8] Niedre M J, Yu C S, Patterson M S, Wilson B C 2005 *Br J Cancer.* **92** 298-304.

[9] Andersen L K, Gao Z, Ogilby P R, Poulsen L, Zebger I 2002 *J. Phys. Chem. A.* **106** 8488-8490

[10] Zebger I, Snyder J W, Andersen L K, Poulsen L, Gao Z 2004 *Photochem. Photobiol.* **79** 319-322

[11] Niedre M J, Patterson M S, Giles A, Wilson B C 2005 *Photochem. Photobiol.* **81** 941-943

[12] Breitenbach T, Kuimova M K, Gbur P, Hatz S, Schack N B, Pedersen B W and Ogilby P R 2009 *Photochem. Photobiol. Sci.* **8** 442–452

[13] Huang G L, Deng C, Zhu J, Xu S K, Han C, Song X B, Yang X Y 2008 *J. Biomed. Opt.***13** 034006

[14] Iinuma S, Schomacker K T, Wagnieres G, Rajadhyaksha M, Bamberg M, Momma T 1999 *Cancer Res.* **59** 6164–6170

[15] Baier J, Maisch T, J Regensburger, Loibl M, Vasold R, W Bäumler, 2007 *J. Biomed Opt.* **12** 064008

[16] Snyder J W, Skovsen E, Lambert J D C, Poulsen L, Ogilby P R 2006 *Phys. Chem. Chem. Phys.* **8** 4280–4293

[17] He Y H, Xing D, Yan G H, Ueda K 2002 *Cancer Letters. 182* 141–145

[18] Fingar V H, Wieman T J, Wiehle S A, Cerrito P B 1992 *Cancer Res.* **52** 4914-4921

[19] Henderson B W, Busch T M, Vaughan L A, Frawley N P, Babich D, Sosa T A *et al* 2000 *Cancer Res.* **60** 525–529

# Antibiofouling polymer coated gold nanoparticles as a dual modal contrast agent for X-ray and photoacoustic imaging

**Guojia Huang, Yi Yuan and Da Xing***
MOE Key Laboratory of Laser Life Science & Institute of Laser Life Science, College of Biophotonics, South China Normal University, Guangzhou 510631, China

E-mail: xingda@scnu.edu.cn

**Abstract.** X-ray is one of the most useful diagnostic tools in hospitals in terms of frequency of use and cost, while photoacoustic (PA) imaging is a rapidly emerging non-invasive imaging technology that integrates the merits of high optical contrast with high ultrasound resolution. In this study, for the first time, we used gold nanoparticles (GNPs) as a dual modal contrast agent for X-ray and PA imaging. Soft gelatin phantoms with embedded tumor simulators of GNPs in various concentrations are clearly shown in both X-ray and PA imaging. With GNPs as a dual modal contrast agent, X-ray can fast detect the position of tumor and provide morphological information, whereas PA imaging has important potential applications in the image guided therapy of superficial tumors such as breast cancer, melanoma and Merkel cell carcinoma.

## 1. Introduction

X-ray is one of the most useful diagnostic tools in hospitals in terms of frequency of use and cost. The X-ray imaging modality can obtain high resolution anatomical images based on X-ray's attenuation when X-ray passes through tissue. It has been chosen as the morphological imaging modality for small animals because of the accuracy of the anatomical information [1]. PA imaging is an emerging non-ionizing, noninvasive imaging modality which can overcome the limitations of purely optical imaging methods that result from the intense optical scattering in soft tissue, and combines the spectroscopic capability of light with the spatial resolution of ultrasound may lead to a new medical imaging technique. PA imaging is interesting for some materials in the body having different optical absorption coefficients, which is a wavelength dependent tissue property. Thus, oxygenated or deoxygenated hemoglobin, water or melanin etc, can be distinguished based on their absorbed spectrum. These merits make PA imaging can be used to identify different functional activities of tissues that may be indistinguishable in other imaging modalities. PA imaging has been employed to image vasculature structure, detect breast tumors, track contrast agent, monitor glucose, measure oxygen saturation, perform molecular and function imaging, count cell with photoacoustic flow cytometry [2-7].

Current contrast agents for X-ray are based on iodinated small molecules, because among nonmetal atoms, iodine has a high X-ray absorption coefficient [8-9]. Iodinated compounds, however, allow only very short imaging time due to rapid clearance by the kidney, which can also cause them to have renal toxicity [10-11]. Instead, PEG coated GNPs can overcome these limitations [12-13]. Furthermore, based on high optical absorption coefficient, GNPs have been used as a contrast agent for PA imaging [14-15].

In this study, for the first time, we use GNPs as a dual-modal contrast agent for X-ray and PA imaging. Soft gelatin phantoms with embedded tumor simulators of GNPs in various concentrations are clearly shown in both X-ray and PA imaging. X-ray can fast detect the position of tumor and provide morphological information, whereas PA imaging has important potential applications in the image guided therapy of superficial tumors such as breast cancer, melanoma and Merkel cell carcinoma.

## 2. Materials and imaging system

### *2.1 Gold nanoparticles*

GNPs were synthesized by citrate reduction of auric chloride ($HAuCl_4$) following the procedure of Frens [16]. The various factors controlling the size, size distribution and shape of the GNPs in the citrate reduction method have been studied in detail [17-19]. In this paper, for the synthesis of 30 nm GNPs, 100 mL of 0.01 wt % $HAuCl_4$ was added to a 250 mL round-bottom and the solution was brought to a boil. 2 mL of 1 wt % sodium citrate was added quickly, which resulted in a color change from blue to burgundy and then the solution was allowed to boil for an additional 15 min. After the solution was cooled to room temperature, 1 mg mPEG-SH was added and stirred for 12 h to covalently modify the surface of the GNPs with PEG. The resulting PEG-coated GNPs were collected by centrifugation at 16,800g for 30 min and washed twice with distilled water.

### *2.2 X-ray imaging system*

A X-ray system (piXarray 100 digital specimen Radiography system, Bioptics, Inc, America) were used. This system incorporates a CCD camera with a phosphor as the X-ray detector that is irradiated with a low-power X-ray source, with image acquisition, system control, and image display and reconstruction performed on the standard desktop computer. Imaged area is 50 mm × 50 mm. The detector of the system consists of a scintillating screen fiber-optically coupled to two 1024 × 1024 CCD image arrays. The tube current of this system is 0.5 mA fixed, and the tube potential from 5 to 45 kVp adjustable. The resolution of this system is 10 lp/mm (50 microns) in contact mode, and 14 lp/mm (36 microns) in 1.4:1 magnification mode (50 micron focal spot).

### *2.3 Photoacoustic imaging system*

For all experiments, a mechanical scanning PA system with a single acoustic transducer to collect the acoustic signals was utilized. An optical parametric oscillator (OPO) (Vibrant 532 I, Opotek, Carlsbad, Calif) was used for light illumination. The OPO operated at 532 nm with pulse duration of 10 ns and pulse repetition rate of 15 Hz. The laser was expended by a concave lens and then homogenized by a piece of ground glass, so that the incident energy density was controlled below 20 mJ/cm$^2$ accordingly to the ANSI standard. A needle hydrophone (Precision Acoustics Ltd, Dorchester, UK; diameter, 1 mm; sensitivity, 850 nV/Pa; frequency response, 200 KHz to 15 MHz), which was pointed to the center of the samples and controlled by a precision stepper to scan circularly around the sample, and receive the PA signals. Total of 200 steps with a constant 1.8° interval were taken for covering a $2\pi$ receiving angle. The PA signals were recorded by the computer through the amplifier and a digital oscilloscope (TDS3032, Tektronix, USA); the sampling rate of the oscilloscope was 250 MSamples/s.

### 3. Results and discussion

The GNPs were prepared by the citrate reduction of the gold ions. Particle size and morphology were determined by TEM. Figure 1(a) shows the representative TEM picture of the 30 nm GNPs. UV–vis absorption spectra of gold nanoparticles showed that the absorption peak was 527 nm.

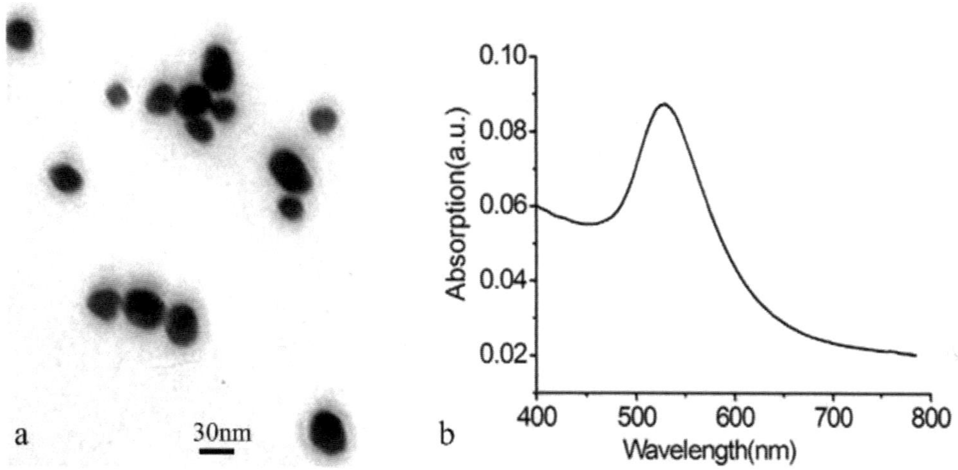


**Figure 1.** (a) Representative TEM picture of 30 nm GNPs (scale bar 30 nm). (b) Optical absorption of 30 nm nanoparticles used in the experiments showing maxima to be 527 nm.

To verify the ability of the GNPs as the X-ray and PA imaging contrast, we constructed a phantom made of 5% gelatin in which 4 inclusions with different concentrations of nanoparticles were embedded (Fig. 2). Each cylinder was 3 mm in diameter, 5 mm in deep, where the nanoparticles concentrations were 10, 5, 2.5, and 1.25 mg/mL respectively. The X-ray imaging is demonstrated in Fig. 2(a), for a cylinder containing 4 inclusions of different GNPs concentrations. The inclusion with a GNPs concentration of 2.5 mg/mL can hardly be recognized from the background, leading us to conclude that this X-ray system has a detection sensitivity of the order of 5 mg/mL. In the corresponding PA image (Fig.2 b), the lowest concentration still supplies good contrast from the background which may owe to the high sensitivity of PA imaging. The X-ray signals are quantified as a function of concentration in Fig2. (c). As the concentration mounts, the X-ray signals increase linearly. Figure2 (d) shows a linear increase of the PA signals as function of concentration. These results strongly support the dual functionality of PEG coated gold nanaoparticles for X-ray and PA imaging.

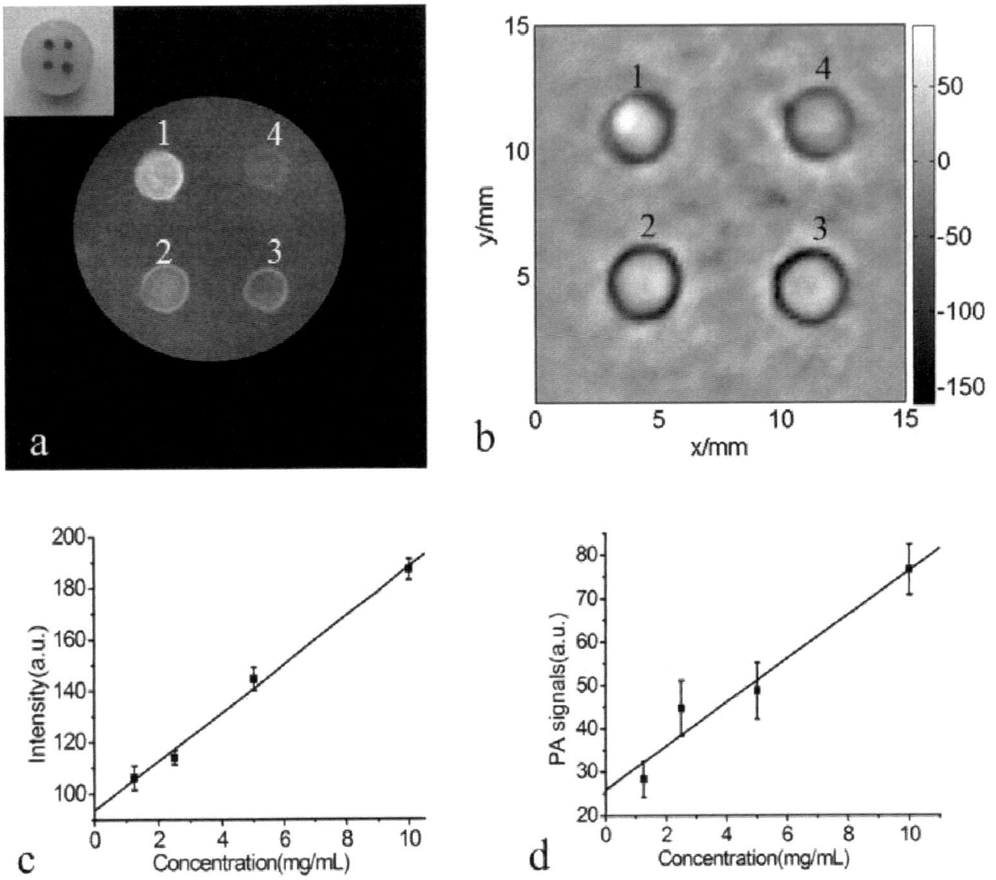


**Figure 2.** X-ray (a) and PA (b) image of PEG coated GNPs at various concentrations, 10, 5, 2.5, and 1.25 mg/mL, respectively. The quantification of the X-ray intensity (c) and PA signals (d) at the corresponding concentrations of GNPs.

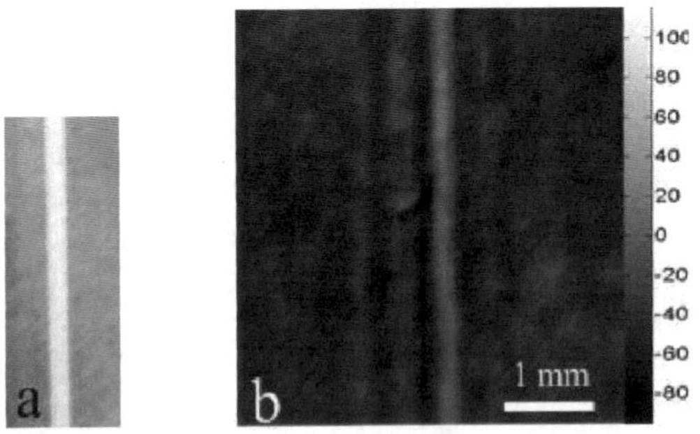


**Figure 3.** (a) the corresponding X-ray and (b) PA imaging of the simulate vascular (10mg/mL GNPs filled in a 0.5-mm-diameter tube) below 5 mm of chicken breast tissues.

To study the quality of GNPs in animal tissues, a 0.5-mm-diameter tube filled with 10 mg/mL GNPs as simulate vascular was embedded in chicken breast tissues. With a 5-mm thick layer of chicken breast tissue, the tube was clearly shown for X-ray imaging in Fig.3 (a) and also had a good contrast for PA imaging in Fig.3 (b), which may take advantage of PA imaging with high resolution.

Due to X-ray's strong penetrating ability and high X-ray absorption coefficient of gold, with deeper chicken breast tissue, tube could still be visible using this X-ray system in our experiments. These results imply that X-ray imaging is potentially to fast detect the position of target and provide morphological information, while PA imaging has the potential to monitor the change of target.

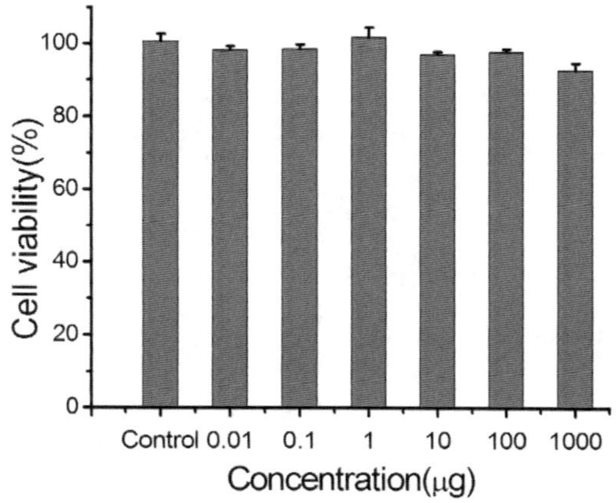


**Figure 4.** Cell toxicity of GNPs measurement. Cell viability of tumor cells after 12 h of incubation with increasing amounts of the PEG coated GNPs. Cell viability was measured using Cell Counting Kit-8（CCK8）.

Cell cytotoxicity assay was performed with a colorimetric tetrazolium salt-based assay, Cell Counting Kit-8 (CCK8). To determine the cytotoxicity of GNPs, tumor cells (103 per well) were cultured in a 96-well microplate for 24 h and then co-incubated with PEG coated GNPs of different concentrations for 12 h, rinsed with PBS, and incubated for another 72 h. OD450, with an absorbance value of 450 nm, was read with a 96-well plate reader (Infinite M200, Tecan, Switzerland) to determine the viability of the cells. The viability of cells was calculated as: cell viability (% of control) = ODTre/ODCon × 100% (where ODTre was the absorbance value at 450 nm of treated cells and ODC on was the absorbance value at 450 nm of control cells). Celltoxicity tests (see Fig. 4) show that PEG coated GNPs display satisfactory biosafety even at 1mg/mL, which is probably a much higher concentration than encountered in vivo.

## 4. Conclusion

In conclusion, we successfully tested the feasibility of PEG coated GNPs as a dual-modal contrast agent for X-ray and PA imaging. We demonstrated that the nanoparticles have high X-ray intensity and PA signal due to elevated optical absorption. The combined imaging methods are complementary and together they provide the rich information needed for the reliable detection and diagnosis. This dual-modality imaging has the potential to become a useful tool in detecting early stage tumor and monitoring the progress of anti-angiogenic therapy. By treating the surface of GNPs to target specific molecules, the detection of tumor such as breast cancer and image guided therapy will be more accurately and the applications of these nanoparticles can be broaden to molecular imaging of primary and metastatic tumor.

## 5. Acknowledgments

This research is supported by the National Basic Research Program of China (Grant No. 2010CB732602), the Program for Changjiang Scholars and Innovative Research Team in University

(Grant No. IRT0829), the National Natural Science Foundation of China (Grant Nos. 30627003 and 30870676), and the Natural Science Foundation of Guangdong Province (Grant No. 7117865).

**References**

[1] Chen W R, Liu H, Yue W, Wang J P, Nordquist R E 2001 Opt. Eng. 40(7) 1249-1254
[2] S. Yang, D. Xing, Q. Zhou, L. Xiang, and Y. Lao 2007 Med. Phys. 34 3294-3301
[3] S. Yang, D. Xing, Y. Lao, D. Yang, L. Ming 2007 Appl. Phys. Lett. 90 243902
[4] Yeqi Lao, Da Xing, Sihua Yang and Liangzhong Xiang 2008 Phys. Med. Biol. 53 4203-4212
[5] Liangzhong Xiang, Yi Yuan, Da Xing, Zhongmin Ou, Sihua Yang, Feifan Zhou 2009 J. Biomed. Opt. 12(2) 021008
[6] X. Wang, X. Xie, G. Ku, L. V. Wang, and G. Stoica 2006 J. Biomed. Opt. 11 024015
[7] Cunguang Lou and Da Xing 2010 Appl. Phys. Lett. 96 211102
[8] Krause, W.and Schneider, P. 2002 Springer: Heidelberg, Germany
[9] Krause W. 1999 Adv. Drug. Deliv. Rev. 37 159-173
[10] Hizoh I, Haller C. 2002 Invest Radiol. 37 428-434
[11] Haller C and Hizoh I. 2004 Invest Radiol. 39 149-154
[12] Kohler N, Fryxell GE, Zhang MQ. 2004 J. Am. Chem. Soc. 126 7206-7211
[13] Lee H., Lee E., Kim D.K., Jang N.K., Jeong Y.Y. and Jon S. 2006 J. Am. Chem. Soc. 128 7383-7389
[14] Copland J A, Eghtedari M, Popov V L, Kotov N, Mamedova N, Motamedi M and Oraevsky A A 2004 Mol. Imaging Biol. 6 341-349
[15] Sharma P, Brown S C, Bengtsson N, Zhang Q, Walter G A, Grobmyer S R, Santa S, Jiang H, Scott E and Moudgil B 2008 Chem. Mater. 20 6087-6094
[16] Frens,G. 1973 Nat. Phys. Sci. 241 848-851
[17] Ji X, Song X, Li J, Bai Y, Yang W and Peng X 2007 J. Am. Chem. Soc. 129 13939-13948
[18] Kimling J, Maier M, Okenve B, Kotaidis V, Ballot H and Plech A 2006 J. Phys. Chem. B 110 15700-15707
[19] Pantungwasa W and Hodak J H. 2007 Mater.Chem. Phys. 108 45-54

# Ultrasmall volume molecular isothermal amplification in microfluidic chip with advanced surface processing

**Guoliang Huang [1, 2,*], Li Ma[2], Xiaoyong Yang [1, 2] and Xu Yang[2]**

[1] Department of Biomedical Engineering, the School of Medicine, Tsinghua University, Beijing, 100084, China

[2] National Engineering Research Center for Beijing Biochip Technology, Beijing, 102206, China

[*]Corresponding author: Guoliang Huang (Email: tshgl@tsinghua.edu.cn)

**Abstract**. In this paper, we developed a metal micro-fluidic chip with advanced surface processing for ultra-small volume molecular isothermal amplification. This method takes advantages of the nucleic acid amplification with good stability and consistency, high sensitivity about 31 genomic DNA copies and bacteria specific gene identification. Based on the advanced surface processing, the bioreaction assays of nucleic acid amplification was dropped about 392nl in volume. A high numerical aperture confocal optical detection system was advanced to sensitively monitor the DNA amplification with low noise and high power collecting fluorescence near to the optical diffraction limit. A speedy nucleic acid isothermal amplification was performed in the ultra-small volume microfluidic chip, where the time at the inflexions of second derivative to DNA exponential amplified curves was brought forward and the sensitivity was improved about 65 folds to that of in current 25μl Ep-tube amplified reaction, which indicates a promising clinic molecular diagnostics in the droplet amplification.

## 1. Introduction

Nucleic acid amplification[1-7] is one of most important tools in life science fields and clinic molecular diagnostics, where specific nucleic acid sequence can be amplified with more than $10^6$-$10^9$ copies within an hour and detected with high sensitivity and specificity. Based on nucleic acid amplification, the single molecular diagnostics has always been a dream of human for clinic application, because of the surface adsorption of bioreactor to DNA molecules, the amplified double stranded DNA to be unfastened under high amplified temperature, the photo-bleaching [8] from the excited light and the background of material of bioreactor and surroundings, etc. The current nucleic acid amplification in Ep-tube, 96 or 384 well microplate usually uses a 25μl bioreaction assays with more than $10^3$ copies of genomic DNA to make enough more amplified double stranded DNA to ignore the above negative effects, and obtain a relative steady exponential mode amplified signal and good consistency to repeated detection.

Polymerase chain reaction (PCR) is most early current nucleic acid amplification process described by Saiki RK (1985) [1], but PCR needs a precise temperature cycling control to keep good specificity of amplification, there is a low efficiency in amplification cycling where only the extend phase can produce new strands and the valid time for real DNA amplification is about 50%. Loop-mediated isothermal amplification (LAMP) is a novel high effective nucleic acid amplification method introduced by Notomi (2000) [2], where nucleic acid amplification is performed at a constant temperature, and there is a high efficiency of amplification about 100%. Compared with PCR, LAMP has some obviously advantages:

Firstly, the whole amplification reaction can be performed at an isothermal condition, and the temperature control is relatively easy to design and realize with low cost. Secondly, the amplification efficiency is significantly improved, reaches to 100%, because no extra time is needed to wait for thermal changes in LAMP reaction. Thirdly, the reaction is highly specific to the amplification target where six specific primers with complete match are required to startup amplification.

The polycarbonate (PC), polymethylmethacrylate (PMMA), polydimethylsiloxane (PDMS), Glass and silicon are current materials used for nucleic acid amplification reactors [3-7], because of adsorption on the coarse surfaces of bioreactor and the limits of detection methods, which general requires at least 20μl reaction assays, otherwise it is difficult to perform a steady exponential mode amplification and good consistency for many repeated detections. Ideal materials could be used for nucleic acid amplification in micro-nanoliter assay systems is absent.

In this paper, we developed a metal micro-fluidic chip with advanced surface processing for ultra-small volume molecular isothermal amplification. A high numerical aperture confocal optical detection system was advanced to sensitively monitor the DNA amplification with low noise and high power collecting fluorescence near to the optical diffraction limit. Nucleic acid amplification in microenvironment was studied for good stability and consistency, high sensitivity. A promising clinic molecular diagnostics in the droplet amplification was performed.

## 2. Materials and methods

### *2.1. Materials*

Oligonucleotide primers were synthesized by Invitrogen Corporation (Shanghai, China). *Bst* DNA Polymerase, Large Fragment and ThermoPol Reaction Buffer were from New England Biolabs (Beijing) Ltd (Beijing, China). $MgSO_4$ was from Beijing Chemical Reagents Company (Beijing, China). dNTPs were from TaKaRa Biotechnology Co., Ltd (Dalian, China). dUTP and Uracil-DNA Glycosylase were from Fermentas Inc (Burlington, Canada). Betaine and BSA were from Fluka Sigma-Aldich Inc (Missouri, USA). EvaGreen was from Biotium Inc (California, USA). Other reagents were analytical grade.

### *2.2. Design of primers*

According to the sequence of the specific *mip* gene from GenBank (Accession No. DQ896988), oligonucleotide primers used for the isothermal amplified assay of *Legionella pneumophila* were designed, which is a bacterial species of pneumonia. Six primers, composing of two outer (F3 and B3), two inner (FIP and BIP) primers and two loop (LF and LB) primers were designed. FIP consists of a complementary sequence of F1 and a sense sequence of F2. Similarly, BIP consists of a complementary sequence of B1 and a sense sequence of B2. The schematic representation and sequence of primers are shown in Figure 1 and Table 1.

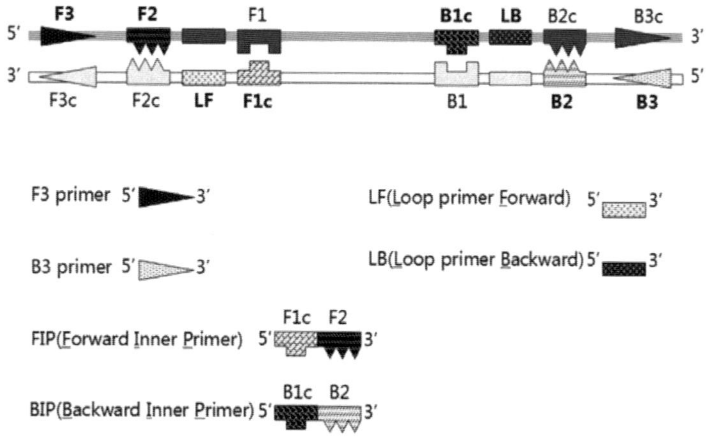


**Figure 1**. Schematic representation of primers used in this study.

**Table 1**. The sequences of the six specific oligonucleotide primers of *L. pneumophila*.

| Primers | Sequences |
|---|---|
| **F3** | GCAAGACGCTATGAGTGG |
| **B3** | TGATTACTTTGTATTGCAAACCA |
| **FIP** | GCCATCAAATCTTTCTGAAACTTGT-CTCAATTGGCTTTAACCGAAC |
| **BIP** | GCGGATGAAAATAAAGTAAAAGGGG-CTTGGCAATACAACAACGC |
| **LF** | TAAGAACGTCTTTCATTTGCT |
| **LB** | CTGAAAACAAAAACAAGCCAG |

*2.3. Bacterial DNA preparation*

Peking University People's Hospital (Beijing, China) kindly provided *L. pneumophila* strain (LP-1). Genomic DNA of LP-1 was extracted and purified by using the QIAamp DNA Mini Kit (Qiagen Inc., CA, USA). DNA concentration was measured by using NanoDrop Spectrophotometer (ND-1000, NanoDrop Technologies, Inc., DE, USA). The number of genomic copies was calculated assuming a molecular size of 3.4 Mbp.

*2.4. Microfluidic chip fabrication*

At first, the microfluid chips with different structure and microenvironment 50nL→7μL are designed by using the machine design software *Solidworks* 2006, and fabricated with metal aluminium materials in thickness 1.5mm by using the Computer Numerical Control Machining Center (JT-M960L, Jiatai Numerical Control Co., Quanzhou, China). An example of masking diagram of microfluid chip was shown in Figure 2(a), where the chip is 60mm in diameter. The center hole is used to fix the chip to detection instrument. The diameter of inlet hole and outlet hole is both of 1.2mm, which fits for Eppendorf tips to inject samples and reactants into the chip. The bioreactor cells were fabricated in different sizes (0.5-3.0mm in diameter and 0.2-1.0mm in depth) to obtain microliter to nanoliter volume ranges. The channel was fabricated with different sizes (0.1-0.5mm in both of width and depth) to limit reagents and sample consumption. All buffer cells are 3.0mm in diameter and 0.5mm in depth for collecting air bubble displaced during sample injection. The curved areas of tunnel near to the center hole was used to block the connection of bioreactor cells by heated press molding, and limit diffusion of reactants from one cell to others during the isothermal amplification.

Secondly, the microfluidic chip was further processed as the processes in Figure 2(b) to make surface smooth and inert to biological molecules, where an advanced inert surface processing was listed as following steps:

a). Clean oil and dirt, in the acidic solution of 50g/L H2SO4, 8g/L Emulsifier OP-10 for five minutes at 25°C.

b). Erode the surface, in the alkaline solution of 50g/L NaOH and 10g/L C6H11O7Na for ten minutes at 60°C.

c). Polish the surface, in solution of 280g/L H3PO4, 100g/L HNO3 and 40g/L H2SO4 for five minutes at 25°C.

d). Anodic oxidation, in electrolytic solution of 300g/L H2SO4, 2g/L HOOC-COOH, 15g/L HOCH2COOH and 15g/L C3H4O4 for thirty minutes at 25°C. The electronic current density is $30mA/cm^2$.

e). Electrolytic coloring, in a black coloring solution of 30% SnSO4, 30% NiSO4 and 15% CuSO4 for twenty minutes at 25°C. The AC voltage is about 6-10V.

f). Seal the surface, in a pH 6 solution of 4.5g/L NiSO4, 0.8g/L CoSO4 and 1.2g/L NaF for forty minutes at 25°C.

g). Clean the chip, in a three aqueous washes to remove un-reacted chemicals.

After the advanced inert surface processing, the chips is encapsulated tightly by a thin PC film from ABI Corporation for use.

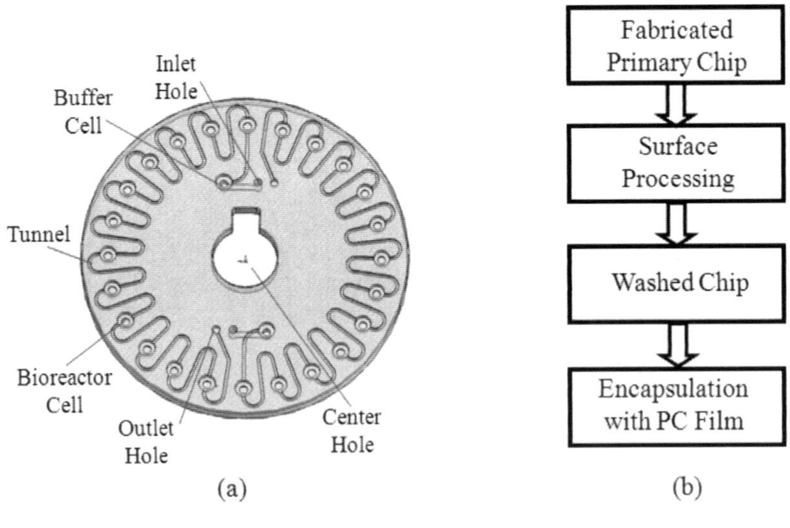


**Figure 2.** The fabrication of microfluidic chips

*2.5. Microfluidic chip isothermal amplification detection system*

To sensitively detect in real time the fluorescent signal of DNA amplification, a confocal optical system was developed, which can detect the bioreactor cell in a limited spherical area with high excited power, while effectively limiting the background. This system used a high numerical aperture of objective NA = 0.72 to collect fluorescence on a focal plane near to the diffractive limit as shown in Figure 3, where there is a high efficiency of 92% at the focal plane. When the target is at the off focus position of 30μm, the efficiency of fluorescence signal collected is very low, about 0.96%. In the microfluidic chip detection, both the diameter and depth of bioreactor cell is 100μm or more, the collected fluorescence is positioned within the range of 50-60μm off the bottom of bioreactor, the background from the material of bioreactor and surrounding is approximated to 0 and can be ignored. Therefore, our developed confocal optical system can effectively restrain the background signals from the materials of bioreactor and surroundings.

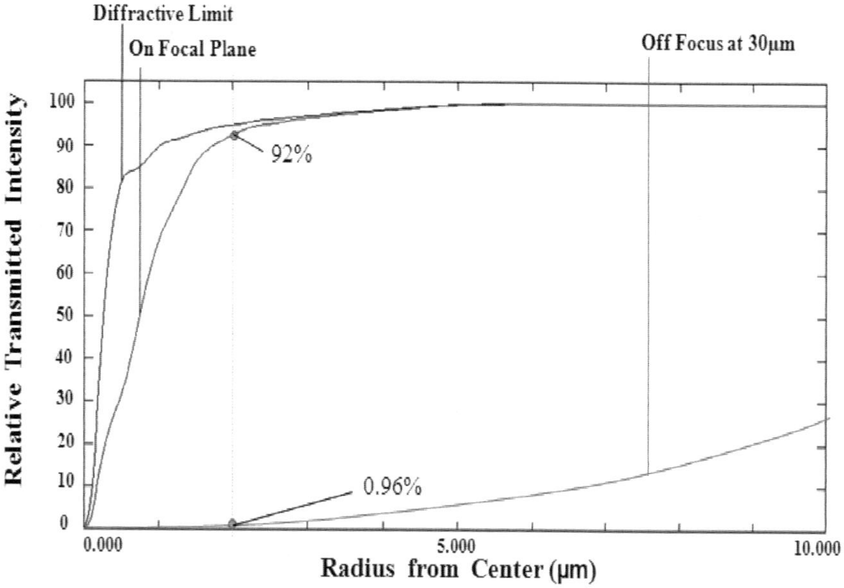


**Figure 3.** The analysis of fluorescent collecting efficiency in developed confocal optical detection system

*2.6. Amplification processes*

The process of DNA amplification in micro-nanoliter fluidic chips (50nL→7μL) is as table 2.

**Table 2**. The process of DNA amplification.

| Temperature(℃) | Time(min) | Remark |
|---|---|---|
| 37 | 5 | for ready |
| 65 | 40-45 | for real isothermal amplification |
| 80 | 5 | for limit life activity |

When the microfluidic chip is used, the inlet and outlet holes are opened by using the Eppendorf tip with light pressure, the mixtures of the circular probes, reagents and DNA samples were injected into the bioreactor cells of microfluidic chip from the inlet hole by the Eppendorf. And then, the inlet and outlet holes were sealed by thin PC film, the channel between any two adjacent bioreactor cells was blocked by heating press to make all bioreactor cells free from others. Finally, the microfluidic chip was setup into our developed confocal optical detection system for DNA isothermal amplification and specific gene identification.

## 3. Results and discussion

The fluorescent detection of nucleic acid amplification usually is affected by many factors, including the surface adsorption of bioreactor, the amplified double stranded DNA, the photo-bleaching, the unfastened double stranded DNA under amplified temperature and the background, etc. Except the amplified double stranded DNA make a positive detection response, while the unfastened double stranded DNA, the surface adsorption and the photo-bleaching all cause a negative detection response and low measuring sensitivity or make the detection response of amplified signal late. The difference from the unfastened double stranded DNA, the surface adsorption and the photo-bleaching usually causes the detection response of nucleic acid amplification unstable. And the background of surroundings and bioreactor material causes a high original value of dynamical amplified signal and low amplified efficiency. Therefore, current nucleic acid amplification in Eppendorf tube, 96 or 384 well microplate usually use a 25μl reaction assays or more, to make enough amplified double stranded DNA to ignore all other negative factors, and get a relative stable exponential amplified curves and good repeated detection.

When the reaction assay is less than 10μl, above negative factors are significant to the nucleic acid amplification as Figure 4. In figure 4, a PMMA microfluidic chip was used, where there are 16 positive amplified cells and 2 negative control cells, DNA template concentrations is 1.3 fM ($10^{-15}$M) and reaction mixtures is 7μl in every bioreactor cells. The maximum fluorescent signal of nucleic acid amplification in 16 bioreactor cells is about 40000, and the minimum fluorescent signal of nucleic acid amplification in 16 bioreactor cells is about 8500, the percent deviation of fluorescent signal was about 129%. The times at the second derivative inflexions of DNA exponential amplified curves for 16 bioreactor cells is from 12'th minute to 23'th minute, the maximum of time difference at these inflexions was about 11 minutes.

The contrast experiment was performed in our advanced inert processing metal microfluidic chip with same DNA template concentration of 1.3 fM and reaction volume of 7μl in every bioreactor cells, as Figure 5. In Figure 5, the maximum fluorescent signal of nucleic acid amplification in 16 bioreactor cells is about 46000, and the minimum fluorescent signal of nucleic acid amplification in 16 bioreactor cells is about 38000, the percent deviation of fluorescent signal was about 19%. The times at the second derivative inflexions of DNA exponential amplified curves for 16 bioreactor cells is from 11.42'th minute to 12.11'th minute, the maximum of time difference at these inflexions was about 0.7 minutes.

From Figure 4 and Figure 5, it is obvious that our developed inert processing metal microfluidic chips can outstanding improve the detection response of nucleic acid amplification in less than 10μl reaction assays.

In order to reduce further the reagents and sample consumption, isothermal nucleic acid amplification experiments were undertaken in different micro-nanoliter reaction volumes of 7.07μl, 3.14μl, 785nl and 392nl, each with the same concentrations of template DNA of 1.3 fM. The real time curves of DNA isothermal nucleic acid amplifications were obtained in Figure 6. It is obvious that the time at the inflexions of second derivative to the DNA exponential amplified curves is nearly the same in the different micro-nanoliter bioreactor cells, and the CVs of time at these inflexions is about 1.96%.

After the denatured double stranded DNA, the surface adsorption of bioreactor, the photo-bleaching and the background from the bioreactor material and surroundings are reduced, a fast response to the amplified double stranded DNA can be detected. With the same reaction mixtures and DNA template concentration, and differences only in reaction volume, nucleic acid amplification was performed respectively in PCR tube with reaction assay of 25μl and in our advanced metal microfluidic chip with a reaction assay of 7.07μl.

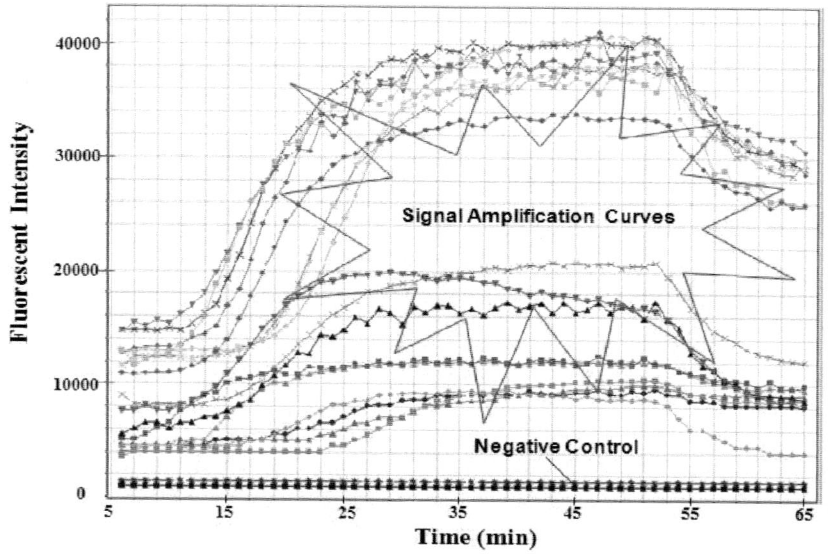


**Figure 4.** The DNA isothermal amplification in primary metal microfluidic chip with 18 bioreactor cells of 7μl in volume.

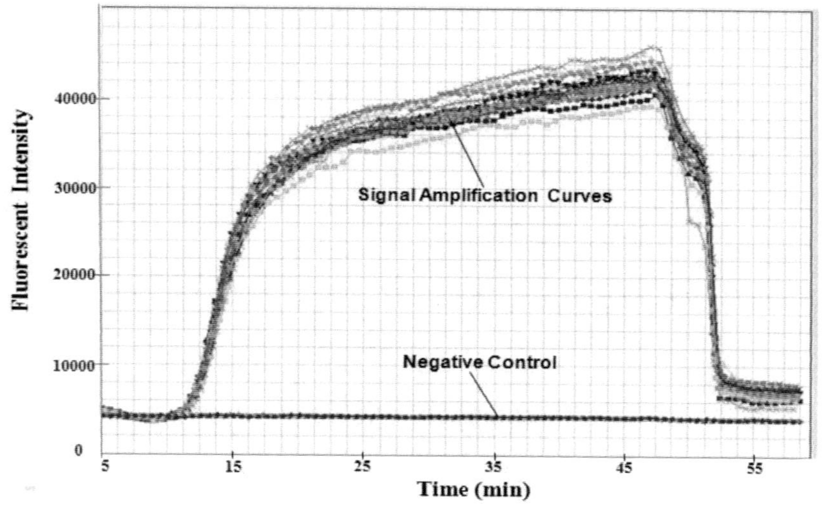


**Figure 5.** The DNA isothermal amplification in advanced inert processing metal microfluidic chip with same DNA template concentrations and reaction volume.

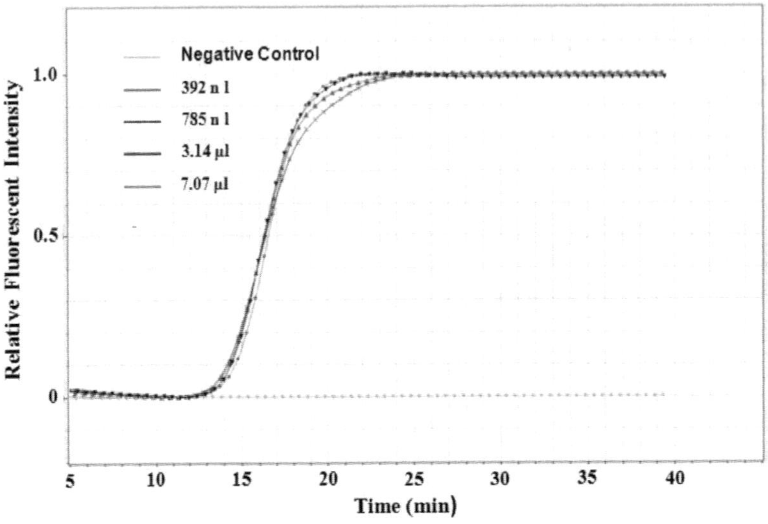


**Figure 6.** The microfluidic chip isothermal amplification in different micro-nanoliter reaction volumes

Figure 7 is the curves of isothermal nucleic acid amplification in the PCR tubes with a reaction volume of 25μl by using the ABI Fast Real-Time PCR System, where the DNA template copies were $2\times10^6$ copies, $2\times10^5$ copies, $2\times10^4$ copies, $2\times10^3$ copies and $2\times10^2$ copies, and the time at the second derivative inflexions of DNA exponential amplified curves for five different DNA template concentrations is 13'th minute, 15'th minute, 17'th minute, 20'th minute and without amplification to 45'th minute. When all amplifications were repeated 4 times, the CVs of the time at these inflexions were about 1%, 1%, 2% and 10% respectively.

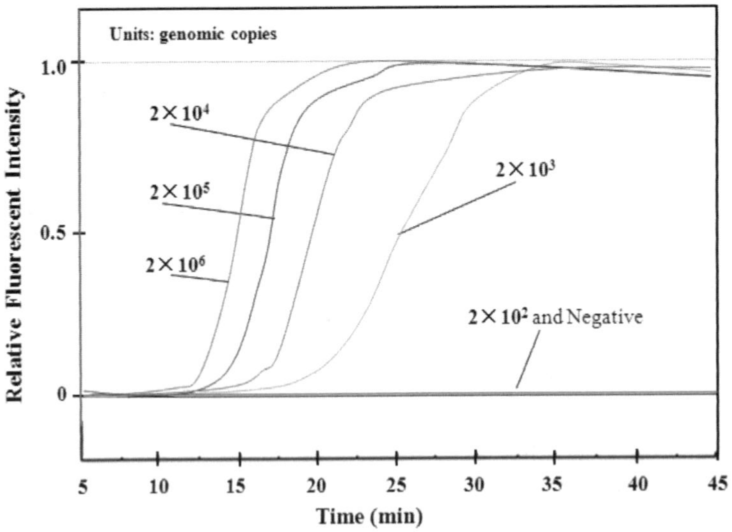


**Figure 7.** The isothermal amplification in 25μl PCR tubes

Figure 8 corresponds the real curves of nucleic acid amplification in the advanced metal microfluidic chip by using our developed confocal optical detection system, where the DNA template copies of $5.7\times10^4$ copies, $5.7\times10^3$ copies, $5.7\times10^2$ copies, $5.7\times10^1$ copies and $3.1\times10^1$ copies were used, and the time at the second derivative inflexions of DNA exponential amplified curves of five different DNA template copies is 12'th minute, 14'th minute, 16'th minute, 19'th minute and 22'th minute. When all amplifications were repeated 5 times, the CVs of inflexion times are respectively 0.2%, 0.3%, 0.8%, 1.1% and 0.9%. From Figure 7 and Figure 8, it is clear that the time at the second derivative inflexions of DNA exponential amplified curves in the advanced metal microfluidic chip have been advanced significantly, the less the DNA template copies is, and the more the time at the second derivative inflexions was advanced. The detection sensitivity in advanced metal microfluidic chip was raised about 65 folds compared to that of in 25μl PCR-tube.

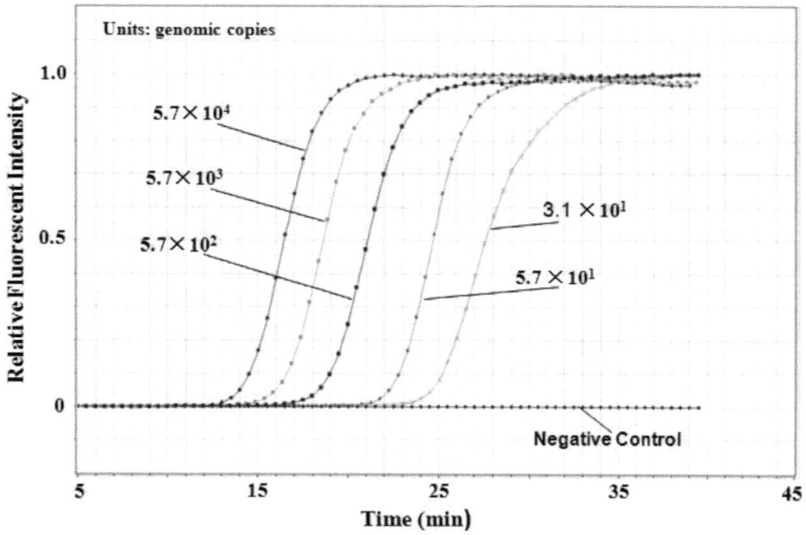


**Figure 8.** The isothermal amplification in advanced metal microfluidic chip with a reaction assay of 7.07μl

From Figure 5, Figure 6 and Figure 8, when using the advanced metal microfluidic chip and confocal optical detection system, a droplet nucleic acid isothermal amplification could be realized for promising clinic molecular diagnostics.

## 4. Conclusion

We have developed a metal micro-fluidic chip with advanced surface processing for micro-nanoliter volume DNA isothermal amplification. This method takes advantages of the nucleic acid amplification with good stability and consistency, high sensitivity about 31 genomic copies and bacteria specific gene identification. A high numerical aperture confocal optical detection system was developed to sensitively monitor the DNA amplification with low noise and high power collecting fluorescence near to the optical diffraction limit. Based on the advanced processing microfluidic chip and confocal optical detection system, the reaction assays of nucleic acid amplification could be reduced to nanoliter in volume. An accelerated nucleic acid isothermal amplification was detected in the ultra-small volume microfluidic chip, where the time at the second derivative inflexions of DNA exponential amplified curves was advanced and the sensitivity was improved about 65 folds to that of in 25μl Ep-tube amplified reaction. The method and system in this paper indicates a promising clinic molecular diagnostics in the droplet amplification.

## References


[1] Saiki RK, Scharf S, Faloona F, Mullis KB, Horn GT, Erlich HA and Arnheim N 1985 Enzymatic amplification of beta-globin genomic sequences and restriction site analysis for diagnosis of sickle cell anemia *Science* **230** 1350–54.

[2] Notomi T, Okayama H Masubuchi H, *et al.* 2000 Loop-mediated isothermal amplification of DNA *Nucleic Acids Res* **28** e63

[3] Lebl M, Burger C, Ellman B, Heiner D, Ibrahim G, Jones A, Nibbe M, Thompson J, Mudra P, Pokorny V, Poncar P and Zenisek K 2001 Fully automated parallel oligonucleotide synthesizer *Coll. Czechoslovak Chem. Commun.* **66** 1299–1314.

[4] Lashkari DA, Hunicke-Smith SP, Norgren RM, Davis RW and Brennan T 1995 An automated multiplex oligonucleotide synthesizer: development of high-throughput, low-cost DNA synthesis *Proc. Natl Acad. Sci. USA*, **92** 7912–15.

[5] Xiaochuan Zhou, Shiying Cai, Ailing Hong, Qimin You, Peilin Yu, Nijing Sheng, Onnop Srivannavit, Seema Muranjan, Jean Marie Rouillard, Yongmei Xia, Xiaolin Zhang, Qin Xiang, Renuka Ganesh, Qi Zhu, Anna Matejko, Erdogan Gulari, and Xiaolian Gao 2004 Microfluidic PicoArray synthesis of oligodeoxynucleotides and simultaneous assembling of multiple DNA sequences *Nucleic Acids Res.* **32** 5409-17.

[6] Lagally, E.T. and Mathies, R.A 2001 Integrated PCR-CE system for DNA analysis to the single molecule limit. Proceedings of Micro Total Analysis Systems. M*onterey*, CA, USA, 21–25 October, 117–118.

[7] Woolley, A.T., Hadley, D., Landre, P., deMello, A.J., Mathies, R.A., Northrup, M.A 1996 Functional Integration of PCR Amplification and Capillary Electrophoresis in a Microfabricated DNA Analysis Device. *Anal. Chem.* **68** 4081–86.

[8] Huang Guoliang, Zhu Jiang, Yang Yang, Xiao Ming, Dong Zhonghua and Deng Cheng 2006 Fluorescence photobleaching properties for microarray chips. *Chinese Journal of Luminescence*. **27** 259-264. (In Chinese)


## Acknowledgements


This subject is Supported by the National Foundation of High Technology of China (2006AA020701、2006AA020803), National Program on Key Basic Research Projects 973 of China (2006CB705700、2011CB707700) and National Natural Science Foundation of China (60978064).

# The application of Silver nanoparticle based SERS in diagnosing thyroid tissue

**Zufang Huang[1], Zuanfang Li[2], Rong Chen[1*], Guannan Chen[1], Duo Lin[1], Gangqin Xi[1] Yongjian Chen[1], Hongxin Lin[1] and Jinping Lei[1]**

1. Key Laboratory of Optoelectronic Science and Technology for Medicine, Ministry of Education, Fujian Normal University, Fuzhou 350007, China

2. Academy of Integrative Medicine, Fujian University of Traditional Chinese Medicine, Fuzhou 350108, China

E-mail: chenr@fjnu.edu.cn

**Abstract**. Surface-enhanced Raman scattering (SERS) is proved to be a powerful analytical tool for investigation of biological tissue. In this study, SERS based on Ag nanoparticles was used to investigate the normal and cancerous thyroid tissue. Preliminary results indicated that Raman peaks and the spectra profile from both normal and cancerous tissues showed a basic similarity, obvious differences are that, first, Raman peaks $563cm^{-1}$, $1449cm^{-1}$ and $1587cm^{-1}$ in cancerous tissue decreased obviously compared with the normal thyroid tissue. Besides, Raman peaks $1004cm^{-1}$ and $1128cm^{-1}$ might be specific peaks for normal thyroid tissue, whereas $1294cm^{-1}$ might attribute to specific peak for cancerous thyroid tissue. In addition, some peaks in normal thyroid tissue appeared to have shifted in cancerous tissue. Intensity ratio of $656cm^{-1}$ vs. $725cm^{-1}$ in normal tissue are significantly different from cancerous tissue ($P<0.005$), and it can be a reference for spectroscopic diagnostics of thyroid tissue. This study demonstrates that SERS can be used to monitor the changes at molecular level as well as a complementary tool in thyroid histopathology.

## 1. Introduction

Thyroid is the largest endocrine organ in human body, and there are increasing cases involved thyroid cancer, although thyroid cancer is one of the least deadly cancers, but it is mainly discovered in young adults. The pathological diagnosis of thyroid usually occurs in the following order: computerized tomography, magnetic resonance imaging and fine-needle aspiration (FNA). FNA is currently the most accepted procedure for diagnosing thyroid tissue, however, sensitivity of this procedure for the thyroid is at times poor with a high rate of false-negative results[1]. In addition, variations of the thyroid diseases present difficulties for the accurate morphological diagnosis, for example, differentiation of hyperplastic nodule, follicular adenoma or follicular carcinoma is difficult[2]. Therefore, new methods which can provide information of biochemical composition and structure, as to improve the diagnosis of thyroid diseases is quite essential.

* To whom any correspondence should be addressed.

One of the possible alternatives to the current methods is Raman spectroscopy, which can provide a molecular fingerprint information of the sample. Thus Raman spectroscopy acts as a power analytical tool with many advantages for characterization of biological molecules and structures due to its unique fingerprint feature; In addition, the characteristics of low sensitivity to water, easy sample preparation, making the application of Raman spectroscopy attract more and more attention of being used in biological research and biomedical diagnosis in recent years. The application of Raman spectroscopy covers a wide range in various human organs, such as brain, breast, gastrointestinal tract, larynx, lung, nasopharynx, and skin[3-9]. However, a major limitation of conventional Raman spectroscopy is its inherently weak signals, which presents great challenges for instrumentation and restricted its clinical application.

To overcome the low efficiency of Raman spectroscopy, a called surface-enhanced Raman scattering was found with ability of greatly enhancing the scattering effect when the analyte of interest is adsorbed onto or close proximity to a rough surface of suitable metals such as gold, silver, or copper. Research results showed several orders of magnitude enhancement effect and then it was rapidly applied to the study of single molecule and tissue. Hereafter Applications of SERS at the tissue level have been reported very recently[10-11].

In this work, we report here the application of silver nanoparticles in diagnosis of thyroid tissue. Tentative assignments of major Raman peaks and possible identities of the thyroid tissue were made. Meanwhile, attempts have been made to distinguish normal thyroid tissue from cancerous thyroid tissue based on difference SERS spectrum as well as the intensity ratio of specific Raman peaks.

## 2. Materials and Methods

### 2.1. Preparation of silver nanoparticles

Colloidal silver was prepared by the aqueous reduction of silver nitrate with hydroxylamine hydrochloride using a simple and fast method reported by[12]. Briefly, Nicolae and Bernhard.12 4.5 mL sodium hydroxide (0.1 mol/L) was added to 5 ml hydroxylamine hydrochloride (6×10-2 mol/L) and then the mixtures were added rapidly to 90 mL silver nitrate (1.11×10-3 mol/L) with vigorous shaking. The maximum absorption band of the silver colloid located at 416 nm.

### 2.2. Sample preparation

Thyroid tissue samples excised from patients were kindly provided by Fujian Provincial Tumour Hospital. Following surgical removal of the thyroid tissues, specimens were sliced and then immediately transported to the lab and kept at 0℃, standard Ag colloidal solution was directly droped onto the thin cross-sectioned pieces which were then dried at the room temperature, finally both normal and cancerous thyroid tissue slices were selected for spectral measurement. All samples were finished within 3h after removal from the patients, then the samples were processed for routine histological examination. Informed consent was obtained from patients who participated in the study, we strictly conformed to the institutional rules governing clinical investigation of human subjects in biomedical research.

### 2.3. Raman Instrumentation

All SERS spectra were recorded with Renishaw InVia Raman microscope system (Renishaw Plc., New Mills, Wotton-under- Edge, UK)which equipped with 785 nm diode laser and 514nm argon-ion laser. The SERS spectra were recorded using 785 nm excitation. The laser power on the sample was about 3 mW, and the exposure time was 10s. A 20x (N.A.=0.75) objective was used for focusing the excitation beam and collecting the backscattered signals. The detection of Raman signal was accomplished with a Peltier cooled charge-coupled device (CCD) camera and the Raman instrument was calibrated using a silicon wafer by centering the band at 520cm$^{-1}$. The WIRE 2.0 software was used for spectra acquisition and analysis.

## 3. Results and discussion

Figure 1 shows the mean spectra from normal and papillary thyroid tissue and the corresponding difference spectrum in the region of $400^{-1}800cm^{-1}$. Prior to other treatment the spectra were preprocessed by sequential 5-point smoothing method to reduce the noise[13]. It can be noted that most Raman peaks as well as the spectra profile from both tissues show a rough similarity, which means that the biochemical compositions are somewhat similar in this two type. Relatively strong bands of $570cm^{-1}$, $656cm^{-1}$, $725cm^{-1}$, $895\text{-}912cm^{-1}$, $963cm^{-1}$, $1094cm^{-1}$, $1331cm^{-1}$, $1374cm^{-1}$, $1449cm^{-1}$, $1587cm^{-1}$ were observed in both tissue. Obvious differences are that, first, in cancer thyroid tissue Raman peaks $563cm^{-1}$, $1449cm^{-1}$, $1587cm^{-1}$ which can be attributed to nucleic acid, collagen and lipids decreased obviously compared with the normal thyroid tissue, besides, Raman peaks $1004cm^{-1}$ and $1128cm^{-1}$ might be specific peaks for normal thyroid tissue, whereas $1294cm^{-1}$ might attribute to specific peak for cancer thyroid tissue. In addition, some peaks such as $1374cm^{-1}$, $1449cm^{-1}$, $1587cm^{-1}$ in the normal thyroid tissue appeared to have shifted in comparison with cancer thyroid tissue. The tentative assignment for the observed SERS bands were listed in table 1.

The preliminary results indicated that SERS spectra can characterize the normal and cancerous thyroid tissue due to their intrinsic differences of biochemical components and molecular structure. Raman peaks at $563cm^{-1}$ and $1449cm^{-1}$ which can be assigned to DNA were easily found to be different in their intensity between normal and cancerous thyroid tissue, and this coincided with the basic concept that DNA content variates during the carcinogenesis process. Meanwhile, we found a peak at $1587cm^{-1}$ which attributed to lipids decreased significantly in cancer tissue compared with the normal tissue. Recent research results showed that obvious difference was found in lipid content between normal tissue and malignant tissue, and they ascribed this phenomenon to the change in energy distribution of lipids[14].

Figure 2 shows a scatter plot for the intensity ratio of $I_{656}$ vs. $I_{725}$ for each tissue sample, previous studies have shown a simple but effective algorithm of intensity ratio based on Raman peaks can be used to classify tumor vs. normal tissue in brain breast, colon and lung[15-17]. In our preliminary study, the mean ratio of $I_{656}$ vs. $I_{725}$ (mean ± SD) for normal tissue (0.409±0.0.082) was significantly different from the mean ratio for cancer tissue (0.900±0.362), (unpaired Student's t-test, $P<0.005$), and this intensity ratio can be a potential diagnostic parameter for detecting and/or classifying different type of thyroid malignancies, and the next work is to increase the number of sample measurements.

## 4. Conclusion

In this preliminary study, we have analyzed and compared the spectra obtained from normal and cancerous thyroid tissue. Although some peaks appear in both two tissue type, but significant spectral differences for cancerous thyroid compared to normal thyroid tissue were found at specific peaks, further calculation based on intensity ratio of $I_{656}$ vs. $I_{725}$ for each tissue sample gives a potential and valuable spectroscopic reference for discriminating normal thyroid tissue from cancerous thyroid tissue. Furthermore, we are working on the development of new statistical methods based on SERS spectra which reflects the changes at molecular level for diagnosis and pathologic analysis of thyroid tissue.

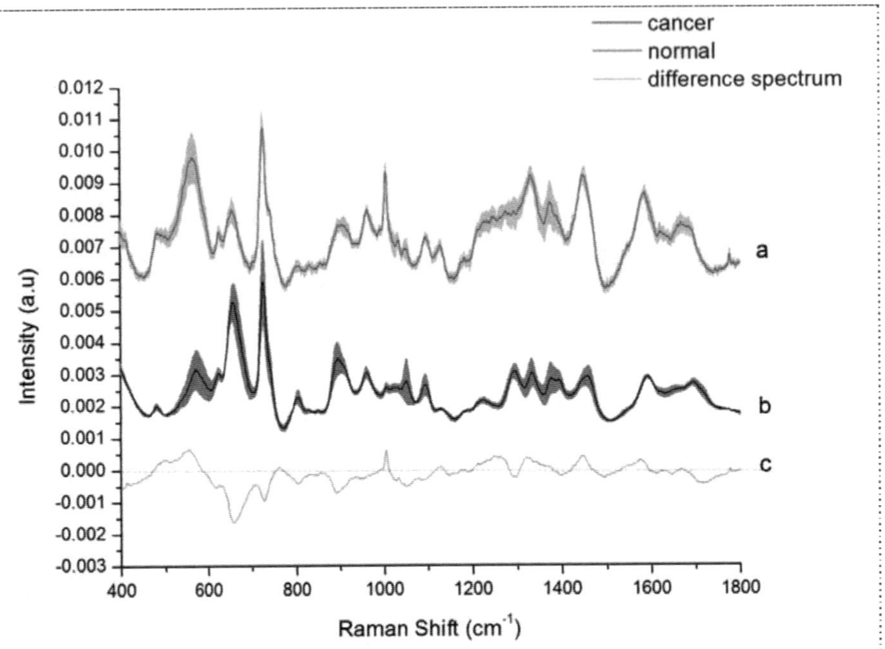


**Figure 1.** Comparison of Mean Raman spectra between normal (a, n=11) and cancerous (b, n=8) thyroid tissue, Shaded area represents the standard deviations (SD), Also shown at the bottom is the difference spectrum(c).

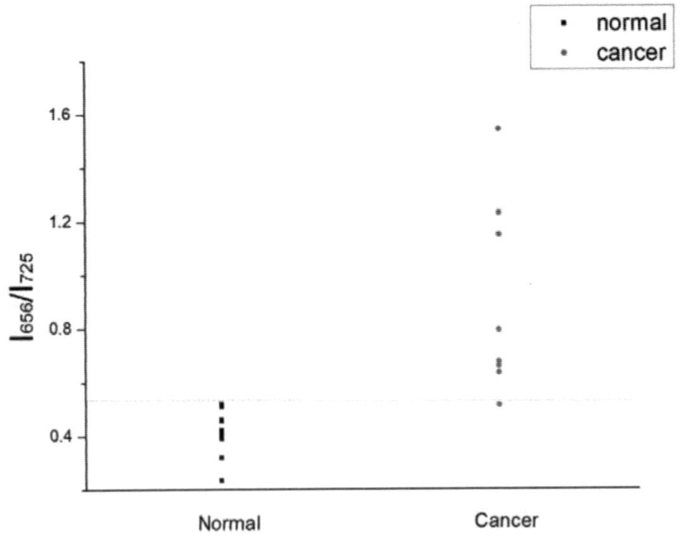


**Figure 2.** Scatter plot of the intensity ratio of Raman peak at 656cm-1 vs. 725cm-1, The mean ratio (0.409±0.0.082) for normal tissue is significantly different (P<0.005)from the mean value (0.900±0.362) for cancer tissue.

Table 1 The peak positions and tentative band assignment of Raman spectra

| Normal thyroid tissue | Cancerous thyroid tissue | components |
|---|---|---|
| 563 | 570 | Nucleic acids, Nucleotide |
| 656 | 658 | Tryptophan |
| 725 | 725 | Adenine, Coenzyme A |
| 803 | 803 | RNA |
| 895-912 | 897 | DNA backbone |
| 963 | 960 | Citric acid |
| 1004 | - | Phenylalanine, Tyrosine |
| 1050 | 1050 | Tryptophan |
| 1094 | 1094 | Lipid, Nucleic acid |
| 1128 | - | Protein |
| - | 1294 | Lipids |
| 1331 | 1333 | Nucleic acid bases |
| 1374 | 1376 | Nucleic acids |
| 1449 | 1458 | Collagen, Phospholipid |
| 1587 | 1591 | Lipids |

## 5. Acknowledgements

This work was supported by the National Natural Science Foundation of China (Nos. 60778046 and 60910106016), the Science and Technology Project of Fujian Province (Nos. 2008J0016 ), the Project of Science Foundation of Ministry of Health and United Fujian Provincial Health and Education Project for Tackling the Key Research (No.WKJ2008-2-046).

## References

[1] Rout P and Shariff S 1999 Diagnostic value of qualitative and quantitative variables in thyroid lesions. *Cytopathology.* **10** 171-9

[2] Teixeira C S, Bitar R A, Martinho H S, Santos A B, Kulcsar M A, Friguglietti C U, da Costa R B, Arisawa E A and Martin A A 2009 Thyroid tissue analysis through Raman spectroscopy *Analyst.* **134** 2361-70

[3] Haka A S, Shafer-Peltier K E, Fitzmaurice M, Crowe J, Dasari R R and Feld M S 2002 Identifying microcalcifications in benign and malignant breast lesions by probing differences in their chemical composition using Raman spectroscopy *Cancer Res.* **62** 5375-80

[4] Jyothi Lakshmi R, Kartha V B, Murali Krishna C, R Solomon J G, Ullas G and Uma Devi P 2002 Tissue Raman spectroscopy for the study of radiation damage: brain irradiation of mice *Radiat Res.* **157** 175-82

[5] Magee N D, Beattie J R, Carland C, Davis R, McManus K, Bradbury I, Fennell D A, Hamilton P W, Ennis M, McGarvey J J and Elborn J S 2010 Raman microscopy in the diagnosis and prognosis of surgically resected nonsmall cell lung cancer *J Biomed Opt.* **15** 026015

[6] Teh S K, Zheng W, Ho K Y, Teh M, Yeoh K G and Huang Z 2008 Diagnosis of gastric cancer using near-infrared Raman spectroscopy and classification and regression tree techniques *J Biomed Opt.* **13** 034013

[7] Wang H, Huang N, Zhao J, Lui H, Korbelik M and Zeng H 2010 Depth-resolved in vivo micro-Raman spectroscopy of a murine skin tumor model reveals cancer-specific spectral biomarkers *J Raman Spectrosc.* (Preprint DOI 10.1002/jrs.2677)

[8] Huang Z, Zeng H, Hamzavi I, McLean DI and Lui H 2001 Rapid near-infrared Raman spectroscopy system for real-time in vivo skin measurements *Opt Lett.* **26** 1782-4

[9] Lau D P, Huang Z, Lui H, Man C S, Berean K, Morrison M D and Zeng H 2003 Raman spectroscopy for optical diagnosis in normal and cancerous tissue of the nasopharynx-preliminary findings *Lasers Surg Med.* **32** 210-4

[10] Liu YN, Zou ZQ, Liu YQ, Xu XX, Yu G and Zhang CZ 2007 Surface-enhanced Raman spectra of natural tissue and cancerous tissue of lung *Spectrosc spect anal.* **27** 2045-8

[11] Pinzaru S C, Andronie L M, Domsa I, Cozar O and Astilean S 2008 Bridging biomolecules with nanoparticles: surface-enhanced Raman scattering from colon carcinoma and normal tissue *J Raman Spectrosc.* **39** 331-4

[12] Nicolae Leopold and Bernhard Lendl 2003 A New Method for Fast Preparation of Highly Surface-Enhanced Raman Scattering (SERS) Active Silver Colloids at Room Temperature by Reduction of Silver Nitrate with Hydroxylamine Hydrochloride *J. Phys. Chem. B* **107** 5723-7

[13] Zhao J, Lui H, McLean DI and Zeng H 2007 Automated autofluorescence background subtraction algorithm for biomedical Raman spectroscopy *Appl Spectrosc.* **61** 1225-32

[14] Gao Z H, Hu B, Ding C and Yu J G 2010 Micro-Raman spectra for lipids in colorectal tissue *Spectrosc spect anal.* **30** 692-6

[15] Huang Z, McWilliams A, Lui H, McLean D I, Lam S and Zeng H 2003 Near-infrared Raman spectroscopy for optical diagnosis of lung cancer *Int J Cancer.* **107** 1047-52

[16] Liu C H, Das B B, Sha Glassman W L, Tang G C, Yoo K M, Zhu H R, Akins D L, Lubicz S S, Cleary J, Prudente R, Celmer E, Caron A and Alfano R R 1992 Raman, fluorescence, and time-resolved light scattering as optical diagnostic techniques to separate diseased and normal biomedical media *J Photochem Photobiol B* **16** 187-209

[17] Utzinger U, Heintzelman D L, Mahadevan-Jansen A, Malpica A, Follen M and Richards-Kortum R 2001 Near-Infrared Raman Spectroscopy for in Vivo Detection of Cervical Precancers *Appl Spectrosc.* **55** 955-9

# Single-shot parallel full range complex Fourier-domain optical coherence tomography

Bingjie Huang, Peng Bu, Nan Nan, Xiangzhao Wang
Shanghai Institute of Optics and Fine Mechanics, Chinese Academy of Sciences
Shanghai 201800, China

**Abstract**

We present a method of parallel full range complex Fourier-domain optical coherence tomography (FDOCT) that is capable of acquiring an artifacts-free two-dimensional (2-D) cross-sectional image, i.e. a full range B-scan tomogram, by a single shot of 2-D CCD camera. This method is based on a spatial carrier technique, in which the spatial carrier-frequency is instantaneously introduced into the 2-D spectral interferogram registered in parallel FDOCT by using a grating-generated reference beam. The spatial-carrier-contained 2-D spectral interferogram is processed through Fourier transformation to obtain a complex 2-D spectral interferogram. From the 2-D complex spectral interferomgram, a full range B-scan tomogram is reconstructed. The principle of our method is confirmed by imaging an onion sample.

**Key words:** optical coherence tomography, Fourier domain detection, spatial carrier frequency.

## 1、Introduction

Fourier-domain optical coherence tomography (FDOCT) [1, 2] has been demonstrated to hold more clinical applicability than conventional Time domain OCT (TDOCT), benefiting from its advantages of high acquisition speed and sensitivity [3]. However, FDOCT suffers from the so-called complex conjugate, dc and autocorrelation artifacts, of which the first reduces the imaging depth range by half, and the last two limit the signal to noise ratio. These disturbing artifacts are due to the straightforward Fourier transformation of the detected real-valued spectral interferogram. To eliminate these artifacts, various methods to reconstruct the complex spectral interferogram have been proposed over the last few years. First, a technique based on phase-shifting interferometry was used [4], where the complex spectral interferogram was retrieved from multiple phase-shifted spectral interferograms. However, several interferograms are required to be measured sequentially for reconstructing a full range axial line image (A-scan), resulting in longer measurement time. In addition, definite phase step between phase-shifted interferograms require a high stability of the optical system and the sample, which precludes its application to in vivo measurement. To improve this method, simultaneous detection of multiple phase-shifted interferograms using $3\times3$ fiber couplers [5] ,polarization-based optical components [6] , and recently dual-reference detection [7] were proposed, but the system's complexity and cost are considerably increased.

Alternatively, carrier-frequency technique is another method to reconstruct the complex interferogram [8]. Recently, carrier-frequency-based methods [9, 10, and 11] have been employed to realize real-time full range complex FDOCT. In these methods the reference beam was phase modulated during lateral scanning of the probing beam to introduce a spatial carrier-frequency into the two-dimensional (2-D) spectral interferogram, from which the complex interferogram can be obtained by Fourier transform or Hilbert transform processing. These methods require only one measurement to acquire a full range FDOCT A-scan image, but the spatial carrier frequency was generated by time-dependent phase modulation, thus any phase error during the sequential phase modulation due to the motion of the measured sample may degrade the suppression ratio of the complex conjugate artifact, especially in Doppler imaging [12, 13]. Therefore a spatial carrier technique, where a spatial carrier frequency is generated by spatially dependent phase modulation at an instant of time, is more suitable for time-varying objects measurement [14].

We have previously reported a one-shot parallel complex full range FDOCT [15] by combining a spatial carrier technique with parallel FDOCT [16], where a tilted reference mirror was used to generate a spatial carrier-frequency into the 2-D spectral interferogram registered in parallel FDOCT. Although the use of tilted reference mirror as a linear spatial phase modulator is simple, the lateral imaging range is limited. An alternative approach to generate linear spatial phase modulation is to use a reflective grating. This grating-based method was previously used as a reference delay line to achieve a scan-free coherence microscopy [17], which was later extended to a high-speed TDOCT [18]. Here, for the first time, we use a grating-generated reference beam to introduce instantaneously a spatial carrier-frequency into the 2-D spectral interferogram registered in parallel FDOCT, to implement a single-shot parallel full range complex FDOCT. The 2-D complex spectral interferogram is reconstructed through Fourier transformation from the spatial-carrier-contained 2-D spectral interferogram with a single shot of the 2-D CCD camera. From the 2-D complex spectral interferomgram, a full range B-scan tomogram is obtained. Therefore our method is more insensitive to the sample motion-induced phase error than the other proposed carrier-frequency methods.

## 2、Method

The schematic diagram of the proposed single-shot parallel full range complex FDOCT system is shown in Fig.1. This system is similar to that described previously in Ref. [15] except that a reflective grating is used as a reference instead of a tilted mirror. In brief, a parallel FDOCT system was adopted to record a 2-D spectral interferogram with a single shot of 2-D CCD camera. The reflective grating G1 was positioned with the Littrow configuration in the reference arm of the interferometer to introduce a spatial carrier-frequency into the 2-D spectral interferogram instantaneously. We define a coordinate system with the z axis along the direction of propagation of light, the x axis the spectral direction, and y axis the lateral direction.

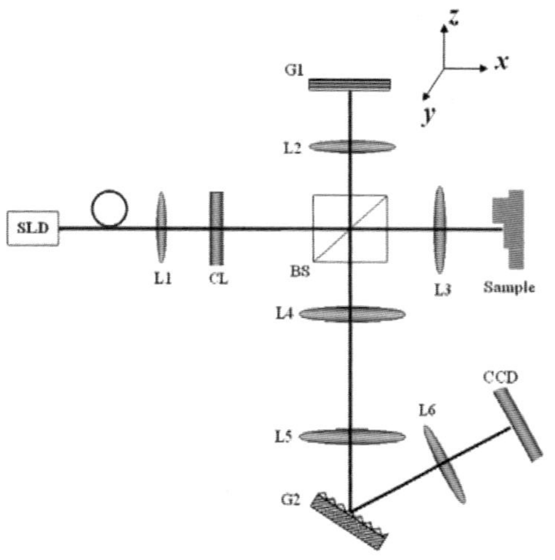


Fig. 1. The schematic diagram of the single-shot parallel full-range complex FDOCT
SLD, super-luminescent diode; L1, the collimator objective lens; L2, L3, L4, L5, L6 denote lenses of focal length 100, 100, 100, 40, 100 mm, respectively; CL, cylindrical lens of focal length 200 mm; BS, cube beamsplitter; G1, G2 denote grating with 600, 1800 lines/mm, respectively. The grating G1 is positioned with Littrow configuration in y-z plane, which is not denoted in the figure.

The grating G1 installed with Littrow configuration was performed in y-z plane, resulting in a tilted reference wavefront. The Littrow angle $\theta$ is determined by

$$\theta = \sin^{-1}(m\lambda_0 / 2p), \tag{1}$$

where $\lambda_0$ is the center wavelength of light, $p$ is the grating period, and $m$ is the diffraction order. When the first-order diffracted light is selected, a spatial carrier frequency $f_0$ is introduced into the 2D spectral interferogram in the lateral direction, which is given by

$$f_0 = 2tg\theta / \lambda, \tag{2}$$

where $\lambda$ is the wavelength of light. So the spatial-carrier-contained 2-D spectral interferogram can be expressed as

$$g(k,y) = g_0(k,y) + 2\sum_n S(k)\sqrt{\alpha_n(y)\beta_0}\cos\left[2k\left(z_n(y) + y \cdot tg\theta\right)\right], \tag{3}$$

where $k = 2\pi / \lambda$ is the wave number; $y$ is the lateral position; $g_0(k,y)$ includes the dc and autocorrelation terms; $S(k)$ is the spectrum of the light source; $\alpha_n(y)$ is the reflectivity of the $n$'th interface within the sample with respect to lateral position $y$; $\beta_0$ is the first-order diffracted efficiency of the grating G1; $z_n(y)$ is the one-pass optical path difference between the probing beam reflected by the $n$'th interface within the sample and the reference beam with respect to lateral position $y$. Equation (3) can be rewritten as

$$g(k,y)=g_0(k,y)+\sum_n b_n(k,y)\exp(i2\pi f_0 y)+\sum_n b_n^*(k,y)\exp(-i2\pi f_0 y), \quad (4)$$

where $b_n(k,y)=S(k)\sqrt{\alpha_n(y)\beta_0}\exp[i2kz_n(y)]$ and $*$ denotes the complex conjugate operator. Fourier transformation of interferogram $g(k,y)$ with respect to lateral position $y$ for each $k$ yield

$$G(k,f_y)=G_0(k,f_y)+\sum_n B_n(k,f_y-f_0)+\sum_n B_n^*(k,f_y+f_0), \quad (5)$$

where $G$ and $B$ are the Fourier transformation of $g$ and $b$, respectively; $f_y$ is the spatial frequency along y direction. When the spatial carrier frequency $f_0$ is large enough, the three spectra on the right-hand side of equation (5) can be separated from each other. Then by band-pass selecting the spectrum of $\sum_n B_n(k,f_y)$ in positive frequency region and inverse Fourier transforming to the original spatial domain, we can obtain the complex 2-D spectral interferogram

$$g_{comp}(y,k)=\sum_n b_n(k,y)=\sum_n S(k)\sqrt{\alpha_n(y)\beta_0}\exp[i2kz_n(y)]. \quad (6)$$

Above-mentioned Fourier transform processing is indeed the same as the Hilbert transform method used by Ref.[11]. Finally, taking inverse Fourier transformation of the complex 2D spectral interferogram with respect to $k$ for each $y$, a complex function $\tilde{I}(y,z)=\sum_n\sqrt{\alpha_n(y)\beta_0}\Gamma(z-2z_n(y))$ is obtained, where $\Gamma$ is the first-order electric field correlation function. Taking the amplitude of $\tilde{I}(y,z)$, we can obtain the full range B-scan tomogram.

## 3、Results and Discussions

The experimental setup shown in Fig. 1 was constructed. A super-luminescent diode (SLD) with a central wavelength ($\lambda_0$) of 840 nm and a FWHM bandwidth of 50 nm was used as a light source. The home-built spectrometer consisted of a reflective grating (G2, 1800 lines/mm), a lens L6 with focal length of 100 mm, and a 2-D CCD camera (Sony XCHR-50, 9-bit, 648×494 pixels, 60 frames per second).

We measured a biological sample. Figure 2(a) shows the cross-sectional image of an onion sample with our proposed method, where the complex conjugate, dc and autocorrelation artifacts are eliminated, compared with the image obtained by a straightforward FDOCT as shown in Fig. 2(b). The suppression ratio of the complex conjugate artifacts was measured to be 28dB. The lateral measurement range was ~1 mm determined by the linear imaging range of CCD camera The available full depth range in air was ~1.8 mm determined by the spectrometer resolution. The full range B-scan rate of our method is determined by the frame rate of CCD camera (60Hz), which suggests a real-time full range FDOCT B-scan imaging is possible.

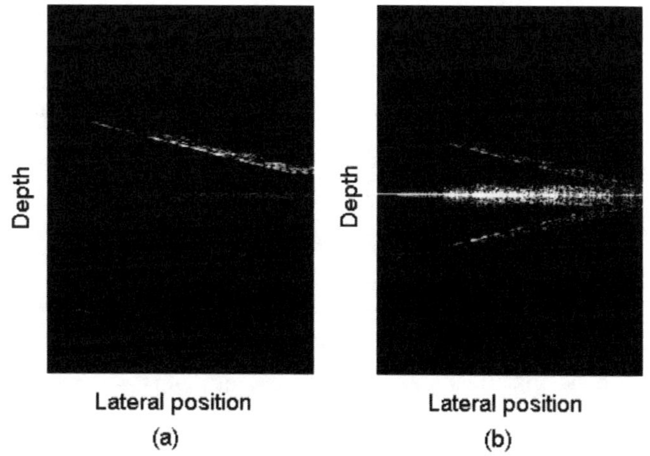


Fig. 2. Cross-sectional image of onion sample obtained by (a) single-shot parallel full range complex FDOCT and (b) straightforward FDOCT.

The system sensitivity was measured to be 78 dB with a partial reflector as a sample. A sensitivity roll off with respect to depth ($z$ axis) was observed. This fall-off of sensitivity was primarily due to the limited resolution of the home-built CCD spectrometer. The numerical error of the wavelength to wavenumber rescaling could also strengthen the fall-off. In our experiment, this fall-off of sensitivity also appeared along the lateral direction. This degradation may be due to the Gaussian intensity profile of the line illumination and vignetting effect of the CCD. In addition, because an SLD is a spatially coherent light source, coherent crosstalk among the lateral points may cause additional degradation of the sensitivity. This coherent crosstalk can be suppressed by using spatially incoherent light source.

The imaging resolution of our system is different for each axis. Along the $x$ axis, the resolution is defined as the focal spot size of the probe beam, and theoretically estimated as 33.6 $\mu m$, which in practice was degraded by the lateral coherent crosstalk effect. The resolution in $y$ axis was 3.0 $\mu m$ depending on the pixel size of the CCD camera. The depth resolution was 6 $\mu m$ in air determined by the coherence length of SLD.

## 4、Conclusion

In summary, we have demonstrated a single-shot parallel complex FD-OCT by using a grating-generated spatial carrier-frequency. The method is insensitive to the phase error due to the motion of the measured object. With this method, a parallel full range complex FDOCT system can be used for high sensitivity B-scan-based Doppler imaging because it is free of motion artifacts while maintaining relatively a long integration time compared to scanning FD-OCT system.

## Acknowledgments

This work was supported by grants from the National Natural Science Foundation of China (60878029 and 60938003) and the International Cooperation Program Project of Shanghai Municipal Science & Technology Commission (08520704200).

## References


[1] Fercher A F, Hitzenberger C K, Kamp G and EL-Ziat S Y, Measurement of intraocular distances by backscattering spectral interferometry, Opt. Commun. 117, 43–8 (1995).

[2] Hausler G, Lindner MW, "Coherence radar" and "spectral radar" — New tools for dermatological diagnosis, J. Biomed. Opt. 3, 21–31 (1998).

[3] R. A. Leitgeb, C. K. Hitzenberger, and A. F. Fercher, Performance of Fourier domain vs. time domain optical coherence tomography, Opt. Express 11, 889–894 (2003).

[4] M. Wojtkowski, A. Kowalczyk, R. Leitgeb, and A. F. Fercher, Full range complex spectral optical coherence tomography technique in eye imaging, Opt. Lett. 27, 1415–1417 (2002).

[5] M. Sarunic, M. A. Choma, C. Yang, and J. A. Izatt, Instantaneous complex conjugate resolved spectral domain and swept-source OCT using 3×3 fiber couplers, Opt. Express 13, 957–967 (2005).

[6] B. J. Vakoc, S. H. Yun, G. J. Tearney, and B. E. Bouma, Elimination of depth degeneracy in optical frequency-domain imaging through polarization-based optical demodulation, Opt. Lett. 31, 362–364 (2006).

[7] Kye-Sung Lee, Panomsak Meemon, William Dallas, Kevin Hsu, and Jannick P. Rolland, Dual detection full range frequency domain optical coherence tomography, Opt. Lett. 35, 1058-1060 (2010).

[8] Jun Zhang, J. Stuart Nelson, and Zhongping Chen, Removal of a mirror image and enhancement of the signal-to-noise ratio in Fourier-domain optical coherence tomography using an electro-optic phase modulator , Opt. Lett. 30, 147-149 (2005).

[9] Y. Yasuno, S. Makita, T. Endo, G. Aoki, M. Itoh, and T. Yatagai,Simultaneous B-M-mode scanning method for real-time full-range Fourier domain optical coherence tomography, Appl. Opt. 45, 1861–1865 (2006).

[10] R. A. Leitgeb, R. Michaely, T. Lasser, and S. C. Sekhar, Complex ambiguity-free Fourier domain optical coherence tomography through transverse scanning, Opt. Lett. 32, 3453–3455 (2007).

[11] Ruikang K. Wang, In vivo full range complex Fourier domain optical coherence tomography, Applied Physics Letters 90, 054103 (2007).

[12] F. Jaillon, S. Makita, M. Yabusaki, and Y. Yasuno, Parabolic BM-scan technique for full range Doppler spectral domain optical coherence tomography, Opt. Express 18, 1358-1372 (2010).

[13] P. Meemon, K.-S. Lee, and J. P. Rolland, Doppler imaging with dual-detection full-range frequency domain optical coherence tomography, Biomed. Opt. Express 1, 537-552 (2010).

[14] M. Takeda and M. Kitoh, Spatiotemporal frequency multiplex heterodyne interferometry, J. Opt. Soc. Am. A 9, 1607–1614 (1992).

[15] Peng Bu, Xiangzhao Wang, and Osami Sasak, One-shot parallel complex Fourier-domain optical coherence tomography using a spatial carrier frequency, Optical Engineering 47(5), 050502 (2008).

[16] B. Grajciar, M. Pircher, A. F. Fercher, and R. A. Leitgeb, Parallel Fourier domain optical coherence tomography for in vivo measurement of the human eye, Opt. Express 13, 1131–1137 (2005).

[17] I. Zeylikovich, A. Gilerson, and R. R. Alfano, Nonmechanical grating-generated scanning coherence microscopy, Opt. Lett. 23, 1797-1799 (1998).

[18] Yuuki Watanabe, Fumitoshi Sajima, Toshiki Itagaki, Kei Watanabe, and Yuuki Shuto, High-spped linear detection time domain optical coherence tomography with reflective grating-generated spatial reference delay, Applied Optics 48, 3401-3405 (2009).

# The spectra character of photodegraded the pyridinoline cross-links by Hypocrellin B

**Jucheng Zhang, Rui Chen, Wei Liu, Zhuo Chen, Lidan Shu, Yingji Liu**

[a] Key Laboratory of Natural Pharmaceutical & Chemical Biology of Yunnan Province, Honghe University, Mengzi, Yunnan Province, P. R. China; [b] school of sicence, Honghe University, Mengzi, Yunnan Province, P. R. China.

Correspondence author Email: hhusci@163.com

**Abstract**. Pyridinoline cross-links is one of the cross-link formation in collagen which in cell matrix, many research shown that this cross-link cause the fibrosis. Hypocrellin B (HB) is one of the nature photosensitizers, this work investigated the pyridinoline cross-link in collagen was photodegraded by HB. The result shown HB can degrade the pyridinoline cross-link with photo. This is to say, HB may be use as the photodynamic reagent to study the fibrosis.

## 1. Introduction

Hypocrellin B (HB), the structure as figure 1, was quantitatively produced by the dehydration of HA which was extracted from the fungus of *Hypocrella bambusae* and crystallized twice from chloroform-petroleum ether before use [1]. It is well known for the high photodynamic activity to tumors and other biological targets but very low dark-toxicity [2-4].

The pyridinoline cross-link is a fluorescence group in collagen and gelatin, and the pyridinoline cross-link formation in collagen so it can't degrade by metalloprotease, then collagen over-amount accumulated in tissues which cause fibrosis [5, 6]. Some research found the Photodynamic therapy (PDT) could prevent the vascular restenosis after surgery and effectively control the postoperative fibrosis in human glaucomatous eyes [7]. It was also reported that the pyridinoline cross-link in the fibrotic skin could be photo-degraded by psoralen under the ultraviolet light [8, 9].

In this work, the spectrum character of the photoreaction of collagen with HB was investigated, the results shown the pyridinoline cross-link in collagen and gelatin could be photo-degraded by hypocrellin B, the reaction system was saturated by $O_2$, Air and Ar respectively.

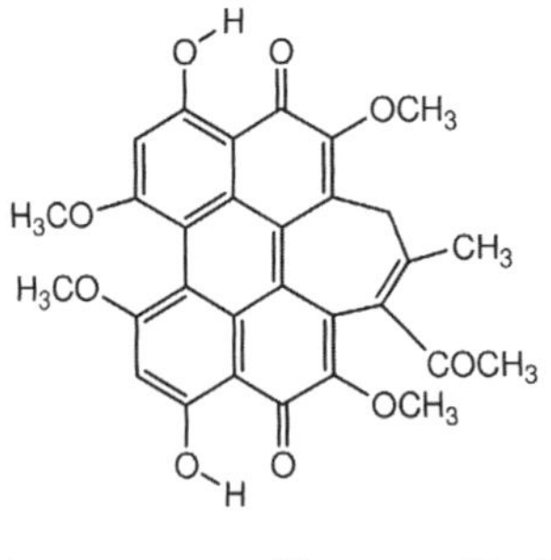


Fig. 1 the structure of hypocrellin B (HB)

## 2. Materials and methods

2.1. Materials

HB was got by Hypocellin A which was extracted from the fungus of *Hypocrella bambusae* and crystallized twice from chloroform–petroleum ether before use. Collagen (type I, c9879) was purchased from Sigma and was used as the received. The phosphate-buffered saline (PBS) of pH = 7.4 was prepared by using 50 mmol/L $KH_2PO_4$ and 50 mmol/L $K_2HPO_4$. The working solutions were prepared shortly before use.

2.2. The photoreaction

The HB-collagen solution was oxygen saturated or bubbled with nitrogen according to the experimental requirements, and then irradiated in a 1 cm×1 cm quartz cuvette. A 450 W mercury vapor lamp equipped with a filter cutting off the wavelengths shorter than 450 nm was used for irradiation. The fluence rate was about 10 mW/cm$^2$ measured by a UV-radiometer.

2.3. Spectroscopy measurements

The absorption spectra were recorded with a 4802s UV/Vis spectrophotometer (Shanghai Unico, China). The F-7000 fluorescence (Hitachi, Japan) was employ to investigate the fluorescence. Samples were dark-adapted at room temperature for 5 min prior to measurement.

## 3. Results

3.1. The spectrum of Collagen in solution

The collagen was been solved in PBS (pH=7.4), the UV-Vis and Fluorescence Spectroscopy of the type I collagen in aqueous were detected in this paper. The results were shown in figure 2.

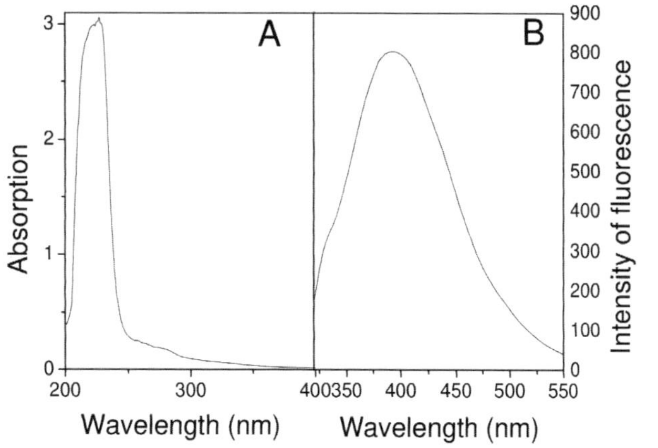


Fig. 2 The spectra character of the collagen A is UV spectrum, B is the Fluorescence spectrum.

The UV spectrum shown the type I collagen had the strong absorption at 227nm. The collagen solution was excited at 295nm, the fluorescence spectrometer detected the 395nm fluorescence emission. The 295/395nm fluorescence is the characteristic of pyridinoline cross-link in type I collagen. So the 395nm fluorescence peak could be as the fluorescent probe to detect the pyridinoline cross-link in collagen.

3.2. The photoreaction of HB with collagen

The photoreaction of hypocrellin B with collagen irradiated by 450W mercury vapor lamp was researched by UV-Vis Spectroscopy and F-7000 fluorescence spectrometer. The experiment found that, one hand, the pyridinoline cross-links can be degraded by the photosensitivity of HB, on the other hand, the pyridinoline cross-links was photodegraded faster when HB and $O_2$ all exist in the system than one of them absent. The ability of photodegraded of HB was shown in figure 3.

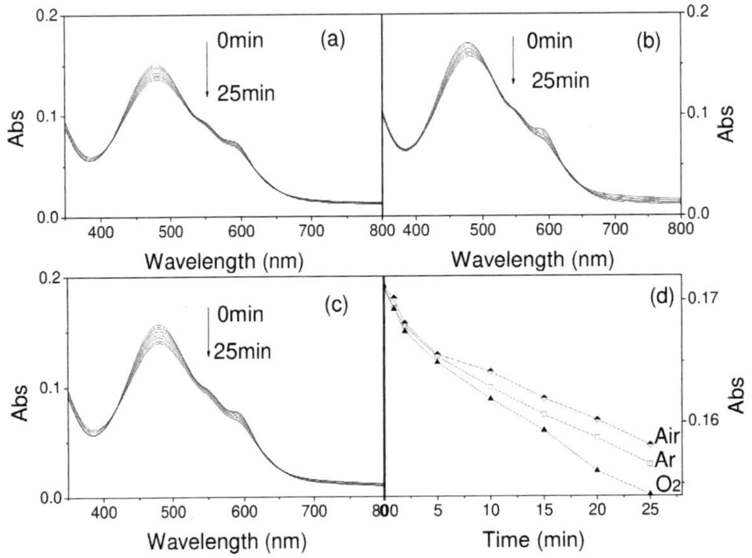


Fig.3 The Absorption spectra of the photo-reaction of collagen with irradiation of HB for 0,1,2,5,10,15,20 and 25min in solution saturated by argon (a) , air(b) and oxygen(c) respectively. (d) Plot of the 395nm fluorescence intensity to irradiation time.

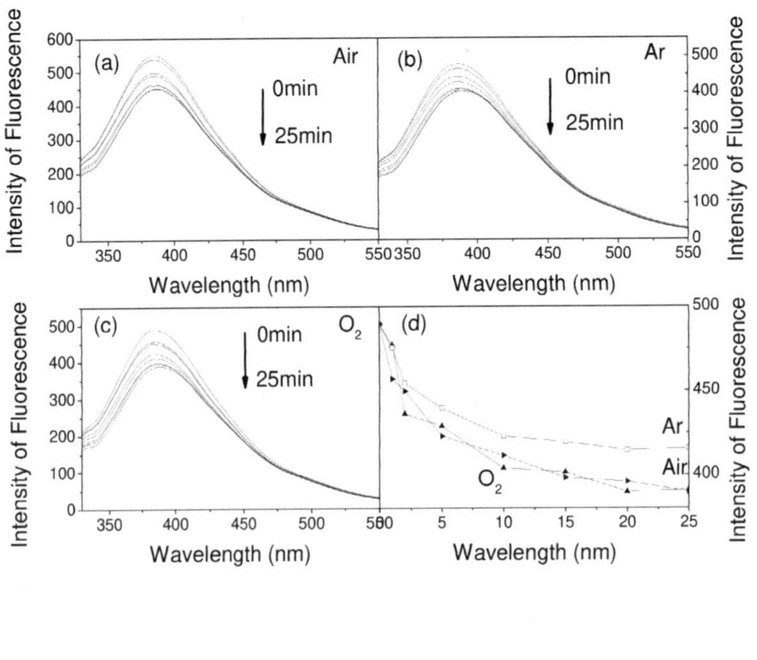

Fig.4 The fluorescence spectra of the photo-reaction of collagen with irradiation of HB for 0,1,2,5,10,15,20 and 25min in solution saturated by air (a) , argon(b) and oxygen(c) respectively. (d) Plot of the 395nm fluorescence intensity to irradiation time.

Figure 3 shows the Absorption of HB in collagen solution with the irradiation. This figure suggested the photoreaction mechanism is main via the type I mechanism. The existence of collagen can 'protect' the HB from being photobleached.

Figure 4 shows changes of the fluorescence of pyridinoline cross-link in collagen with HB during the irradiation. The result shown the pyridinoline cross-link was photodegraged when the HB existed in the system with irradiation. It can be seen that the 395nm fluorescence quenching, reflecting degradation of pyridinoline cross-link, was much more effective in the presence of oxygen. So the photo-degradation speed of pyridinoline cross-link dependent on oxygen.

The ROS are the most important reactive species for the photoreaction (Type II mechanism) when the oxygen presence. The type I mechanism is the main way to degrade this cross-link when the oxygen absent from the reaction system.

## 4. Conclusions

In this paper, the 295/395nm fluorescence was ascribed to the pyridinoline cross-link formation in collagen. The result shown the level of pyridinoline cross-link in type I collagen could be reduce by HB photodynamic action under irradiation by mercury vapor lamp. The existence of collagen can 'protect' the HB from being photobleached. This suggesting the HB may use as photodynamic reagent for research the fibrosis in future.

**Acknowledgement**: This research was supported by the Key Discipline Foundation of Yunnan province and the Scientific Research Project of Honghe University (Project XJ1Y0902).

*References*

[1] Jiang, L. J. (1990) Chin. Sci. Bull. 21, 1608–1616.
[2] Y.Z. Hu, J.Y. An, L. Qin, L.J. Jiang, J. Photochem.Photobiol., A Chem. 78 (1994) 247–251.
[3] M.J. Fehr, S.L. Carpenter, Y. Wannemuehler, J.W. Petrich, Biochimie 34 (1995) 15845– 15848.
[4] J. Moan, K. Berk, Photochem. Photobiol. 55 (1992) 931–948.
[5] Trojanowska M, LeRoy E C, Krieg B E T. J Mol Med, 1998, 76: 266—274
[6] Dennis A H, David R E. J Biol Chem, 1996, 271: 26508—26516
[7] Brinckmann J, Neess C M, Gaber Y, et al. The Journal of Investigative dermatology, 2001, 117(2):269-273
[8] Diestelhorst M, Grisanti S. Archives of ophthalmology, 2002, 120(2): 130-134
[9] Zhang J C, Liu W, Yi Z Z, et al.Chinese Sci Bull, 2009, 54: 2230-2234.

# Ultrafast fluorescence spectroscopy of biologically relevant chromophores using type II difference frequency generation

**J. Léonard, T. Gelot, K. Torgasin and S. Haacke**

Institut de Physique et Chimie des Matériaux de Strasbourg, Department of Ultrafast Optics and Nanophotonics, UMR 7504 Strasbourg University – CNRS, F-67034 Strasbourg, France

Corresponding author: stefan.haacke@ipcms.u-strasbg.fr

**Abstract**. A novel femtosecond fluorescence experiment based on type II difference frequency mixing is demonstrated. This approach is particularly interesting for near-UV emitting biological chromophores like amino acids and nucleotides, as the fluorescence is converted into the spectral range where CCD have their highest quantum efficiencies. The method is implemented with a 5-kHz amplified Ti:Sapphire laser system and first results obtained with 2,5-diphenyloxazole (PPO) in ethanol are reported.

## 1. Introduction

In the past years, much research has been conducted on the ultrafast fluorescence dynamics of proteins and oligonucleotides or complexes thereof with the aim of elucidating their structural dynamics. Examples range from the investigation of spatially restricted solvation dynamics [1] to intra-protein energy and charge transfer [2, 3], leading to static and/or collisional quenching. In particular, in the case of static fluorescence quenching the distance-dependence of photo-oxydation or non-radiative resonance energy transfer bears interesting information e.g. about stacking interactions in DNA or in protein/DNA complexes [4, 5], and allows the distinction of different binding motifs [6].

Femtosecond fluorescence up-conversion due to sum frequency generation [7-10] or Kerr gating [11-13] have been the methods of choice for most of these studies. In the present contribution, we explore the use of type II difference frequency mixing (DFG) in a fluorescence down-conversion scheme. DFG is advantageous for the biologically relevant near-UV emitting fluorophores, like tryptophan, tyrosine or 2-aminopurine, in particular when broadband recording of time-resolved spectra with a CCD is aimed for [9, 14, 15]. When using an intense 800 nm gate pulse, most of the DFG signal is in the visible, i.e. in the wavelength range of highest CCD sensitivity.

## 2. Experiment and methods

Type II DFG is characterized by the following energy conservation and phase-matching conditions

$$\omega_D = \omega_{fl} - \omega_G \qquad \overrightarrow{k_D^o} = \overrightarrow{k_{fl}^e} - \overrightarrow{k_G^e} \tag{1}$$

UV fluorescence (*fl*) and gate pulse (*G*) are both linearly polarized along the extraordinary (*e*) axis of the non-linear medium, BBO in our case, and the difference frequency mixing yields a VIS signal polarized along the ordinary (*o*) axis. It is spectrally resolved with a 25-cm focal length spectrometer and detected with a Peltier-cooled low-noise CCD camera (Spec-10, Princeton Instruments). The latter operates in 10-pixel horizontal binning, such that one channel integrates ≈3.5 nm of the DFG signal. As background rejection is a major concern in these experiments, it is worth mentioning that the *e*-polarized gate beam does not fulfill phase-matching conditions for frequency doubling, and that the fundamental close to 800 nm, the gate beam, can be attenuated over several orders of magnitude by a combination of band pass filters and a linear polarizer.

The home-made set-up is sketched in figure 1. A solution of molecules is circulated through a quartz capillary (S) of section 0.5x0.5 mm$^2$. The UV fluorescence is excited at 5 kHz, by sub-100 fs, 266-nm pump pulses generated by Third Harmonic Generation (THG) from the fundamental 800 nm beam of a regenerative Ti:Sapphire amplifier. P1 and P2 are two parabolic mirrors of 3-inch diameter, and focal lengths of 5 cm and 30 cm, respectively. The downconversion process occurs in a BBO crystal mounted on a motorized rotary stage by spatially overlapping the UV fluorescence with a ≤ 35 μJ gate pulse at 800 nm. Both fluorescence and gate beam spots are matched in size and have 0.5-0.6 mm diameter. Rotating the BBO crystal around the vertical axis selects the spectral detection window by optimizing the phase matching condition (1). The gate beam is horizontally polarized, i.e. along the crystal axis, and the type II condition down-converts fluorescence with the same horizontal linear polarization. The 266-nm excitation beam is likewise horizontally polarized. The gate beam can be delayed by up to 2 ns.
With a 0.20 mm-thick BBO crystal, we achieve a 200-fs time resolution similar to what was observed in up-conversion experiments with near-UV emitting chromophores [16, 17]. The time resolution is determined as the FWHM of the time-resolved signal obtained by frequency mixing of the 266 nm pump light scattered from a diluted milk solution. While in the 0.2-mm crystal group velocity mismatch affects the time resolution [16], recent experiments with thinner crystals indicate that it is presently limited by wave front misalignment of fluorescence and gate in our set-up. Data displayed below have been measured with a 0.4-mm BBO crystal.

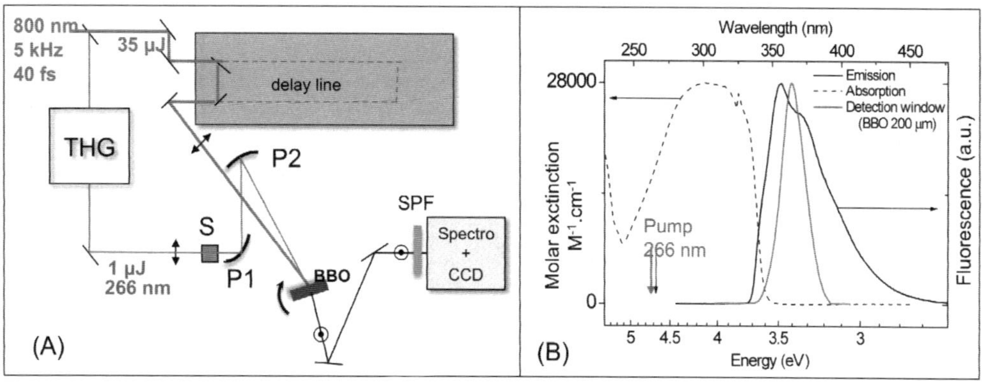


**Figure 1 A.** Layout of the down-conversion experiment. S = sample, THG = third harmonic generation, P1/P2 = parabolic mirrors, SPF: short-pass filter. The gate pulse (red line) and the center of the fluorescence cone (purple) form a ≈7°-angle. Arrows indicate linear polarizations of fluorescence, gate and DFG signal beams. **B.** Steady-state absorption and emission spectra of 2,5-diphenyloxazole (PPO) in ethanol used for preliminary tests of the experimental set-up. The red curve is the DFG spectrum (transposed into the UV spectral region) at 10-ps time delay.

We have tested the set-up performances by time-resolving the fluorescence of 2,5-diphenyloxazole (PPO) in ethanol (see Fig. 1.1-B for the stationary absorption and emission spectra). Although we aim at producing the highest signal-to-noise ratio, both fluorescence excitation conditions and gate beam intensities are kept low enough to operate in a regime of DFG signal linearity. This prevents the BBO crystal to be damaged in a long-term run. In the present contribution we report results for DFG wavelengths shorter than 730 nm only. As a matter of fact, residual (depolarized) scattered light from the intense gate pulse is eliminated by a shortpass filter (Thorlabs FES0750). Hence, any fluorescence emitted around 400 nm (and downconverted around 800 nm) remains undetectable.

## 3. Results and discussion

Fig. 1.2 displays an overview of the results of the experiment performed with PPO at a fixed BBO angular position, adjusted here for maximum DFG signal at 665 nm. This selects the near-UV fluorescence wavelength range around 363 nm, i.e. close to the maximum of the steady-state spectrum. Note that the spectral width of the DFG signal is 75 nm, translating into a near-UV spectral bandwidth of 20-22 nm. As the steady-state spectrum is almost constant around 363 nm, this value can be taken as the crystal's acceptance bandwidth in the present conditions.

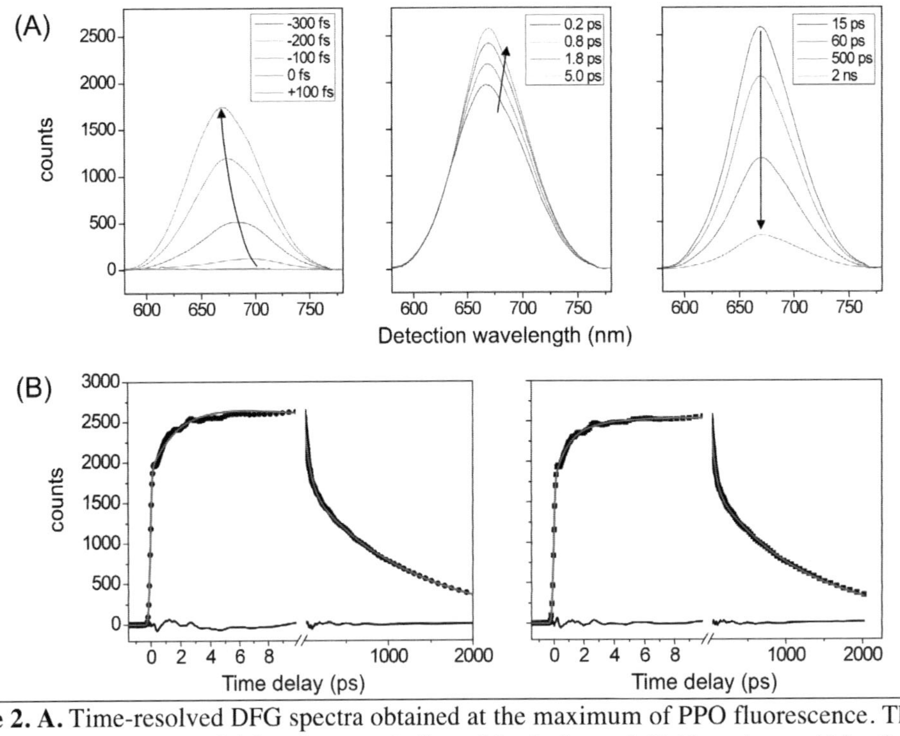


**Figure 2. A.** Time-resolved DFG spectra obtained at the maximum of PPO fluorescence. Three panels illustrate different ranges of delay times, as indicated in the legend. **B.** Experimental kinetic traces at 665 nm (squares) fitted with a sum of 3 (left) or 4 exponentials (right), convoluted with a 200-fs Gaussian representing the instrument response function (red line). Fit residuals (bottom line). Note the break at 10 ps on the time axis.

The first 300 fs display a dramatic spectral blue-shift due to group velocity differences (chirp) of the fluorescence. The only dispersive elements in the set-up are the 0.5 mm ethanol solution and the 0.25-mm fused silica capillary output window. The signal rise and slight red-shift observed over the first 5 picoseconds (cf. fig. 2A center) are attributed to intramolecular or solvent relaxation. Kinetic traces recorded in the range of 580 to 720 nm (fig. 3) underscore this gradual red-shift, as the rise times observed in the long-wavelength traces appear as decay components in the short-wavelength counterparts. This initial ≈10-ps phase is followed by a monotonous decay without any significant shift. We expect both anisotropy decay and the excited state lifetime to be responsible for the signal decay.

The signal-to-noise ratio is as good as 2000:1, as can be seen from the kinetic traces in fig. 2B. These have been obtained after ≈30 min data collection and averaging (0.4-mm BBO). The background noise level is defined by the rms of the spurious signal detected at negative times. Note that these traces are produced from a single CCD channel. Further binning and averaging would lead to an even higher dynamic range. The present data quality thus allows for a very accurate determination of the lifetimes in a multi-exponential fitting procedure. Fig. 2B-left shows a tri-exponential fit with two decay times and a single 2-ps rise time. But it is obvious that the recorded short delay time data deviate from this single exponential rise. The fidelity is significantly improved ($\chi_2$ reduced by 40 %, and better residuals) with a fit including a second rise time. Whether the signal rise really is biexponential, possibly indicating two different relaxation mechanisms, or is simply non-exponential (as may be the case for relaxation processes) is beyond the focus here. The point is rather to illustrate the very good signal-to-noise ratio of the time-resolved fluorescence data, and our ability to refine the quantitative analysis by including up to four time constants. For the signal decay we find a 60-ps time constant attributed to rotational diffusion of the PPO molecules in ethanol, as excitation and detection are linearly co-polarized, followed by a time constant of 1.32±0.1 ns. Even though a precise determination of the latter would ideally require a larger time span to be covered, it is in very good agreement with the well-characterized fluorescence lifetime of PPO in ethanol (1.4 ± 0.2 ns, [4]).

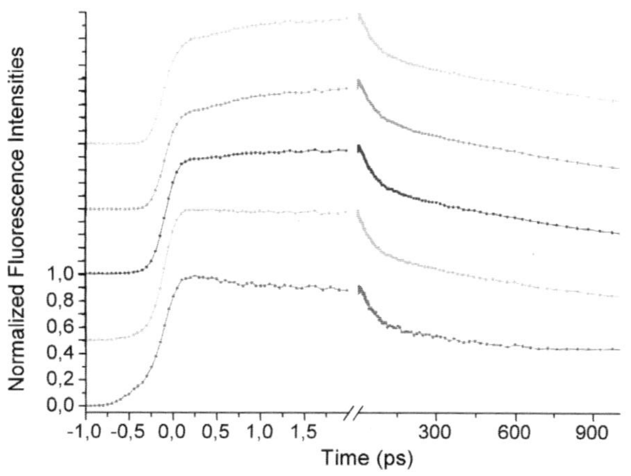


**Figure 3:** Selection of kinetic traces for PPO. DFG signal from 580 nm (red) to 720 m (cyan). The wavelength dependence of the initial sub-5 ps phase supports the interpretation of vibrational and or dielectric relaxation inducing a progressive red-shift of the fluorescence.

## 4. Conclusions

In summary, we have shown that type II difference frequency mixing is a new way of time-resolving near-UV fluorescence at wavelength < 400 nm with 200-fs resolution and an excellent signal-to-noise ratio. It is an interesting alternative to the wide-spread type I "up-conversion" technique, in particular in view of investigating local structural dynamics of proteins or oligonucleotides. Indeed, when multi-

exponential decay components need to be resolved, the present approach optimized for signal detection using a CCD affords excellent signal-to-background ratio.
Further developments of the experiment involve selective near-UV excitation for Tryptophan or Aminopurine and the use of a near-infrared OPA as gate pulses, in view of recording complete time-resolved DFG spectra in the 500- 900 nm range.


**Acknowledgements**

The authors are grateful to O. Crégut and G. Dekyndt for important technical assistance. The work was funded by the French "Agence National pour la Recherche sur le SIDA". T.G. is grateful for a PhD grant from "Région d'Alsace".


**References**


[1] Peon J, Pal S K and Zewail A H 2002 Hydration at the surface of the protein monellin: Dynamics with femtosecond resolution *Proc. Nat. Acad. Sci.* **99** 10964-9

[2] Qiu W, Li T, Zhang L, Yang Y, Kao Y-T, Wang L and Zhong D 2008 Ultrafast quenching of tryptophan fluorescence in proteins: Interresidue and intrahelical electron transfer *Chem. Phys.* **350** 154

[3] Qiu W, Wang L, Lu W, Boechler A, Sanders D A R and Zhong D 2007 Dissection of complex protein dynamics in human thioredoxin *Proc. Nat. Acad. Sci. USA* **104** 5366-71

[4] Xia T, Becker H-C, Wan C, Frankel A, Roberts R W and Zewail A H 2003 The rna-protein complex: Direct probing of the interfacial recognition dynamics and its correlation with biological functions *Proc. Nat. Acad. Sci. USA* **100** 8119-23

[5] Xia T, Wan C, Roberts R W and Zewail A H 2005 Rna-protein recognition: Single-residue ultrafast dynamical control of structural specificity and function *Proc. Nat. Acad. Sci. USA* **102** 13013-8

[6] Bernacchi S, Stoylov S, Piémont E, Ficheux D, Roques B P, Darlix J L and Mély Y 2002 Hiv-1 nucleocapsid protein activates transient melting of least stable parts of the secondary structure of tar and its complementary sequence *Journal of Molecular Biology* **317** 385

[7] Mahr H and Hirsch M D 1975 An optical up-conversion light gate with picosecond resolution *Opt. Comm.* **13** 96-9

[8] Shah J 1988 Ultrafast luminescence spectroscopy using sum frequency generation *Ieee Journal of Quantum Electronics* **24** 276-88

[9] Haacke S, Taylor R A, Bar-Joseph I, Brasil M J S P, Hartig M and Deveaud B 1998 Improving the signal-to-noise ratio of femtosecond luminescence up-conversion by multichannel detection *JOSA B* **15** 1410-7

[10] Schanz R, Kovalenko S A, Kharlanov V and Ernsting N P 2001 Broad-band fluorescence up-conversion for femtosecond spectroscopy *App. Phys. Letters* **79** 566-8

[11] Kinoshita S, Ozawa H, Kanematsu Y, Tanaka I, Sugimoto N and Fujiwara S 2000 Efficient optical kerr shutter for femtosecond time-resolved luminescence spectroscopy *Rev. Scient. Instr.* **71** 3317-22

[12] Schmidt B, Laimgruber S, Zinth W and Gilch P 2003 A broadband kerr shutter for femtosecond fluorescence spectroscopy *Applied Physics B: Lasers and Optics* **76** 809-14

[13] Arzhantsev S and Maroncelli M 2005 Design and characterization of a femtosecond fluorescence spectrometer based on optical kerr gating *Appl. Spectrosc.* **59** 206-20

[14] Zhao L J, Lustres J L P, Farztdinov V and Ernsting N P 2005 Femtosecond fluorescence spectroscopy by upconversion with tilted gate pulses *Physical Chemistry Chemical Physics* **7** 1716-25

[15] Zgrablić G, Voïtchovsky K, Kindermann M, Haacke S and Chergui M 2005 Ultrafast excited state dynamics of the protonated schiff base of all-trans retinal in solvents *Biophys. J.* **88** 2779-88

[16] Cannizzo A, Bräm O, Zgrablic G, Tortschanoff A, Oskouei A A, van Mourik F and Chergui M 2007 Femtosecond fluorescence upconversion setup with broadband detection in the ultraviolet *Opt. Lett.* **32** 3555-7

[17] Onidas D, Markovitsi D, Marguet S, Sharonov A and Gustavsson T 2002 Fluorescence properties of DNA nucleosides and nucleotides:,ää a refined steady-state and femtosecond investigation *The Journal of Physical Chemistry B* **106** 11367-74

# Detection and analysis of diamond fingerprinting feature and its application

**Xin Li [1], Guoliang Huang [1, 2,*], Qiang Li[1] and Shengyi Chen[1]**

[1] Department of Biomedical Engineering, the School of Medicine, Tsinghua University, Beijing, 100084, China

[2] National Engineering Research Center for Beijing Biochip Technology, Beijing, 102206, China

[*]Corresponding author: Guoliang Huang (Email: tshgl@tsinghua.edu.cn)

**Abstract.** Before becoming a jewelry diamonds need to be carved artistically with some special geometric features as the structure of the polyhedron. There are subtle differences in the structure of this polyhedron in each diamond. With the spatial frequency spectrum analysis of diamond surface structure, we can obtain the diamond fingerprint information which represents the "Diamond ID" and has good specificity. Based on the optical Fourier Transform spatial spectrum analysis, the fingerprinting identification of surface structure of diamond in spatial frequency domain was studied in this paper. We constructed both the completely coherent diamond fingerprinting detection system illuminated by laser and the partially coherent diamond fingerprinting detection system illuminated by led, and analyzed the effect of the coherence of light source to the diamond fingerprinting feature. We studied rotation invariance and translation invariance of the diamond fingerprinting and verified the feasibility of real-time and accurate identification of diamond fingerprint. With the profit of this work, we can provide customs, jewelers and consumers with a real-time and reliable diamonds identification instrument, which will curb diamond smuggling, theft and other crimes, and ensure the healthy development of the diamond industry.

## 1. Introduction

Diamond is one of the world's most precious jewels. The production of diamond is rare and not well-distributed in nature so that diamond industries in many countries rely on import which create a capacious market for diamond smuggling. High worth, small size and easy carry make people take risks for diamond smuggling. How to curb these crimes has become an onerous but urgent problem.

At present, researches of diamond identification technology at home and abroad can be divided into three levels. The first level is the most primitive way which is relatively simple, including thermal conductivity tests, rigidity verification, growing point examination, reflected light observation, similar chemical composition tests etc. The second level is the common methods which are relatively reliable, including birefringence character, refractive index measurement, characteristic spectrum etc. The third level is more advanced which includes cathode luminescence technology, Raman spectrum technology, neutron activation analysis technology and X-Ray diffraction technology [1].

Indeed, there have been numerous reports on diamond identification technologies. Most of these works focus on the diamond material characters. However, researches on diamond fingerprint still remain deficient. The only literature report about space frequency is delivered by Weiszmann Institute of Sciences in Israel. Technologies in this report have been patented and the "Gemstone Registration System" [2] has been applied into real life. In this system, tilt collection of fingerprint will cause image distortion and affect the accuracy of fingerprint analysis.

In this study, parallel light beam irradiates the diamond table board vertically and excites surface reflected light, which can produce a space spectrum as Fingerprinting Map, and has good specificity and space-invariance. Recording diamond fingerprint and binding to other information (origin, production date, trading resume, present owner, carat weight etc.) as a "Diamond File" can improve management of diamond imports and exports for customs and effectively strike diamond smuggling crimes.

## 2. Theoretical analysis of diamond fingerprinting map

### *2.1. Optical Fourier Transform Implementation*

In order to obtain the space spectrum of diamond surface reflected light, we need to process Fourier Transform. The mathematical expression is demonstrated as follows.

$$F\{E(x,y)\} = \int_{-\infty}^{\infty}\int E(x,y)\exp[-i2\pi(f_x x + f_y y)]dxdy \tag{1}$$

In the optical field, Fourier Transform can be implemented by Fourier lens which has incomparable parallel processing speed comparing with computer. The lens plays an important role in the process of 2D Fourier Transform, the lens could produce a space frequency spectrum on the behind focal plane, which is related to the space frequency of the diamond table board reflected light wave.

Analysis is executed in the case of plane illumination light wave as shown in figure 1. Diffraction plane is located in the distance of $L_0$ from the lens and is illuminated by monochromatic plane wave. In this case, the relationship between the complex amplitude in back focus plane $\tilde{E}_1(x_1, y_1)$ and the complex amplitude in diffraction plane $\tilde{E}(x, y)$ is demonstrated as follows [3].

$$\tilde{E}_1(x_1, y_1) = \frac{1}{i\lambda f}\exp\left[\frac{ik}{2f}\left(1-\frac{L_0}{f}\right)\left(x_1^2 + y_1^2\right)\right]F\left\{\tilde{E}(x, y)\right\} \tag{2}$$

Where $k$ represents wave number, $\lambda$ is wave length, $f$ represents focal length, $L_0$ corresponds to the distance from the diffraction plane to the lens. The relationship between spatial frequency coordinate $(f_x, f_y)$ and back focus plane coordinate $(x', y')$ is shown as follows.

$$x' = \lambda f f_x, y' = \lambda f f_y \tag{3}$$

Formula (2) indicates that the complex amplitude distribution on back focus plane consists of the product of the Fourier transform of the complex amplitude distribution on diffraction plane $F\left\{\tilde{E}(x, y)\right\}$ and a phase factor $\frac{1}{i\lambda f}\exp\left[\frac{ik}{2f}\left(1-\frac{L_0}{f}\right)\left(x_1^2 + y_1^2\right)\right]$.

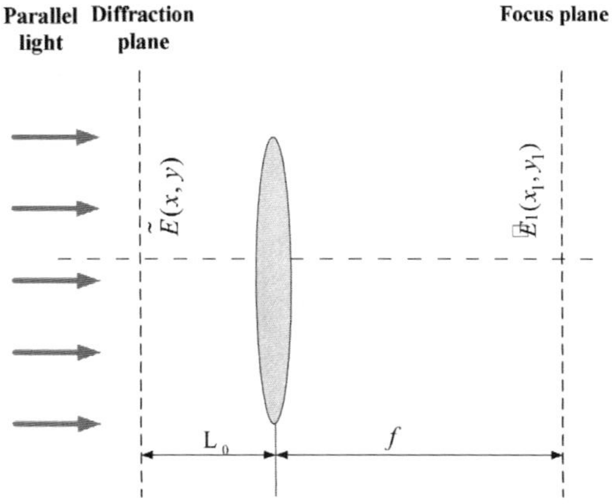


**Figure 1.** Optical Fourier Transform setup of lens.

*2.2. Theoretical Analysis of Diamond Fingerprinting Map*

In this study, parallel light beam irradiates the diamond surface vertically and excites surface reflected light beams which have discrete spatial frequencies. The superposition of light beams produce specific complex amplitude distribution in space domain which is processed by 2D Fourier transform by the lens .The space spectrum is displayed in the form of complex amplitude distribution in the back focus plane. In fact, the CCD can only detect light intensity instead of complex amplitude distribution. Therefore it's necessary to calculate the light intensity distribution $|\tilde{E}'(x', y')|^2$ in frequency domain .The light intensity distribution is demonstrated as follows.

$$|\tilde{E}'(x',y')|^2=\sum_{i=1}^{i=n}A_i^2\left[\delta(\frac{x'}{\lambda f}-f_{x_i})\delta(\frac{y'}{\lambda f}-f_{y_i})\right]^2 \qquad (4)$$

In the Formula (4), $A_i$ represents amplitude of reflected light beams, $\lambda$ represents wave length of light source, $f$ represents lens' focal length, $(f_{x_i}, f_{y_i})$ represents spatial frequency coordinate of different light beams, $(x', y')$ is the screen coordinate.

Formula (4) is the mathematic model of the diamond fingerprinting map, which indicates that the light intensity distribution in the receiving screen is comprised of a series of the square of discrete impact functions which are visually a set of bright spots. The number of the spots n is equal to the number of diamond surfaces. Besides, the position of the spots corresponds to the spatial frequency value. The relationship has been shown as Formula (3). Consequently, if the coordinate $(x', y')$ of the spot is given, we could calculate the spatial frequency $(f_{x_i}, f_{y_i})$ of diamond surface geometry structure character. On the contrary, if the geometry structure of diamond is given, we can calculate the position of the corresponding spatial frequency spot as well.

### 3. Construction of diamond fingerprint detecting platform

In order to realize the detection of diamond fingerprint and the related researches, the diamond fingerprinting map detecting system is constructed. For the first time, a coaxial symmetry light path is adopted in figure 2.

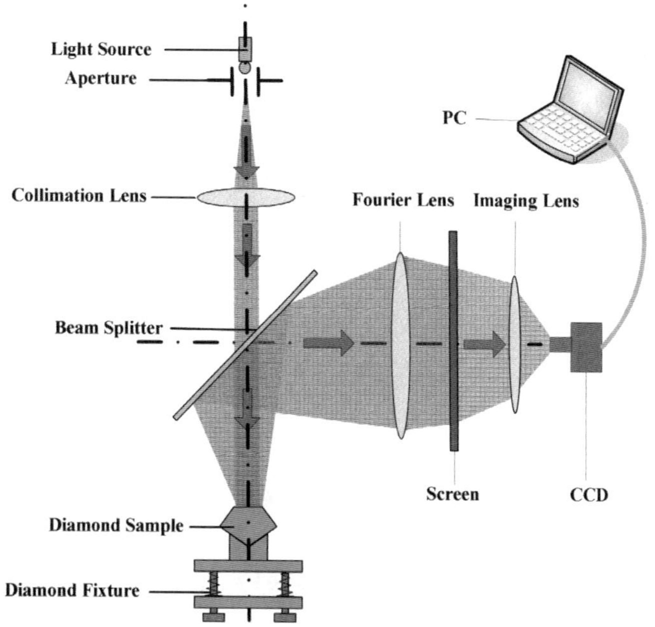


**Figure 2.** Diamond fingerprinting map detecting system.

The focal length of lens will affect the distances from beam splitter to the pin hole and to the screen. If we use lens of short focal length to decrease the volume, some mechanical components which help to connect or fasten and some optical components such as the splitter, would prevent the light beams and lead to deficiency of space spectrum map. Conversely, If we choose the long one, the area of the spectrum would be very large and a tremendous receiving screen would be needed, will result in unexpected big system structure in volume. In order to gain the best focal length, we did experiments in several different focal lengths of 33mm, 50mm, 100mm, 150mm. Experimental results show that the focal length of 50 mm has the best effect.

The concave-convex type of lens affects the imaging performance as well. Based on the experiments using double-convex and concave-convex lens, the results indicate that in the condition of lens at 25mm in diameter, concave-convex lens achieve better convergence of the high spatial frequency beams.

In the system, VC2003 (Visual Studio 2003) linking to the OpenCV library is used as the software development environment. Coding is completed with functions of background-noise-removed real-time image collection, image preview, saving and analysis. Due to factors such as coherence of the light source and reflected light of components, there is heavy background noise in experiment. Background-noise-removed real-time image collection can help users observe more clear and legible fingerprint. The software interface of diamond fingerprinting spectrum analysis is shown in figure 3.

Diamond fingerprinting map is analyzed accurately by "Radius Comparison Method", the main process is shown as follows.

(1) Detect and calculate the center pixel coordinate of each spot in the fingerprinting spectrum image.

(2) Compare the amount of spots in each spectrum images. Analysis ends if the amount is not equal.

(3) Calculate the distances from the point of zero spectrum to all other spectrum spots (referred to as radius below) and sort them in a sequence.

(4) Compare the two sequences and calculate the mean square error and the maximum error.

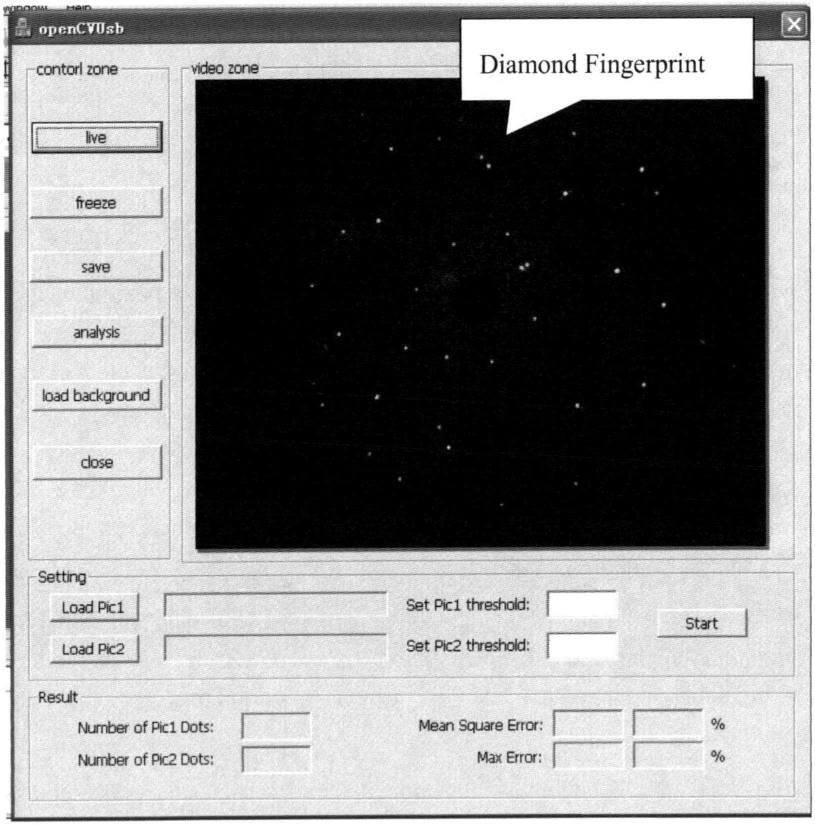


**Figure 3.** Diamond fingerprinting map detecting system.

## 4. Results and discussion

### *4.1. The light coherence impact to the system.*

In this paper, Image Signal to Noise Ratio (ISNR) is used to analyze the light coherence impact to the system. ISNR formula is shown as follows [4].

$$R_{SNR} = 10\lg\frac{I_{max} - B}{\sigma} \quad (5)$$

In the formula, $R_{SNR}$ represents ISNR. $I_{max}$ is the maximum gray value of the spectrum spot signal. $B$ represents the gray value of the background, which can be estimated from the average value of the image background of all pixels. $\sigma$ represents the noise standard deviation.

In the condition of the same mechanical structure and light path, we respectively did experiments based on LED (partial coherent source) and laser (complete coherent source). Firstly we turned off the light source and collected a background image, the average gray value $B$ was obtain. Secondly we turned on the light source and collected a background image to calculate the noise standard deviation $\sigma$. Finally we placed the diamond into the system and collected the diamond fingerprinting spectrum images and estimated the maximum gray value $I_{max}$. The experimental data is shown in table 1.

**Table 1.** Fingerprinting map analysis based on different coherence sources

| | $I_{max}$ | $B$ | $\sigma$ | $R_{SNR}$ (dB) |
|---|---|---|---|---|
| **LED** | 176 | 16 | 9.81 | 12.12 |
| **Laser** | 254 | 16 | 11.85 | 13.03 |

Data in the table 1 illustrates that coherence source can enhance the Signal Noise Ratio of diamond fingerprinting image.

*4.2. Feasibility analysis of real-time accurate identification*

In fact, when the same diamond was repeatedly detected, we can't guarantee that the position of diamond is completely the same. Therefore, it is prerequisite that fingerprinting radius information is invariable under center translation and axis rotation. We call them translation invariance and rotation invariance.

In this article, the two parameters "Mean Square Error" (MSE) and "Maximum Error" (ME) are used to estimate the difference between two fingerprint images. When the precondition of the same spots amount is met, the distances from zero-point to all the spots are sorted, MSE the mean square deviation of the radius in the same sequence number indicates the similarity of fingerprint images. The less the mean square deviation of the radius is, the higher the similarity of fingerprint images is. ME is the maximum value of all the subtractions of the radius in the same sequence number. Theoretically, both of the MSE and ME are translation invariance and rotation invariance. However, error exists in the actual system and these two parameters can be used to evaluate the accuracy of the diamond fingerprinting map detecting system.

In the illumination of laser, we did experiment for center translation invariance and axis rotation invariance, recorded and analyzed five groups of fingerprinting maps. The analysis results are shown in table 2 and table 3. Further more, we did experiments for another sample and analyzed its fingerprints as well. The analysis results are shown in table 4. The MSE and ME in these tables are presented in the form of percentage, which is ratio of the real MSE or ME to the maximum radius in the maps.

**Table 2.** Center translation error analysis of fingerprint

| | 1 | 2 | 3 | 4 | 5 | Average |
|---|---|---|---|---|---|---|
| **MSE** | 0.76% | 0.72% | 0.98% | 1.16% | 0.70% | 0.86% |
| **ME** | 1.55% | 1.63% | 1.30% | 2.89% | 1.02% | 1.67% |

**Table 3.** Axis rotation error analysis of fingerprint

| | 1 | 2 | 3 | 4 | 5 | Average |
|---|---|---|---|---|---|---|
| **MSE** | 0.72% | 0.89% | 0.82% | 1.55% | 1.88% | 1.17% |
| **ME** | 1.12% | 1.97% | 2.58% | 3.42% | 3.62% | 2.54% |

**Table 4.** Fingerprinting map error analysis of different samples

| **MSE** | **ME** |
|---|---|
| 3.27% | 9.18% |

The comparison of the three tables demonstrates that there is identifiable difference in different samples. Both the MSE and ME of the fingerprint of the same sample are small compared to that of different samples, which certifies the feasibility of real-time accurate identification.

Error produced by center translation is due to large distortion when high spatial frequency beams pass through lens. The essence of axis rotation is the movement of each diamond surface. Therefore one of the error sources of axis rotation is the same as the center translation. Besides, it's obvious that the error of axis rotation is a little larger than that of center translation, the reason is that lens, receiving screen and CCD are not completely coaxial. In a word, the error of axis rotation can be regarded as the superposition of the error of center translation and incomplete co-axis.

## 5. Conclusions

In summary, we have constructed the theoretical model of diamond fingerprint and developed a diamond fingerprint detecting system, the fingerprinting maps of diamond are collected and the validity of theoretical model is proved. Based on our method and system, the coherence impact of light source to the system is analyzed and the feasibility of real-time accurate identification of diamond fingerprinting maps is verified.

There are two innovations in this paper. Firstly, co-axial symmetry light path detecting structure is adopted in our diamond fingerprinting map detecting system, which overcomes the existed disadvantage in reference 2 where CCD tilt imaging leads to distortion of collecting image and improves the accuracy of fingerprint analysis. Secondly, the software system realizes fingerprinting map preview, saving, analysis as well as real-time background-noise-removed collection, which is more simple and efficient than the mechanical and optical methods.

## References


[1] Shen K Y, Wang C and Ma W X 2001 Diamond research progress *Gems & Gemology*,**2** 38-41.

[2] Vanier D. J., Wallner H. F. and Leydon M 1998 Gemstone registration system US:5828405.

[3] Yu D Y, Tan H Y 2006 *Engineering Optic* ( Beijing: Machinery industry Press) p 380-385.

[4] Bar Shalom Y 1978 Tracking methods in a multi-target environment *IEEE Trans Automat Control* **4** 618-626.


## Acknowledgements


This subject is Supported by the National Foundation of High Technology of China (2006AA020803), National Program on Key Basic Research Projects 973 of China (2006CB705700、2011CB707700) and National Natural Science Foundation of China (60978064).

# High-speed optical coherence tomography signal processing on GPU

**Xiqi Li**[1,2,3]**, Guohua Shi**[1,2,3]**, and Yudong Zhang**[1,2]

[1]The Laboratory on Adaptive Optics, Institute of Optics and Electronics, Chinese Academy of Sciences, Chengdu 610209, China

[2]The Key Laboratory on Adaptive Optics, Chinese Academy of Sciences, Chengdu 610209, China

[3]Graduate School of the Chinese Academy of Sciences, Beijing 100039, China

E-mail: lixiqi@yahoo.cn

**Abstract.** The signal processing speed of spectral domain optical coherence tomography (SD-OCT) has become a bottleneck in many medical applications. Recently, a time-domain interpolation method was proposed. This method not only gets a better signal-to noise ratio (SNR) but also gets a faster signal processing time for the SD-OCT than the widely used zero-padding interpolation method. Furthermore, the re-sampled data is obtained by convoluting the acquired data and the coefficients in time domain. Thus, a lot of interpolations can be performed concurrently. So, this interpolation method is suitable for parallel computing. An ultra-high optical coherence tomography signal processing can be realized by using graphics processing unit (GPU) with computer unified device architecture (CUDA). This paper will introduce the signal processing steps of SD-OCT on GPU. An experiment is performed to acquire a frame SD-OCT data (400A-lines×2048 pixel per A-line) and real-time processed the data on GPU. The results show that it can be finished in 6.208 milliseconds, which is 37 times faster than that on Central Processing Unit (CPU).

## 1. Introduction

Optical coherence tomography (OCT) is a non-invasive, cross-sectional optical imaging technique that allows for high-sensitivity, high-resolution imaging of backscatter samples [1,2,3]. Because of system simplicity and economical design, spectral domain optical coherence tomography (SD-OCT) has undergone rapid development. SD-OCT is based on spectral interferometer and detected by a line scan charge coupled device (CCD). The interference signal I(*k*) can be expressed as:

$$I(k) = S(k)\left(1 + 2\int_0^{\infty} a(z)\cos(2knz)dz + \int_0^{\infty}\int_0^{\infty} a(z)a(z')e^{-i2kn(z-z')}dzdz'\right) \qquad (1)$$

Where I(*k*) denotes the detected spectrum; S(*k*) denotes the spectral intensity distribution of light source; a(*z*) is the scattering amplitude. The a(*z*) can be obtained by Fourier Transformation of the detected spectrum I(*k*). It is obvious that the intensity information is coded in the wave number *k*. Unfortunately, the CCD detects the signal evenly in wavelength . It need resample the data from wavelength space to wave number space, or else, it lead to coherence envelop broadening and fall-off degradation in SD-OCT systems [4]. Fig.1 shows the signal processing procedure. The intensity data $S_{ou}t(x)$ of the spectrometer is acquired by the linescan CCD. The interference spectrum $S_{int}(x)$ is obtained after reducing the dc terms and the low-frequency noises by the background subtraction procedure. The re-sampled data $S(k)$ is obtained by a non-uniformly sampled, which distributed linearly in wave number spaced.

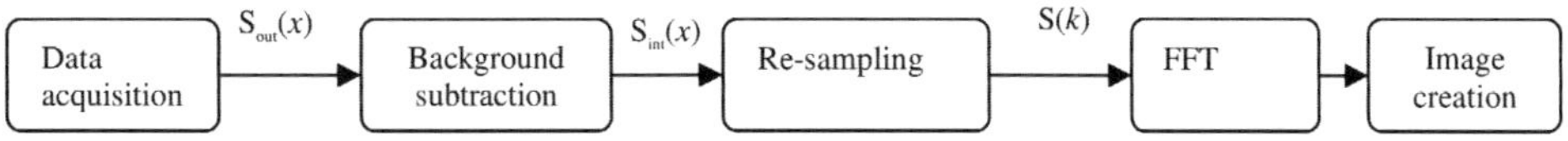


**Fig.1**. Flow chart of SD-OCT structure imaging procedure.

Re-sampling is an important part in the signal processing step. Because $k$ and  has a relationship of $k$=2 / and $k$=2 / $^2$ and the frequency oscillation period is as small as two pixels in the spectrometer, resulting any local interpolation scheme (such as cubic spline) is bound to fail [4]. Zero-padding interpolation together with linear interpolation is widely used to resample in SD-OCT imaging processing. This method needs huge FFT operations and a lot of logical operations, resulting in the low processing speed. Thus, a lot of SD-OCT systems are post processing. Real-time optical coherence tomography processing by using Digital Signal Processing (DSP) or field-programmable-gate-array-based (FPGA) was proposed [5-8], which increase the system complexity and cost because additional hardware inclusions are needed. A real-time swept source polarization-sensitive OCT system by using OpenMP model has been reported [9], but that needs expensive multi-core central processing unit (CPU).

In the last ten years, graphics processing unit (GPU) becomes more and more powerful in parallel computing [10,11]. Its special architecture is well suitable for the problems that can be parallel executed. Commodity GPU like NVIDIA's GTX 295 has 480 processing cores and can achieve 1784 G float point operation per second (FLOPS) of computational horsepower, which is much faster than that of the current CPU. But traditional tools for computing on GPU were too complex to program, limiting the general purpose use of GPU. Recently, NVIDIA Corporation develops a tool for computing on GPU, which is named as the compute unified device architecture (CUDA) programming model [10,11]. With the help of CUDA model, a general purpose computing application can be developed easily. A real-time 4D signal processing and visualization using GPU on a regular nonlinear-k Fourier-domain OCT system is reported [12]. But the linear spline interpolation (LSI) they used is barely improve the signal-to-noise ratio (SNR) of the SD-OCT system. A high speed and high accuracy interpolation method is needed.

A time domain interpolation for Fourier domain optical coherence tomography was reported [1]. This method not only obtains a better SNR but also gets a faster computing speed than commonly used zero-padding interpolation [1]. Furthermore, by this interpolation method, the linear k space data can be obtained by convoluting the original data detected in CCD with interpolation coefficients [1].Thus, only the data and coefficients in cut-off window are processed. So, this interpolation method is suitable for parallel computing.

This paper will briefly introduce the time-domain interpolation. A CUDA program is developed to utilize the real-time parallel computing capability of the GPU. This program follows the steps as Fig.1 shows. These steps are processed step by step, but every step is processed by a lot of threads concurrently, which can accelerate the processing speed of SD-OCT significantly.

## 2. Time domain interpolation

In SD-OCT system, interference signal $x_1(n)$ is acquired by the pixels of the linescan CCD, whose corresponding wavelengths are linearly distributed in wavelength space. We assumed that there has a sequence of wavelengths linearly distribute in $k$ space, whose interference data are $x_2(s)$. The virtual position index $s$ is obtained by mapping this sequence to CCD pixel index $n$. The interpolated data can be obtained by using the follow equation [2]:

$$x_2(s) = \frac{1}{N+1}\sum_{n=0}^{N-1} x_1(n)\left\{1+2\sum_{j=1}^{\frac{N}{2}}\cos\frac{2\pi}{N}j(s-n)\right\} \qquad (2)$$

The range of index $n$ and $s$ is determined by the pixels of the CCD $N$ and the resolution of the OCT system. The coefficients are defined as:

$$C(\Delta) = \frac{1}{N+1}\left(1+2\sum_{j=1}^{\frac{N}{2}}\cos\left(\frac{2\pi}{N}j\Delta\right)\right) \qquad (3)$$

•$s$-$n$. Eq. 3 shows that the coefficients do not relate to the data. It can be computed separately if $s$ and $n$ known. The virtual CCD position of the linear $k$ space can be obtained by calibrating the spectrometer of the SD-OCT system.

It has reported that only a small number ranges of C( ) contribute to the interpolation. Therefore, a narrow window can be used to cut off the C( ) with an acceptable accuracy to reduce the computation time, whose cut off width is $L$[2]. The larger $L$ can get a better accuracy, but it would consume more computing time. Thus, the cut off width $L$ is determined by a trade-off between the accuracy and the computation time [2]. The interpolation can be finally expressed as:

$$x_2(s) = \sum_{n=Min}^{Max} x_n H_{N\times N}(n, s_n)W(n) \qquad (4)$$

Where Min and Max are determined by the cut-off length and the position $s$. $H_{N\times N}$ is the coefficients matrix. $W(n)$ is the cut-off windows. A rectangle window is suitable. Eq.4 shows that $x_2(s)$ are only relate to several original data and the coefficients range in the cut-off window. So the sequence $x_2(s)$ can be computed out by a lot of parallel streams.

### 3. Signal Processing on GPU

Figure.2 shows the setup of SD-OCT. The light source is a SLD (Superlum, Russia, SLD-371-HP1) with a bandwidth of 45 nm FWHM at 840 nm. The spectrometer consists of a line scan camera (AVVIVA, SM2 CL2015, 28 KHz at 2048 pixels,), a transmission grating [Wasatch Photonics, 1200 (lines×pairs)/mm at 830 nm], and an achromatic lens (f=150 mm). Two galvo mirrors are driven by a waveform generator (PCI 6221, National Instrument) to control the scan light. The interference is detected by the CCD and transmits to computer after it was grabbed by the image acquisition device (IMAQ) (PCIe 1430, National Instrument). A computer with a signal-core CPU AMD sempron 3600+ at 2.0GHz and a GPU NVIDIA GTX295 is used as the data acquisition and processing computer. The operating system was Microsoft Windows XP pro SP2 and the software developing platform was Microsoft Visual Studio 2005 SP1.

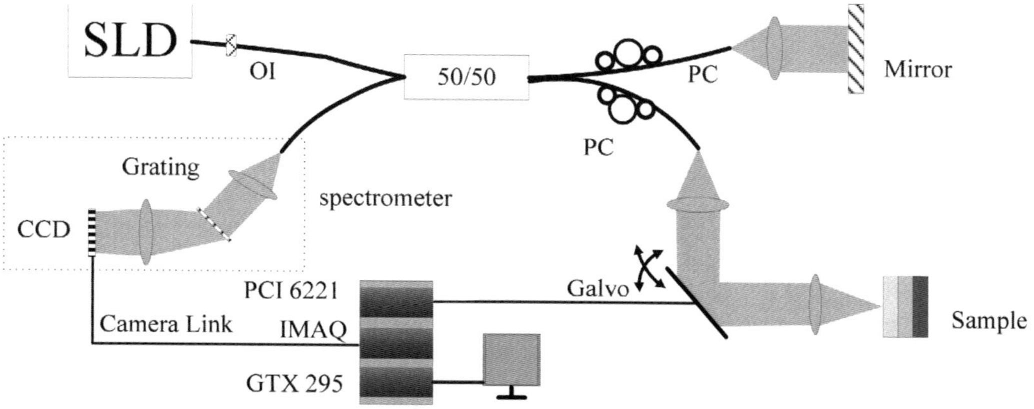


**Fig.2.** The setup of a SD-OCT system. SLD: SLD, superluminescent diode; PC, polarization controller; OI, optical isolator; CCD: Charge Coupled Device.

In order to utilize capability of the GPU and speed up the imaging processing time, a NVIDIA GTX 295 GPU is used to do the data processing instead of CPU. In CUDA model, GPU is treated as a dedicated coprocessor of the host CPU. CUDA allows the same code to be simultaneously run on different cores of GPU as threads [10,11]. All threads are organized into thread blocks, which are executed concurrently on one Single-Instruction Multiple-Data (SIMD) stream multiprocessor (SM). A GPU has two types of memory: global memory and SM on-chip memory. Each SM has four different types of on-chip memory, namely, constant cache, texture cache, registers, and shared memory [10,11]. Different type's memory has different properties. In order to gain optimal performance when utilizing CUDA, the user must determined the thread blocks, the shared memory, registers, and the global memory usage.

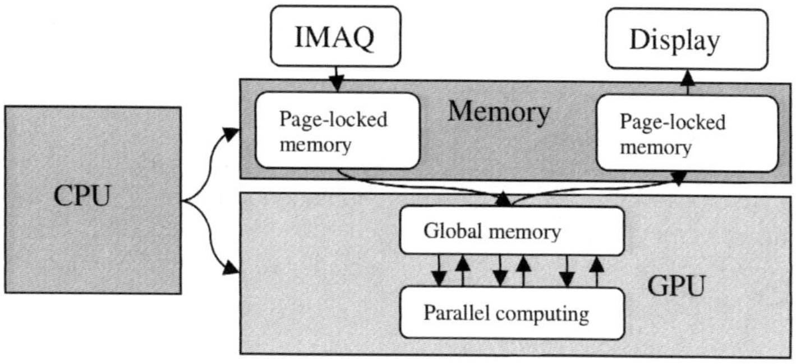


**Fig.3**. Data flow chart of GPU computing.

Figure.3 shows data flow of GPU computing for SD-OCT. Data is acquired by IMAQ and transmitted to a bulk page-locked memory, in which system can get a high transmission performance. And then, the data is transmitted to the global memory of the GPU for parallel computing. After the data is processed by GPU, it is transmitted to page-locked memory again. Finally, it is showed by the display screen. All the operations are scheduled by CPU. In order to utilize the capability of the GPU, the parallel computing on GPU follows the steps as Fig.4 shows.

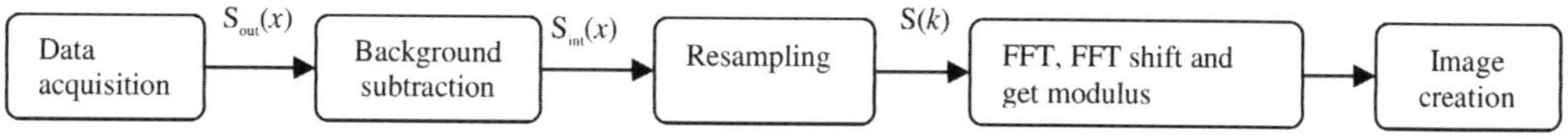


**Fig.4.** Data processing step on GPU

### *3.1. Background subtraction*

x'=x-DC

Symbol x' stands for the interference data of the SD-OCT system. The symbol DC stands for the spectral intensity distribution of the light source. It is the sum of the all lines' acquired data and then divided by the line number. This step is not suitable for parallel computing.

### *3.2. Re-sampling*

In CUDA architecture, global memory read function would consume a lot of time, usually 400 to 600 clock cycles [10,11]. As discuss above, one point's interpolation only relates to the acquired data and coefficients in the cut-off window. A lot of interpolations use the same acquired data and a lot of lines use the same coefficients. If all the data stored in global memory, it would consume a lot of read memory time. Additionally, if a lot of blocks read the same memory, it would result in block conflicts [10,11]. So, it's better to share these data and coefficients by reading them into every block's shared memory before interpolation take place. The read speed of the shared memory can reach the read of the registers, which is the least communication latency in threads. It can be realized like this:

```
__shared__ float data [block size][begin to end];

__shared__ float coefficient [block size][ cut-off length];
```

By using the shared memory, every block read the acquired data and the coefficient once, reducing the read latency and block conflict. After this step, interpolation is operated by multiply-add functions as Fig. 3 shows. A point can be operated by a thread. So a lot of points can be operated concurrently, reducing the computing time.

### *3.3. FFT, FFT shift and get modulus*

Depth information can be retrieved by performing an inverse FFT algorithm. The NVIDIA CUDA fast Fourier transform library (CUFFT) provides a simple interface for computing parallel FFT on an NVIDIA GPU, which allows users to leverage the floating-point power and parallelism of the GPU without having to develop a custom, GPU-Base FFT implementation [13]. By setting the types, transform directions and size of the application programming interface (API), FFT can be operated automatically. The data obtained from FFT operations are complex quantities and the sequence is not distributed well. So, it needs the FFT shift and get modulus operations to do the pre-processing for the image creation.

### *3.4. Image creation*

The demodulated data from FFT is complex, so it needs to map to structure image by Eq.5 [2]. A lot of intensity also can be operated concurrently in this step. A pixel of the image is obtained by combining two pixels of the acquired data and restored in 8 bit in step.

$$Intensity = Contrast \times \left(10 \times \log 10(x_2(s')) + Brightness\right) + 255 \qquad (5)$$

The resolution of GPU computing supports 32 bit and 64 bit [10,11]. The 64 bit's computing will consume much more time than that of the 32 bit's. Furthermore, the error between the accuracy of 64 bit and 32 bit is acceptable for OCT applications [9]. Thus, 32 bit's computing is adopted in this program.

## 4. Result and discussion

In order to compare the processing capability of the GPU and CPU, the computing time of the steps is measured by CUDA event function. A CUDA event is started and stopped at the beginning and ending, respectively. The computing time of a program is obtained by recording the time between the two events, whose resolution can be reach to microseconds (us). A speed up ratio (R) is defined as $R=t_c/t_g-1$, where $t_c$ and $t_g$ is the computing time on CPU and on GPU, respectively.

An Experiment is performed to obtain the computing time. One frame data contained 400 (A-lines per frame) × 2048 (pixels per lines) ×12 (bits per pixel) is acquired by the image acquisition device in SD-OCT. Table.1 shows every step's computing time on GPU and that of on CPU. A single core CPU is adopted to compare the parallel capability of GPU. The difference between the time for copy data to GPU and the time for copy data to host is that the image data is compressed and stored in 8 bit per pixel. So, the time for copy image data to host is less than the time for copy acquired data to GPU. The R of the background subtraction is less than others is because it is The Computing time on GPU is much faster than that of on CPU, which has folded more than 37 times computing time. It should be noted that the R depends the capability of the CPU and GPU. Different CPU and GPU result in different R. It's easy to realize real-time video rate SD-OCT imaging by this software program. The computing capability can be achieved 60,000 A-lines per second, which is much faster than the acquisition speed of the linescan camera. Thus, the acquisition speed becomes the imaging limitation of SD-OCT system.

**Table 1.** Computing time of different methods[a].

| | GPU time(ms) | CPU time (ms) | R |
|---|---|---|---|
| **Copy acquired data to GPU** | 0.463 | | |
| **Background subtraction** | 2.047 | 31.42 | 14.35 |
| **Interpolation**[b] | 2.060 | 87.18 | 42.32 |
| **FFT** | 0.854 | 52.63 | 60.63 |
| **Image creation** | 0.689 | 67.09 | 96.37 |
| **Copy image data to Host** | 0.095 | | |
| **Total time** | 6.208 | 238.32 | 37.90 |

[a]Note:the computer platform:CPU, AMD sempron 3600+ 2.0 GHz; Memory: 2G DDR2 800;GPU, NVIDIA GTX 295.

[b]Note: the cut-off length is L=21.

In conclusion, we have described a new way to accelerate the signal processing speed of SD-OCT. The computing time of time-domain interpolation on GPU can be 42-folded times less than that on CPU. The total imaging processing time of SD-OCT has been improved by more than 37 times. The total signal processing of SD-OCT can be finished in 6.208. So, with the help of the GPU, real-time SD-OCT data processing is realized. Furthermore, CUDA model supports multi GPUs. The user can get a higher computing capability by using several GPUs, which can process more A-lines per frame in real-time.

## 5. References


[1] Yudong Zhang, Xiqi Li, Ling Wei, Kai Wang, Zhihua Ding, and Guohua Shi, *Opt. Lett.* **34** 1849

[2] James G. Fujimoto, in Handbook of Optical Coherence Tomography, Brett E. Bouma and Guillermo J. Tearney, eds, (Marcel Dekker, Inc. 2002) Chap.1.

[3] R. Leitgeb, C. Hitzenberger, and Adolf Fercher, *Opt. Express.* **11** 889

[4] Christophe Dorrer, Nadia Belabas, Jean-Pierre Likforman, and Manuel Joffre, *J. Opt. Soc. Am. B.* **17** 179

[5] A. W. Schaefer, J. Joshua Reynolds, D. L. Marks, and S. A. Boppart, *IEEE Trans. Biomed. Eng.* **51** 186

[6] S. Yan, D. Piao, Y. Chen, and Q. Zhu, *J. Biomed. Opt.* **9** 454

[7] J. Su, J. Zhang, L. Yu, H. G. Colt, M. Brenner, and Z. Chen, *J. Biomed Opt.* **13** 030506

[8] T. E. Ustun, N. V. Iftimia, R. D. Ferguson, and D. X. Hammer, *Rev. Sci. Instrum.* **79** 114301
[9] Gangjun Liu, Jun Zhang, Lingfeng Yu, Tuqiang Xie, and Zhongping Chen, *Appl. Opt.* **48** 6365
[10] NVIDIA CUDA. http://developer.nvidia.com/object/cuda.html.
[11] NVIDIA CUDA computer Unified Device Architecture: Programming Guide, Version 2.0beta2, Jun.2008.
[12] Kang Zhang and Jin U. Kang, *Opt. Express.* **18** 11772
[13] NVIDIA CUDA CUFFT Library, Version 1.1, Oct, 2007.

# Study on the Relationship between Delayed Fluorescence and Photosynthetic Capability at Elevated Temperature in Higher Plants

**Ying Li[1], Wenhai Xu[1,3], Junsheng Wang[1], Da Xing[2]**
[1] College of Information Science Technology, Dalian Maritime University, Dalian 116026, China
[2] MOE Key Laboratory of Laser Life Science & Institute of Laser Life Science, South China Normal University, Guangzhou 510631, China

E-mail: ly86950888@163.com

**Abstract.** With the continuous elevation of the global temperature, high-temperature stress has been a major environmental factor that affects plant growth and productivity. Effects of short-term heat temperature stress on light-induced delayed fluorescence (DF) decay kinetic curve, intensity and emission spectrum have been investigated in C3 soybean (Jing Huang No.3) and C4 maize (Yun Xi No.5081) species. The temperature responses of DF decay kinetic curve from two different species show that decay rate characteristics are affected by high temperature. The spectroscopy measurements indicate that heat stress influence the shape of DF emission spectra of two species, especially the peak intensities at 685nm and 730nm. Moreover, our results clearly demonstrate that DF intensity of each plant positively correlated with F730/F685 of DF emission spectra at elevated temperatures. In addition, the net photosynthetic rate (*P*n) of samples has the same temperature response with DF intensity and F730/F685. Based on these results, we can conclude that there is an excellent correlation between F730/F685 of DF emission spectra, DF intensity and *P*n in both C3 and C4 plants. Therefore, we proposed that the F730/F685 of DF emission spectrum can be used to measure the photosynthetic capability of higher plants to heat stress.

## 1. Introduction

There are many factors to restrict the growth and production of crops and forest plants, such as salinity, UVB radiation, drought, low or high temperature, etc. With the predicted greenhouse-induced rise in global surface temperature possibly already under way, the extremely temperature has been a normal weather in many cities. Accordingly, the effects of elevated temperatures on plant growth and ecosystem function have become a major area of concern. As we know, the high temperature affects the photosynthesis of plant, changes the structure and function of the photosynthetic apparatus, and causes various damages in thylakoid membranes [1]. There are some biochemical methods to examine the plant responses to increasing temperature. At present, the chlorophyll fluorescence techniques and photosynthetic capability detection are the main approaches for measuring plant stress responses. However, photosynthetic capability detection based on gas exchange technique is easy to interference

[3] Correspondence author: Wenhai Xu, Tel: +86-411-84723177, Fax: +86-411-84723177, Email:whxu@newmail.dlmu.edu.cn

from environmental factors [2]. Although chlorophyll fluorescence technique is a fast method of detection of plant stress response, there are some difficulties in practical application [3]. In recent years, some researchers proposed that delayed fluorescence (DF) can be a useful and effective method to detect the plant stress response, such as high temperature, salt and UVB stress, etc [4,5]. Accordingly, the analysis of DF behaviour can be useful in assessing the state of the photosynthetic processes.

In 1951, Strehler and Arnold firstly reported the DF, also referred as "delayed luminescence (DL)" and "delayed light emission (DLE)", which is a long-live low level photon emission by plant after being illuminated with light and put into darkness [6]. After that, many scholars were eager to find out the characteristics of DF, the relationship between DF and environmental stress, and the biochemical representation of DF.

For years, researchers have been reported some DF characteristics such as origination of the peak at 730nm in DF emission spectrum [7], correlation between DF intensity and net photosynthetic rate (*P*n) [4,5] and the environmental stress of plant based on DF [5,8], and so on. However, the responses of decay characteristics of DF signal and emission spectrum to stress has not been researched.

The objectives of this work were to investigate the variation of decay characteristics of DF decay signal and the DF emission spectrum at elevated temperatures. Moreover, the relationships between DF intensity, peak intensity of DF emission spectra and *P*n of two different samples were researched. Experimental results indicated that the variation of decay characteristics of DF decay signals leads to the variation of DF intensity at elevated temperatures. The high temperature varied the peak intensities of DF emission spectrum. Additionally, results demonstrated that peak ratio (F730/F685) of DF emission spectrum, DF intensity and *P*n response to temperature in a same way in both C3 and C4 samples.

## 2. Materials and Methods

### *2.1. Plant material and elevated temperature*

Seeds of soybean (*Glycine max* (L.) Jing Huang No.3) and maize (*Zea May* L. Yun Xi 5081) were grown in a plant growth chamber (Conviron, model E7/2, Winnipeg, Canada) under a photoperiod of 14 hours with a light intensity of 400μmol photons $m^{-2}s^{-1}$. The growth relative humidity (RH) or temperature was controlled as 70/80% (day/night) or 26°C/22°C (day/night), respectively. After a week, the seedlings were thinned out in the uniform size. Two to four plants were maintained per pot. The elevated temperatures were carried by water bath. The aluminium cuvette containing the detached leaves of samples were dipped into a water bath, and after incubation for 30min, DF intensity, *P*n, and DF spectrum was measured at the temperatures indicated, respectively. The experimental temperatures were set as 25°C, 30°C, 35°C, 40°C, 45°C and 50°C. For all experiments, the fourth or fifth leaf was used.

### *2.2. Measurements of delayed fluorescence (DF) decay signal and intensity*

DF decay signals and intensities were measured by home-made portable DF detection system in our laboratory [4]. The excitation light was a set of LED (λ=660nm), and the light intensity was controlled within the range between 0 and 3000μmol photons $m^{-2}$ $s^{-1}$. The DF decay signal was collected by an ultra-high-sensitive Channel Photomultiplier DC-Module (MP963, Perkin-Elmer, Wiesbaden, Germany). In this paper, the DF intensity was integrated within 1 seconds of DF decay signal and registered as count per second (cps). The technical details of the system are described in reference [4].Each experiment was replicated at least five times.

### *2.3. Measurements of net photosynthesis rate(Pn)*

The plant net photosynthetic rate (*P*n), according to traditional interpretation, is used to represent plant photosynthetic capability. In this study, *P*n was measured by a portable photosynthesis system (Model: LI-COR 6400, Lincoln, Nebraska, USA). In the experiment, a fixed concentration of $CO_2$

(1250~1350ppm) was chosen, which is saturated for C3 and C4 plants [9]. The temperature or humidity of system's chamber was controlled at 26±0.5°C or 75%, respectively. Each experiment was replicated at least five times.

*2.4. Measurements of delayed fluorescence (DF) spectrum*

The DF emission spectra of samples were measured by a fluorescence spectrometer LS55 (Perkin-Elmer, USA). The DF emission spectrum measurements were carried out under phosphorescence model. DF emission spectrum was excited at 660nm and recorded from 670 to 760nm. Each experiment was replicated at least five times.

## 3. Results and Discussion

*3.1. DF decay signals response to elevated temperature*

Figure 1 shows the responses of DF decay signals to elevated temperatures. As shown in figure 1, with increasing of temperature, the initial intensity and decay rate of DF decay signal decreased obviously, especially at higher temperature.

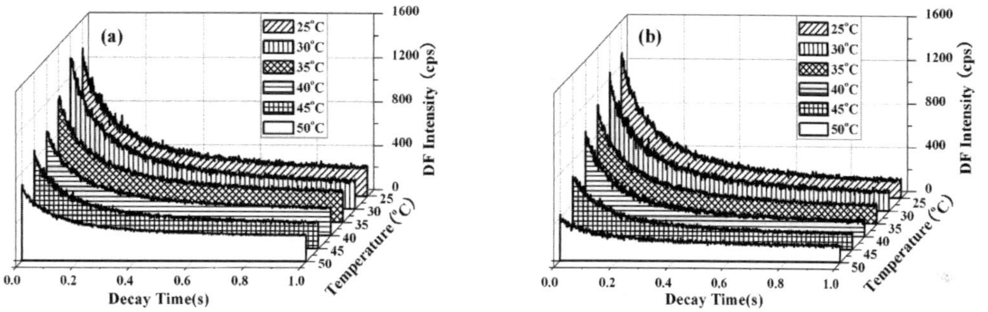


**Figure 1**. The responses of DF decay signals to elevated temperature in maize (Yun Xi No. 5081) (a) and soybean (Jing Huang No.3) (b) leaves

To better research the DF decay characteristics at elevated temperature, we used curve fitting to analyse the DF decay signals in figure 1. In our recent report, we proposed that DF decay signal can be simplified as a three components decay kinetic model was shown as following [10]:

$$I_{DF}(t) = I_1 e^{\frac{t}{\tau_1}} + I_2 e^{\frac{t}{\tau_2}} + I_3 e^{\frac{t}{\tau_3}} \quad (1)$$

In equation (1), $I_1 e^{\frac{t}{\tau_1}}$, $I_2 e^{\frac{t}{\tau_2}}$, $I_3 e^{\frac{t}{\tau_3}}$ are corresponding to the electron reverse of three components, $Q_A^-$, $Q_B^{2-}$, and $F_d^-$, respectively. Although $I_3 e^{\frac{t}{\tau_3}}$ comes from the cyclic electron flow around PSI, the research testified that the third component is almost a constant [10]. Accordingly, the equation (1) can be further simplified as a double exponential:

$$I_{DF}(t) = I_1 e^{\frac{t}{\tau_1}} + I_2 e^{\frac{t}{\tau_2}} + C \quad (2)$$

$\tau_1$, which is related to the electron recombination of $P_{680}^+$ and $Q_A^-$, is called decay rate constant of the fast decay component. $\tau_2$, which is related to the electron recombination of $Q_A^-$ and $Q_B^{2-}$, is called decay rate constant of slow decay component. $I_1$ and $I_2$ are portion to $[Q_A^-]$ and $[Q_B^{2-}]$. And C represents the total of the DF component which comes from the cyclic electron flow around PSI and not negligible noise. We fitted the DF decay signals which are shown in figure 1 using the equation (2), and the fitting results are listed in table 1.

**Table 1.** Fitting results of DF decay signals at elevated temperature of two samples

| Sample | Temperature (°C) | Fitting curve function: $I_{DF}(t)=I_1e^{\frac{t}{\tau_1}}+I_2e^{\frac{t}{\tau_2}}+C$ | | | | | |
|---|---|---|---|---|---|---|---|
| | | $R^2$ | $C$ | $I_1$ | $\tau_1$ | $I_2$ | $\tau_2$ |
| Maize (Yun Xi No. 5081) | 25 | 0.989 | 251.831 | 480.923 | -0.066 | 506.669 | -0.271 |
| | 30 | 0.994 | 245.733 | 590.932 | -0.064 | 524.979 | -0.267 |
| | 35 | 0.994 | 277.839 | 392.429 | -0.066 | 458.082 | -0.267 |
| | 40 | 0.994 | 243.855 | 325.485 | -0.060 | 366.380 | -0.262 |
| | 45 | 0.989 | 230.114 | 288.555 | -0.058 | 304.001 | -0.270 |
| | 50 | 0.993 | 236.696 | 166.315 | -0.060 | 254.496 | -0.263 |
| Soybean (Jing Huang No.3) | 25 | 0.994 | 151.664 | 519.067 | -0.058 | 602.125 | -0.263 |
| | 30 | 0.989 | 166.856 | 480.932 | -0.066 | 566.661 | -0.271 |
| | 35 | 0.988 | 136.670 | 415.831 | -0.069 | 500.958 | -0.288 |
| | 40 | 0.991 | 169.383 | 332.630 | -0.050 | 349.992 | -0.273 |
| | 45 | 0.987 | 136.989 | 290.833 | -0.071 | 217.259 | -0.291 |
| | 50 | 0.979 | 140.368 | 113.802 | -0.070 | 132.497 | -0.310 |

It is clearly that two components fit for the DF decay signals at elevated temperatures of two samples were very excellent ($R^2 \geqq 0.979$). Moreover, the results listed in table 1 shown that values of $I_1$ and $I_2$ decreased when temperature is above 25°C for soybean sample. $\tau_1$ and $\tau_2$ are slightly different as the noise. The results indicated that under high temperature, the electron recombination of $P_{680}^{+}$ and $Q_A^-$, $Q_A^-$ and $Q_B^{2-}$ are greatly restrained. The inhibition leaded to the decrease of the $\lfloor Q_A^- \rfloor$ and $\lfloor Q_B^{2-} \rfloor$, accordingly deceased the values of $I_1$ and $I_2$. According to equation (2) and table (1), we concluded that the decreasing of values of $I_1$ and $I_2$ are the main factor which lead to the deceasing of $I_{DF}(t)$. The similar results were observed in maize sample. As is known, C4 plant species have a higher temperature optimum for photosynthesis than C3 plants. Accordingly, maximum values of $I_1$ and $I_2$ on elevated temperature is observed at 30°C for maize sample and at 25°C for soybean sample, respectively.

*3.2. DF spectra responses to elevated temperature*

The DF spectra clusters of two samples are shown in figure 2(a) and (b). Data were normalized at the peak of 685nm.

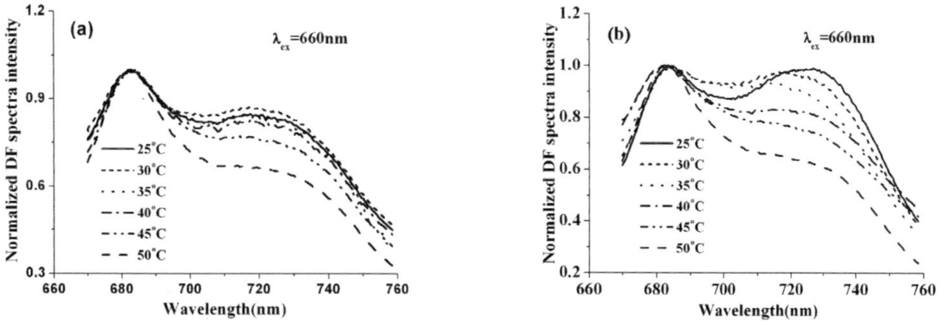


**Figure 2.** The responses of DF emission spectra to elevated temperature in maize (Yun Xi No. 5081) (a) and soybean (Jing Huang No.3) (b) detached leaves

In room temperature, the DF emission spectrum, in turn, exhibits two maxima in the 685nm and 730nm region. Figure 2(a) shows the changes in DF emission spectrum of maize samples exposed to elevated temperatures. As shown in figure 2(a), the peak intensity at 730nm attained its maximum when the temperature was 30°C. Above 30°C, with increasing temperature, the peak intensity at 730nm continued to decrease. However, as shown in figure 2(b), the temperatures higher, the peak intensity at 730nm lower.

*3.3. Correlation between DF intensity, Pn and DF emission spectrum*

From the results of 3.1 and 3.2, we observed a surprised phenomenon that the response of DF intensity and emission spectra to elevated temperature is similar. For the maize sample, when temperature was 30°C, the DF intensity and peak intensity at 730nm respectively attained the maximum. However, for the soybean sample, with increasing temperature, both them decreased. For better analyse the response of DF emission spectrum to elevated temperature, we researched the ratio of peak at 730nm to 685nm, which was marked as F730/F680. We hypothesized that DF intensity and F730/F685 of emission spectrum has some correlation. To verify the hypothesis, the correlations between DF intensity, F730/F685 and *Pn* of two samples at elevated temperature were researched. As shown in figure 3(a) and 3(b), the DF intensity exhibited a good agreement with the F730/F685 for both maize and soybean samples in response to elevated temperatures. Moreover, the results showed that the F730/F685 and *Pn* responded to elevated temperatures in a same way in both two samples (figure 3(a) and 3(b) insert maps). Similar experiments were performed for the leaves of two soybean species (Ke Feng No.1 and You Chun No.4) and three maize species (Jin Dan No.36, Ji No.853 and Yun Xi No. 422). The similar results have been found in the five different samples. All the results suggested that F730/F685 has the same response with DF intensity and *Pn* to elevated temperature in both C3 and C4 plants.

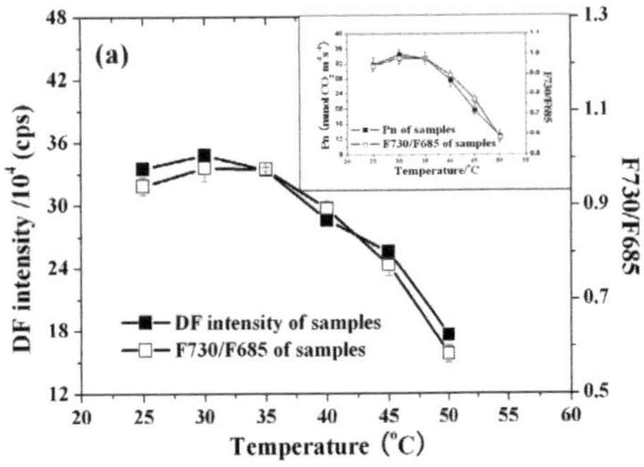


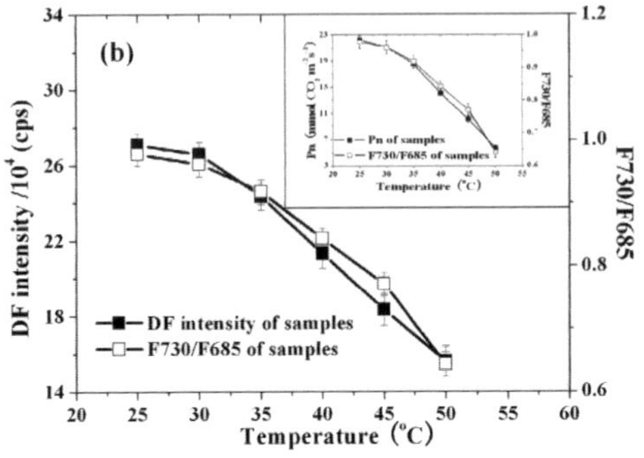


**Figure 3.** Effect of elevated temperature on DF intensity and F730/F685 in maize (a) and soybean (b) leaves. Effect of elevated temperature on *P*n and F730/F685 in maize and soybean samples is respectively shown in the inset.

## 4. Conclusion

With the detached leaves of maize and soybean as testing samples, we have studied the variations of decay characteristics of DF decay signals and peak intensity of DF emission spectra under high temperature conditions. The results clearly show that, the high temperature inhibit the electron recombination of $P_{680}^{+}$ and $Q_A^-$, $Q_A^-$ and $Q_B^{2-}$, accordingly decrease the amplitudes of fast and slow decay components. Therefore, the DF intensity decreased with increasing temperature. The spectroscopic analysis results demonstrated that the high temperature affect the peak intensity of DF emission spectra. Moreover, the response of F730/F685 of DF emission spectra to elevated temperature is the same as that of DF intensity and *P*n. Since the correlation between DF intensity and *P*n under stress has been reported before. Therefore, we proposed that the F730/F685 of DF emission spectrum can be used as an effective test for the sate of photosynthetic apparatus and truly reflects the plant photosynthetic ability under high temperature stress.


## Acknowledgments

This research is supported by the National Natural Science Foundation of China (60774056), and the Department of Education New Teacher Foundation of China (200801511013).


## Reference


[1] Martin Cajanek, Michal Stroch, Irena Lachetova, Jiri Kalina and Vladimir Spunda 1998 *J. Photochem. Photobiol. B: Biol.* **47** 39

[2] C. L. Wang, D. Xing and Q. Chen 2004 *Biosens. Bioelectron.* **20** 454

[3] A. Baker, R. Inverarity, M. Charlton and S. Richmond 2003 *Environ. Pollut.* **124** 57

[4] Junsheng Wang, Da Xing, Lingrui Zhang and Li Jia 2006 *Biosens. Bioelectron.* **22** 2861

[5] Lingrui Zhang, Da Xing and Junsheng Wang 2007 *Sensors.* **7** 52

[6] B. L. Strehler and W. Arnold 1951 *J. General Physiology.* **34** 809

[7] Lizhang Zeng, Da Xing, Lingrui Zhang, Qiang Li and Chenglong Wang 2007 *Luminescence.* **22** 387

[8] Zhang Lingrui, Xing Da and Wen Feng 2009 *Chinese Science Bulletion.* **54** 4009

[9] Fracheboud. Y. and Ridaut. J.M. 2002 *J. Exp. Bot.* **53** 1967

[10] WANG Jun-sheng, XU Wen-hai, ChENG Wei-shan and LI Jian 2007 *JACAT LASER BIOLOGY SINICA..* **16** 668

# Laser homeostatics on delayed onset muscle soreness

**T C Y Liu[1], D R Fu, X G Liu and Z X Tian**
Lab Laser Sports Medicine, South China Normal University, University Town, Guangzhou, GD 510006, China

E-mail :liutcy@scnu.edu.cn

**Abstract.** Delayed onset muscle soreness (DOMS) and its photobiomodulation were reviewed from the viewpoint of function-specific homeostasis (FSH) in this paper. FSH is a negative-feedback response of a biosystem to maintain the function-specific fluctuations inside the biosystem so that the function is perfectly performed. A stressor may destroy a FSH. A stress is a response of a biosystem to a stressor and may also be in stress-specific homeostasis (StSH). A low level light (LLL) is so defined that it has no effects on a function in its FSH or a stress in its StSH, but it modulate a function far from its FSH or a stress far from its StSH. For DOMS recovery, protein metabolism in the Z-line streaming muscular cell is the essential process, but the inflammation, pain and soreness are non-essential processes. For many DOMS phenomena, protein metabolism in the Z-line streaming muscular cell is in protein metabolism-specific homeostasis (PmSH) so that there are no effects of LLL although the inflammation can be inhibited and the pain can be relieved. An athlete or animal in the dysfunctional conditions such as blood flow restriction and exercise exhaustion is far from PmSH and the protein metabolism can be improved with LLL.

## 1. Introduction

Skeletal muscle is one of the most vulnerable tissue in the extremities. Exercise-induced muscle damage (EIMD) is commonly experienced following either a bout of unaccustomed physical activity or following physical activity of greater than normal duration or intensity, which manifests as prolonged muscle dysfunction, delayed onset muscle soreness (DOMS), and leakage of muscle proteins into circulation. The mechanistic factor responsible for the initiation of EIMD is not known; however, it is hypothesised to be either mechanical or metabolic in nature. This model states that

[1] To whom any correspondence should be addressed.

damage initiation may be either metabolic or mechanical, or a combination of both, depending on the mode, intensity and duration of exercise and the training status of the individual[1]

EIMD/DOMS is a well-known phenomenon in exercise training and physical activities. Effective therapy to reduce skeletal muscle injury associated with severe or eccentric exercise is needed. Many therapeutic approaches have been used to speed recovery in view of previous theories proposed for its mechanism, such as lactic acid, muscle spasm, connective tissue damage, muscle damage, inflammation, the enzyme efflux theories, and an integration of two or more theories, but few of them were effective [2]. In this paper, EIMD/DOMS phenomena and its photobiomodulation (PBM) was reviewed from the viewpoint of function-specific homeostasis (FSH).

**2. Function-specific homeostasis**

Homeostasis is a classic concept in physiology [3]. However, it is too obscure to be studied so that it has been developed as FSH[4-5]. FSH is a negative-feedback response of a biosystem to maintain the function-specific fluctuations inside the biosystem so that the function is perfectly performed.

The quality of a FSH includes function complexity and performance stability. An elite athlete has a sport-specific homeostasis (SpSH) to maintain sport performance. For a 100 m sprint runner, the 100 m sprint time represents the function complexity. The less the running time, the more complicated the run. The coefficient of variation (CV) of the sprint time represents the performance stability. The higher the CV, the lower the performance stability.

There are many subsystems to maintain a SpSH, SpSH-essential subsystems (SESs) and SpSH-non-essential subsystems (SNSs), which are in their respective SES-specific homeostasis (SESH) and SNS-specific homeostasis (SNSH), respectively [4]. In order to realize the transition from low quality SpSH1 into high quality SpSH2, there also are two kinds of training, extraordinary training (ET) and ordinary training (OT). ET establishes the SESHs of SpSH2 by destroying SpSH1. OT maintaining the SESHs has two phases, OTA during which the SNSHs of SpSH2 and then SpSH2 are established, and OTB during which SpSH2 is maintained.

A stress is a function of a biosystem[6]. A function-specific stressor (FSSR) is so defined that it disrupts the FSH of the function. A function-specific stress (FSS) of a biosystem is defined to be a response of the biosystem to a FSSR and may also be in FSS-specific homeostasis (FSSH) in which a FSS disrupt FSH1 but establish FSH2. A factor is called low level factor (LLF) if there is no effects of the factor on a function in its FSH or a FSS in its FSSH, but modulate a function far from its FSH or a FSS far from its FSSH. Obviously, novel eccentric (lengthening contraction) exercise (ECC) is a sport-specific stressor, and ET-OTA is in sport-specific stress-specific homeostasis because it realized the transition from SpSH1 into SpSH2. ECC can increase the optimum lengths of both the knee extensors and knee extensors flexors during the preseason in professional soccer[7]. Only 30 min of eccentric exercise per week for eight weeks was sufficient to improve health risk factors [8].

ET results in EIMD/ DOMS. The physiological stress associated with ECC induced muscle damage impairs insulin stimulation of Insulin-stimulated insulin receptor substrate (IRS)-1, phosphatidylinositol (PI) 3-kinase, and Akt (protein kinase B)-kinase, presumably leading to decreased insulin-mediated glucose uptake[9]. Using high density oligonucleotide based microarrays, MacNeil et al. [10]screened for differences in mRNA expression caused by 17beta-estradiol (E2) and 150 single-leg

ECC, and found that the signaling and regulation of early Ras homologue gene family, member A (RhoA) and nuclear factor of activated T-cells (NFAT) are altered following the ECC for muscle remodeling and repair.

Increases in the mRNA expression of interleukin (IL)-6, IL-8, and cyclooxygenase 2 (COX2) occur in the vastus lateralis as a result of damaging ECC in young, recreationally trained males [11].

Inflammation induced by high-force ECC in skeletal muscle is greater when a high carbohydrate (CHO) compared to a low CHO diet is consumed during recovery [12]. Even grounding the body to the earth alters measures of immune system activity and pain [13].

Unaccustomed strenuous exercise that includes lengthening contraction (LC) often causes DOMS, a kind of muscular mechanical hyperalgesia. Augmentation of the mechanical response in muscle thin-fiber sensory receptors might be related to the muscle tenderness in DOMS after ECC[14]. Nerve growth factor (NGF) upregulation through activation of B(2) bradykinin receptors is essential (though not satisfactory) to mechanical hyperalgesia after exercise[15].

DOMS has three subsequent phases: Z-line streaming, proteolysis of damaged proteins and protein synthesis, among which the last two phases is just protein metabolism that is the essential process of DOMS recovery [16]. The DOMS normally recovers in about seven days if the protein metabolism is in its protein metabolism-specific homeostasis (PmSH). For EIMD/DOMS, there are oxidative stress, inflammation, pain/soreness and protein metabolism. All the other processes might be far from their respective FSH, oxidative-process-specific homeostasis (OSH), inflammation-specific homeostasis (ISH) and pain/soreness -specific homeostasis (PsSH), but protein metabolism might be in protein-metabolism-specific homeostasis (PmSH). The factor effects on EIMD/DOMS would then be discussed.

## 3. DOMS in PmSH

Fu [17] has studied the effects of niacin supplement at low level on the inflammation of skeletal muscle, level of serum CK and lactate dehydrogenase (LDH), protein level of sirtuin 1(SIRT1), p65 and 20S protease of skeletal muscle in rats following a bout of ECC. She found niacin supplement could reduce inflammatory of EIMD/DOMS, which indicated the process of inflammatory may be far from ISH. Niacin supplement increased SIRT1 protein level, reduced p65 protein level and then inhibited inflammation for a bout of ECC induced EIMD / DOMS. Niacin supplement did not affect the expression of 20S protease in soleus and the levels of serum CK and LDH, respectively, which indicated that protein metabolism may be in PmSH, and the niacin supplements did not affect the metabolic processes.

Any LLFs can not affect normal DOMS recovery. DOMS, creatine kinase (CK) activity, and isometric and concentric torque all changed over the 72-hour period after ECC; however, oral beta-hydroxy-beta-methylbutyrate (HMB) and alpha-ketoisocaproic acid (KIC) had no significant effect on any of the indices of muscle damage[18]. Early RhoA and NFAT signaling and regulation are altered following the ECC for muscle remodeling and repair, but are not affected by E2[10].

Eight healthy male volunteers performed 200 maximal eccentric contractions with each leg. Although Non-steroidal anti-inflammatory drugs (NSAIDs) inhibit the exercise-induced satellite cell proliferation, Mikkelsen et al.[19]observed myofibrillar and collagen protein synthesis were unaffected by the local NSAID infusion. Five hours post-exercise, the mRNA expression of COX2 was sixfold higher in the NSAID leg compared with the unblocked leg, but the expression of growth factors and matrix-related genes were unaffected by NSAID. All subjects completed 4 sets of 12 repetitions of eccentric elbow flexion with their non-dominant arm. Goldfarb et al.[20]reported that 4 weeks of pretreatment with a fruit/berry/vegetable concentrate (FVC) can attenuate blood oxidative stress markers induced by ECC but had no significant impact on the functional changes related to pain and muscle damage. These phenomena indicated that oxidative process is far from oxidative-process-specific homeostasis (OSH), but protein metabolism is in PmSH. As McGinley et al. [21] have reviewed, there is little evidence to support a role for vitamin C and/or vitamin E in protecting against muscle damage although there is some evidence to show that both antioxidants can reduce indices of oxidative stress, and antioxidant supplementation may actually interfere with the cellular signalling functions of reactive oxygen species (ROS), thereby adversely affecting muscle performance. Since the potential for long-term harm does exist, the casual use of high doses of antioxidants by athletes and others should perhaps be curtailed[21].

Forty-five repetitions of eccentric elbow flexion were performed with 90% of one maximum repetition by each subject. The exercises were done as three sets of 15 repetitions with 3 minutes rest between sets. Afroundeh et al. [22] found carbohydrate increased the inflammatory (IL-6) response following resistance exercise, but had no effects on C-reactive protein (CRP) and CK. Consumption of theaflavin-enriched black tea extract led to improved recovery and a reduction in oxidative stress and DOMS responses to acute anaerobic intervals, but it had no significant effects on either peak power or mean power during the 30 s Wingate Anaerobic Test (WAnT). These phenomena indicated that inflammation is far from inflammation-specific homeostasis (ISH), but protein metabolism is in PmSH.

Branched-chain amino acid (BCAA) supplementation may attenuate muscle soreness, but it does not ameliorate eccentric exercise-induced decrements in muscle function or increases in reputed blood markers of muscle damage, when consumed before exercise and for 3 d after an ECC bout[23]. Prophylactic supplementation of N-acetyl-cysteine (NAC) and epigallocatechin gallate (EGCG) blunted soreness ratings, but significant increases in muscle levels of bax and bcl-2 were observed in all groups with no significant differences between groups [24]. These phenomena indicated that soreness is far from pain/soreness -specific homeostasis (PsSH), but protein metabolism is in PmSH. The PmSH can be upgraded if there is a beneficial stress in stress-specific homeostasis so that the normal DOMS recovery can be promoted. Protease supplementation seems to attenuate muscle strength losses after eccentric exercise[25]. Subjects performed two sets of 20 maximal eccentric elbow flexion exercises with one arm. Supplementation with ellagitannins from pomegranate extract significantly improves recovery of isometric strength 2-3 d after the damaging eccentric exercise [26]. The initial downhill run of mice at 8 wk of age consisted of 150 min of running on a motorized treadmill. The pretreatment with a food pellet containing 10 mg of curcumin powder blunted the increase in plasma creatine kinase 24 h after the EE[27]. Adenosine $A_3$ receptor stimulation induces protection of skeletal muscle from the EIMD induced by the EE [28]. Calcium entry blocker nifedipine by gavage (2 mg/kg/day in two equal doses) attenuated EE induced muscle damage caused due to its beneficial effect on EE impaired muscle microcirculation[29]. Consuming milk-based carbohydrate-protein (CHO-P) after muscle-damaging exercise is more beneficial in attenuating decreases in muscle performance and increases in active DOMS at 48 h than ingestion prior to exercise [30], but post-exercise milk-based CHO-P supplementation at lower dose has no influences of on muscle glycogen replacement, inflammation, or muscle function[31], and the addition of whey protein isolate to a dietary carbohydrate supplement did not attenuate systemic indices of muscle damage or inflammation, nor did it restore muscle function more rapidly than when the carbohydrate fraction was ingested alone [32-33]. BCAA supplementation (isoleucine:leucine:valine = 1:2.3:1.2) at about 100 mg/kg body weight may suppress muscle damage[34-35], but BCAA supplementation at lower dose does not ameliorate eccentric exercise-induced decrements in muscle function or increases in reputed blood markers of muscle damage [23], and adding leucine to carbohydrate beverages did not affect acute muscle recovery and squat performance during both initial testing and during a subsequent exercise bout 72 hours later in resistance trained subjects[36].

### 4. DOMS far from PmSH

The recovery of DOMS in PmSH is normal recovery. The normal recovery can be impaired by many factors so that DOMS is far from PmSH. After exhaustive endurance exercise, muscle damage can be produced by metabolic disturbances associated with ischaemia [37]. The blood flow restriction (BFR) resulted in more soreness than exercise without BFR and greater reductions in pain-pressure threshold and maximal voluntary contraction [38].

DOMS is far from PmSH due to ageing. A reparative capability of muscle was reduced in response to a single exhaustive bout of heavy resistance weight lifting in old (>120 wk of age) compared with young (14-20 wk of age) rats[39]. Older individuals (age 66±2 yr) did not exhibit the compensatory increase in beta-cell secretion observed among young individuals (age 22±1 yr) after ECC [40]. Thus, with increasing age, the pancreatic beta-cell may be less responsive to the physiological stress associated with ECC.

DOMS is far from PmSH due to insulin resistance. Increased phosphorylation of the 70-kDa ribosomal S6 kinase (p70S6k) signaling is strongly correlated with the degree of muscle adaptation following exercise. Katta et al.[41] have compared the phosphorylation of p70S6k, Akt, and mammalian target of rapamycin (mTOR) in the tibialis anterior (TA) muscles of lean and obese Zucker rats following a bout of eccentric exercise (EE). EE increased p70S6k (Thr389) phosphorylation at 0, 1 and 3 h after the EE in the lean TA and at 3 h in the obese TA Zucker rats. mTOR (Ser2448) phosphorylation was elevated in the lean TA immediately after EE but remained unaltered in the obese TA. EE increased Akt (Thr308) and Akt (Ser473) phosphorylation in the lean but not the obese TA. The subjects completed exhaustive downhill running protocol on a motorized treadmill. Allicin supplementation had significantly lower plasma levels of CK, muscle-specific CK and IL-6, and reduced perceived muscle soreness after exercise[42].

### 5. Photobiomodulation

PBM is a modulation of laser irradiation or monochromatic light (LI) on biosystems, which stimulates or inhibits biological functions but does not result in irreducible damage [4-5]. The LI used in PBM is always low intensity LI (LIL), ~10 mW/cm$^2$. However, moderate intensity LI (MIL), $10^{2\sim3}$ mW/cm$^2$, is of PBM if the radiation time is not so long that it damages organelles or cells. LIL is a kind of LLF, and MIL is also a kind of LLF if the radiation time is short enough. Both the two kind of LI are called low level LI (LLL). LLL has been used to treat DOMS after EIMD.

LLL can promote the recovery of DOMS far from PmSH. Exhaustive downhill running was used to induce muscle injury in rat gastrocnemius muscle. LIL significantly reduced serum CK activity at 48 h after exercise[43]. The load-resistance swimming test forced adult male Wistar rats to swim until exhaustion. 40 s MIL treatment lowered CK activity and muscular apoptosis [44].

LLL can not promote the recovery of DOMS in PmSH. The experimenter raised the subjects elbow position to 135° of flexion (upright position) and the subject was asked to lower this one-repetition maximum (1-RM) weight eccentrically for 3 sec. 80 s MIL treatment significantly decreased pain, but had no effects on girth and resting angle [45]. Isokinetic exercise was performed to induce pain related to DOMS. LIL treatment significantly decreased pain, but had no effects on isokinetic peak torque[46].

## 6. Discussion

PBM can be also used to treat DOMS before EIMD. After MIL treatment with 810 nm diodes at 1724 $mW/cm^2$ for 30 s or placebo treatment, subjects performed 75 maximal knee extensors eccentric contractions (five sets of 15 repetitions; velocity = 60 degrees seg(-1); range of motion = 60 degrees ). Baroni et al. [47] found that MIL treatment before ECC was effective in terms of attenuating the increase of muscle proteins in the blood serum and the decrease in muscle force. The DOMS is in PmSH, but MIL pretreatment can promote its recovery. In this case, MIL might be a beneficial stress in stress-specific homeostasis, but it needs to be further studied.

## 7. Acknowledgement

This work was supported by National Science Foundation of China (60878061).

## 8. References

[1] Tee J C, Bosch A N and Lambert M I 2007 Metabolic consequences of exercise-induced muscle damage. *Sports Med.* **37** *827*

[2] Cheung K, Hume P and Maxwell L 2003 Delayed onset muscle soreness : treatment strategies and performance factors. *Sports Med.* **33** *145*

[3] Cannon W B 1932 The Wisdom of the Body. *WW Norton, New York.*

[4] Liu T C Y, Liu R, Zhu L,Yuan J Q, Hu M and Liu S H 2009 Homeostatic photobiomodulation. *Front Optoelectron China.* **2** *1*

[5] Liu C Y and Zhu P (Eds) 2009 Intranasal Low Intensity Laser Therapy. Beijing: *People's Military Medical Press.* **50** *413.*

[6] Liu T C Y, Li F H, Zhu L and Liu S H 2010 Photobiomodulation on stress. *Chin. J. Laser. (to be published).*

[7] Brughelli M, Mendiguchia J, Nosaka K, Idoate F, Arcos A L and Cronin J 2010 Effects of eccentric exercise on optimum length of the knee flexors and extensors during the preseason in professional soccer players. *Phys. Ther. Sport.* **11** *50*

[8] Paschalis V, Nikolaidis M G, Theodorou A A, Panayiotou G, Fatouros I G, Koutedakis Y and Jamurtas A Z 2010 A Weekly Bout of Eccentric Exercise Is Sufficient To Induce Health-Promoting Effects. *Med. Sci. Sports Exerc.*

[9] Del Aguila L F, Krishnan R K, Ulbrecht J S, Farrell P A, Correll P H, Lang C H, Zierath J R and Kirwan J P 2000 Muscle damage impairs insulin stimulation of IRS-1, PI 3-kinase, and Akt-kinase in human skeletal muscle. *Am. J. Physiol. Endocrinol Metab.* **279** *E206*

[10] MacNeil L G, Melov S, Hubbard A E, Baker S K and Tarnopolsky M A 2010 Eccentric exercise activates novel transcriptional regulation of hypertrophic signaling pathways not affected by hormone changes. *PloS. One.18;***5** *e10695.*

[11] Buford T W, Cooke M B, Shelmadine B D, Hudson G M, Redd L and Willoughby D S 2009 Effects of eccentric treadmill exercise on inflammatory gene expression in human skeletal muscle. *Appl. Physiol. Nutr. Metab.* **34** *745*

[12] Depner C M, Kirwan R D, Frederickson S J and Miles M P 2010 Enhanced inflammation with high carbohydrate intake during recovery from eccentric exercise. *Eur. J. Appl. Physiol.* **109** *1067*

[13] Brown D, Chevalier G and Hill M 2010 Pilot study on the effect of grounding on delayed-onset muscle soreness. *J. Altern. Complement Med.* **16** *265.*

[14] Taguchi T, Sato J and Mizumura K 2005 Augmented mechanical response of muscle thin-fiber sensory receptors recorded from rat muscle-nerve preparations in vitro after eccentric contraction. *J. Neurophysiol.* **94** *2822*

[15] Murase S, Terazawa E, Queme F, Ota H, Matsuda T, Hirate K, Kozaki Y, Katanosaka K, Taguchi T, Urai H and Mizumura K 2010 Bradykinin and nerve growth factor play pivotal roles in muscular mechanical hyperalgesia after exercise (delayed-onset muscle soreness). *J. Neurosci.* **30** *3752*

[16] Liu T C Y, Huang P, Liu X G, Chen X Y, Liu J, Wang S X, Cui L P, Xu X Y Guo H, Jin H, Deng S X and Ji L L 2006 Delayed Onset Muscle Soreness: Three-Phase Hypothesis and Its Clinical Applications. *Med. Sci. Sport Exer.* **38** *S124*

[17] Fu D R 2010 The molecular mechanism of exercise induced skeletal muscle damage and its niacin effects. *Ph. D. thesis, South China Normal University.*

[18] Nunan D, Howatson G and van Someren K A 2010 Exercise-induced muscle damage is not attenuated by beta-hydroxy-beta-methylbutyrate and alpha-ketoisocaproic acid supplementation. *J. Strength Cond. Res.* **24** *531*

[19] Mikkelsen U R, Schjerling P, Helmark I C, Reitelseder S, Holm L, Skovgaard D, Langberg H, Kjær M and Heinemeier K M 2010 Local NSAID infusion does not affect protein synthesis and gene expression in human muscle after eccentric exercise. *Scand. J. Med. Sci. Sports.*

[20] Goldfarb A H, Garten R S, Cho C, Chee P D and Chambers L A 2010 Effects of a Fruit/Berry/Vegetable Supplement on Muscle Function and Oxidative Stress. *Med. Sci. Sports Exerc.*

[21] McGinley C, Shafat A and Donnelly A E 2009 Does antioxidant vitamin supplementation protect against muscle damage? *Sports Med..* **39** *1011*

[22] Afroundeh R, Siahkouhian M and Khalili A 2010 The effect of post-exercise carbohydrate ingestion on inflammatory responses to short time, high-force eccentric exercise. *J. Sports Med .Phys. Fitness.* **50** *182*

[23] Jackman S R, Witard O C, Jeukendrup A E and Tipton K D 2010 Branched-chain amino acid ingestion can ameliorate soreness from eccentric exercise. *Med. Sci. Sports Exerc.* **42** *962*

[24] Kerksick C M, Kreider R B and Willoughby D S 2010 Intramuscular adaptations to eccentric exercise and antioxidant supplementation. *Amino. Acids.* **39** *219*

[25] Buford T W, Cooke M B, Redd L L, Hudson G M, Shelmadine B D and Willoughby D S 2009 Protease Supplementation Improves Muscle Function after Eccentric *Exercise.Med. Sci. Sports Exerc.* **41** *1908*

[26] Trombold J R, Barnes J N, Critchley L and Coyle E F 2010 Ellagitannin consumption improves strength recovery 2-3 d after eccentric exercise. *Med. Sci. Sports Exerc.* ***42*** *493*

[27] Davis J M, Murphy E A, Carmichael M D, Zielinski M R, Groschwitz C M, Brown A S, Gangemi J D, Ghaffar A and Mayer E P 2007 Curcumin effects on inflammation and performance recovery following eccentric exercise-induced muscle damage. *Am. J. Physiol .Regul. Integr. Comp. Physiol.* **292** *R2168*

[28] Wang R, Urso M L, Zambraski E J, Rader E P, Campbell K Pand Liang B T 2010 Adenosine A(3) receptor stimulation induces protection of skeletal muscle from eccentric exercise-mediated injury. *Am. J. Physiol. Regul. Integr. Comp. Physiol.* **299** *R259*

[29] Heap S J, Fulgenzi G L and Hudlicka O 2006 Microcirculation in rat soleus muscle after eccentric exercise: the effect of nifedipine. *Eur. J. Appl. Physiol.* **97** *687*

[30] Cockburn E, Stevenson E, Hayes P R, Robson-Ansley P and Howatson G 2010 Effect of milk-based carbohydrate-protein supplement timing on the attenuation of exercise-induced muscle damage. *Appl. Physiol. Nutr. Metab.* **35** *270.*

[31] Wojcik J R, Walber-Rankin J, Smith L L and Gwazdauskas F C 2001 Comparison of carbohydrate and milk-based beverages on muscle damage and glycogen following exercise. *Int. J. Sport Nutr. Exerc. Metab.* **11** *406.*

[32] White J P, Wilson J M, Austin K G, Greer B K, St John N and Panton L B 2008 Effect of carbohydrate-protein supplement timing on acute exercise-induced muscle damage. *J. Int. Soc .Sports Nutr.* **5** *5*

[33] Betts J A, Toone R J, Stokes K A and Thompson D 2009 Systemic indices of skeletal muscle damage and recovery of muscle function after exercise: effect of combined carbohydrate-protein ingestion. *Appl. Physiol. Nutr. Metab.* **34** *773*

[34] Negro M, Giardina S, Marzani B and Marzatico F 2008 Branched-chain amino acid supplementation does not enhance athletic performance but affects muscle recovery and the immune system. *J. Sports Med. Phys. Fitness.* **48** *347*

[35] Shimomura Y, Inaguma A, Watanabe S, Yamamoto Y, Muramatsu Y, Bajotto G, Sato J, Shimomura N, Kobayashi H and Mawatari K 2010 Branched-chain amino acid supplementation before squat exercise and delayed-onset muscle soreness. *Int. J. Sport Nutr. Exerc. Metab.* **20** *236*

[36] Stock M S, Young J C, Golding L A, Kruskall L J, Tandy R D, Conway-Klaassen J M and Beck T W 2010 The effects of adding leucine to pre and postexercise carbohydrate beverages on acute muscle recovery from resistance training. *J. Strength Cond. Res.* **24** *2211*

[37] Ebbeling C B and Clarkson P M 1989 Exercise-induced muscle damage and adaptation. *Sports Med.* **7** *207*

[38] Umbel J D, Hoffman R L, Dearth D J, Chleboun G S, Manini T M and Clark B C 2009 Delayed-onset muscle soreness induced by low-load blood flow-restricted exercise. *Eur. J. Appl .Physiol.* **107** *687*

[39] Tamaki T, Uchiyama S, Uchiyama Y, Akatsuka A, Yoshimura S, Roy R R and Edgerton V R 2000 Limited myogenic response to a single bout of weight-lifting exercise in old rats. *Am. J.Physiol. Cell Physiol.* **278** *C1143*

[40] Krishnan R K, Hernandez J M, Williamson D L, O'Gorman D J, Evans W J and Kirwan J P 1998 Age-related differences in the pancreatic beta-cell response to hyperglycemia after eccentric exercise. *Am. J. Physiol.* **275** *E463.*

[41] Katta A, Karkala S K, Wu M, Meduru S, Desai D H, Rice K M and Blough E R 2009 Lean and obese Zucker rats exhibit different patterns of p70s6 kinase regulation in the tibialis anterior muscle in response to high-force muscle contraction. *Muscle Nerve.* **39** *503*

[42] Su Q S, Tian Y, Zhang J G and Zhang H 2008 Effects of allicin supplementation on plasma markers of exercise-induced muscle damage, IL-6 and antioxidant capacity. *Eur. J. Appl. Physiol.* **103** *275*

[43] Liu X G, Zhou Y J, Liu T C and Yuan J Q 2009 Effects of low-level laser irradiation on rat skeletal muscle injury after eccentric exercise.*Photomed. Laser Surg.* **27** *863*

[44] Sussai D A, Carvalho Pde T, Dourado D M, Belchior A C, dos Reis F A and Pereira D M 2010 Low-level laser therapy attenuates creatine kinase levels and apoptosis during forced swimming in rats. *Lasers Med. Sci.* **25** *115*

[45] Douris P, Southard V, Ferrigi R, Grauer J, Katz D, Nascimento C and Podbielski P 2006 Effect of phototherapy on delayed onset muscle soreness. *Photomed. Laser Surg.* **24** *377*

[46] Vinck E, Cagnie B, Coorevits P, Vanderstraeten G and Cambier D 2006 Pain reduction by infrared light-emitting diode irradiation: a pilot study on experimentally induced delayed-onset muscle soreness in humans. *Lasers Med. Sci.* **21** *11*

[47] Baroni B M, Leal Junior E C, De Marchi T, Lopes A L, Salvador M and Vaz M A 2010 Low level laser therapy before eccentric exercise reduces muscle damage markers in humans. *Eur. J. Appl. Physiol.*

# Characterizing lamina propria of human gastric mucosa by multiphoton microscopy

**Y C Liu[1*], H Q Yang[1*], G Chen[2*], S M Zhuo[1], J X Chen[1] and J Yan[3]**
1. Institute of Laser and Optoelectronics Technology, Fujian Provincial Key Laboratory for Photonics Technology, Key Laboratory of OptoElectronic Science and Technology for Medicine of Ministry of Education, Fujian Normal University, Fuzhou 350007, China
2. Department of Pathology, Fujian Provincial Tumor Hospital, Fuzhou, 350014, China
3. Department of Surgery, Fujian Provincial Tumor Hospital, Fuzhou, 350014, China

E-mail: chenjianxin@fjnu.edu.cn (J X Chen) or ynjun@yahoo.com (J Yan)

**Abstract.** Lamina propria (LP) of gastric mucosa plays an important role in progression of gastric cancer because of the site at where inflammatory reactions occur. Multiphoton imaging has been recently employed for microscopic examination of intact tissue. In this paper, using multiphoton microscopy (MPM) based on two-photon excited fluorescence (TPEF) and second harmonic generation (SHG), high resolution multiphoton microscopic images of lamina propria (LP) are obtained in normal human gastric mucosa at excitation wavelength $\lambda_{ex} = 800$ nm. The main source of tissue TPEF originated from the cells of gastric glands, and loose connective tissue, collagen, produced SHG signals. Our results demonstrated that MPM can be effective for characterizing the microstructure of LP in human gastric mucosa. The findings will be helpful for diagnosing and staging early gastric cancer in the clinics.

## 1. Introduction

Due to poor diagnosis, the stomach cancer is the most second common cause of death [1]. Most stomach cancers develop in the gastric mucosa (inner layer or lining). The gastric mucosa consists of epithelial tissues, lamina propria (LP), and muscularis mucosae. The histologists and pathologists have been mainly focusing on epithelial tissue and LP which is the site that inflammatory reactions occur [2]. LP is composed of a delicate network of collagenous, reticular fibers and contains numerous capillaries, lymphocytes, some plasma cells, eosinophils, and mast cells. Because of the branching of the gland, crowding of the cells and the obscuring of the glands, it is often difficult to observe LP in human gastric mucosa.

Multiphoton microscopy (MPM) has many advantages over traditional optical imaging, including higher resolution, deeper penetration depth, inherent optical sectioning and less phototoxicity [3, 4]. In the nonlinear spectral imaging of the biosciences, MPM based on TPEF and SHG can be also used [3]. MPM was employed for characterizing human gastrointestinal mucosa at the cellular level without the

[*] Equally contributed to this work

need for fluorescent dye [5]. To the best of our knowledge, few studies were used to observe LP of human gastric mucosa by MPM. In this paper, the microstructure of LP in normal human gastric mucosa is investigated by using MPM.

## 2. Materials and Methods

In our study, the specimen was provided by the Surgical Department of Fujian Provincial Tumor Hospital. The fresh and normal gastric mucosa specimen was excised from the normal part of gastrectomy for the gastric cancer patient. In order to obtain a complete transverse cross-section of the LP in each section (5 μm in thickness) of normal human gastric mucosa, the mucosa specimen was excided perpendicular to the gastric layer. Theses sections were sandwiched between the microscope slide and the cover glass. To avoid dehydration or shrinkage during the whole multiphoton imaging, a little phosphate-buffered saline solution (PBS solution, PH=7.4) was dipped into the section specimen. Written informed consent was obtained from the patient prior to gastrectomy and our study protocol was approved by the ethical committee of Fujian Provincial Tumor Hospital.

The experimental setup of MPM has been described previously [6-8]. The excitation light source is a mode-locked femtosecond Ti: sapphire laser (110fs, 76MHz) tuning from 700 to 980 nm (Coherent Mira 900-F, Coherent Inc., Santa Clara, CA, USA). It was equipped through a laser scanning microscope (Zeiss LSM 510 META) with a Plan-Apochromat 63 × (N.A. = 1.4) oil immersion objective (Zeiss), which was used to focus the excitation bean and collect the backward signals. The polarization direction of the laser light is the horizontal polarization. The laser intensity attenuation was implemented by an acousto-optic modulator (AOM, Carl Zeiss Inc., Thornwood, NY, USA). By using a galvanometer-driven optical scanner (Zeiss), the excitation laser beam was scanned. A main dichroic beam splitter (MDBS, Zeiss, HFT KP 650) was employed for reflecting the incident excitation laser beam to the specimen and directing the TPEF and SHG signals to the META detector, which was used to detect the backward signals. The META detector is composed of a high-quality refection grating as a dispersive element and an optimized 32-channel photomultiplier tube (PMT) array capable of collecting the dispersed output signal in 10.7 nm bandpass. All 32-channel PMTs cover a spectral width of approximately 340 nm across the visible spectrum (377-716 nm). In this work, the laser excitation wavelength is 800 nm. The META detector has eight detecting channels, and two independent channels (SHG and TPEF) from eight channels were used. One channel (387-409 nm) was used to detect SHG signal of collagen. Another channel (430-708 nm) was employed to detect TPEF signal of cells in gastric glands. Moreover, no photobleaching and photodamage were observed in a series of multiphoton imaging for LP within the normal human gastric mucosa. The images are 12-bit pixel depth.

## 3. Results and Discussion

Combining TPEF and SHG, high contrast and high resolution multiphoton images of LP in normal human gastric mucosa are obtained. Figures 1(a)-(c) display multiphoton images of LP in normal human gastric mucosa based on TPEF and SHG at excitation wavelength $\lambda_{ex} = 800$ nm. In Fig. 1(a), the cells of gastric glands in gastric mucosa can be visualized via TPEF (430-708 nm bandpass channel). The source of tissue TPEF signal is mainly the reduced pyridine nucleotides (NADH and NADPH; hereafter NAD(P)H) and oxidized flavoproteins (FPs) in the cells of gastric glands. As shown in Fig. 1(b), the connective tissue of LP can be imaged alone by SHG of collagen (387-409 nm bandpass channel). Overlaying two channels (SHG and TPEF), a higher contrast SHG/TPEF image of LP was obtained. As can be seen from the Fig. 1(c), the microstructure of LP in gastric mucosa can be readily observed. The gastric glands are separated by loose connective tissue bands. Furthermore, the glands are not quite straight and not quite parallel to one another.

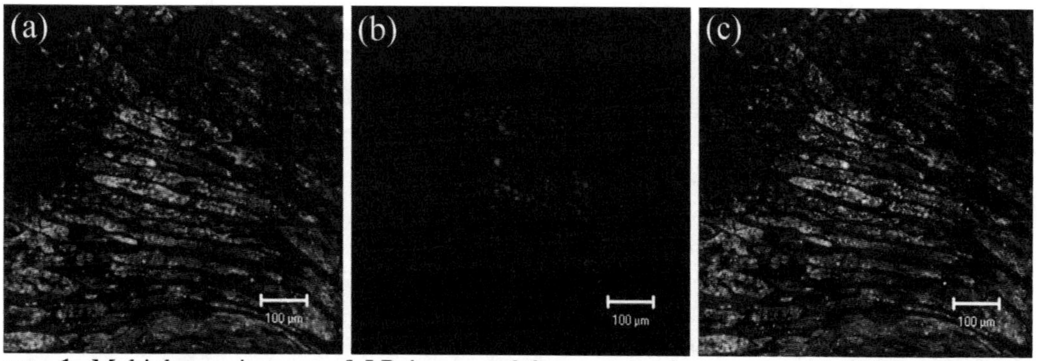

**Figure 1.** Multiphoton images of LP in normal human gastric mucosa. (a) TPEF image (green color-coded, 430-708nm). (b) Representative SHG image (red color-coded, 387-409nm). (c) SHG/TPEF image was obtained by overlaying two channels.

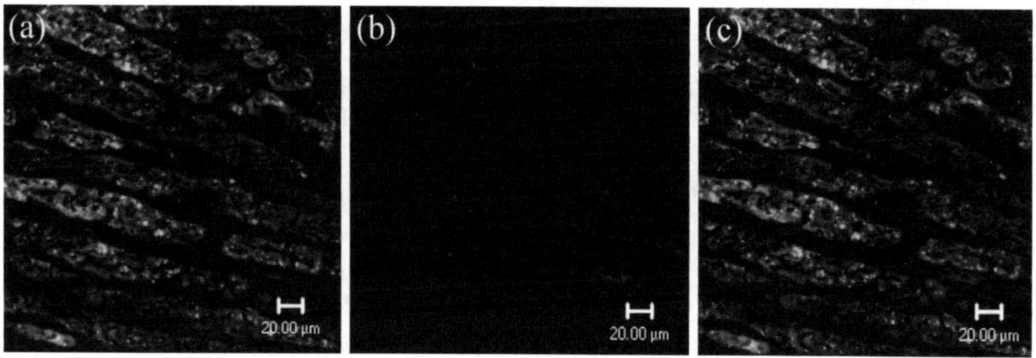

**Figure 2.** Magnified multiphoton microscopic images. (a) TPEF image (green color-coded, 430-708nm). (b) Representative SHG image (red color-coded, 387-409nm). (c) SHG/TPEF image was obtained by overlaying two channels.

To further characterize the LP in normal human gastric mucosa, the magnified multiphoton images were obtained, shown in Fig. 2. As can be seeen, the microstrucute of LP is more clearly observed. The green-fluorescing thick rope-like structures correspond to gastric glands in the gastric mucosa, whereas the red color signal orignates from collagen of loose connective in the LP.

## 4. Conclusion

In conclusion, MPM based on TPEF and SHG was used to investigate the microstructure of LP in normal human gastric mucosa and high contrast and resolution multiphoton images of LP were obtained at excitation wavelength $\lambda_{ex} = 800$ nm. The results showed MPM is very effective to examine LP of normal human gastric mucosa at high contrast and resolution, without the need for additional fluorescent probe. Moreover, the method will be further used to examine the variation of LP in human gastric mucosa. Our results will be helpful for deciding the clinical diagnosis and the clinicopathological stage of gastric cancer. With the advance of endomiroscopy, MPM can provide the accurate and comprehensive information for the clinical research of human gastric tissue.

## Acknowledgements

This project was supported by the National Natural Science Foundation of China (60908043 and 30970783), the Program for New Century Excellent Talents in University (NCET-07-0191), the Natural Science Funds for Distinguished Young Scholar in Fujian Province (2009J06031), Natural Science Foundation of Fujian Province (2010J01136).

## References


[1] Boyle P and Levin B 2008 *World Cancer Report 2008* (Geneva: WHO Press) p 42

[2] Arai K, Ota H, Hidaka E, Hayama M, Sano K, Sugiyama A, Akamatsu T and Katsuyama T 2004 Histochemical, ultrastructural, and three-dimensional observation of smooth muscle cells in human gastric mucosa *Histochem. Cell Biol.* **121** 229–37

[3] Zipfel W R, Williams R M and Webb W W 2003 Nonlinear magic: multiphoton microscopy in the biosciences *Nat. Biotechnol.* **21** 1369–77

[4] Williams R M, Zipfel W R and Webb W W 2001 Multiphoton microscopy in biological research *Curr. Opin. Chem. Biol.* **5** 603–8

[5] Rogart J N, Nagata J, Loeser C S, Roorda R D, Aslanian H, Robert M E, Zipfel W R and Nathanson M H 2008 Multiphoton imaging can be used for microscopic examination of intact human gastrointestinal mucosa ex vivo *Clin. Gastroenterol. Hepatol.* **6** 95–101

[6] Chen J X, Zhuo S M, Chen R, Jiang X S, Xie S S and Zou Q L 2007 Depth-resolved spectral imaging of rabbit oesophageal tissue based on two-photon excited fluorescence and second harmonic generation *New J. Phys.* **2** 212

[7] Zhuo S M, Chen J X, Luo T S and Zou D S 2006 Multimode nonlinear optical imaging of the dermis in ex vivo human skin based on the combination of multichannel mode and Lambda mode *Opt. Express* **14** 78107820

[8] Zhuo S M, Chen J X, Xie S S, Hong Z B and Jiang X S 2009 Extracting diagnostic stromal organization features based on intrinsic two-photon excited fluorescence and second-harmonic generation signals *J. Biomed. Opt. Lett.* **14** 020503

# Parallel scan hyperspectral fluorescence imaging system and biomedical application for microarrays

**Zhiyi Liu[1], Suihua Ma[1], Yanhong Ji[2], Le Liu[1], Jihua Guo[1], Yonghong He[1*]**

1 Laboratory of Optical Imaging and Sensing, Graduate School at Shenzhen, Tsinghua University, Shenzhen, 518055, P. R. China

2 MOE Key laboratory of Laser Life Science & Institute of Laser Life Science, South China Normal University, Guangzhou, 510631, P. R. China

* Corresponding author email: heyh@sz.tsinghua.edu.cn

**Abstract:** Microarray research offers great potential for analysis of gene expression profile and leads to greatly improved experimental throughput. A number of instruments have been reported for microarray detection, such as chemiluminescence, surface plasmon resonance, and fluorescence markers. Fluorescence imaging is popular for the readout of microarrays. In this paper we develop a quasi-confocal, multichannel parallel scan hyperspectral fluorescence imaging system for microarray research. Hyperspectral imaging records the entire emission spectrum for every voxel within the imaged area in contrast to recording only fluorescence intensities of filter-based scanners. Coupled with data analysis, the recorded spectral information allows for quantitative identification of the contributions of multiple, spectrally overlapping fluorescent dyes and elimination of unwanted artifacts. The mechanism of quasi-confocal imaging provides a high signal-to-noise ratio, and parallel scan makes this approach a high throughput technique for microarray analysis. This system is improved with a specifically designed spectrometer which can offer a spectral resolution of 0.2 nm, and operates with spatial resolutions ranging from 2 to 30 μm. Finally, the application of the system is demonstrated by reading out microarrays for identification of bacteria.

## 1. Introduction

The advent of DNA microarrays has revolutionized the biomedical research and development in genetic engineering [1-2]. Microarray research, which offers great potential for analysis of gene expression profile, and leads to greatly improved experimental throughput, has been an advanced technology for biomedical applications since the end of the $20^{th}$ century [3-6].

Although microarray technology is of growing importance in various fields of research, it is still young and in need of improvement in many aspects. The operation of the microarray scanners has generated considerable interest in the current study, and a number of methods have been developed for the analysis of microarrays, such as chemiluminescence [7], surface plasmon resonance [8], and fluorescence markers. Fluorescence imaging is popular for the readout of microarrays. Most fluorescence scanners detect the band-integrated emission corresponding to the band-pass filter utilized for each fluorophore [9-11]. In these scanners one filter corresponds to one kind of fluorophore. This approach is a direct imaging of sample plane, however, there are some drawbacks existing in this measurement approach. First, as different fluorescent tags are excited with different wavelengths, and the fluorescence signals are distinguished by band-pass filters, the chosen fluorescent tags should have no spectral overlap of their emission spectra and minimal overlap of their absorption spectra. Second, for these filter-based scanners, it is difficult to determine whether the measured data have been corrupted by some extraneous emission sources. For example, the spot-localized contamination due to spotting chemistry and the emission from substrates affect the determination of genes expression [12], especially for weakly expressed genes. These drawbacks aforementioned limit this approach in reliable and high-accuracy analysis of microarrays.

Hyperspectral imaging overcomes these limitations by recording the entire emission spectrum for every voxel within the imaged area in contrast to recording only fluorescence intensities of filter-based scanners. Hyperspectral approach could detect mixed emission from several fluorescent tags with overlapping emission spectra with only one filter. Multivariate data exploitation is then used for data analysis [13-14]. The wealth of information that is available from spectrally resolving an image makes hyperspectral imaging an extremely attractive technique, and hyperspectral instruments have been reported over the past years [15-18]. In most of these papers, hyperspectral fluorescence imaging systems are employed mainly for imaging biological samples, and some efforts have been devoted to microarray research. While, we attempt to develop a system suitable and optimized for microarrays readout.

We have previously developed a quasi-confocal, parallel scan fluorescence imaging system, which is a filter-based scanner for microarray analysis [11]. Different from laser confocal scanners which are point-focusing of the laser light, quasi-confocal method focuses the laser light to a line, and the whole region of interest is detected with parallel scan of the line light along the direction vertical to the line. Quasi-confocal method provides a high signal-to-noise ratio, and parallel scan offers a high throughput which is important for microarray analysis. Considering the advantages offered by quasi-confocal parallel scan architecture and the great potential of hyperspectral imaging, in this paper we improve the former system and develop a quasi-confocal, multichannel parallel scan hyperspectral system for analysis of microarrays. This system is improved with a specifically designed spectrometer, which can offer a high effective spectral resolution. Two lasers are presently employed in the system, and new laser sources could be easily added to this system for multichannel analysis. The spatial resolution and sensitivity of this system are discussed in detail. Finally, the application of this system is demonstrated by reading out microarrays for identification of bacteria.

## 2. Methods and materials

### *2.1. Constitution of the hyperspectral scanner*

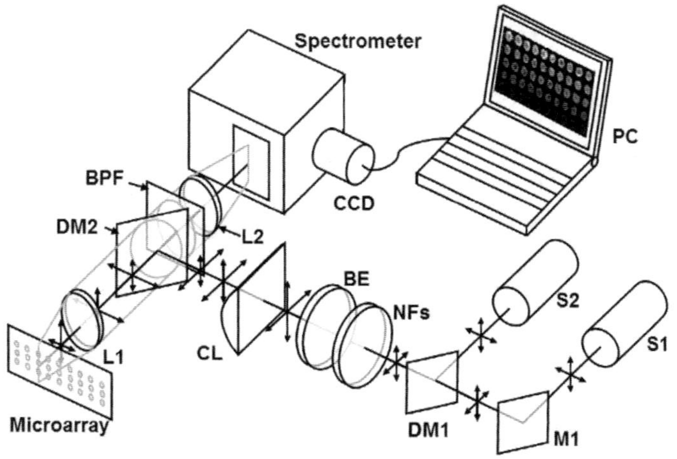


**Figure 1.** Layout of the system. S1: red laser; S2: blue laser; DM1: dichroic mirror 1; NFs: neutral-density filters; BE: beam expander; CL: cylindrical lens; DM2: dichroic mirror 2; L1: scan lens; BPF: band-pass filter; L2: imaging lens.

Figure 1 shows the optical layout of the system. A He-Ne laser with a wavelength of 632.8 nm and a power of 12.5 mW, a solid state laser with a wavelength of 473 nm and a power of 12.5 mW, are used as light sources. Two lasers are present here, however, new lasers could be easily added to the system to enlarge spectral ranges for multichannel detection. The laser fluence is adjusted to fit experimental conditions by use of a filter wheel populated with a set of neutral-density filters. The laser light is broadened by a beam expander and directed through a cylindrical lens (f = 30 mm) for the desired line illumination. Then a dichroic mirror which is characterized by high reflectivity at the laser wavelength and high transparency at longer wavelengths, is used to direct the laser light through the scan lens (f = 50 mm). For higher resolution and fluorescence collection efficiencies, apochromatic objectives with high numerical apertures (NA) are used.

The scan lens focuses the incident laser light to a line approximately 5 mm long on the surface of the microarray mounted on the scan stage, which is capable of 80 mm of travel in *x* direction and 60 mm of travel in *y* direction. Fluorescence excited by the focused line light is collected by the scan lens and directed through the dichroic mirror. A band-pass filter is used to further eliminate unwanted scattered and reflected laser light. An imaging lens (f = 80 mm) is then used to focus the fluorescence signals for a line image, which is relayed to the entrance slit of the spectrometer for the acquisition of the spectrally dispersed image. The spectrometer, which is able to offer an effective spectral resolution of 0.2 nm, is specifically designed for this system. A CCD array (Canon 500D) records these spectrally dispersed images for further data analysis. Then the whole plane is detected with parallel scan along the direction vertical to the focused line. The system achieves so-called "quasi-confocal" imaging because some light from adjacent voxels along the illuminated line could mix at the CCD array. While vertical to the illuminated line, light from adjacent voxels is efficiently rejected.

### 2.2. *Fabrication of microarrays*

Throughout this study, we prepare microarrays ourselves to test the characterization and the imaging capability of the system. For the fabrication of the DNA-arrays, the magnetron sputtering gold films are used as substrates. The fluorophores labeled DNA strands are thiolated for covalent bonding to substrates.

The gold film needs to be cleaned thoroughly before the printing of sample spots. Firstly, the gold film is cleaned with a solution consisting of $H_2O_2$ (30%), $NH_3$ (30%) and deionized water in a 1:1:5 ratio for 10 min [19]. Then, the film is dipped into acetone and ethanol for 10 min oscillating successively. After that, the film is thoroughly rinsed by deionized water and then dried and stored until further use.

The fluorescence labeled DNA is dissolved with ultra pure water to prepare solutions with different concentrations. When all above are ready, the gold film is put into a wet box, which provides a wet environment for the whole process of array printing. In our study, printing of the array is carried out manually. After that, the gold film, kept in the wet box, is put into the electric heating constant temperature incubator (Shanghai permanent science and technology the Limited company, DHP-9052) for 2h incubation at 37 °C. Then the DNA array is ready for optical scan.

Equation (1) shows the calculation of sample density ( $D_s$ ) for the printed spots:

$$D_s = V_p \times M_p \times N_A / A_o , \qquad (1)$$

where $N_A$ is the Avogadro's constant, $V_p$ is the extracted volume for one spot, $M_p$ is the concentration of fluorescent dye with the unit of μmol/L, and $A_o$ is the area of one spot on the sensing plane. Thus the unit of sample density is the number of fluorescence molecules per square micro ( $\text{fluors}/\mu\text{m}^2$ ). The spots with variant sample densities can be acquired by adjusting concentrations of fluorescent dye.

### 2.3. *Hyperspectral data analysis*

Analysis of hyperspectral data is crucial to the fluorescence image reconstruction and gene expression. Prior to analysis of the hyperspectral data, calibration data frames of the system are obtained by use of mercury argon calibration source (Ocean Optics). The acquired spectral data can be used to calibrate the wavelength scale and remove the slight influence of image curvature due to the spectrometer with the linear interpolation method.

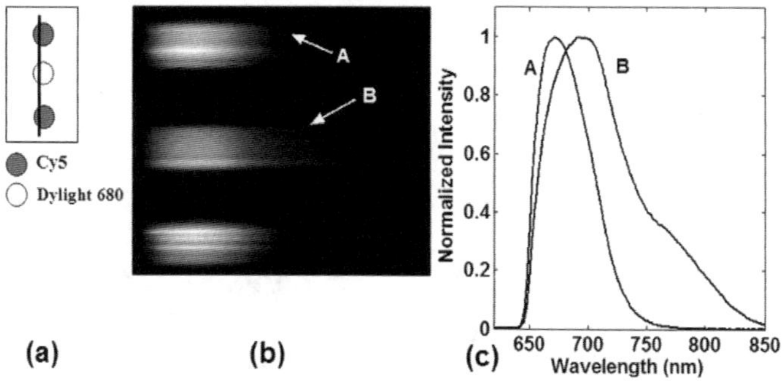


**Figure 2.** Hyperspectral data analysis. (a) Design of the dye labeled spots. (b) A single frame of data corresponding to the line in (a). (c) Intensity curves of rows marked in (b).

To illustrate the hyperspectral data analysis, a sensing model of manually printed sample spots is prepared, as shown in figure 2(a). Cy5 (absorption/ emission: 649/ 667 nm) and Dylight 680 (absorption/ emission: 680/ 715 nm) are employed as fluorescent tags, which are corresponding to red laser excitation. Figure 2(b) shows one frame of data corresponding to the vertical line in figure 2(a). Each row represents the spectrum of one certain point in the line region, thus the entire picture of figure 2(b) is the spectra of illuminated line region. Figure 2(c) shows the intensity curves of the rows marked in figure 2(b), which are normalized for comparison. As can be seen from figure 2(c), the normalized spectra of these two dyes are overlapped. For every voxel on the array plane, the intensity of each emission source is calculated by integration of corresponding spectrum. The whole image of the array is acquired by analysis of all the data frames obtained during optical scan.

## 3. System Characterization

### *3.1. Performance of the hyperspectral scanner*

The spatial resolution of this hyperspectral system is characterized using a U. S. Air Force standard resolution target (Edmund Optics) with broadband illumination. As mentioned above, this system is designed to enable simple changes of scan lens to obtain different resolutions for variant applications. Figure 3(a) (left) is an image of the resolution target obtained by an achromatic lens with the focus length of 50 mm. As can be seen from this image, element 1 of group 4 of the resolution target, corresponding to 16.0 line paris/mm (~31 μm), is resolved in both the horizontal and the vertical direction. For higher resolution, a 10×, NA 0.25 objective is also available, and the corresponding image is shown in figure 3(a) (right). In this case, element 4 of group 5, corresponding to 45.3 line paris/mm (~11 μm), can be resolved. The spot diameters and center-to-center spacings of commercial microarrays are of the order of ~100 μm, thus the 10~30 μm resolution is good and adequate for microarray analysis. A 40×, NA 0.65 objective is used for an even higher resolution of ~2 μm and higher fluorescence collecting efficiencies than 10× objective. Thus the resolution of this hyperspectral system ranges from 2 to 30 μm. However, as the length of the scan line, corresponding to the throughput within a single scan, decreases with the use of high magnification objectives, we should choose the spatial resolution according to the size of the array.

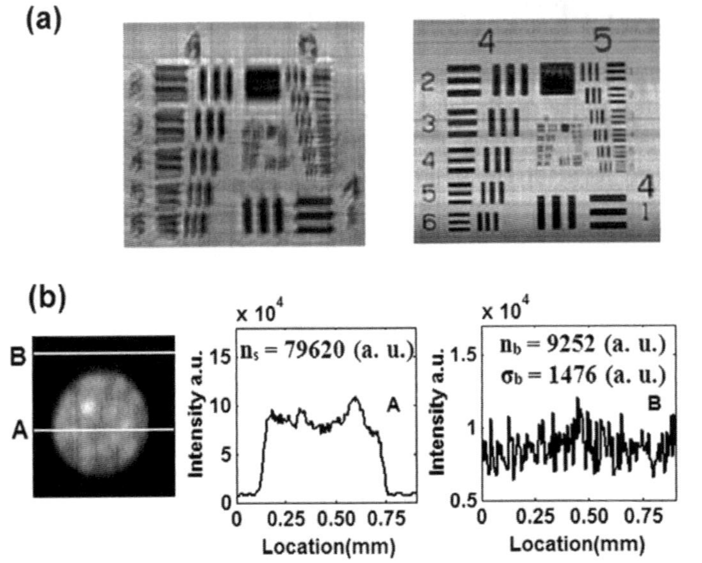


**Figure 3.** The spatial resolution and sensitivity of the system. (a) Scanning images of the standard resolution target. Left: 30- μm resolution; right: 10- μm resolution. (b) Left: the scanning image of a spot; middle and right: intensity graphs of the marked rows A and B, respectively. The values of $n_s$, $n_b$ and $\sigma_b$ are given in the graphs.

The sensitivity of fluorescence imaging is calculated as follows [20-21]:

$$S_d = D_s / SNR, \qquad (2)$$

$$SNR = (n_s - n_b) / \sigma_b, \qquad (3)$$

where $S_d$ is the detection sensitivity, $D_s$ is the sample density and $SNR$ is the signal-to-noise ratio. Eq. (3) shows the calculation of $SNR$ [22], where $n_s$ is the average intensity obtained from the spot of interest, $n_b$ is the average intensity of background, and $\sigma_b$ is the standard deviation of background. We use a spot with the sample density of 30 fluors/$\mu m^2$ for illustration, as shown in figure 3(b) (left). To demonstrate the signal intensity and background noise level, intensity graphs of the marked rows A and B are plotted to the right of the spot image, and the values of $n_s$, $n_b$ and $\sigma_b$ are given in the graphs. Calculated by Eq. (3), an $SNR$ of approximately 48 is derived, thus the detection sensitivity is 0.63 $\text{fluors}/\mu m^2$ according to Eq. (2). The sensitivity can be further improved by pixel binning method [11]. The sensitivity achieved from this hyperspectral fluorescence imaging system is comparable to that of commercial microarray scanners, thus it is adequate for analysis of most microarrays.

### *3.2. multi-channel fluorescence imaging*

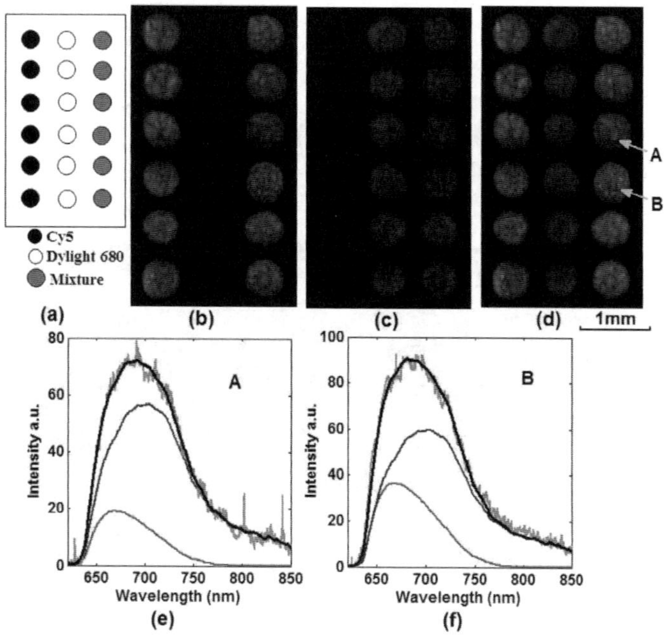


**Figure 4.** Hyperspectral imaging of spectrally overlapping fluorescent dyes. (a) Design of an array. (b) Image of Cy5. (c) Image of Dylight 680. (d) Image of the whole sample plane. (e) and (f) Multivariate data analysis for the determination of the ratios of each dye. Red curve: the weighted spectrum for Cy5; blue curve: the weighted spectrum for Dylight 680; brown curve: the raw spectrum for a certain point of the mixture spots; Black curve: the calculated spectrum as a sum of these two weighted spectra.

To investigate the multichannel, hyperspectral imaging performance of the system, we design a sensing model for demonstration, as shown in figure 4(a). For all the spots in the first column, only Cy5 labeled DNA is used, and only Dylight 680 labeled DNA is applied in the second column. These two cases correspond to the multicolor labeling strategy which is generally utilized in gene expression research [23]. In order to mimic heterozygotes [24] and eliminate unwanted artifacts, such as emission from substrates and spot-localized contamination, a mixture of these two dyes is applied in the third column. Figure 4(b) and (c) show the fluorescence images of Cy5 and Dylight 680, respectively. The whole image of the sample plane is acquired by combining the concentration maps of these two dyes, as shown in figure 4(d). The raw emission spectra (brown color) for two points A and B in figure 4(d) are shown in figure 4(e) and (f). Although the same Cy5-Dylight 680 ratio is applied for spots in the third column, the spectra of the points within one spot vary from each other due to the inhomogeneities of distributions of fluorescence molecules. With multivariate data analysis, the weight of each dye is determined. In figure 4(e) and (f), the red and blue curves correspond to the weighted spectra of Cy5 and Dylight 680, respectively, and the black curve is a sum of these two weighted spectra. As can be seen from figure 4(e) and (f), the sum curves (black) correspond well to the raw spectra (brown) of the two randomly chosen points, indicating that the entire hyperspectral image stack could be modeled using linear admixtures of these two dyes. For presentation, only two dyes are applied here corresponding to a single laser, the scanning image demonstrates the ability of this hyperspectral scanner to identify multiple fluorescent labels with spectral analysis.

## 4. Biomedical Application of Microarrays

The identification of bacteria using gene chips is performed by this quasi-confocal, parallel scan hyperspectral fluorescence imaging system as an example of biomedical application of microarrays.

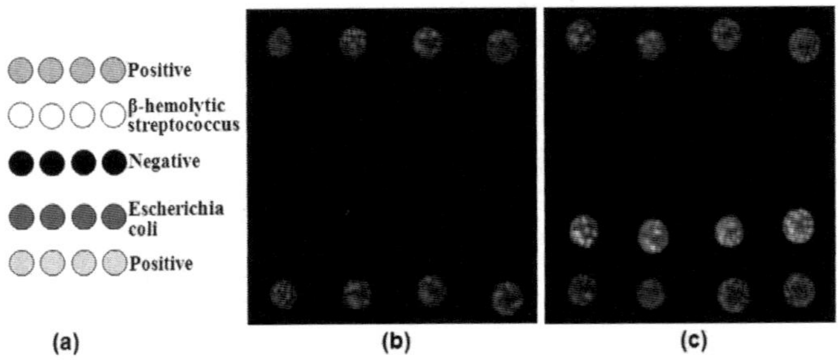


**Figure 5.** Identification of bacteria as an example of biomedical application of microarrays. (a) Design of the microarray. Scanning results with (b) no microbe; (c) only Escherichia coli in the sample.

A gold film covered quartz glass is used as the substrate, which is carefully cleaned before the printing of probes. Then two $5\times4$ probe arrays are printed manually on the substrates. All the probes are thiolated for covalent bonding to gold films. As shown in figure 5(a), the array comprises one row of Escherichia coli probes, one row of β-hemolytic streptococcus probes, one row of negative probes as quality report marks of the chip hybridization, and two rows of positive probes. For the validation of results on the chip, the positive probes should show a high fluorescent intensity signal, and the negative probes should show a low (or zero) signal.

Throughout this study, Cy5 is used for positive and β-hemolytic streptococcus DNA fragments, and Dylight 680 is for Escherichia coli. Figure 5(b) shows the detection results with no microbe present in the sample. As expected, the positive probes report a high fluorescent intensity signal, while the negative probes show a blank signal. Figure 5(c) shows the detection results with only Escherichia coli present in a sample. Besides positive probes, only the row of Escherichia coli probes shows a high signal, corresponding well with expectation. The experimental results indicate that the target genes are selectively hybridized to the probes. This hyperspectral system generates high quality images and can be successfully applied in biomedical research.

## 5. Conclusion

In this paper, we have proposed a quasi-confocal, multichannel parallel scan hyperspectral fluorescence imaging system. This system employs a pushbroom scan architecture with the line focusing of the excitation laser and records the spectrum for every voxel of the excited line region, and the whole sample plane is detected with parallel scan vertical to the scanning line. Coupled with multivariate data analysis, this system is capable of quantitative analysis of the concentrations of all the emitting species for any pixel on the sample plane. The spatial resolution of this system ranges from 2 to 30 $\mu m$. Although the quasi-confocal mechanism limits the system in extreme high-resolution microscopic imaging, the spatial resolution of this system is particularly suitable for imaging of microarrays. The sensitivity of this system is $0.63\ \text{fluors}/\mu m^2$, which is adequate and good for microarray analysis. The application of this system is demonstrated by reading multicolor microarrays for identification of bacteria. We believe that the proposed hyperspectral system can be used to build a commercially valuable device for microarrays analysis application.

**Acknowledgements**

This research was made possible with the financial support from the 863 project, China (grant 2006AA06Z402), NSFC China (grants 30770592 and 60608019).

**References:**

[1] Fodor S P, Read J L, Pirrung M C, Stryer L, Lu A T and Solas D 1991 *Science* **251** 767-73

[2] Fodor S P, Rava R P, Huang X C, Pease A C, Holmes C P and Adams C L 1993 *Nature* **364** 555-6

[3] Duggan D J, Bittner M, Chen Y, Meltzer P and Trent J M 1999 *Nat. Genet.* **21**(Suppl.) 10-4

[4] Brown P O and Botstein D 1999 *Nat. Genet.* **21** 33–7

[5] MacBeath G and Schreiber S L 2000 *Science* **189** 1760–3

[6] Huang Y, Joo S, Duhon M, Heller M, Wallace B and Xu X 2002 *Anal. Chem.* **74** 3362–71

[7] Wolter A, Niessner R and Seidel M 2008 *Anal. Chem.* **80** 5854-63

[8] Homola J 2008 *Chem. Rev.* **108** 462-93

[9] Ruckstuhl T, Walser A, Verdes D and Seeger S 2005 *Biosens. Bioelectron.* **20** 1872-7

[10] Huang G L, Deng C, Zhu J, Xu S K, Han C, Song X B and Yang X Y 2008 *J. Biomed. Opt.* **13** 034006

[11] Liu Z, He Y, Liu L, Ma S, Chong X, Hu Z, Ma H and Guo J 2010 *Opt. Lasers Eng.* **48** 849-55

[12] Martinez M J, Aragon A D, Rodriguez A L, Weber J M, Timlin J A, Sinclair M B, Haaland D M and Werner-Washburne M 2003 *Nucl. Acids Res.* **31** e18

[13] Haaland D M, Sinclair M B, Timlin J A, Van Benthem M H, Martinez M J, Aragon A D and Werner-Washburne M 2003 *Proc. SPIE* **4959** 55–67

[14] Bro R and DeJong S 1997 *J. Chemometr.* **11** 393–401

[15] Sinclair M B, Haaland D M, Timlin J A and Jones H D T 2006 *Appl. Opt.* **45** 6283-91

[16] Tsurui H, Nishimura H, Hattori S, Hirose S, Okumura K and Shirai T 2000 *J. Histochem. Cytochem.* **48** 653–62

[17] Sinclair M B, Timlin J A, Haaland D M and Werner-Washburne M 2004 *Appl. Opt.* **43** 2079–88

[18] Stimson M J, Haralampus-Grynaviski N and Simon J D 1999 *Rev. Sci. Instrum.* **70** 3351-4

[19] Wang R H, Sara T, Maria M, Maria M S and Marco M 2004 *Biosens. Bioelectron.* **20** 967–74

[20] Nicolini C, Malvezzi A M, Tomaselli A, Sposito D, Tropiano G and Borgogno E 2002 *IEEE. Trans. Nanobiosci.* **1** 67–72

[21] Nagarajan R and Peterson C A 2002 *IEEE. Trans. Nanobiosci.* **1** 78–84

[22] Schena M 2000 *Microarray Biochip Technology* (Natick, MA: Eaton)

[23] Bender K 2005 *Methods Mol. Biol.* **297** 243–52

[24] Lovmar L, Ahlford A, Jonsson M and Syvanen A C 2005 *BMC Genomics* **6** 35

# Microstructure imaging of human rectal mucosa using multiphoton microscopy

**N R Liu[1★], G Chen[2★], J X Chen[1], J Yan[3], S M Zhuo[1], L Q Zheng[1] and X S Jiang[1]**
1. Institute of Laser and Optoelectronics Technology, Fujian Provincial Key Laboratory for Photonics Technology, Key Laboratory of OptoElectronic Science and Technology for Medicine of Ministry of Education, Fujian Normal University, Fuzhou, China
2. Department of Pathology, Fujian Provincial Tumor Hospital, Fuzhou, China
3. Department of Surgery, Fujian Provincial Tumor Hospital, Fuzhou, China

★ Equally contributed to this work
Send correspondence to: Prof. J X Chen or Dr. J Yan
Email: chenjianxin@fjnu.edu.cn (J X Chen) or ynjun@yahoo.com (J Yan)

**Abstract.** Multiphoton microscopy (MPM) has high resolution and sensitivity. In this study, MPM was used to image microstructure of human rectal mucosa. The morphology and distribution of the main components in mucosa layer, absorptive cells and goblet cells in the epithelium, abundant intestinal glands in the lamina propria and smooth muscle fibers in the muscularis mucosa were clearly monitored. The variations of these components were tightly relevant to the pathology in gastrointestine system, especially early rectal cancer. The obtained images will be helpful for the diagnosis of early colorectal cancer.

## 1. Introduction

It is well known that the rectal carcinoma has the high mortality and is the second most common gastrointestinal malignancy. With the western patterns of diet and physical inactivity, the incidence rate of colorectal cancer is increasing. The greatest increase in the incidence of colorectal cancer is in Asia[1]. Patients with stage I colorectal cancer have a 5-year survival of ≥90%, whereas patients with stage Ⅱ, Ⅲ, Ⅳ colorectal cancer have a 5-year survival of 82.5%, 59.5%, 8.1% respectively[2]. So, reducing the mortality of colorectal cancer relies on early detection. Unfortunately, the routine endoscopy is difficult to identify the early colorectal cancers, especially flat adenomas and carcinomas in situ, mainly due to the absence of typical morphological structures. Therefore, much more progress has been made towards developing new optical image-systems that improve the predictability of dysplastic and malignant colorectal lesions, such as photodynamic diagnosis[3-5], fluorescence imaging systems[6-7], and laser-induced spectroscopy[8-9] .

Multiphoton microscopy (MPM) based on two-photon excited fluorescence and second harmonic generation has the ability to obtain the images of tissues at the cellular and subcellular level[10-12]. It has some advantages in imaging unstained tissue samples and has been used to evaluate and monitor morphological structure and functional state of epithelial tissues[10-11,13]. MPM has the potential to become a powerful tool for noninvasively early diagnosis of cancers in colorectal system[14]. The variations of mucosa and submucosa layers are important to early diagnoses of rectal cancer. So, in this study, the microstructures of human rectal mucosa are investigated using multiphoton microcopy. The obtained images will be helpful for the diagnosis of early colorectal cancers.

## 2. Materials and methods

In this study, the *ex-vivo* human rectum tissues were examined. The fresh specimens were provided by Fujian Provincial Tumor Hospital, originated from five patients aged 40-60 years old. Every patient had his consent to participate in the study. The samples were cut into 5 $\mu m$ thickness sections by the freezing microtome and stored in a low temperature refrigerator (-86℃) before they were used. Each section was surely comprised of a complete transverse cross-section of the mucosa layers and sandwiched between the microscope slide and cover glass. To avoid dehydration or shrinkage during the whole imaging process, the specimen was sprinkled with PBS solution (PH 7.4). In this study, we are chiefly interested in the rectal mucosa layer.

The multiphoton microscopic imaging system used in this study has been described in previous publications[9,15]. It was mainly an Axiovert 200 microscope (Zeiss LSM 510 META) equipped with a mode-locked femtosecond Ti: sapphire laser (Coherent Mira 900-F), tunable from 700 nm to 980 nm (110 fs, 76 MHz). A Plan-Apochromat 63× (NA=1.4) oil immersion objective (Zeiss) was employed for focusing the excitation beam and for collecting the backward signals. The multichannel mode can achieve TPEF and SHG images, respectively, which separated by the dichroic mirror in the detection path. In this study, the excitation wavelength of 800 nm was used. So, one channel was corresponding to the wavelength range of 387 - 409 nm to show microstructure of tissue component from SHG signals (red color-coded), whereas another channel covered the wavelength range from 430 to 708 nm in order to present the morphology of tissue component from TPEF signals (green color-coded ). An optional HRZ 200 fine-focusing stage (HRZ 200 stage, Carl Zeiss) was used to obtain a large-area image of rectum tissue and record the focus position.

## 3. Results and discussions

Figure 1 shows a large-area TPEF/SHG image (877.5 ×585.0$\mu m$) of human rectal mucosa. As can be seen from figure 1a, a little collagen ( ↓ white arrow) with mesh morphology of lamina propria in mucosa layer can be imaged alone by one channel (red color-coded). The thin red band surrounds individual glands (showed in figure 1c), which likely represents the basement membrane and portions of myofibroblastic sheath[14]. Similarly, in figure 1b the absorptive cells, goblet cells, and a large number of intestinal glands can be isolated from collagen using another channel (green color-coded).

The content and distribution of absorptive cells, goblet cells, intestinal glands and some scattered smooth muscle bundles ( ↓ white arrow) between the intestine glands, smooth muscle fibres in the muscularis mucosae can be clearly visible in figure 1b. Overlaying two channels yields a high-contrast TPEF/SHG image of mucosa, as shown in figure 1c, indicates that a little collagen located between two intestinal glands. Imaging results of normal rectal mucosa are in excellent agreement with histology as showed in reference 16.

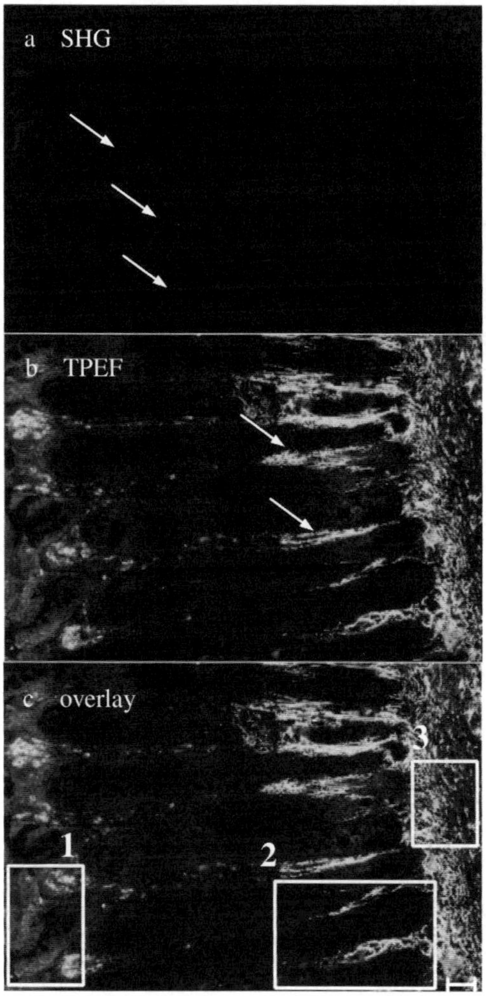


**Figure 1.** A large area SHG/TPEF image of human rectum mucosa: (a) SHG image of collagen (↓white arrow); (b) TPEF image of epithelial cells in the epithelium, absorptive cells, goblet cells, crypts of Lieberkuhn in the lamina propria, and some scattered smooth muscle bundles (↓white arrow) between the intestine glands, smooth muscle fibers in the muscularis mucosae; (c) overlay of the SHG/TPEF image. Scale bar is 50 $\mu m$.

In order to clearly present the microstructure of mucosa layer, figure 2 shows the epithelium image of the mucosa, white box 1 in figure 1c. The imaging field of view is 140.0 × 213.1 $\mu m$. As can be seen, the absorptive cells ( ↓ white arrow ①) and goblet cells ( ↓ white arrow ②) in figure 2b can be distinctly identified and show the fluorescent cytoplasmatic granules of mitochondria. The non-fluorescent nuclei displayed darkly on the optical image and surrounded by the fluorescent mitochondria, and the clear cellular boundary.

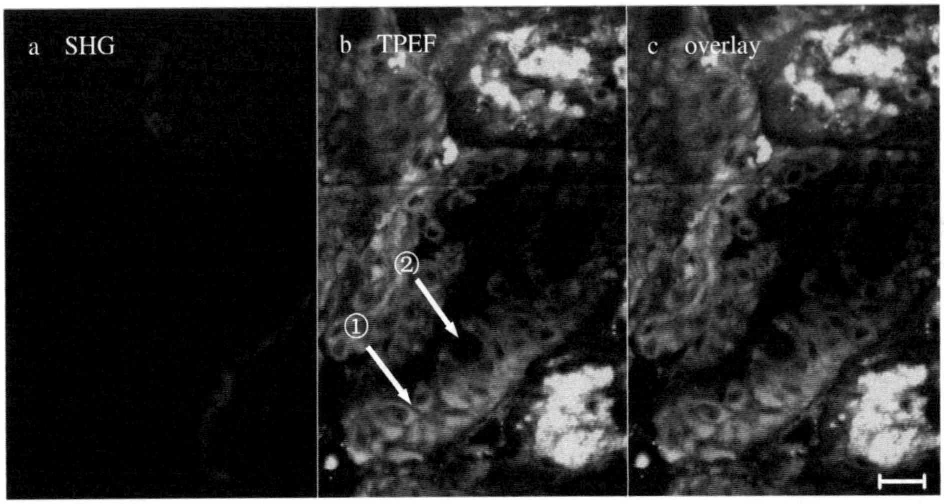


**Figure 2.** The microstructure of the mucous epithelium in figure 1c (white box 1), showing SHG image of collagen in figure 2a, the absorptive cells (↓white arrow ①) and goblet cells (↓white arrow ②) in figure 2b,. Scale bar is 20 μm

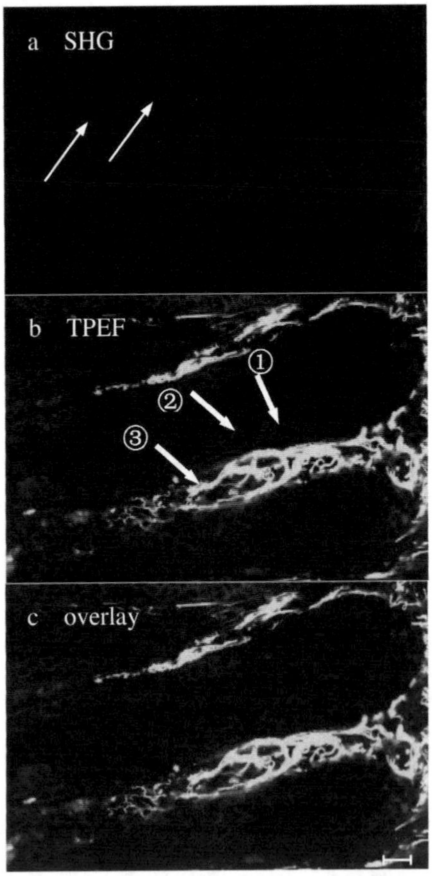


**Figure 3.** SHG/TPEF image of the lamina propria of human rectum, regions of interest (ROI) in figure 1c (white box 2): (a) SHG image of collagen (↓white arrow); (b) TPEF images of absorptive cells (↓white arrow ①), goblet cells (↓white arrow ②), scattered smooth muscle bundles (↓white arrow ③) in figure 3b; (c) SHG/TPEF image was obtained by overlaying two channels. Scale bar is 20 μm

Displayed in Figurefigure 3 are the TPEF/SHG images of the lamina propria of human rectal mucosa. Imaging field of view is 296.5 × 192.5 *μm*. Figure 3a shows a little collagen ( ↓ white arrow) with mesh morphology of lamina propria in mucosa layer surrounding the intestine glands can be imaged alone by one channel (red color-coded) just like in figure 1a. Similarly, figure 3b reveals the two-photon excited fluorescence images (green color-coded) of absorptive cells ( ↓ white arrow ①), goblet cells ( ↓ white arrow ②) in the intestine glands, scattered smooth muscle bundles ( ↓ white arrow ③) between the intestine glands. The high-contrast TPEF/SHG image is shown in figure 3c.

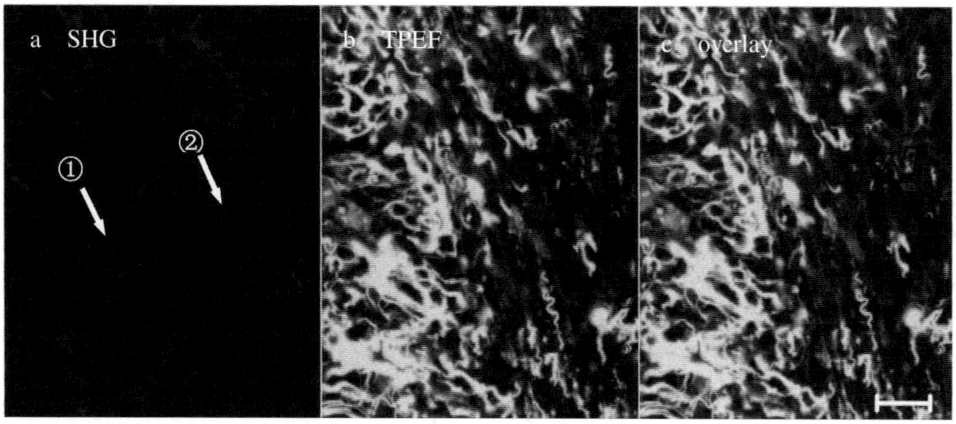


**Figure 4.** The microstructure images of the muscularis mucosae of human rectum, regions of interest (ROI) in figure 1c (white box 3), showing the more detailed structural information of smooth muscle: (a) SHG image of collagen, inner circular (↓white arrow①) and outer longitudinal (↓white arrow ②); (b) TPEF images of smooth muscle fibres; (c) SHG/TPEF image was obtained by overlaying two channels. Scale bar is 20 μm.

The morphology of smooth muscle fibres in the muscularis mucosae is also displayed in figure 4 (imaging field of view was 119.4 × 151.4 *μm*). SHG signals (red color-coded) of fibers are of inner circular layer ( ↓ white arrow①) and outer longitudinal layer ( ↓ white arrow ②), as shown in the figure 4a. Similarly, figure 4b also presents the two-photon excited fluorescence images (green color-coded) of an inner circular and outer longitudinal layer of smooth muscle in the corresponding location. The high-contrast TPEF/SHG image is shown in figure 4c.

The morphology and function of the intestinal glands and cells in rectum mucosa may alter with disease progression of early rectum cancers[17]. The thickness of mucosa[18] and the morphology of smooth muscle fibres in the muscularis mucosae also changed when cancer occurs. So, the microscopic imaging of rectal mucosa layer may further develop a new method for histological evaluation in the early diagnosis of rectum diseases.

## 4. Conclusions

In conclusion, we demonstrate that multiphoton microscopic imaging technique has the potential to present the microstructures of human rectal mucosa. And the imaging results of normal rectal mucosa are in excellent agreement with histology. These imply that the multiphoton microscopic imaging technique has the potential to non-invasively diagnose early rectal cancer.

**5. Acknowledgments**

The work was supported by the National Natural Science Foundation of China (60908043 and 30970783), the Program for New Century Excellent Talents in University (NCET-07-0191), the Natural Science Funds for Distinguished Young Scholar in Fujian Province (2009J06031), the Science and Technology Program of Education Department in Fujian Province (JB09073), Natural Science Foundation of Fujian Province (2010J01136).

**References**

[1] Garcia M, Jemal A, Ward E M, Center M M, Hao Y, Siegel R L and Thun M J 2007 *Global Cancer Facts & Figures 2007* (Atlanta, GA: American Cancer Society).

[2] O'Connell J B, Maggard M A and Clifford Y. Ko 2004 Colon Cancer Survival Rates With the New American Joint Committee on Cancer Sixth Edition Staging *Journal of the National Cancer Institute* **96** 1420-25.

[3] Endlicher E and Messmann H 2004 Photodynamic diagnosis in the gastrointestinal tract *Gastrointestinal endoscopy clinics of North America* **14** 475-85.

[4] Mayinger B, Reh H, Hochberger J and Hahn E G 1999 Endoscopic photodynamic diagnosis: oral aminolevulinic acid is a marker of GI cancer and dysplastic lesions *Gastrointestinal endoscopy* **50** 242-6.

[5] Brown E, McKee T, diTomaso E, Pluen A, Seed B, Boucher Y and Jain R K 2003 Dynamic imaging of collagen and its modulation in tumors in vivo using second-harmonic generation *Nat Med* **9** 796 - 800.

[6] Brand S, Stepp H, Ochsenkühn T, Baumgartner R, Baretton G, Holl J, Von Ritter C, Baumgartner G and Sackmann M 1999 Detection of colonic dysplasia by light-induced fluorescence endoscopy: a pilot study *International journal of colorectal disease* **14** 63-8.

[7] Polglase A L, McLaren W J, Skinner S A, Kiesslich R, Neurath M F and Delaney P M 2005 A fluorescence confocal endomicroscope for in vivo microscopy of the upper-and the lower-GI tract *Gastrointestinal endoscopy* **62** 686-95.

[8] Mayinger B, Jordan M, Horner P, Gerlach C, Muehldorfer S, Bittorf B R, Matzel K E, Hohenberger W, Hahn E G and Guenther K 2003 Endoscopic light-induced autofluorescence spectroscopy for the diagnosis of colorectal cancer and adenoma *Journal of Photochemistry and Photobiology B: Biology* **70** 13-20.

[9] Chen J X, Zhuo S M, Luo T S, Jiang X S, Zhao J J 2006 Spectral characteristics of autofluorescence and second harmonic generation from ex vivo human skin induced by femtosecond laser and visible lasers *Scanning* **28** 319–26.

[10] Chen J X, Zhuo S M, Chen R, Jiang X S, Xie S S and Zou Q L 2007 Depth-resolved spectral imaging of rabbit oesophageal tissue based on two-photon excited fluorescence and second harmonic generation *New Journal of Physics* **9** 212-23.

[11] Zhuo S M, Chen J X, Jiang X S, Xie S S, Chen R, Cao N, Zou Q L and Xiong S Y 2007 The layered-resolved microstructure and spectroscopy of mouse oral mucosa using multiphoton microscopy *Phys Med Biol* **52** 4967–80.

[12] Zipfel W R, Williams R M and Webb W W 2003 Nonlinear magic: multiphoton microscopy in the biosciences *Nat Biotechnol* **21** 1369–77.

[13] Skala M C, Squirrell J M, Vrotsos K M, Eickhoff J C, *Gendron- Fitzpatrick A, Eliceiri* K W and Ramanujam N *2005 Multiphoton* microscopy of endogenous fluorescence differentiates normal, pre*cancerous, and* cancerous squamous epithelial tissues *Cancer Res* **65** 1180–6.

[14] Rogart J N, Nagata J, Loeser C S, Roorda R D, Aslanian H, Robert M E, Zipfel W R and Nathanson M H 2008 Multiphoton Imaging Can Be Used for Microscopic *Examination* of Intact Human Gastrointestinal Mucosa Ex Vivo *Clinical Gastroenterology and Hepatology* **6** 95-101.

[15] Zhuo S M, Chen J X, Luo T S, Zou D S and Zhao J J 2006 Multimode nonlinear optical imaging of the dermis in ex vivo human skin based on the combination of multichannel mode and lambda mode *Opt Express* **14** 7810–20.

[16] http://www.ipathology.cn/news_display.asp?id=4259

[17] Wang T D, Friedland S, Sahbaie P, Soetikno R, Hsiung P L, Liu J T, Crawford J M and Contag C H 2007 Functional imaging of colonic mucosa with a fibered confocal microscope for real-time in vivo pathology *Clinical Gastroenterology and Hepatology* **5** 1300-5.

[18] Huang Z W, Zheng W, Xie S S, Chen R, Zeng H S, David I. M ans Harvey L 2004 Laser-induced autofluorescence microscopy of normal and tumor human colonic tissue *International journal of oncology* **24** 59-64.

# Chloride ion-dependent surface-enhanced Raman scattering study of biotin on the silver surface

**Fangfang Liu[1], Huaimin Gu[1], Xiaojuan Yuan[1], Yue Lin[2] and Xiao Dong[1]**

[1] MOE Key laboratory of Laser Life Science & Institute of Laser Life Science, College of Biophotonics, South China Normal University, Guangzhou, China
[2] School of Education Science, South China Normal University, Guangzhou, China

E-mail: guhm@scnu.edu.cn

**Abstract.** In the present paper, the surface enhanced Raman scattering (SERS) technique was employed to study the SERS spectra of biotin molecules formed on the silver surface. The adsorption geometries of biotin molecules on the silver surface were analyzed based on the SERS data. It can be found that most vibration modes show a Raman shift in silver sol after the addition of sodium chloride solution. In addition, The Raman signals of biotin become weaker and weaker with the increase of the concentration of sodium chloride. This may be due to that the interaction between chloride ions and silver particles is stronger than the interaction between biotin molecules and silver particles. When the concentration of sodium chloride in silver colloid is higher than 0.05mol/L, superfluous chloride ions may form an absorption layer so that biotin can not be adsorbed on silver surface directly. The changes in intensity and profile shape in the SERS spectra suggest different adsorption behavior and surface-coverage of biotin on silver surface. The SERS spectra of biotin suggest that the contribution of the charge transfer mechanism to SERS may be dominant.

## 1. Introduction

Surface enhanced Raman scattering (SERS) effect was first observed by Fleischmann et al in pyridine adsorbed on the roughened silver electrode in 1974 [1]. SERS is a powerful and convenient analytical tool in structural analysis of materials, which can greatly enhance the weak normal Raman signals of molecules and materials.

Generally, there are two theories on SERS accepted widely. One is electromagnetic enhancement, and the other one is charge transfer enhancement [2, 3]. The first mechanism is based on the strong enhancement of the electromagnetic field on the metal surface, as a consequence of the plasmon excitation induced by the incident radiation on a roughed surface. The second consists of a resonant Raman effect caused by a charge-transfer process from the metal to the adsorbate [4]. In most of the SERS systems, the electromagnetic and charge transfer enhancement happen simultaneously, and the ratio of their enhancement varies in different SERS systems.

In the course of SERS study, many works have focused on the influence of halides, mainly $Cl^-$, in the adsorption and SERS effects [5, 6]. A number of works have reported that $Cl^-$ anion is able to create a high degree of roughness in order to stabilize the surface active sites, attributed to the chemical mechanism of SERS enhancement [7, 8], as a large charge-transfer between the metal surface and the adsorbate may occur.

Biotin, which is also known as Vitamin H, is a water-soluble vitamin and generally classified as a B-complex vitamin. Biotin has always intrigued biochemists for it participates as a cofactor in gluconeogenesis, fatty acid synthesis and branched chain amino acid catabolism [9]. In addition, biotin regulates gene expression [10] and is involved in some systemic processes such as development [11] and metabolism [10]. It was reported that biotin deficiency or biotin excess affect reproductive functions [12].

In this study, the SERS spectrum of biotin in silver colloid was recorded under the treatment of low laser power and the adsorption geometries were analyzed based on the SERS data. The relation between adsorption and SERS of biotin on borohydride-reduced Ag colloids upon the addition of $Cl^-$ was investigated. Through analysis of the SERS spectra, it is evident that the surface-coverage and adsorption orientation of biotin may be easily influenced by the added $Cl^-$.

## 2. Experimental

### *2.1. Chemicals*

The biotin in the solid form was purchased from Sigma-Aldrich Chemical Company (USA) with a stated purity of greater than 99.5% and it was used without any further purification. Silver nitrate and sodium borohydride were used as received. All aqueous solutions were prepared using doubly distilled water.

The collidal silver nanoparticles were prepared by the chemical reduction of $AgNO_3$ with $NaBH_4$ according to the standard procedure [13]. Briefly, 8.5 mg $AgNO_3$ powder dissolved in 50 mL doubly distilled water were added dropwise to 150 mL of ice-cold aqueous solution of $2\times10^{-3}$ mol/L $NaBH_4$ with vigorous stirring. The resulting yellowish colloidal solution was found to be stable over several days at room temperature and has a surface Plasmon absorption maximum measured at 392 nm (data not shown).

Samples for SERS measurement were prepared by adding 2 mL saturated biotin solutions to 1.5 mL silver sol. In the final step, 0.05 mL water or aqueous NaCl solutions of different concentrations were added in order to characterize the influence of $Cl^-$. The final concentration of biotin in all mixed systems was $5\times10^{-5}$ mol/L. After the addition of different concentrations of aqueous NaCl, the final concentrations of $Cl^-$ in the mixed systems were 0.01, 0.04 and 0.05 mol/L, respectively.

The final solution was sealed in a capillary with one hour incubation for the SERS measurement. In the measurement of SERS on silver surface, a droplet (0.02 mL) of incubated solution was evenly dropped on a clean glass slide at room temperature.

*2.2. Instrumentation*

Raman spectra were acquired using a triple grating Raman micro-spectrometer (Acton Spectro @2300i, Princeton Acton, USA) equipped with a Ar ion laser ($\lambda$ = 514.5 nm) and a liquid nitrogen cooled coupled charge device(CCD) detector. The spectral resolution was set at 2 $cm^{-1}$ and the laser power focused on the sample was about 5 mW.

## 3. Results and Discussion

*3.1. Molecular geometry*

Figure 1 shows the chemical structure of biotin. The molecule contains a bicyclic ring system whose mirror symmetry is broken by a valeric acid side chain. The top ring of the bicyclic system is an ureido ring, and the bottom ring containing sulfur atom, is a tetrahydrothiophene ring. Biotin further possesses three contiguous stereocenters on the tetrahydrothiophene ring in the all-*cis* configuration [14].

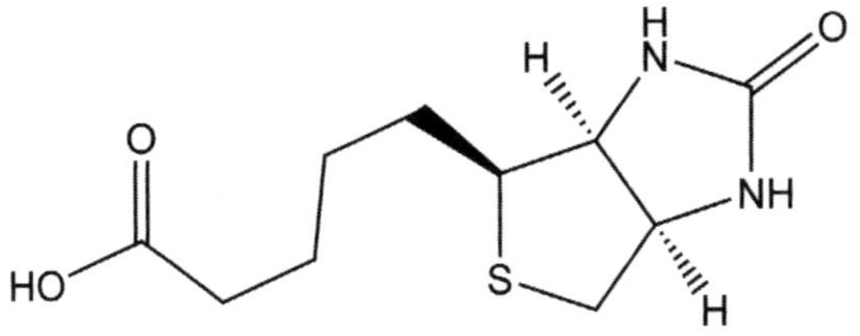


**Figure 1.** Chemical structure of biotin.

*3.2. SERS of biotin in silver sol*

Figure 2 displays the Normal Raman spectroscopy of biotin in aqueous saturated solution, in which the Raman bands are sparse and the intensity is low, especially in the higher wavenumber range. Figure 3 shows the SERS of biotin in silver colloid. The Raman bands assignments of the NRS and SERS in silver colloid are given in Table 1. In making assignments, several former literatures have been used [14-18].

The density functional theory (DFT) and Hartree-Fock calculational frequencies of biotin performed by Emami et al [14] suggest that O-H in-plane bending and O-H out-of-plane bending vibrations mixes with C-C stretching and C-H out-of-plane bending vibrational modes, respectively. Our experimental bands of 1318 $cm^{-1}$ and 795 $cm^{-1}$ in the NRS show excellent agreement with theoretically values, respectively. The strong band at 1635 $cm^{-1}$ which generally appears around 1662 $cm^{-1}$ comes from the ureido ring stretching of biotin [18].

The relative intensities of the SERS spectra are expected to differ significantly from that of the NRS owing to specific surface selection rules [19]. The surface selection rule suggests that when a molecule adsorbed flat on the silver surface, its out-of-plane vibrational modes will be more enhanced than its in-plane vibrational modes, and vice versa when it is adsorbed perpendicular to the surface [19, 20]. Further, vibrations involving atoms that are close to the silver surface will be enhanced. When the wavenumber difference between Raman bands in normal and SERS spectra is not more than 5 $cm^{-1}$, the molecular plane is expected to be perpendicular to the silver surface [21].

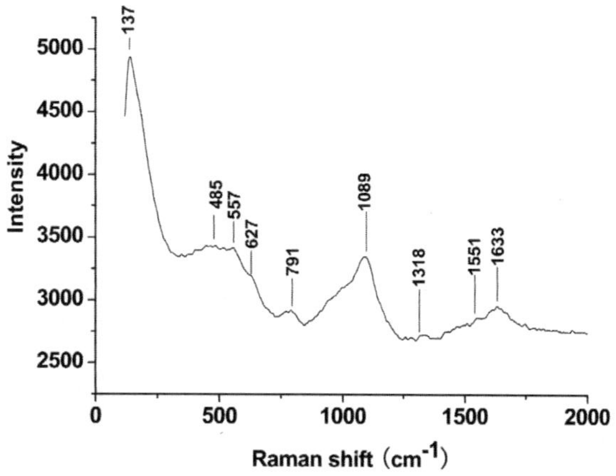


**Figure 2.** Normal Raman spectrum of biotin in aqueous saturated solution.

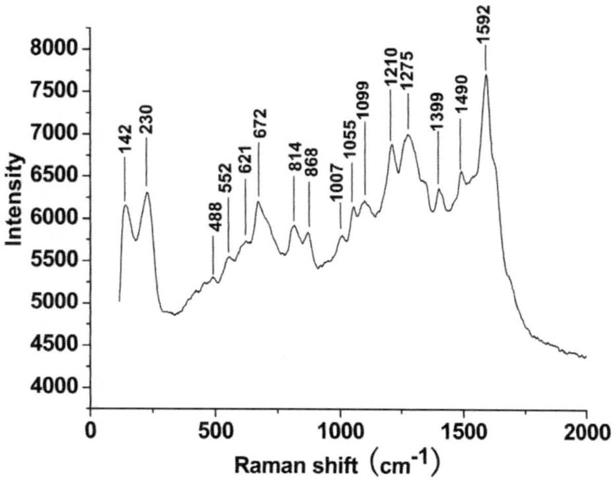


**Figure 3.** SERS spectrum of biotin in silver colloid.

In the SERS spectrum of the title compound, prominent bands are observed at 1592, 1275, 1210, 814, 672, 230 and 142 $cm^{-1}$. The ureido ring stretching vibration mode is observed as a very strong band at 1592 $cm^{-1}$ in the SERS spectrum which is at 1551 $cm^{-1}$ in the Raman spectra. It suggests that the ureido ring is not perpendicular to the silver surface. The medium band at 814 $cm^{-1}$ in the SERS spectrum and the weak band at 795 $cm^{-1}$ in the NRS have been attributed to the C-H out-of-plane deformation. The weak band at 1007 $cm^{-1}$ in the SERS spectrum has been attributedto the C-H in-plane bending vibrations. The comparison of the above two different C-H vibrational modes suggests that the molecular of biotin is parallel to the silver surface. Further, The C-H out-of-plane bending is obtained in the SERS spectrum at 1490 $cm^{-1}$ and the absence of this vibration here in the NRS indicates again the parallel orientation of the molecule at the metal surface.

**Table 1.** Frequencies and assignments of Normal Raman and SERS bands of biotin.

| NRS ($cm^{-1}$) | SERS ($cm^{-1}$) | Assignment | References |
|---|---|---|---|
| 137 vs | 142 vs | CCCC torsion | [17] |
| | 230 vs | C-S-C out-of-plane deformation, Ag-O stretching | [15,16] |
| 485 m | 488 vw | | |
| 559 m | 552 vw | O-H out-of-plane deformation | [17] |
| 627 vw | 621 vw | N-H in-plane bending | [14] |
| | 672 s | N-H wagging | [14] |
| 795 w | 814 m | C-H out-of-plane deformation | [16] |
| | 868 m | | |
| | 1007 vw | C-H in-plane bending | [14] |
| | 1055 m | | |
| 1089 vs | 1099 s | C-O stretching | [14] |
| | 1210 vs | C-O stretching, C-H in-plane bending, C-C stretching, O-H in-plane bending | [16,17] |
| | 1275 vs | C-O stretching | [14] |
| 1318 m | | C-H out-of-plane bending, CH-$CH_2$ deformation, C-C stretching | [16,17] |
| | 1399 s | $CH_2$ stretching | [14] |
| | 1490 m | C-H out-of-plane bending | [14] |
| 1551 s | 1592 vs | Ureido ring stretching | [18] |
| 1635 s | | Ureido ring stretching | [18] |

Abbreviations: vw, very weak; w, weak; m, medium; s, strong; vs, very strong.

The presence of $CH_2$ mode at 1399 $cm^{-1}$ in the SERS spectrum indicates the closeness of $CH_2$ group with metal surface. Emami et al [14] suggested that the vibrations of -$CH_2$- group (scissoring, wagging, twisting and rocking) contribute to several normal modes in the low frequency region and the wagging, twisting and rocking modes are distributed in a wide range. The experimental band at 1210 $cm^{-1}$ supports this argument. The strong band at 1275 $cm^{-1}$ is originated by the C-O stretching which is one of the characteristic carboxylic group vibrations. Also medium C-O stretching is found at 1099 $cm^{-1}$ in the SERS spectrum with a blue shift of 10 $cm^{-1}$ compared with this band in NRS. The bands at 672 $cm^{-1}$ and 621 $cm^{-1}$ are probably originated by N-H wagging and N-H in plane bending, respectively. The Raman band at 559 $cm^{-1}$ in NRS assigned to O-H out-of-plane deformation vibration mode appears at 552 $cm^{-1}$ in the SERS spectrum with a shift of 7 $cm^{-1}$. This pronounced shift in the spectrum of the hydrogen-bonded species demonstrates that the OH group vibrations are likely to be most sensitive to the environment. The appearance of a strong band at 230 $cm^{-1}$ in the SERS spectrum in colloid ascribed to the Ag-O stretching indicates that the biotin molecule is adsorbed on the silver surface in colloid by the chemisorption of oxygen atom. Also, the band at 230 $cm^{-1}$ is strengthened by the C-S-C out-of-plane deformation mode resulting from the tetrahydrothiophene ring vibration. According to the surface selection rule, it is supposed that the C-O and N-H vibration modes have strong interaction with the silver surface. The tetrahydrothiophene ring and the ureido ring of biotin are close to the silver particle surface in the adsorption geometry.

*3.3. Chloride effect on SERS of biotin*

Figure 4 displays the SERS spectra of biotin after addition of different concentrations of $Cl^-$. The presence of $Cl^-$ shows the following main effects: (1) decrease in the relative intensity of most SERS bands; (2) the spectral profile changes with shifts and splitting of selected bands; (3) decrease of a Ag-adsorbate bending band below 200 $cm^{-1}$; and (4) the elimination of bands appearing at 488, 1007, 1055, 1399 and 1490 $cm^{-1}$.

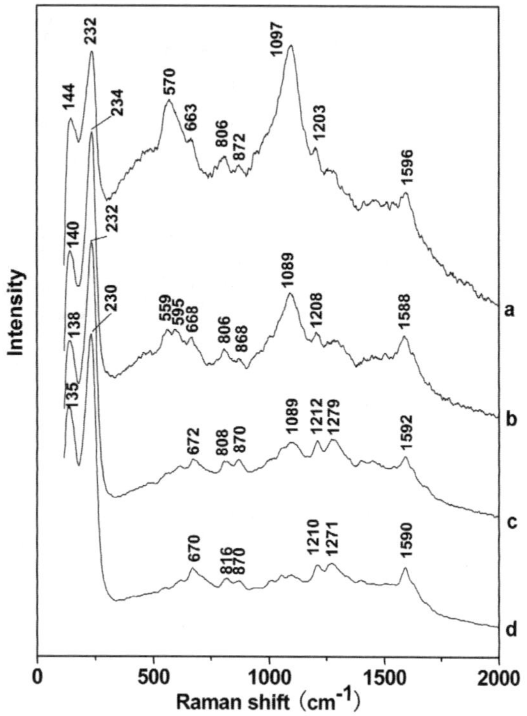


**Figure 4.** Effect of $Cl^-$ in the SERS of biotin with the final concentration of $Cl^-$ in the mixed systems is: (a) 0.01 mol/L; (b) 0.02 mol/L; (c) 0.04 mol/L; (d) 0.05 mol/L.

The effect of $Cl^-$ on the SERS intensity is not well understood, since there are some adsorbates whose SERS signals are positively enhanced by the halide [22], while others are negatively decreased by it [23]. It is believed that these differences can be attributed to the nature of the adsorbate and the double contribution of the halide on both the EM and CT mechanisms, thus making difficult the analysis of $Cl^-$ effect. The SERS intensity quenching observed may be attributed to a surface-coverage restriction imposed by the co-adsorption of the halide [23], effectively competing with biotin for adsorption sites on the metal surface. When the concentration of sodium chloride in silver colloid is higher than 0.05mol/L, superfluous chloride ions may form an absorption layer so that biotin can not be adsorbed on silver surface directly.

The above phenomena may be attributed to the changes in adsorption orientation and surface-coverage [24]. In addition, these changes are assumed to be correlated with the aggregation of Ag sol in the absence of $Cl^-$. A correlation between the UV–Vis absorption spectra and the morphologies observed in the TEM micrographs indicates that $Cl^-$ induces changes in the particle sizes with the formation of both bigger and smaller particles due to the aggregation and diffusion of the initial particles in the presence of the $Cl^-$ [25]. Due to the aggregation effect, large average radii of new particles leads to a decrease of the available surface and, a resultant increase of the adsorbate/surface ratio. It may be concluded that the SERS spectra obtained after adding $Cl^-$ may be due to multilayer surface-coverage. In this case, the gradual decrease of SERS intensity with increasing $Cl^-$ can account for the strong contribution of the CT mechanism, where absorbed molecule always shows weak SERS intensity at dense surface coverage [25].

On the basis of the increase of the adsorbate/surface ratio in the presence of halide, it may be deduced that the negative effect of surface concentration induced by the co-adsorption of halide, which competes with biotin for the adsorption on the surface, is compensated by the multilayer surface coverage and the aggregation of Ag sol created by the $Cl^-$. The intensity, frequency and shape of the bands may also change due to the adsorbate reorientation.

After addition of $Cl^-$ to the biotin-Ag system, the adsorption intensity and spectral profile differ greatly from that of both aqueous biotin and biotin-Ag system, in the absence of $Cl^-$, such that one cannot consider the deviations to be attributed to simple desorption of biotin from the silver surface. The effect of $Cl^-$ is also noted in the <300 $cm^{-1}$ region. Figure 4 shows the SERS of biotin with different concentration of $Cl^-$. The existence of an intense band attributed to Ag-adsorbate stretching vibration at 230 $cm^{-1}$, in the absence of $Cl^-$, demonstrates the adsorption of biotin on the metal, which may be produced by the electron transition between biotin and silver surface. With the increase of the concentration of $Cl^-$, this band is strengthened, indicating the existence of adsorption competition between $Cl^-$ and biotin and the increased possibility of CT interaction between adsorbate and metal. The changes in intensity and profile shape in the SERS spectra suggest different adsorption behavior and surface-coverage of biotin on silver surface.

**4. Conclusion**

The NRS and surface-enhanced Raman scattering spectra of biotin were obtained with very low laser power and the assignment of the vibrational wavenumbers was accomplished with the help of several former reports. The adsorption geometries of biotin molecules were analyzed based on the SERS data.

The $Cl^-$ effect studies of the SERS spectra of biotin on silver surface suggests the existence of a first-layer effect and changes of adsorption behavior in the presence of $Cl^-$. The changes of SERS intensity and spectral profile in presence of $Cl^-$ are governed by the CT mechanism and co-adsorption of anions and molecules.

SERS is a powerful and convenient analytical tool for the investigation of the molecular structure information of biotin and its interaction with metal surface.

**Acknowledgements**

This research is supported by the National Natural Science Foundation of China (60678050).

**References**

[1] Fleischmann M, Hendra P J and McQuillan A J 1974 *J.Chem.Phys.Lett.* **26** 163

[2] Otto A, Mrozek I, Grabhorn H and Akemann W 1992 *J. Phy. Condens. Mat.* **4** 1143

[3] Schatz G C 1984 *Acc. Chem. Res.* **17** 370

[4] Persson B N J 1981 *Chem. Phys. Lett.* **82** 561

[5] Wetzel H and Gerischer H 1980 *Chem. Phys. Lett.* **76** 460

[6] Cortes S S and Ramos J V G 2001 *Surf. Sci.* **473** 133

[7] Li Y S, Cheng J and Wang Y 2000 *Spectrochim. Acta.* A **56** 2067

[8] Du Y and Fang Y 2004 *Spectrochim. Acta.* A **60** 535

[9] Gravel R A and Narang M A 2005 *J. Nutr. Biochem.* **16** 428

[10] Dakshinamurti K 2005 *J. Nutr. Biochem.* **16** 419

[11] Watanabe T 1996 *Experientia.* **52** 149

[12] Mock D M 2005 *J. Nutr. Biochem.* **16** 435

[13] Creighton J A, Blatchford C G and Albrecht M G 1979 *J. Chem. Soc. Faraday Trans.* **75** 790

[14] Emami M, Teimouri A and Chermahini A N 2008 *Spectrochim. Acta.* A **71** 1516

[15] Panicker C Y, Varghese H T and Philip D 2006 *Spectrochim. Acta.* A **65** 802

[16] Li P W, Zhang J, Zhang L and Mo Y J 2009 *Vib. Spectrosc.* **49** 2

[17] Leopold N, Pinzaru S C, Baia M, Antonescu E, Cozar O, Kiefer W and Popp J 2005 *Vib. Spectrosc.* **39** 169

[18] Swamy M J, Heimburg T and Marsh D 1996 *Biophys. J.* **71** 840

[19] Creighton J A 1988 *Spectroscopy of Surfaces-Advances in Spectroscopy* ( New York: John Wiley and Sons Inc) p 37

[20] Gao X, Davies J P and Weaver M J 1990 *J. Phys. Chem. A* **94** 6858

[21] Levi G, Patigny J, Massault J P and Aubard J 1992 *13nd. Int. Conf. on Raman Spectroscopy* (Wurzburg) p 652

[22] Grochala W, Kudelki A and Bukowska J 1998 *J. Raman Spectrosc.* **29** 681

[23] Aubard J, Bagnasco E, Pantigny J, Ruasse M F, Levi G and Byren E W 1995 *J. Phys. Chem.* **99** 7075

[24] Zhang A P, Fang Y 2006 *Chem. Phys. Lett.* **429** 518

[25] Cortes S S, Ramos J V G, Morcillo G and Tinti A 1995 *J. Colloid. Interf. Sci.* **175** 358

# Potable NIR spectroscopy predicting soluble solids content of pears based on LEDs

Yande Liu*, Wei Liu, Xudong Sun, Rongjie Gao,Yuanyuan Pan, Aiguo Ouyang,

School of Mechatronics Engineering, East China Jiaotong University, Changbei Open and Developing

District, Nanchang, 330013, China

## ABSTRACT

A portable near-infrared (NIR) instrument was developed for predicting soluble solids content (SSC) of pears equipped with light emitting diodes (LEDs). NIR spectra were collected on the calibration and prediction sets (145:45). Relationships between spectra and SSC were developed by multivariate linear regression (MLR), partial least squares (PLS) and artificial neural networks (ANNs) in the calibration set. The 45 unknown pears were applied to evaluate the performance of them in terms of root mean square errors of prediction (RMSEP) and correlation coefficients (r). The best result was obtained by PLS with RMSEP of 0.62°Brix and r of 0.82. The results showed that the SSC of pears could be predicted by the portable NIR instrument.

**Keywords:** Near-infrared spectroscopy, Artificial neural network, Soluble solids content, Partial least squares, Portable, LEDs

## 1. INTRODUCTION

Pear is one of major fruits in our daily life. Brix is one of the major factors that affect taste in pears. Traditionally, pear quality is determined quantitatively by using destructive testing methods, which are both waste and time-

* correspondence: Email: jxliuyd@163.com; Telephone: 0086-0791-7046173; Fax: 0086-0791-7046173

consuming. Some papers have been published on the NIR analysis of brix value in pears [1]. Even if these instruments are highly precise, their application to field research, such as monitoring of chemical changes of developing pears on trees, are limited by their large size and weight. Fortunately, the availability of low cost miniaturized spectrometers has opened up the possibility of portable devices which can be used for brix prediction. However, before these technology can be successfully transferred to the fruit industry and ,especially, implemented on a large scale , further research is required into solve these problems: the device must be low cost, be quick-response and deliver robust performance .

Presently, there are several commercially available portable NIR instruments offered for determining Brix value [2-3], such as (Camps et al. (2009); Cayuela., 2008; Liu et al., 2007; Lu et al., 2006; Gómez et al., 2006; McGlone et al., 2003; Miyamoto et al., 1995; Saranwong et al. (2003)). Temma et al. showed that the portable NIR instrument developed by their research centre had excellent potential in determining Brix value of intact apples and it is being successfully implemented in various fruits, such as apples (Liu, Ying, Fu, & Lu, 2007; Lu & Ariana, 2002); cherries (Lu, 2001); mangoes (Schmilovitch, Mizrach, Hoffman, Egozi, & Fuchs, 2000); peaches (Ying, Liu, Wang, Fu, & Li, 2005) [4]. However, the illuminant of a portable NIR instrument when compared to a research one is still not stable. The high-power light emitting diode (LED) is a semiconductor device usually adapted on portable photometers (LED-photometers) as radiation source because it confers some advantages for optical instruments as simplicity, easy operation, light stability and low power and cost [5-6].

The objective of this study was to research the feasibility of rapid measurement of the SSC of intact pear fruit with NIR instrument equipped with LEDs. Investigate the performance for determining SSC of pear quality nondestructively and discusses the developed high-power LED spectrometer and presents the test results characterizing the developed NIR module.

## 2. MATERIALS AND METHODS

### 2.1. Samples preparation and destructive analyses

A total of samples of 190 'crystal' pears (from Shandong province, China) were purchased from a local market. The pears were stored two days at 20°C and 60% relative humidity before spectra collection. Immediately after spectra collection, each pear was mixed using a juice centrifuge and the obtained juice was filtered using a filter paper. Corresponding to each pear, the SSC value of filtered juice was determined using a hand-held refractometer

(LB32T, MinRui, Guangdong, China) and expressed in °Brix. 190 samples were used to analyze and develop Vis-NIR models in laboratory. In terms of the testing samples, as selected by fairly normally distributing around the mean value, 45 pears were used for prediction set. The remaining 145 samples were used for calibration set. To ensure the adaptability of the calibration models, the samples with high and low values were put in the calibration set and the other samples selection inside each group were performed manually.

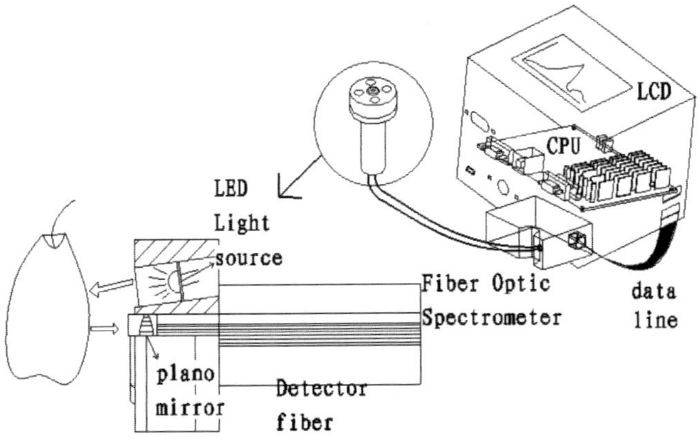


Fig.1. space diagram of the portable NIR instrument

### 2.2. Spectral Collection

As shown in figure 1, the portable NIR instrument consisted of a hand-held probe, a detector, system microcontroller, display unit and data processing software. The handheld probe consisted of a Plano mirror, a fiber-optics, four LEDs. The Plano mirror is located in the center of the round metal bottom cover. It can take more rays of light in the fiber-optics. The length of the fiber-optics is 2 meters and the fiber core diameter is 400 um. Four different types of 3W high power LEDs (FZ-P003ORDX-XXXP, Ark Lighting, Guangdong, China) made up the light source, centered at the wavelengths 620, 850, 880, and 940 nm. The LEDs were arranged in a round metal bottom cover equably [7]. The length of the handle is 120 mm, minor diameter is 40 mm and major diameter is 76 mm [8]. A convex in the center of metal bottom cover was able to secure and tight attachment the surface of pears, thus the reflected light could be prevented and the pure diffuse reflection spectrum is detected by the probe and transmit the optical fiber to the spectrometer. Diffuse transmission spectra were collected using a fiber spectrometer equipped with a 3648-element linear silicon CCD array detector (TCD1304AP, Toshiba, Japan), which covers a spectral range of 345–1040nm at a sampling interval of 1.5nm. The system microcontroller adopted a single board computer (PCM-9361, Advantech, China). The measurement result was displayed on a LCD screen. Seen from Figure 1, all modules, except for the fiber-optics probe, were housed in a mater case

(30×20×15cm). The digitalized absorbance data send by the control unit was displayed on the screen. Therefore, the parameters can be set just one touching pen [9].

One hundred and ninety samples of pears were used to register the blank signal. The measurements with the portable NIR spectrometry were performed in a room temperature. The round metal bottom cover was tight attachment the surface of the sample. Adjusted the pear and let the center of the round metal bottom cover aim the fruit shoulder (form the fruit stem axis). When the light sources, spectrometer and energized, the diffuse reflection spectra data automatically on the LCD screen. A dark scan was carried out, and a white Teflon tile was used for white calibration at the beginning of each measurement session. For each fruit, spectra were taken at four equidistant positions around the equator (approximately $90^{o}$). Then connect the portable NIR spectrometry to a computer with the USB port and the spectra data can be transmitted by USB wire to the computer. The data that were processed was finally used to verify feasibility of the portable NIR spectrometry.

**2.3. Data Analysis**

Before the calibration step, the spectra in reflectance(R) were converted to absorbance (log (1/R)) values to obtain linear correlations of the NIR spectra with the SSC measurement values. Then, the log (1/R) spectra were pre-treated by using multiplicative scatter correction (MSC). Chemometric analysis was performed using the Unscrambler software v9.5 (CAMO AS, Trondheim, Norway) and Matlab 7.0 (The MathWorks, USA). The Unscrambler program was used to establish partial least squares (PLS) model and regression, multivariate linear regression (MLR) model and the Matlab 7.0 (The MathWorks, USA) was used for developing the ANNs models. PLS were carried out to perform linear models of prediction between spectral data and the values obtained from the destructive tests. In order to get efficient and quick models, the optimal number of latent variables for the PLS models was obtained by using the leave-one-out cross-validation technique.

ANNs are the most widely used mathematical algorithms for overcoming non-linearity in calibration model which were more and more widely applied during the past several years [10-11]. Back-propagation training algorithms are usually preferred for content applications and were used in this study [12]. ANNs comprise several layers. PCA was performed firstly to extract information from the whole spectra region as principle components, which were used to be the neurons of network input layer.

The statistic correlation coefficient (r), root mean squares error of calibration (RMSEC) and root mean squares error of prediction (RMSEP) were used to evaluation of the fit of a mathematical model to the experimental data.

Bias denotes the average of the differences between estimated and measured values. The correlation coefficient of validation is calculated using Eq. (1) from the samples for validation 。 r, RMSEP, Bias and RMSEC are defined as follows[13]:

$$r = \sqrt{\sum_{i=1}^{n}\left(\hat{y}_i - y_i\right)^2} \Big/ \sqrt{\sum_{i=1}^{n}\left(\hat{y}_i - y_m\right)^2} \quad \text{(1)}$$

$$RMSEC, RMSEP = \sqrt{\frac{1}{n}\sum_{i=1}^{n}(\hat{y}_i - y_i)^2} \quad \text{(2)}$$

$$Bias = \frac{1}{n}\sum_{i=1}^{n}(\hat{y}_i - y_i) \quad \text{(3)}$$

Where, $\hat{y}_i$ = predicted value of the i-th observation; $y_i$ = measured value of the i-th observation; $n$ = number of observations in the calibration or prediction set; $y_m$ = mean value of the calibration or the prediction set.

## 3. RESULTS AND DISCCUSSION

### 3.1. SSC distribution

The mean SSC was 10.3 °Brix. The portable device was applied on the analysis of 190 samples of pears in order to determinate the standard deviations (SD) of measurements. In this study, the SSC measurements (n=190) were approximately normally distributed around the mean (max=12.7 °Brix, min=8.2 °Brix). An overview of SSC distributions of pears samples in the calibration and validation sets were presented in Table 1.

Table 1 characteristics of calibration and validation sample sets of pear used

| Items | Quantity | Range(°Brix) | Mean(°Brix) | SD (°Brix) |
|---|---|---|---|---|
| Calibration set | 145 | 8.2-12.7 | 10.4 | 1.05 |
| Validation set | 45 | 8.6-12.3 | 10.1 | 0.98 |

### 3.2. Spectral pre-processing analyses

The NIR spectra often require mathematical pretreatment to remove interference of specula radiation, such as smooth, multiplicative scatter correction (MSC) and first derivative. The pretreatment methods were attempted to eliminate the interference by smooth, MSC and first derivative, respectively. For results in Table 2, the RMSEP, RMSECV and R were very close by different pretreatment methods. This indicated that the major information in pear spectra could be reserved while noise was removed by 3-average smoothing. Consequently, the PLS methods were used to develop the multivariate calibration models, where twenty factors (latent variables) were necessary to produce reasonable results for the spectra with 3-average smoothing spectral pretreatments, respectively [14].

Table 2 Prediction results and number of factors of PLS models with different spectral pre-processing

| Pre-processing | RMSEC | RMSEP | R |
|---|---|---|---|
| None | 0.54 | 0.60 | 0.85 |
| 3-average smoothing | 0.58 | 0.58 | 0.88 |
| MSC | 0.59 | 0.64 | 0.87 |
| Fist-derivative | 0.53 | 0.60 | 0.85 |

### 3.3. The establishment of models

PLS was used to build the prediction models, because PLS have the potential to estimate not only component concentrations but also chemical and physical properties from their infrared spectra. Due to its simplicity and small volume of calculations, it is used to analyze different data. PLS method itself is a linear method of data analysis. Fig. 2(a) and (b) were the calibration and prediction models. From the plots, the low values were over estimated, while high values were under estimated, and the least squares solution was pulled to the mean. This was known as the "Dunne-effect" for regression on normally distributed data. Hence, PLS was unsuitable for dealing with the nonlinear phenomenon.

In the above discussion of the prediction results using PLS models, no consideration has been given to the contribution of the individual individual wavelength. This was because the PLS model applied linear transforms using the entire wavelength data. As a result, it was difficult to ascertain which individual wavelength was directly related to SSC. However, it would be helpful to examine how SSC levels are related to individual wavelengths, so that a better understanding of spectroscopy of pear, and SSC may be obtained. For that reason, an MLR model was built.

After PLS processing of the 190 samples, cross validation revealed the existence of six LVs. The cumulative reliabilities of the first six LVs accounted for 87% of X and 31% of Y (Fig. 3), suggesting a possible association between the sensitive individual wavelengths and SSC. By choosing spectral wavebands with the highest absolute loading weight values in the spectral region (800–980), the optimal wavelengths were: 812, 825, 831, 846, 856, 872, 891, 904, 918,925, 954, 966and 979nm. Since the individual wavelengths might be particularly important for reducing sugar content, all these wavelengths were selected in the loading weight curves, whether or not they were the real fingerprint spectra. The performance of these wavelengths was then evaluated. MLR model was built based on the above sensitive wavelengths and the calibration and prediction results of the MLR model are shown

Fig. 4(a) and (b). The model displayed high correlation ($r^2 = 0.87, 0.78$), SEC (0.58), and SEP (0.64), respectively, for calibration and prediction.

In the BP-ANN method, BP-ANN consisting of three layers (input layer, hidden layer and output layer) was established. The tansig function was used for the transfer function between the input layer and the hidden layer, and the prelim function for that between the hidden layer and the output layer. PLS was performed firstly to extract information from the whole spectral region as factors (principle components), which were then used to be the neurons of network input layer. At the beginning of a training run, the parameters of ANNs including input nodes, hidden nodes, learning coefficient, momentum and number of iterations were initialized with optional values [15]. Being in operation process, the modifications of the network input nodes (4), hidden nodes (5), learning coefficient (0.12) and momentum (0.05–0.50) were selected by the back-propagation of the errors and the degree of approximation. Because there was only one active compound in established models, the output layer contained one neuron. Figure 5 showed the measured and predicted plots of brix for the BP-ANN model. It is clear that the ANNs model obtained better result of r = 0.83 and RMSEP = 0.57°Brix.

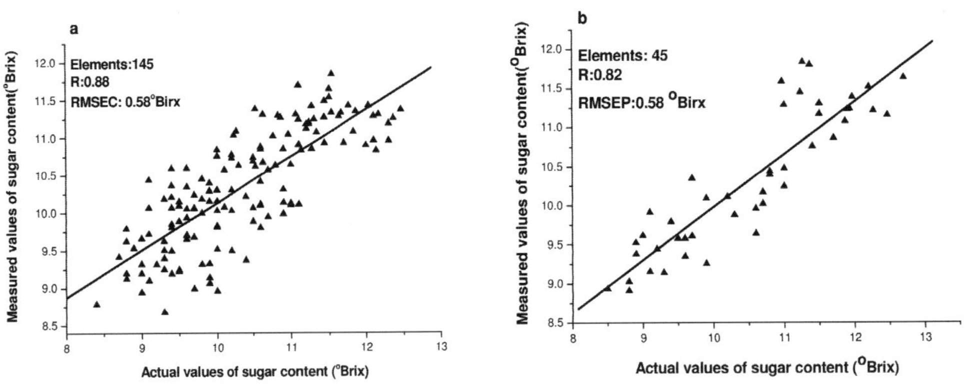


Fig.2.PLS calibration (a) and prediction (b) models

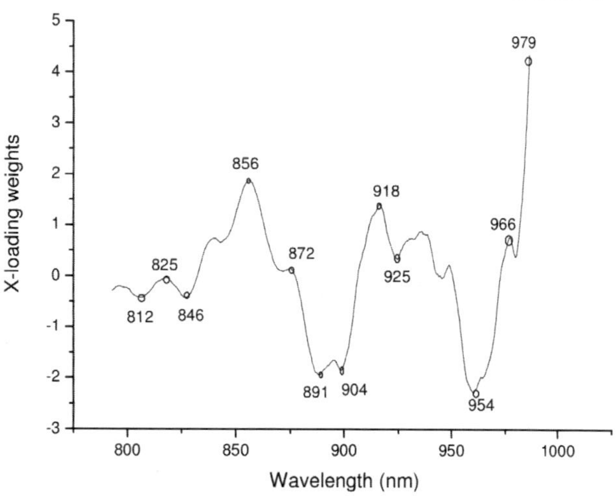


Fig.3. Loading weights of the first six LVs of the PLS model.

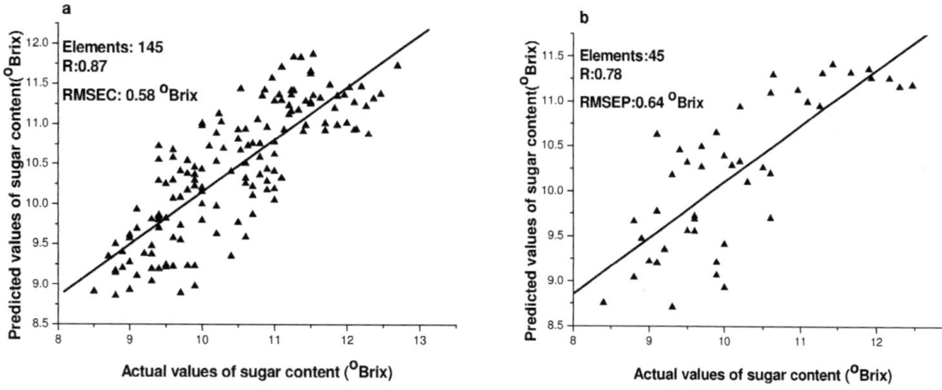


Fig.4.MLR calibration (a) and prediction (b) models

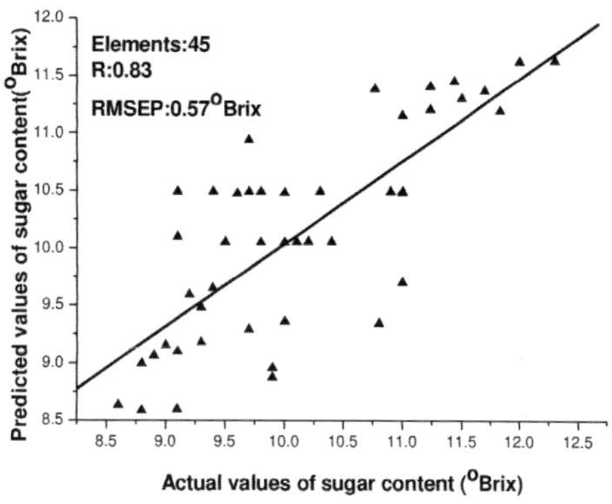

Fig.5. ANN prediction model

### 3.4. Comparison of calibration methods

The quality of the calibration was quantified by SEC, SEP and r. A good model should have a low SEC, a low SEP, a high r but also a small difference between SEC and SEP [16-17]. The PLS model, based on 3-average smoothing spectral pretreatment, could be obtained the best of prediction results with r=0.88 and SEP=0.58°Brix. The plot of relationship between the values of the SSC measurement and the PLS model prediction, with MSC spectral pretreatment, is shown in table 3. Scatter plots between actual and predicted Brix values of the validation sample set are also shown in Figure 4. The result showed that the PLS spectra using wavelength region from 800 to 950nm more than the ANNs and MLR equation.

Table 3 Calibration results for Brix value using spectral data measued with Portable spectrometer

| Methods | Wavelength | Calibration | | Prediction | |
|---|---|---|---|---|---|
| | | $R_c$ | SEC | $R_v$ | SEP |
| PLS | 800-980 | 0.87 | 0.51 | 0.82 | 0.62 |
| MLR | 800-980 | 0.87 | 0.58 | 0.78 | 0.64 |
| ANN | 800-980 | 0.88 | 0.51 | 0.83 | 0.57 |

## 4. CONCLUSION

In this paper, performance of PLS, MLR and ANNs were compared in developing calibration models for brix prediction in pears by the portable near infrared device. ANNs was applied to improve prediction performance by correcting the nonlinearities which limited the performance of the classical linear method. Compared to PLS, the performance of ANNs obtained better result, with higher r and lower RMSEP for brix value. The portable NIR spectrometry uses a NIR high-power LED emitting as portable detector whose noise was kept at low levels by a control system based on thermistor and thermoelectric cooling. Thus, the portable NIR spectrometry showed to be an economical and viable alternative for SSC analysis of pears. This paper successfully demonstrates the application of NIR spectroscopy using the spectral region from 800 nm to 950 nm to the quantitative measurement of pears in PLS models study.

## ACKNOWLEGEMENTS

This study was supported by the Natural Science Foundation of Jiangxi Province (2008GQN0029, 2007GZN0266), National Science and Technology Support Plan (2008BAD96B04), Special Science and

Technology-Support Program for Foreign Science and Technology Cooperation Plan (2009BHB15200) and Technological expertise and academic leaders training plan of Jiangxi Province (2009DD00700).

## REFERENCES

1. T. Temm a, K. Hanamatsu, and F. Shinoki, "Development of a portable near infrared sugar-measuring instrument," J. Near Infrared Spectrosc., vol. 10, pp. 77–83, 2002.
2. Y.B. Ying, Y.D. Liu, J.P. Wang, X.P. Fu, and Y.B. Li, "Fourier transform near infrared determination of total soluble solids and available acid in intact peaches," Transactions of the ASAE, vol. 48 (1), pp. 229-23, 2005.
3. Y.D. Liu, Y.B. Ying, and H.Y. Jiang, "Rapid determination of maturity in apple using outlet detection and calibration model optimization,"Transactions of the ASAE, vol. 49 (1), pp. 91-95, 2006.
4. T. Temma, K. Hanamatsu and F. Shinoki, J.Near Infrared Spectrosc. 10, 77 2002.
5. Chemleds. Telcom Devices datasheet.
6. B.A. Matveev, et al., Mid-infrared (3–5 mm) LEDs as sources for gas and liquid sensors, Sensors and Actuators B 38–39 (1997) 339–343.
7. Pascal Chalus, Serge Walter, and Michel Ulmschneider, "Combined wavelet transform–artificial neural network use in tablet active content determination by near-infrared spectroscopy," Analytica Chimica Acta, vol. 591, PP. 219–224, 2007.
8. Jouko Malinen, Markku Ka¨nsa¨koski, Raimo Rikola, and Colin G. Eddison "LED-based NIR spectrometer module for hand-held and process analyser applications," Sensors and Actuators B, vol. 51, pp. 220–224, 1998.
9. Edvaldo da N´obrega Gai˜ao et al, "An inexpensive, portable and microcontrolled near infrared LED-photometer for screening analysis of gasoline," Talanta, vol. 75, PP. 792–796, 2008.
10. F. Despagne, D.L. Massart, Analyst .157–178.174–180, 1998.
11. D. Perez-Marin, A. Garrido-Varo, J.E. Guerrero, Talanta 72 28–42, 2007.
12. B.N.Peshlov, F.E.Dowell, F.A.Drummond, and D.W.Donshue, "Comparison of three near infrared spectrophotometers for infestation detection in wild blueberries using multivariate calibration models," J. Near Infrared Spectrosc, vol. 17, PP. 203–212, 2009.

13. Lingzhi Zhao , Ying Dou , Ye Guoa, Yonghong Chen, and Yulin Ren, "Artificial neural networks for non-destructive determination of acetylspiramycin powder by short-wavelength NIR spectroscopy," Vibrational Spectroscopy, vol. 47, pp. 21–25, 2008.

14. R. Lu, and D. Ariana, "A near-infrared sensing technique for measuring internal quality of apple fruit," Applied Engineering in Agriculture, vol. 18 (5), pp. 585_590, 2002.

15. Yongni Shao, Yong He, Antihus H ,Gomez , Annia G .pereir ,Zhengjun Qiu, and Yun Zhang, "Visible/near infrared spectrometric technique for nondestructive assessment of tomato'Heatwave'(Lycopersicum esculentum) quality charateristics," Journal of Food Engineering , vol . 81, pp. 672-678, 2007.

16. Y.D. Liu, Y.B. Ying, X.P. Fu, and H.S. Lu, "Experiments on predicting sugar content in apples by FT-NIR technique," Journal of Food Engineering, vol. 80, pp. 986_989, 2007.

17. B.M. Nicolaï, K.I. Theron, J. Lammertyn, and Kernel, "PLS regression on wavelet transformed NIR spectra for prediction of sugar content of appe,"Chemometrics and Intelligent Laboratory Systems, vol. 85 (2), pp. 243-252., 2007.

# Comparison of skin optical clearing by different drug delivery method

**Z. Mao*** [1], **X. Ge** [2], **E. Li**[3]

[1] Sports Health & Science School, Wuhan Sports University, Wuhan, 430079, China.

[2] Department of Fundamental Theories, Shandong Sports University, Jinan, 250102, China.

[3] Institute of Physical Education, Huazhong Normal University, Wuhan, 430079, China.

E-mail: mm73@163.com

**Abstract.** Previous studies have shown that optical clearing techniques have great potential for optical diagnosis and therapy. In order to further discuss the effect by different drug delivery method, diffusion chambers were used to simulate in vivo physiological environment; and the optical clearing agents were acted on the surface of porcine skin, the saline was contacted with the dermal side (epidermal application). Then they were compared with the traditional method- samples immersed in the optical clearing agents (mainly dermal immersion). The transmission intensity of skin was monitored by an integrating sphere system. The changes in thickness were measured before and after experiments. The results showed that the optical clearing effect and the changes in thickness are not coincident under the two different methods. It indicates that the optical clearing effect and mechanism may relate to the drug delivery method.

## 1. Introduction

The optical clearing technique can increase the penetration depth of light in tissue significantly. Among the investigations of the various tissues, skin attracts much attention [1- 5].

At present, the simple drug delivery method for skin optical clearing in vitro is to immerse the samples in the solution [1- 4], which means the optical clearing agents act on both sides of the skin, especially the dermis with large permeability. The corresponding way that drug delivery through in vivo skin is subdermal injection [5- 7], which is traumatic and difficult to operate. The final goal of the optical clearing technique is to explore the in vivo clinical application. In actual clinical application, the desirable and noninvasive way is to apply the drug on the skin surface directly; and let the drug penetrates through the epidermis to dermis where it will be efficient.

Recently, Yeh finds structural modification or dissociation of collagen is the mechanism of skin optical clearing based on in vitro experiments [8, 9]. Our research [7] indicates the mechanism concerning about dissociation of collagen is inconsistent between in vivo in vitro study; even some of the optical clearing agents can not induce the obvious effect on skin in vivo as that of in vitro. Therefore, different experimental methods may lead different role and effect of agents. However, the previous work mainly focused on in vitro study, such as screening efficient optical clearing agents, using physical- medical methods to achieve ideal optical clearing effect and discussing the mechanism.

---

* To whom any correspondence should be addressed.

Though research on in vivo skin is difficult, it is important and necessary before the technique carried out for clinical use.

In order to develop a simple experimental setup that can simulate the in vivo physiological environment, we used horizontal diffusion chambers to imitate in vivo physiological environment; and compare the optical clearing effect with the simple drug delivery method - samples immersed in the optical clearing agents.

## 2. Materials and Methods

*2.1. Preparation of optical clearing agents*

In previous work, we found that glycerol solution with high concentration achieve good clearing effect. Three typical concentrations of glycerol –100%, 75% and 30% were chosen as optical clearing agents. Saline was set as control. The refractive index was measured with a refractometer (Digital Abbe Refractometer, WAY-2S; Shanghai, China). The concentration and refractive indices of the OCAs used in this study were listed in table 1.

**Table1** The refractive indices of the glycerol solutions

| Concentration of glycerol | Refractive indices (n) | Data from references (n) |
|---|---|---|
| 100% | 1.470 | 1.47 |
| 75% | 1.435 | 1.43 |
| 30% | 1.375 | 1.369 |
| saline | 1.335 | 1.34 |

*2.2. Horizontal diffusion chambers to imitate in vivo physiological environment*

We used horizontal diffusion chambers to imitate in vivo physiological environment in our experiment. The diffusion chambers were made of two colorimetric wares. Each chamber was 50 mm × 50 mm ×5 mm, with an aperture of 30 mm in diameter on one side, as shown in Figure 1. The samples were cut into 50mm ×50 mm sections, and clamped tightly between two apertures. One chamber near the epidermis was filled with OCAs, and the other one was filled with isotonic saline. Both sides of the skin were able to react to the solution through the apertures. The osmotic pressure of isotonic saline was similar to in vivo physiological conditions. By this means, the dynamics of drug delivery is similar to the common clinical manner and the skin can be monitored continuously. On the other hand, the chambers are easy to clean and the dose of solution was limited.

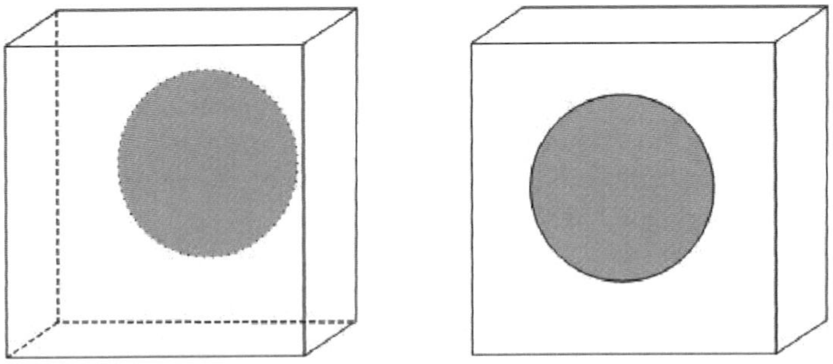

**Figure 1** The sample box: horizontal diffusion chambers.

*2.3. Part 1: measurement of skins by immersion (mainly dermal immersion)*

In this part of our experiment, a double-integrating sphere system was used to measure the optical character parameters; the experimental setup was shown as reference [10]. The samples should be very thin, so rats' skin was chosen and immersed in agents. The detection system is mainly comprised by the reflective ball and transmitted ball. A 1.6 mW, 1mm beam in diameter He-Ne laser ($\lambda$ = 632.8 nm, 5 mW, Melles Griot, U.S.) was chopped mechanically at 1 k Hz (Model SR 540), and then split two beams by a beam splitter. The reflected beam was a small fraction (20%) of laser beam, which irradiated to a reference sphere of 70 mm in diameter. The other (80%) of laser beam irradiated to the sample mounted in which placed in front of the port ($\phi$ = 25 mm) of a big integrating sphere of 210 mm in diameter. The intensity of the transmitted light from the sample including collimated and diffuse components of transmitted radiation was collected by the integrating sphere. The diffuse transmittance and diffuse transmittance were detected; and then, by transport theory and calculation, we can obtain the optical character parameters of the sample.

Because the structure of rats' skin is very similar to the human's, we selected the back skin of rats. The experimental procedures are shown as follows:

(1) The hair was removed after the rats were sacrificed. The skin of back region was taken and cut into 5cm × 5cm square after the subcutaneous fat layer was peeled carefully.

(2) The thickness of sample was measured by using a micrometer.

(3) Measuring the initial reduced scattering coefficient of the samples by the double-integrating sphere system;

(4) Measuring the thickness, reduced scattering coefficient of the skins after they were immersed in solutions for 10 minutes.

(5) We repeated all the procedures above four times. The whole experiment lasted for 60 minutes.

### *2.4. Part 2: measurement of skins with horizontal diffusion chambers to simulate physiological environment (epidermal application)*

In Part 2 of this experiment, we set a drug delivery system, horizontal diffusion chambers to simulate physiological environment, as shown in Figure 1. We chose thick porcine skins to test because the rat skin is too thin to leak when clamped between two chambers. A single integrating sphere system was used to test the transmittance of the samples, as shown in reference [11].

Skin specimens were obtained from the porcine abdomen in a local slaughterhouse within 1 hour postmortem. All samples were measured within 8 h and stored at 4 °C. The stratum corneum and the subcutaneous fat of the whole skin specimens were cleaned before the samples were dissected for measurements. The samples were cut into 50mm ×50 mm sections.

After the thickness was measured, each skin sample was clamped between two diffusion chambers, which were held together. 12 ml of OCAs was added to the donor cell in contact with the epidermal side of the skin, while 12 ml of isotonic saline was added to the receptor cell in contact with the dermal side. The chambers were set up with the receptor cell touching to the port ($\phi$ = 25 mm) of the integrating sphere. In order to determine the OCAs induced optical clearing effect of skin; the dynamical transmission intensity of samples was measured at 0, 1, 2, 3, 4, 5, 10, 15, 20, 25, 30, 40, 50 and 60 min after the application of OCAs.

## 3. results

### *3.1. results of Part 1*

#### *3.1.1. The reduced scattering coefficient of samples*

To trace the influence of the glycerol on skin optical properties, the reduced scattering coefficient of skins was recorded after immersed for every 10mins by each concentration group. Figures 2 shows the ratio of scattering coefficient measured at every point ($\mu$s' (t), t=10, 25, 40, 60min) and the initial value. The results indicate that the reduced scattering coefficient of skins decreases gradually after treatment with the OCAs. In contrast, there is a little increase in the control group alone after treatment with the isotonic saline. Among the agents, 100% glycerol caused the greatest changes in reduced scattering coefficient, and 30% glycerol did the poorest.

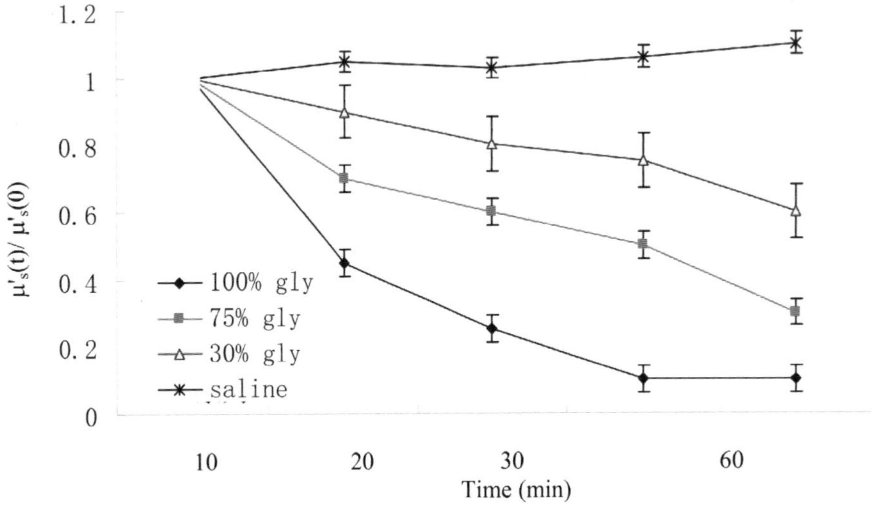


**Figure 2** The reduced scattering coefficient of skins after treated with different concentration glycerol.

*3.1.2. The change in thickness of samples*

Table 2 lists the thickness of samples before and after the experiment, and the corresponding change. SPSS statistical software was used to perform an F- test among all the groups to compare the differences in the thickness changes after the experiment. Within each group, a paired-sample t- test was performed to compare skin thickness before and after an experiment. It is obvious that the skins shrunk by solutions with high concentration. A paired- sample t- test shows that there are significant differences within groups treated by 100% and 75% glycerol ($P< 0.01$). But a little swell in control. And an F- test shows that there are significant differences in thickness change between high concentration groups and control ($P< 0.01$).

**Table 2** Skin thickness before and after experiments and the corresponding changes

| Agents | $d_0$ (mm) | $d$ (mm) | $\Delta d$ (mm) |
|---|---|---|---|
| 100% | 1.02±0.05 | 0.85±0.11** | -0.26±0.10** |
| 75% | 1.06±0.04 | 0.85±0.08** | -0.20±0.09** |
| 30% | 1.10±0.05 | 1.08±0.12 | -0.02±0.12 |
| saline | 1.08±0.10 | 1.15±0.08 | 0.09±0.07 |

($d_0$- thickness before experiments; $d$- thickness after experiments; $\Delta d$- thickness changes after experiments. *means $P<0.05$ ; **means $P\leq0.01$)

*3.2. results of Part 2*

*3.2.1. The relative transmittance of skin*

The relative transmittance can be deduced as the following formula (1) based on the measurements of transmission intensity:

$$T = \frac{I_t}{I_0} \quad (1)$$

$I_t$ and $I_0$ are the transmission intensity measured at time $t$ and initially ($t = 0$), respectively. Figure 3 shows the changes of relative transmittance after treatment with solutions. The relative transmittance of skin was deduced with time. The results indicate that the relative transmittance of skins increases gradually after treatment with the OCAs. In contrast, there is some decrease in the control group alone after treatment with the isotonic saline. Seen from the figure, 100% glycerol achieves the greatest changes in relative transmittance and the best optical clearing effect correspondingly. However, 30% glycerol nearly has no effect on the skin.

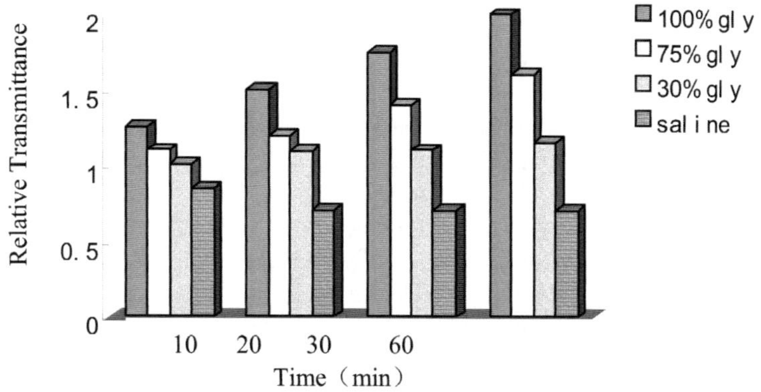


**Figure 3** Relative transmittance of porcine skins after treatment with solutions

*3.2.2. The thickness of skins*

The thickness of samples before and after the experiment, and the corresponding change are listed in Table 3. The skins seem a slight shrinkage treated by high concentration glycerol; but a little swell treated by 30%glycerol and saline. However, a paired- sample t- test shows that there is no significant difference within each group, and an F- test shows that there is no difference in thickness change among groups (P> 0.05).

**Table 3** Skin thickness before and after experiments and the corresponding changes ($\overline{X} \pm$ SD)

| Agents | $d_0$ (mm) | $d$ (mm) | $\Delta d$ (mm) |
|---|---|---|---|
| 100% | 1.58±0.12 | 1.51±0.11 | -0.07±0.11 |
| 75% | 1.61±0.11 | 1.59±0.08 | -0.02±0.07 |
| 30% | 1.60±0.09 | 1.61±0.12 | 0.01±0.05 |
| saline | 1.66±0.10 | 1.74±0.05 | 0.07±0.01 |

($d_0$- thickness before experiments; $d$- thickness after experiments; $\Delta d$- thickness changes after experiments. *means P<0.05 ; **means P≤0.01)

## 4. Discussion

The high scattering character of tissue is mainly because of refractive index mismatch among cellular components: cell membrane, cytoplasma, cell nucleus, cell organelles, melanin granules, and the extracellular fluid [1]. It is well known that turbidity of a dispersive physical system can be effectively controlled by providing matching of refractive indices of the scatterers and the ground material. Optical clearing technique proposed by Tuchin attracted intensive attentions. By immersing tissue with biocompatible, hyper-osmotic and refractive index matching agents, tissue scattering reduces due to refractive index matching of scatterers and ground matter. As a result, the turbidity of biological tissues can be effectively controlled, which increases penetration depth of light in tissues [1-6].

In this study, the experiment was divided into two parts based on the medication method: mainly by dermal immersion and by epidermal application. We used diffusion chambers to imitate in vivo physiological environment in Part 2. The optical clearing effect of glycerol with three different concentrations was compared by the different medication above. We found that 100% and 75% glycerol achieved obvious optical clearing effect, but 30% glycerol almost had no influence on skins by epidermal application. In this work, the changes in thickness of skins treated by glycerol solutions were investigated, under two different medications. The results show that skins shrunk by immersed in solutions. But their thickness would have no change if just the epidermis contacted with OCAs, and the dermis contacted with saline.

Much previous in vitro research found the dehydration and the shrinkage of skin was marked when the OCAs were directly acted to the dermis [9]. In clinical applications, the OCAs would probably be applied to the epidermis, and then diffuse into the dermis. So we apply the diffusion chambers to simulate transdermal OCAs delivery *in vivo*. Then the OCAs can not be effective unless they penetrate from epidermis to dermis; on the other hand, part of the OCAs in the dermis was replaced by saline in the receptor cell for the substances exchange. Therefore, in our Part 2 tests, the effective OCAs in the dermis may be less than that in the previous studies [2, 4, 6, 12], and the corresponding dehydration and shrinkage of skins may also be milder. As a result, there is no significant change in skin thickness after treatment of solutions. It indicates that the effect of optical clearing agents may differ between in vitro and in vivo situations in basis of our different medications. Hence, according to our former research, the mechanism may not identical under the corresponding medications [7, 11].

## 5. **Conclusion**

The optical clearing effect may relate to the drug delivery method. By using diffusion chambers to simulate in vivo transdermal drug delivery, the skins achieve less optical clearing effect and shrinkage than agents' immersion. However, we provide a worthy of learning way to imitate in vivo research.

## Acknowledgments

This study was supported by the Youth fund of the education department of Hubei province, China (Q20093303). The authors are thankful to Dan Zhu, Xiang Wen in Britton Chance Center for Biomedical Photonics, Wuhan, China, for their help.

## References

[1] Tuchin V V 2005 *J. Phys. D: Appl. Phys.* **38** 2497
[2] Jiang J J and Wang R K 2004 *Phys Med Biol.* **49** 5283
[3] Yeh A T, Choi B, Nelson J S, et al. 2003 *J. Invest. Dermatol.* **121** 1332
[4] Xu X and Zhu Q 2007 *Opt. Commun.* **279** 223
[5] Galanzha E I, Tuchin V V, Solovieva A V, Stepanova T V, Luo Q, and Cheng H 2003 *J. Phys.-Lond.-D Appl. Phys.* **36** 1739
[6] Vargas G, Chan E K, Barton J K., et al 1999 *Laser. Surg. Med.* **24** 133
[7] Wen X, Mao Z, Han Z, Zhu D 2010 *J. Biophoton.* **3** No. 1-2, 44
[8] Hirshburg J, Choi B, Nelson J S, and Yeh A T 2006 *J Biomed. Opt.* **11** 040501
[9] Yeh A T and Hirshburg J, 2006 *J. Biomed. Opt.* **11** 014003
[10] Zhu D, Luo Q, Chen J 2003 *Lasers Surg. Med.* **33** 226
[11] Mao Z, Zhu D, Hu Y, Wen X, Han Z 2008 *J. Biomed. Opt.* **13** 021104
[12] Choi B, Tsu L, Chen E, et al 2005 *Lasers Surg. Med.* **36** 72

# Dextran-coated $Fe_3O_4$ magnetic nanoparticles as a contrast agent in thermoacoustic tomography for hepatocellular carcinoma detection

**Huan Qin, Dong Xu and Sihua Yang***
MOE Key Laboratory of Laser Life Science & Institute of Laser Life Science, College of Biophotonics, South China Normal University, Guangzhou 510631, China

E-mail:yangsh@scnu.edu.cn

**Abstract.** Microwave-induced thermoacoustic tomography can provide a novel imaging modality for clinical detection. Significant progress has been made in the past several years in microwave-induced thermoacoustic tomography. In this paper, we investigate the feasibility of using dextran-coated $Fe_3O_4$ magnetic nanoparticles as a contrast agent in thermoacoustic tomography for hepatocellular carcinoma detection. Dextran-coated $Fe_3O_4$ magnetic nanoparticles administered intravenously are phagocytosed by resident Kupffer cells in normal reticuloendothelial system (RES) within the liver, but are not retained in tumor tissue. Consequently, there are significant differences in thermoacoustic signal intensity between normal RES and tumors, which result in increased lesion conspicuity and detectability. This provides the improvement of lesion-to-liver contrast for thermoacoustic tomography. A fast thermoacoustic computed tomography system with a multielement linear transducer array was used to image cancerous liver tissue with circular scanning. The results show that the system can provide molecular imaging with functionalized contrast agents for high-contrast detecting hepatocellular carcinoma and has the potential to become a novel approach for clinical diagnosis in the future.

## 1. Introduction

Pure microwave imaging has the advantage of good imaging contrast but suffers from poor spatial resolution due to the long wavelength of microwave[1-2]. Pure ultrasound imaging, an established medical imaging modality, can yield good spatial resolution but has poor contrast. Microwave-induced thermoacoustic computed tomography which is similar to photoacoustic imaging, combines the merits

* Correspondence author: Tel: +86-20-85211346, Fax: +86-20-85216052, Email: yangsh@scnu.edu.cn

of both microwave imaging and ultrasound imaging, achieves excellent microwave absorption contrast and ultrasound spatial resolution, when compared to competing imaging techniques[3-8]. Various tissues present particular characteristics in their absorption spectra. Photoacoustic imaging can map optical absorption distribution whereas microwave-induced thermoacoustic imaging is related to the dielectric properties in the objects.

Thermoacoustic source wave is a result of microwave-induced thermal effect. A small temperature rise can be produced when a biological tissue is irradiated by a microwave pulse with adequate energy. Subsequently, the heated structure thermally expands and contracts, becoming a source of acoustic wave. By detecting the acoustic signals and via image reconstruction, thermoacoustic thermoacoustic computed tomography can be realized based on the heterogeneity in microwave absorption inside the object. This technique typically uses the endogenous contrast provided primarily by water content and ionic concentration in tissue[9]. While powerful, thermoacoustic tomography would benefit greatly from the development of exogenous contrast agents. Exogenous contrast agents would decouple contrast from penetration depth and open applications in which the endogenous contrast is insufficient, especially in molecular imaging. Our group has reported fabricated iron oxide nanoparticles conjugated with tumor ligands for targeted TAT tumor detection at the molecular level[10]. The results indicate that thermoacoustic molecular imaging with functionalized iron oxide nanoparticles may contribute to targeted and functional early cancer imaging. And modified iron oxide combined with suitable tumor markers may also be used as novel nanomaterials for targeted and guided cancer thermal therapy.

In this paper, we investigate the feasibility of using dextran-coated $Fe_3O_4$ magnetic nanoparticles as a contrast agent in thermoacoustic tomography for hepatocellular carcinoma detection. Dextran-coated $Fe_3O_4$ magnetic nanoparticles administered intravenously would be phagocytosed by Kupffer cells, which are mainly settled in the liver. Furthermore, Kupffer cells will be redistributed when tumor occurs in the liver, and will absent or markedly decrease in number in tumor entity[11-16]. Therefore, liver tumor retains thermoacoustic signal intensity after dextran-coated $Fe_3O_4$ magnetic nanoparticles administration, while signal enhancement in normal liver tissue. This provides the improvement of lesion-to-liver contrast for thermoacoustic tomography. A fast thermoacoustic computed tomography system with a multielement linear transducer array was used to image cancerous liver tissue with circular scanning. The results show that the system can provide molecular imaging with functionalized contrast agents for high-contrast detecting hepatocellular carcinoma and has the potential to become a novel approach for used in clinical diagnosis in the future.

## 2. Data-acquisition system and imaging reconstruction

The data-acquisition system for TAT molecular imaging mainly composes of B-mode digital ultrasound diagnostic equipment (Model CTS-5000B, SIUI, China) with a multi-element transducer (L2L50A) and a high speed digital card (PCI-6541, NI, USA). The 128-element linear transducer with a scanning width of 53 mm was used for thermoacoustic signal acquisition. The multi-element transducer has a center frequency of 2.5 MHz with a nominal bandwidth of 70%. The transducer array, which has a computed tomography capability, can select the thermoacoustic signals which are parallel to its centre cross section in a limited view. The digital data acquisition (DAS) card featured a 12-bit analog-to-digital (D/A) converter with a sampling rate of 25 MHz was employed for collecting digital signals. A microwave generator at 6 GHz (BW-6000HPT, China) transmitted out short microwave pulses adjustable from 0.3 to 1.2 μs, whose maximum trigger repetition rate can reach 500 Hz. The microwave pulses were coupled into a rectangular waveguide in TE10 mode with a cross section of 34.8 × 15.8 mm2. For 6 GHz microwaves, the penetration depths of fat and muscle are 5.2 and 0.7cm respectively. Other soft tissue has a penetration depth between those of fat and muscle. The energy density per volume of the microwave pulse is about 0.25 mJ/cm3, which is within the safety standard for animal and human use.

A custom-built control circuit provided a synchronized clock signal to trigger the DAS card and the microwave generator simultaneously. The TA signals received by the transducer array, after pre-amplification and phase adjustment, were acquired with the DAS card, and then transferred to a personal computer for subsequent data processing. The phantom in the holder can be rotated at several scanning positions for acquisition of multiple image signals with comprehensive tissue information. Generally the signals at 20 circular scanning positions around the object were sufficient to recover a clear and accurate image. Due to the directivity properties of the linear transducer array, an improved limited-field-filtered back-projection (LFBP) is employed to reconstruct the thermoacoustic image. More technical information about the fast data-acquisition system can be found in the previous literatures.

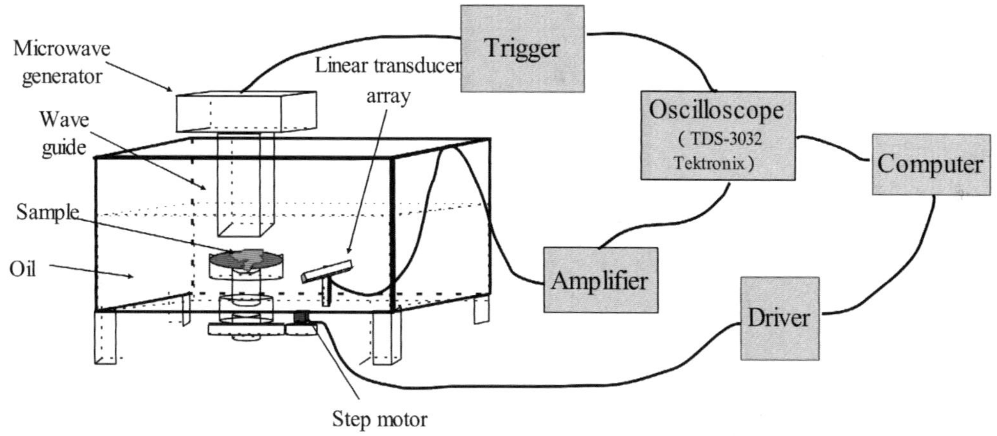


**FIG.1.**Schematic of the experimental setup for microwave-induced thermoacoustic data acquisition system.

### 3. Results

Dextran-coated $Fe_3O_4$ magnetic nanoparticles supplied by Department of Biomedical Engineering, Zhejiang University , the $Fe_3O_4$ particles are almost spherical with diameters ranging from 10 to 20 nm. After coate with dextran, dextran-coated $Fe_3O_4$ magnetic nanoparticles are almost spherical with diameters ranging from 30-50 nm[17].

Dextran-coated $Fe_3O_4$ magnetic nanoparticles for TAT molecular imaging .A stable water-soluble dextran-coated $Fe_3O_4$ magnetic nanoparticles suspension was obtained after the final centrifugation of the solution. In a phantom study, TA signals of dextran-coated $Fe_3O_4$ magnetic nanoparticles at different concentrations as shown in the Eppendorf tubes from left to right in the actual inset picture are detected under the same microwave energy density. It can be seen that the TA signal produced by dextran-coated $Fe_3O_4$ magnetic nanoparticles solution is in linear relationship with their concentrations ($R^2$ = 0.9961) in Fig.2. These data suggest that TA signal of the contrast agent is much higher than water and dextran-coated $Fe_3O_4$ magnetic nanoparticles solution at a concentration of 0.1 mg/ml can achieve effective signal increase. The concentration at the level of 0.1 mg/ml can be used as a reference for the in vivo study.

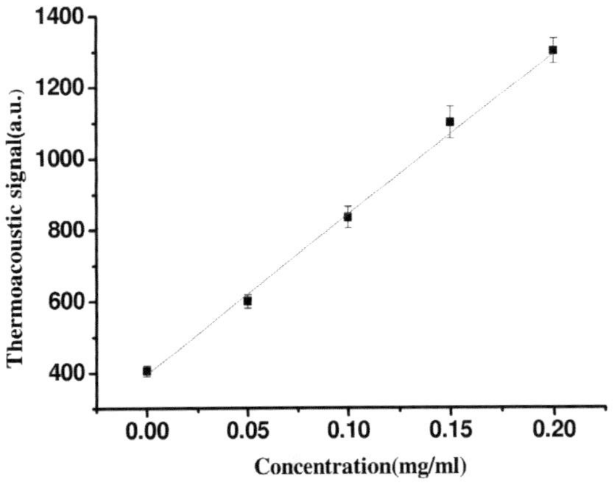


**FIG.2.** Linear fitting of thermoacoustic signals of the contrast agents at different concentrations; the solid line is a linear fit with standard deviations (n=3). Dextran-coated $Fe_3O_4$ magnetic nanoparticles at different concentrations are shown in the Eppendorf tubes from left to right in the actual inset picture of Fig.2. The thermoacoustic signal produced by dextran-coated $Fe_3O_4$ magnetic nanoparticles was observed to be linearly dependent on the concentration ($R^2$ = 0.9965).

Different concentrations of dextran-coated $Fe_3O_4$ magnetic nanoparticles solution mixed with human blood were also investigated to estimate the appropriate concentration for TAT molecular imaging. Four identical plastic tubes with a diameter of 1 mm filled by human blood and dextran-coated $Fe_3O_4$ magnetic nanoparticles solution were embedded at a depth of 10 mm in porky fat tissue, the cross section of the three tubes' placement is shown in the inset picture of Fig. 3(a). The tube marked 'A' was filled by 150 μl dextran-coated $Fe_3O_4$ magnetic nanoparticles of 0.1 mg/ml with 150 μl blood. 'B' was filled by 150 μl dextran-coated $Fe_3O_4$ magnetic nanoparticles solution of 0.075 mg/ml mixed with 150 μl blood, another one marked 'C' was filled by 150 μl dextran-coated $Fe_3O_4$ magnetic nanoparticles solution of 0.05 mg/ml mixed with 150 μl blood; and the other one filled by 300 μl pure blood.The TAT image in Fig. 3(a) obtained by circular scanning at 20 scanning positions agrees well with the actual phantom shown in the inset of Fig. 3(a). The thermoacoustic map of tube 'A' containing dextran-coated $Fe_3O_4$ magnetic nanoparticles and blood has a much higher signal-noise-ratio (SNR) than that of tube 'D' containing blood only due to high microwave absorption of the exogenous contrast agent. Thermoacoustic signals were calculated in the tubes whose boundaries were clearly visualized in the ellipses, and shown in Fig. 3(b).The TA signals can be increased roughly 4 times using a dextran-coated $Fe_3O_4$ magnetic nanoparticles enhanced sample compared to pure blood.

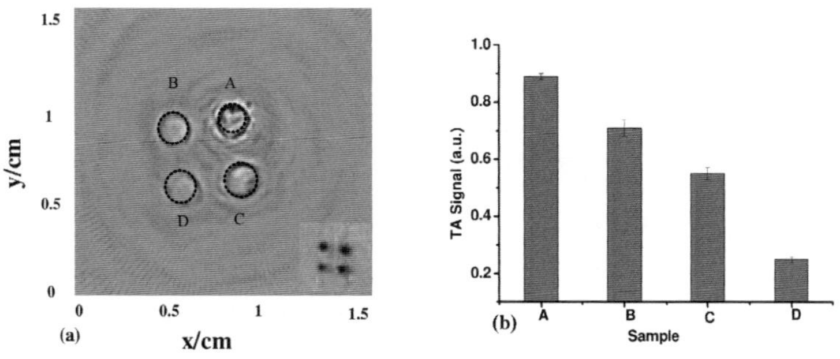


**FIG. 3.** (a) Thermoacoustic imaging of thermoacoustic contrast agents mixed with human blood at four different concentrations in plastic tubes. (b) Thermoacoustic signals were calculated in the tubes whose boundaries were clearly visualized in the dashed ellipses. Data are the mean values of three replicates.

Because dextran-coated $Fe_3O_4$ magnetic nanoparticles administered intravenously would be phagocytosed by Kupffer cells, which are mainly settled in the liver. Furthermore, Kupffer cells will be redistributed when tumor occurs in the liver, and will absent or markedly decrease in number in tumor entity. Therefore, liver tumor retains thermoacoustic signal intensity after dextran-coated $Fe_3O_4$ magnetic nanoparticles administration, while signal enhancement in normal liver tissue. A phantom made from agar simulation the distribution of dextran-coated $Fe_3O_4$ magnetic nanoparticles in the liver with hepatocellular carcinoma after intravenous .the round phantom was made from 15% gelatin , 12.5% milk, and 72.5% water to simulate the electromagnetic and acoustic properties of live tissue. dextran-coated $Fe_3O_4$ magnetic nanoparticles were doped in the phantom but absent in a small round which simulation the tumor entity. The small round is in the middle of the big round phantom. Induced thermoacoustic waves emanated from the distribution of the water and the dextran-coated $Fe_3O_4$ magnetic nanoparticles, the signals at 20 scanning stops around the sample were acquired to reconstruct the images. Fig.4. shows the thermoacoustic image by the LFBP reconstruction algorithm with the real picture shown in the inset of Fig.4.This image clearly show that the signal from dextran-coated $Fe_3O_4$ magnetic nanoparticles is much higher than the water. we can see in the TAT image, the signal only enhance in distribution of the dextran-coated $Fe_3O_4$ magnetic nanoparticles, this result indicate that when dextran-coated $Fe_3O_4$ magnetic nanoparticles administered intravenously ,and after phagocytosed by Kupffer cells, the signal from normal live tissue would be much higher than the tumor entity because Kupffer cells are mainly settled in the liver but absent in tumor entity. This result is successful ex vitro evidence for the possibility of application of dextran-coated $Fe_3O_4$ magnetic nanoparticles as thermoacoustic image for hepatocellular carcinoma detection.

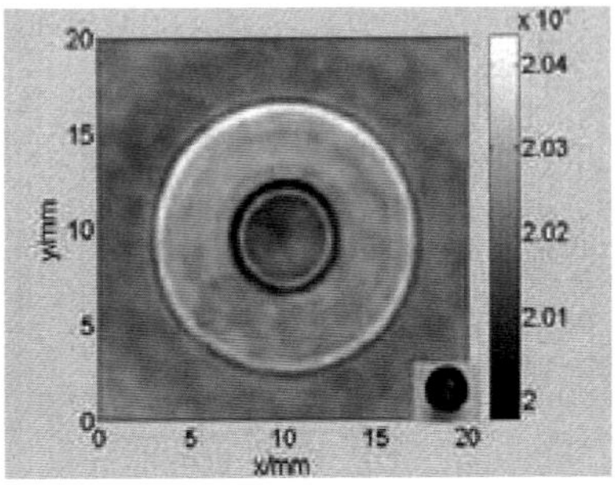


**FIG.4.** Thermoacoustic imaging of a phantom made from agar simulation the distribution of dextran-coated $Fe_3O_4$ magnetic nanoparticles in the liver with hepatocellular carcinoma. The inset picture is photography of the simulated agar. Dextran-coated $Fe_3O_4$ magnetic nanoparticles only distribution in the ring of the big round simulated the normal live tissue and absent in the small round simulated the tumor entity.

## 4. Conclusion

In summary, we have successfully shown that the peak-to-peak TA signal can be increased roughly 4 times using a dextran-coated $Fe_3O_4$ magnetic nanoparticles enhanced sample compared to pure blood. The phantom Simulation the distribution of the liver with hepatocellular carcinoma after intravenous dextran-coated $Fe_3O_4$ magnetic nanoparticles clearly show that the signal from dextran-coated $Fe_3O_4$ magnetic nanoparticles is much higher than the water. This result indicate when dextran-coated $Fe_3O_4$ magnetic nanoparticles administered intravenously ,and after phagocytosed by Kupffer cells, the signal from normal live tissue would be much higher than the tumor entity because Kupffer cells are mainly settled in the liver but absent in tumor entity. We presented our preliminary results that the feasibility of using dextran-coated $Fe_3O_4$ magnetic nanoparticles as a contrast agent in thermoacoustic tomography for hepatocellular carcinoma detection and has the potential to become a novel approach for used in clinical diagnosis in the future.

## References

[1] J. C. Bolomey and C. Pichot, 1990 “Microwave tomography: From theory to practical imaging systems,” Int. J. Imaging Syst. Technol. 2, 144–156.

[2] S. C. Hagness, A. Taflove, and J. E. Bridges, 1998 “Two-dimensional FDTD analysis of a pulsed microwave confocal system for breast cancer detection: Fixed-focus and antenna-array sensors,” IEEE Trans. Biomed. Eng. 45, 1470–1479 .

[3] R. A. Kruger, W. L. Kiser, Jr., D. R. Reinecke, and G. A. Kruger, 2003 “Ther moacoustic computed tomography using a conventional linear transducer array,” Med. Phys. 30, 856–860.

[4] Y. Wang, X. Xie, X. Wang, G. Ku, K. L. Gill, D. P. O‘Neal, G. Stoica, and L. V. Wang, 2004 “Photoacoustic tomography of a nanoshell contrast agent in the in vivo rat brain,” Nano Lett. 4, 1689–1693.

[5] X. Wang, Y. Pang,G. Ku, X. Xie, G. Stoica, and L. V. Wang, 2003 “Noninva sive laser-induced photoacoustic tomography for structural and functional in vivo imaging of the brain,” Nat. Biotechnol. 21, 803–806.

[6] Caijun Cao, Liming Nie, Cunguang Lou and Da Xing,2010 “The feasibility of using

microwave-induced thermoacoustic tomography for detection and evaluation of renal calculi," Phys. Med. Biol. 55 ,5203-5212 .

[7] Songbo Ma, Sihua Yang, and Da Xing,2010 "Photoacoustic imaging velocimetry for flow-field measurement," OPTICS EXPRESS. 18, 9991–10000.

[8] Z. Yuan and H. Jiang, 2007 "Three-dimensional finite-element-based photoacoustic tomography: Reconstruction algorithm and simulations," Med. Phys.34, 538–546 .

[9] Liming Nie, Da Xing,Quan Zhou, Diwu Yang, and Hua Guo , 2008 "Microwave-induced thermoacoustic scanning CT for high-contrast and noninvasive breast cancer imaging , " Med. Phys. 35 , 4026-4032.

[10] Liming Nie, Zhongmin Ou, Sihua Yang, and Da Xing, 2010 "Thermoacoustic molecular tomography with magnetic nanoparticle contrast agents for targeted tumor detection," Med. Phys. 38,4193-4198 .

[11] Patrick Wunderbaldinger,Lee Josephson,and Ralph Weissleder, 2002 "Tat Peptide Directs Enhanced Clearance and Hepatic Permeability of Magnetic Nanoparticles," Bioconjugate Chem. 13, 264−268.

[12] Jerome Segers,Catherine Laumonier, Carmen Burtea, Sophie Laurent, Luce Vander Elst, and Robert N. Muller, 2007 "From Phage Display to Magnetophage, a New Tool for Magnetic Resonance Molecular Imaging," Bioconjugate Chem.18, 1251−1258.

[13] Sophie Laurent,Delphine Forge,Marc Port,Alain Roch,Caroline Robic,Luce Vander Elst,and Robert N. Muller, 2008 "Magnetic Iron Oxide Nanoparticles: Synthesis, Stabilization, Vectorization,Physicochemical Characterizations, and Biological Applications," Chem. Rev. 108, 2064–2110.

[14] Akihiro Tanimoto1, Go Wakabayashi, Hiroshi Shinmoto1, Seishi Nakatsuka1, Shigeo Okuda1,and Sachio Kuribayashi, 2005 "Superparamagnetic iron oxide-enhanced MR imaging for focal hepatic lesions: a comparison with CT during arterioportography plus CT during hepatic arteriography,"J Gastroenterol. 40, 371–380.

[15] Akihiro Tanimoto, Sachio Kuribayashi, 2006 "Application of superparamagnetic iron oxide to imaging of hepatocellular carcinoma," European Journal of Radiology, 58, 200–216.

[16] Gyo rgy Bay, 2009 "Kupffer cells in non-alcoholic fatty liver disease: The emerging view," Journal of Hepatology. 51 ,212–223.

[17] NIU Wanting, ZHANG Luyao, YANG Hao, PAN Min, CHEN Yuquan, 2009 "Preparation and characterization of dextran coated superparamagnetic Iron oxide nanoparticles for medical applications," Technology Review,1000-7857.

# Palmtop spectrophotometer for DNA and protein measurement in micro-nanoliter assays

**Tian Qiu [1, 2], Guoliang Huang [1, 2,*], Xiaoyong Yang[1, 2], Li Ma[2] and Xu Yang[2]**

[1] Department of Biomedical Engineering, the School of Medicine, Tsinghua University, Beijing, 100084, China

[2] National Engineering Research Center for Beijing Biochip Technology, Beijing, 102206, China

[*]Corresponding author: Guoliang Huang (Email: tshgl@tsinghua.edu.cn)

**Abstract**. Spectrophotometer, an important tool in life science, medicine, and analytical fields, usually uses an optical path of 10 mm or more for absorbance measurement of UV light. This corresponds to a sample consumption of ⩾ 50 μL in volume and a narrow measuring range of 0.5-50 ng/μL for nucleic acid samples and 0.05-2 mg/mL for protein samples. Higher concentrations must be diluted for measurement. In this paper, we developed an advanced palmtop spectrophotometer for the measurement of both DNA and protein concentrations in micro-nanoliter assays. We constructed a fiber transmission and a fiber reflection absorbance detection scheme illuminated by either UV-LED or deuterium lamp. The sensitivity of 0.5 ng/μL and a wide measuring range of 0.5-2000 ng/μL in concentrations were obtained for DNA, and the sensitivity of 0.05 mg/mL and a wide measuring range of 0.05-100 mg/mL were also obtained for protein. However, sample consumption is only 1 μL in volume for fiber transmission detection scheme and 500 nL for fiber reflection detection scheme. The linear correlation coefficient of measured concentrations to theoretical concentrations is greater than 0.99. With the profit of this work, a miniaturized spectrophotometer with better sensitivity and wider measuring range can be produced for analytical applications.

## 1. Introduction

Spectrometry is an instrumental analysis method that measures the intensity and wavelength of light emission, absorption, or scattering. Ultraviolet-Visible (UV-Vis) spectrophotometry, based on molecular absorption spectrum caused by electronic transition, is with a wide wavelength range from 200nm to 800nm [1]. It is commonly used in quantitative analysis of organic compounds [2-4]. In ordinary molecular biology experiments, determination of nucleic acid or protein concentration is an essential step. UV-Vis spectrophotometry is the most common, faster and more cost-effective way than other analytical methods [5-6]. Taking microarray as an example, DNA extraction is essential to probe preparation and hybridizing samples preparation. High purity DNA template is the premise of the subsequent steps, which is usually measured by UV-Vis spectrophotometry [7].

Though traditional UV-Vis spectrophotometer [8], using 10mm×10mm quartz cuvette, has high sensitivity, it has the following disadvantages. 1) Large sample consumption, specifically tens of micro-liter or even more volume is required, which loses the suitability for precious biological samples' detection. 2) The linear range is so narrow that high concentration samples have to be diluted and the operation becomes complicated. 3) After each measurement, washing of used cuvette is also time and labor intensive. 4) Traditional instrument, with complicated structures inside, occupies a large area of the already crowded laboratory. All of these problems make the traditional spectrophotometer unsuitable for the high-throughput detection of precious biological samples.

Facing these problems, several micro-UV-Vis spectrophotometers were developed for micro-liter biological samples. Typical commercially available products are Nanodrop ND-1000 [9], GE NanoVue [10]. They take advantage of liquid surface tension to contain samples between two surfaces instead of the traditional cuvette, which not only decreases the sample consumption, but also broadens the measurement range and simplifies the operations. However, some problems still remain unsolved, such as low precision and reproducibility, complicated instrument structure, and high cost of instrument.

Fiber-optic biosensors have attracted great attention because of their potential for sensitive measurements in biological and chemical applications. In our previous work, we have developed several opt-biosensors for detection of interactions between biological moleculars [11-13]. These opt-biosensors are capable of label-free detection of micro-nano-liter biological samples reaction; however, they cannot determine the net concentration of initial samples. The integration of three steps, namely sample preparation, chemical reaction, and results detection on a chip is the ultimate goal to develop a micro-Total-Analysis-System (μTAS) or Lab-on-a-Chip (LOC) for point-of-care diagnostics [14]. As a result, the research of fiber-optic biosensor for measuring initial sample concentration quickly, accurately, and label-freely becomes crucial.

In this paper, we report an advanced palmtop spectrophotometer for UV spectrophotometric detection of micro-nanoliter DNA and protein samples. Taking advantage of surface tension and fiber reflection method, the linear range, detection limit, and reproducibility of the system were proved to be suitable for normal analytical applications. The sample consumption is much less (500 nL minimum) compared with current commercial instruments (2 μL minimum), which is essential for precious biological samples detection. The novel system illuminated by ultraviolet light emitting diode (UV-LED) facilitates simple fabrication process and lower cost. Moreover, without the restriction of UV transparent material, the system is more suitable for integration on μTAS or LOC.

## 2. Materials and methods

### *2.1. Materials*

Single-strand DNA (ssDNA) standard substances, purchased from National Institute of Metrology (Beijing, China), were Salmon Sperm DNA with concentrations of 2050, 998, 501, 100 and 11 ng/μL, respectively. TE buffer (10mM Tris-HCl, pH 8.0, 1mM EDTA, Capitalbio Corporation, Beijing) was used to dilute the standard substances to a series of 19 concentrations ranging from 2000 to 1 ng/μL.

BSA (Bovine Serum Albumin) was used as protein sample. BSA solid powder (Proliant, Ankeny, IA) of 10 g was dissolved in 1 mL PBS (Phosphate Buffered Saline, Four Ring Sunny Bioscience Co., Beijing) to obtain the highest concentration of 100 mg/mL. Then, a series of 15 samples was diluted to concentrations from 100 to 0.05 mg/mL, and the final concentrations were quantified by Nanodrop ND-1000 (Thermo Fisher Scientific, Wilmington, DE).

### *2.2. Apparatus*

A spectrometer (USB2000+, Ocean Optics, Dunedin, FL) was used for the spectral measurement. Micro-deuterium lamp (DT-Mini-2-GS, Ocean Optics) and two UV-LEDs (central wavelength 265 and 285 nm, bandwidth 10 and 10 nm respectively, Sensor Electronic Technology, Columbia, SC), were used as the light source. Quartz fibers with center diameter 115 μm and 400 μm were also purchased from Ocean Optics. Quartz fiber reflective probes were manufactured by Xingyuan Coorperation (Beijing, China). The probe positions 6 fibers around one central fiber with center distances 220 μm and the fibers' diameters are all 200 μm.

*2.3. Construction of the Palmtop spectrophotometer*

Figure 1 shows the fiber reflection detection scheme of the palmtop spectrophotometer, which is mainly a mechanical swing-arm structure. Liquid sample is compressed to a thin film and held still by surface tension between upper surface of the reflective probe (4) and the plane mirror (6). The thickness of the liquid film, nearly half of the optical path length, can be adjusted by screw thread micrometer (2) (GCM4002-TM, Daheng Optics and Electronics, Beijing).

UV Light is emitted from micro-deuterium lamp or UV-LED (13), transmits through incoming fiber (11) and liquid sample (5), and after reflected by plane mirror (6) transmits through the sample (5) once more. The reflective light is collected by six outgoing fibers (10) of reflective probe (4), and finally irradiates into spectrometer (12). Computer receives spectrogram by software (Spectrasuite, Ocean Optics), and the concentrations of samples were calculated by readings of the spectrogram. Magnet (7) is attracted with screw on the fixed arm (3) in order to keep the swing arm (1) and the head of screw thread micrometer (2) in close contact.

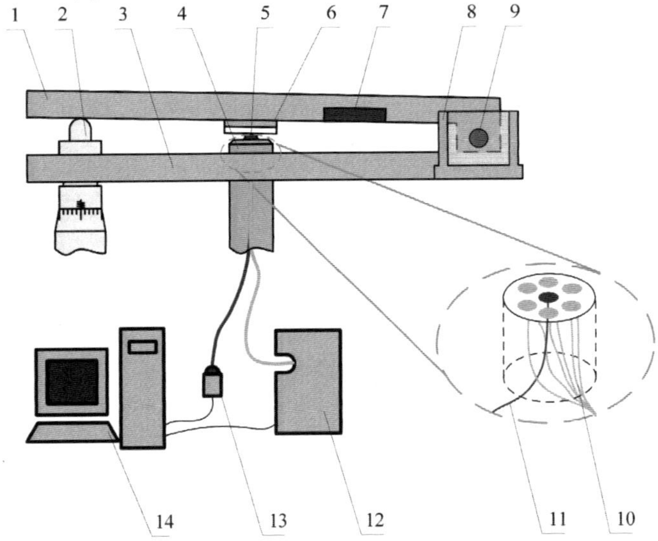


**Figure 1.** Schematic diagram of the miniaturized measurement system
1-swing arm; 2-screw thread micrometer; 3-fixed arm; 4-reflective probe; 5-liquid sample; 6-plane mirror; 7-magnet; 8-pedestal; 9-shaft; 10-outgoing fiber; 11-incoming fiber; 12-micro-spectrometer; 13-UV-LED; 14-computer

**Figure 2.** Photo of the miniaturized measurement system

Another approach of the system, the transmission detection scheme, could be easily transformed by changing connectors of optical fibers, so that the transmitting light emitted from 400 μm fiber below will not be reflected, but received by the upper 115 μm fiber. Figure 2 is a photo of the major part of constructed reflection detection system.

### *2.4. Absorbance measurement*

First, the clearance between the upper surface of the probe and the plane mirror was adjusted to specific distances with a clearance gauge. The deuterium lamp was warmed up. The spectrometer was connected to the computer and sampling parameters were configured. Parameters of spectrophotometer were set up as follows: integration time 0.5 s for deuterium lamp and 30 ms for UV-LED, boxcar 5, average times 3.

After adjustment, the swing arm (1) was swung up, and 0.5-1 μL solvent was added onto the center of the reflective probe (or incoming fiber in transmission scheme). By swinging down the arm to the place as shown in figure 2, a compressed liquid film (5) was formed between the upper surface of the probe (4) and the plane mirror (6), whose thickness, in relation to the measurement path length could be changed by the screw thread micrometer (2).

Background light intensity $I_b$ was measured, after the shutter of the lamp was closed. Then, the shutter was opened and the reflective light intensity of solvent $I_0$ was then measured. After that, the shutter was closed again to protect the user from UV radiation. Both the probe and the mirror (or two fibers' end in transmission scheme) were cleaned by gauze, and then sample solution was added. Subsequently the reflective light intensity of solution $I_1$ was measured by the same steps described above. The path lengths of solution and solvent measurement have to be maintained the same.

In experiments of transmission method, every solution was measured 3 times independently to test the system's feasibility; while in reflection method, every measurement was repeated 15 times independently in order to get a reliable result on the system's performance.

### *2.5. Detection principle and data processing*

Lambert-Beer's law states that absorbance of solution $A$ is proportional to solute concentration and light path length through the solution [15], which is written as

$$A = \varepsilon bc \tag{1}$$

where $\varepsilon$ is absorption coefficient of the substance determined by the solute, $b$ is light path length, $c$ is substance concentration. $A$ is solution absorbance of specific wavelength (260 nm for DNA and 280 nm for protein), which is defined as

$$A = \lg \frac{I_0 - I_b}{I_1 - I_b} \tag{2}$$

where $I_0$ is the transmitted light intensity of solvent; $I_1$ is the transmitted light intensity of solution; $I_b$ is the background intensity, including stray light and dark noise of detector.

When light intensities $I_0$, $I_1$, $I_b$ are measured, the concentration $c$ can be calculated using the following derived formula

$$c = (\lg \frac{I_0 - I_b}{I_1 - I_b}) / (\varepsilon b) \tag{3}$$

where absorption coefficient $\varepsilon$ is 0.03 μL/(ng·cm) for ssDNA and 0.67 mL/(mg·cm) for BSA.

## 3. Result and discussion

### *3.1. Protein measurement using deuterium lamp-fiber transmission method*

First, BSA samples were measured using fiber transmission method with a deuterium lamp. The path length was set to 0.5 mm, and the results are shown in figure 3, where the experimental results are plotted against measurements of commercial instruments. As the fitted line approaches a slope of 1, the measurements more closely match the true concentrations.

The reproducibility is satisfied with standard deviation between measurements lower than ±0.02 mg/mL (in range of 0.2 – 10 mg/mL) and ±1.5% (in range of 10 – 80 mg/mL). High linearity is kept in the range of 1-50 mg/mL, while large negative deviation in high concentration measurement (> 50 mg/mL) was observed. That is because 0.5 mm path length is too large for high concentration detection, in which the intensity of passing light is too low for CCD (charge-coupled devices) to detect.

According to the Lambert-Beer's law (1) and CCD's normal detection range (absorbance is between 0.03-3), 0.2 mm path length can be used to measure protein solutions in range of 2.24-223.88 mg/mL. That represents the main range of normal protein concentrations, hence 0.2mm is theoretically the optimized path length.

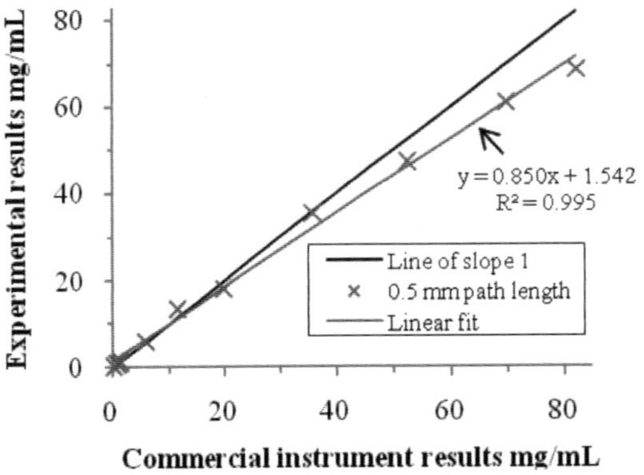


**Figure 3.** Protein measurement result using deuterium lamp-fiber transmission method

*3.2. Protein measurement using LED-fiber transmission method*

In order to fulfill the portability and low cost of a palmtop system, LEDs were tested as light source. As shown in figure 4, standard deviations are lower than ±0.07 mg/mL (in range of 0.2 – 10 mg/mL) and ±0.6% (in range of 10 – 80 mg/mL), which satisfy the basic requirement of spectrophotometry [9-10]. Compared with former experiment illuminated by deuterium lamp, due to the usage of shorter path length (0.2 mm), the reproducibility was improved in high concentration range. Moreover, the large negative deviation at high concentrations was also smaller using 0.2 mm path length, which agrees with the theoretical prediction of 0.2 mm as the optimal length.

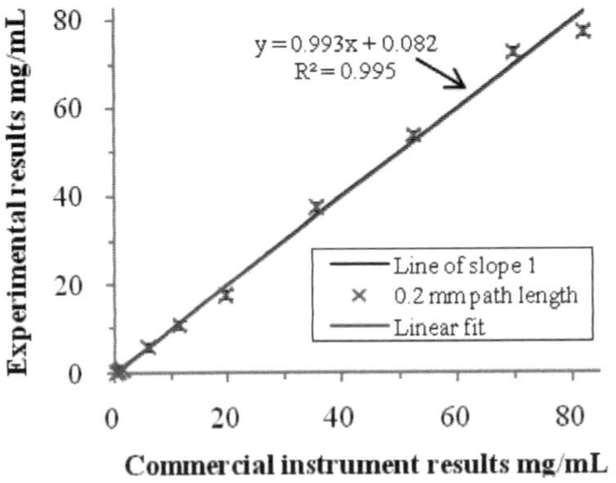


**Figure 4.** Protein measurement result using LED-fiber transmission method

UV-LED was proved to be suitable for absorbance detection in miniaturized system. Because light intensity of UV-LED on 280 nm is much higher than that of deuterium lamp, the integration time of detection was set to 30 ms, less than one tenth of that used for deuterium lamp experiment (500 ms). This not only results in a faster testing time, but also may lead to lower detection limit and higher measurement precision.

*3.3. Protein measurement results using LED-fiber reflection method*

In contrast with the transmission absorbance detection method discussed above, micro sample was illuminated by one fiber in the center, and the reflective light was received by 6 surrounding fibers. 1 mm and 0.2 mm path lengths were both evaluated.

Because of the large concentration range of samples, results of whole measuring range are shown in figure 5(a). In supplement, the commonly used low concentration range 0.05-4 mg/mL is also shown in figure 5(b). High linearity is kept in the whole range (0.05-100 mg/mL) using 0.2mm and 1mm path lengths with correlation coefficient of 0.9995. Standard deviations are lower than ±0.15 mg/mL (in range of 0.05 – 10 mg/mL) and ±4% (in range of 10 – 100 mg/mL), and the detection limit is 0.05 mg/mL.

Positive deviation at high concentration was observed in figure 5. The reason may lie in the fact that the reflective light received by 6 outgoing fibers travels larger path length than that of calculation. According to formula (3), when smaller path length *b* is used in calculation, calculated concentration will reveal a positive deviation.

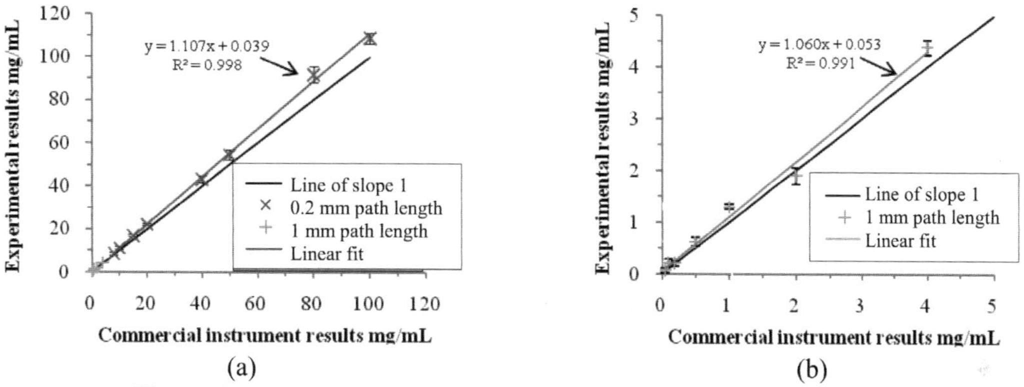


**Figure 5.** Protein measurement results using LED-fiber reflection method
(a) Whole measuring range: 0.05-100 mg/mL; (b) Low concentration range: 0.05-4 mg/mL

*3.4. Nucleic acid measurement results using LED-fiber reflection method*

Similar to protein detection, nucleic acid measurement can also be achieved using the same system. In figure 6(a) and 6(b), the system reveals high reproducibility, high sensitivity, and wide detection range in single-strand DNA standard substances tests. The linear range is 0.5 - 2000 ng/μL with correlation coefficient of 0.9992, and standard deviations are lower than ±2.10 ng/μL (in range of 2 - 100 ng/μL) and ±3% (in range of 100 - 2000 ng/μL). In addition, when measuring the sample of 0.5 ng/μL, the concentration measurement is 0.56±0.12 ng/μL, which means the sensitivity reaches 0.5 ng/μL, better than that of commercial instrument.

Positive deviation in high concentration was also observed in DNA measurement. The major reason is the same, namely the path length error in calculation. The concentration error caused by path length error can be derived as follows basically from formula (1).

$$c_{real} = \frac{A}{\varepsilon \cdot b} \quad (4)$$

When there is a path length error during measurement, the concentration can be stated as

$$c_{meas} = c_{real} + \Delta c = \frac{A}{\varepsilon \cdot (b + \Delta b)} \quad (5)$$

where $c_{real}$ , $c_{cal}$ , $\Delta c$ represent solution's real concentration, measured concentration and concentration deviation respectively.

Use formula (5) minus (4), then $\Delta c$ is derived:

$$\Delta c = \frac{A}{\varepsilon\cdot(b+\Delta b)} - \frac{A}{\varepsilon\cdot b} = \frac{A}{\varepsilon\cdot(b+\Delta b)}\cdot\frac{-\Delta b}{b} = c_{meas}\cdot\frac{-\Delta b}{b} \quad (6)$$

Because $\Delta b$ and $b$ are constant during measurement, it is obvious that the positive deviation caused by path length error is proportional to calculated solution concentration.

In experiments, the linear coefficient of concentration deviation to measured concentration of 0.2 mm and 1 mm path length are 0.99 and 0.91 respectively, so the deviation value can be linearly fitted and corrected. The corrected results after using this linear model arithmetic, shown as red line in figure 6(a), are perfectly matched to the theoretical value (maximum ±2.8 ng/μL deviation in range of 0.5 – 100 ng/μL, ±5% deviation in range of 100 – 2000 ng/μL).

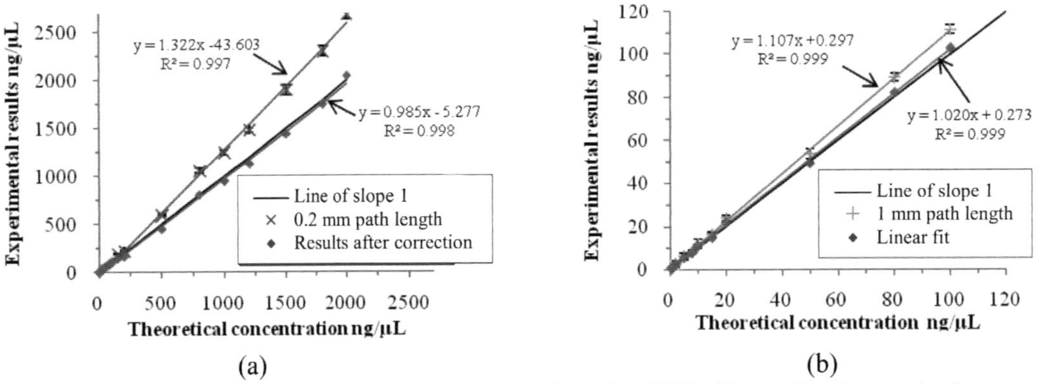


**Figure 6.** Nucleic acid measurement results using LED-fiber reflection method
(a) Whole measuring range: 0.5 – 2000 ng/μL; (b) Low concentration range: 0.5 – 100ng/μL

### 3.5. *Comparison between novel palmtop system and traditional systems*

The novel measurement system is designed to maintain the advantages of micro- spectrophotometer, known as wide test linear range and simple operation. Nevertheless, it offers a different way from traditional systems for spectrophotometric detection. In fiber-reflective system, measuring path length is more than twice as the distance between two solid surfaces, which means that small volume liquid sample is compressed but not stretched to liquid-column so as to reach the path length. Less liquid sample is required for detection (500 nL minimum), which is crucial for precious biological samples detection, and liquid surface tension will not result in non-uniform solute distribution anymore, which facilitates more precise absorbance measurement.

High coaxiality of incoming and outgoing fibers is required in a transmitted measurement system. However, in a reflective system, there is no coaxiality requirement. In addition, calculation by Lambert-Beer's law is based on the portion of relative intensity of solution and solvent, so that perpendicularity between plane mirror and incoming light is also not so strict. As a result, the structure to maintain liquid thin film is simplified, which leads to looser requirement for manufacturing precision and lower cost of the instrument.

In addition, fiber reflection scheme excludes the restriction of UV transparent substrate material. Commonly used materials for biochip fabrication, *i.e.* silicon, gold, aluminum, are all not transparent to UV light, but they are able to reflect UV light. Without the requirement of UV transparency, the reflective system can also work with these materials, so that it has the potential of on-chip absorbance measurements integration in Micro Total Analysis System.

## 4. Conclusion

We developed an advanced palmtop spectrophotometer for both nucleic acid and protein measurement in micro-nanoliter assays. A fiber transmission absorbance detection scheme and a fiber reflection scheme illuminated either by UV-LED or by deuterium lamp were both constructed. With the optimized detection path lengths, the sensitivity of 0.5 ng/μL and a wide measuring range of 0.5-2000 ng/μL were obtained for nucleic acid sample, and the sensitivity of 0.05 mg/mL and a wide measuring range of 0.05-100 mg/mL were also obtained for protein sample. With the profit of this work, a miniaturized spectrophotometer with high reproducibility, sensitivity, and wide measuring range can be produced for analytical applications.

**References**


[1] Li K 2005 *Analytical Chemistry* (Beijing: Peking University Press) p 327

[2] Qi J 2006 *Modern Analytical Measurement Technology* (Shanghai: Tongji University Press) p 244

[3] Waechter H, Bescherer K, Duerr C J, *et al.* 2009 405 nm Absorption Detection in Nanoliter *Anal. Chem.* **81** 9048–54

[4] Xiong Y, Zhu D, Duan C, *et al.* 2010 Small-volume fiber-optic evanescent-wave absorption sensor for nitrite determination *Anal. Bioanal. Chem.* **396** 943–8

[5] Wang C J and Mbi A 2007 A new acetone detection device using cavity ringdown spectroscopy at 266 nm: evaluation of the instrument performance using acetone sample solutions *Meas. Sci. Technol.* **18** 2731–41

[6] Galban J, de Marcos S, Sanz I, *et al.* 2010 CCD detectors for molecular absorption spectrophotometry. A theoretical and experimental study on characteristics and performance *Analyst* **135** 564–569

[7] Xing W and Cheng J 2006 *Experiments in Biochip Technologies* (Beijing: Tsinghua University Press) p 28

[8] Beckman 2002 *DU 800 Spectrophotometer installation and operating instructions*

[9] Thermo Fisher Scientific 2007 *NanoDrop ND-1000 Spectrophotometer V3.5 User's Manual*

[10] GE 2007 *NanoVue User Manual V4.0*

[11] Xu S, Huang G, Deng C, *et al.* 2008 Label-free detection of protein microarray with high throughput surface plasmon resonance imaging (SPRI) *J. Inno. Opt. Health Sci.* **1** 107-14

[12] Zhu J, Deng C, Huang G, *et al.* 2009 Quantitative fluorescence correction incorporating FRET and its use for measurement of hybridization efficiency on microarrays *Anal. Chem.* **81** 1426-32

[13] Li Q, Huang G, Gan W, *et al.* 2009 Dynamic and label-free high throughput detection of biomolecular interactions based on phase-shift interferometry *Proc. SPIE* **7519** 75190H

[14] Cheng J, Kricka L J 2001 *Biochip Technology* (Philadelphia: Harwood Academic Publishers) p 269

[15] Li K 2005 *Analytical Chemistry* (Beijing: Peking University Press) p 335



**Acknowledgments**

This subject is supported by the National Foundation of High Technology of China (2006AA020701、2006AA020803), National Program on Key Basic Research Projects 973 of China (2006CB705700、2011CB707700) and National Natural Science Foundation of China (60978064).

# Effect of halideions on the surface-enhanced Raman spectroscopy of methylene blue for borohydride-reduced silver colloid

**Xiao Dong, Huaimin Gu*, and Fang Liu**
MOE Key Laboratory of Laser Life Science, College of Biophotonics, South China Normal University, 510631, Guangzhou, China

**Abstract.** The surface enhanced Raman scattering (SERS) spectrum of methylene blue (MB) was studied when adding a range of halideions to borohydride-reduced silver colloid. The halideions such as chloride, bromide and iodide were added as aggregating agents to study the effects of halideions on SERS spectroscopy of MB and observe which halideion gives the greatest enhancement for borohydride-reduced silver colloids. The SERS spectra of MB were also detected over a wide range of concentrations of halideions to find the optimum concentration of halideions for SERS enhancement. From the results of this study, the intensity of SERS signal of MB was enhanced significantly when adding halideions to the colloid. Among the three kinds of halideions, chloride gives the greatest enhancement on SERS signal. The enhancement factors for MB with optimal concentration of chloride, bromide and iodide are $3.44\times10^4$, $2.04\times10^4$, and $1.0\times10^4$, respectively. The differences of the SERS spectra of MB when adding different kinds and concentrations of halideions to the colloid may be attributed to the both effects of extent of aggregation of the colloid and the modification of silver surface chemistry. The purpose of this study is to further investigate the effect of halideions on borohydride-reduced silver colloid and to make the experimental conditions suitable for detecting some analytes in high efficiency on rational principles.

## 1. Introduction

Since the phenomenon of surface-enhanced Raman scattering (SERS) was discovered by Fleischman in 1974 [1], and was explained partially by Jeanmaire in 1977 [2], SERS has excited significant interest in many research fields [3-6]. The enhancement of Raman signal in SERS is related to the so-called "SERS-active substrates" which are various metallic structures with sizes on the order of tens of nanometers. It is well known that "surface roughness" is the key factor for such enhancement in SERS. However, the mechanisms giving rise to the surface enhancement is not quite clear yet. In general, the explanation is due to two mechanisms: the electromagnetic mechanism and the charge transfer mechanism [7]. Among the various SERS-active substrates, silver colloids are the most

* Corresponding author. Tel.: +86-20-85216972, *E-mail:* guhm@scnu.edu.cn

commonly used SERS active substrates due to theirs better enhancement effect and simple preparation.

In this paper, borohydride-reduced silver colloid were used as SERS-active substrate for its highly SERS-active features and particle stability. The halideions such as chloride, bromide and iodide were added as aggregating agents to study the effects of halideions on SERS spectroscopy of MB and to observe which halideion gives the greatest enhancement for borohydride-reduced silver colloid. MB was chosen as target molecule because MB has a very strong SERS signal. The SERS spectra of MB were also detected over a wide range of concentrations of halideions to find the optimum concentration of halideions for SERS enhancement. The degree of aggregation and the morphology of particles in colloid could be estimated by the absorption spectra. From this study, some general principles that could be used to guide the choice of conditions which would give the result were found. The purpose of this study is to further investigate the effects of halideions on borohydride-reduced silver colloid and to make the experimental conditions suitable for detecting some analytes in high efficiency on rational principles. In addition, this study will have a profound implication to SERS users about their interpretation of SERS spectra when obtaining these anomalous bands.

## 2. Experimental

### *2.1. Instrumentation*

A 785nm Semiconductor laser was used as excitation source. The Raman system was equipped with a BX-41 Olympus microscope and 45× objective. Spectrometer system (Acton spectro@2300i, Princeton Acton, USA), was used with a 1200 gr/mm holographic grating. An air cooled CCD detector (Pixis 256, Princeton Acton, USA) was used for measuring the Raman signal by an integration method. The signals were detected at concentration of 2.5 mM. The laser power at the sample was 2mW. The concentration of aggregating agents which was added to the colloid was 0.02 M (to final concentration). The Raman spectrum and absorption spectrum were obtained at room temperature. The absorption spectra were detected by spectrophotometer (Lamda 35, PE company, USA). The data processing was operated by using Origin 7.5 software.

### *2.2. Materials*

Silver nitrate ($AgNO_3$), Sodium bohohydride ($NaBH_4$), and Methylene blue (MB) were purchased from Sigma. Sodium chloride (NaCl), sodium iodide (NaI), and sodium bromide (NaBr) were purchased from Sangon. The tri-distilled water was used for all solution preparation.

### *2.3. Silver colloids and Sample preparation*

The silver colloid was prepared by Dirk and Charles borohydride-reduced method [8]. a. 8.5 mg $AgNO_3$ was dissolved in 50 ml $H_2O$ to yield $10^{-3}$ M $AgNO_3$ solution. b. 11.349 mg $NaBH_4$ was dissolved in 150 ml $H_2O$ to yield $2\times10^{-3}$ M $NaBH_4$ solution. c. 50 mL $AgNO_3$ solution was added dropwise to 150 mL of ice-cold $NaBH_4$ solution and the mixture was stirred vigorously. d. The stirring continued for one hour until glassy yellow colour was obtained. The silver colloid was stored in an amber bottle and was laid in dark place. The UV/vis absorption spectrum of the colloid has a maximum absorption band at 392 nm with a full width half-maximum (fwhm) of 50 nm.
The sample that was used for Raman studies was MB. The concentration of MB solution was $1\times10^{-4}$ mM (to final concentration). The solution was stored at ambient temperature. 30 μL aggregating agent and 250 μL MB solution were added to 750 μL silver colloid gradually. The mixture was left for 10 minutes and then the SERS signals were detected.

## 3. Results and discussion

### *3.1. Features of Ag colloid*

The UV-vis spectra of borohydride-reduced silver colloid with different concentration of NaCl are shown in Figure 1. The absorption spectra can provide general informations on the characterization of colloids, such as the average particle size, the shape of particles in colloid, and the aggregation degree of particles in colloid. Because the UV-vis absorption spectrum of colloids is closely related to the localized surface plasmon resonances (LSPR) which are collective metallic nanostructures. From this figure it can be seen clearly that the absorption maximum of this kind of silver colloid is at 492 nm and the full width at half maximum (FWHM) is about 50 nm. From published literature [9], the average size of silver nanoparticles is 15 nm. With the increase of NaCl concentration, the intensity of the surface plasmon absorption band decreases and the FWHM is broadened. The addition of NaCl at low concentrations (0.01 M-0.04 M) does not alter the position of the absorption band dramatically. However, the absorption band is red-shifted when adding NaCl at the highest concentration (0.05 M). It implies that the addition of NaCl at low concentration mainly causes the modification of the silver surface chemistry but causes little aggregation, aggregation occurs when the concentration of NaCl increases to 0.05 mol $dm^{-3}$. The result may be attributed to that most residual ions are displaced by $Cl^-$ and the repulsive force between particles is reduced, which leads to the aggregation of particles.

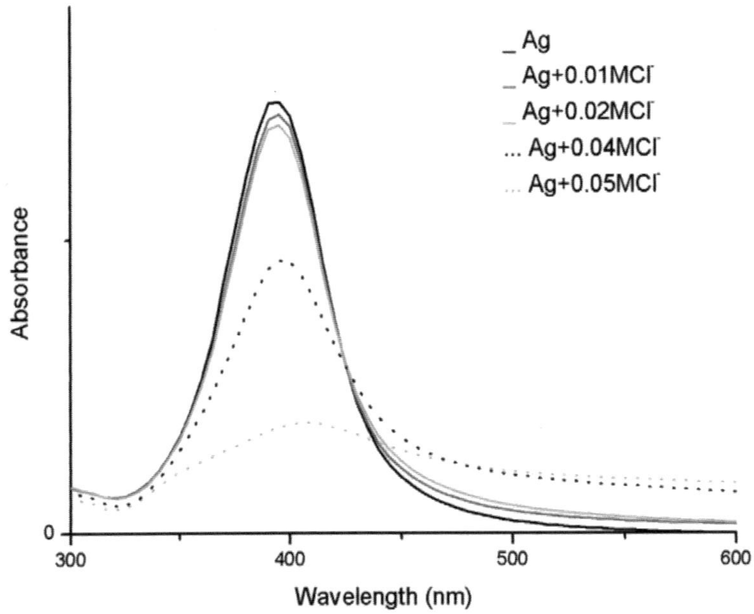


**Figure1.** UV-vis absorption spectrum of the borohydride-reduced silver colloid and the spectra when adding a wide range of concentrations of NaCl, respectivelty.

*3.2. SERS measurements*

Figure 2 shows the SERS spectra of MB on addition of different concentrations of NaCl. Noticeable enhancement is achieved on addition of NaCl at the final concentration of 0.02 mol $dm^{-3}$. As the concentration of NaCl increases, the intensity of SERS signal is reduced. It may be attributed to the reason that the addition of $Cl^-$ at a much lower concentration mainly causes the modification of the silver surface chemistry but causes little aggregation. The modification of silver surface caused by $Cl^-$ changes the binding affinity of MB and it leads to an increase in SERS intensity. When adding more $Cl^-$ to the colloidal system, the SERS intensity of MB drops, and it may be due to the collapse of colloidal particles. Figure 2b shows the normal Raman spectrum of 100 mM MB aqueous solution. The bands are observed at 447, 496, 598, 670, 776, 812, 861,907, 950, 1071, 1163, 1298, 1393, 1474,

1627 cm$^{-1}$. From figure 2a, a new band at 241 cm$^{-1}$ appears when adding $Cl^-$ to the colloid. This band is assigned to Ag-Cl bonds [10]. At the same time, the band of Ag-Cl bonds is in proportion to MB bands. The result is reflected in the 241:1627 cm$^{-1}$ ratio. It can be ascribed to that the MB molecules have positive charge and can easily adsorb on the silver surface which has been treated by $Cl^-$, and the MB molecules can't displace $Cl^-$, which has a high affinity to the silver surface.

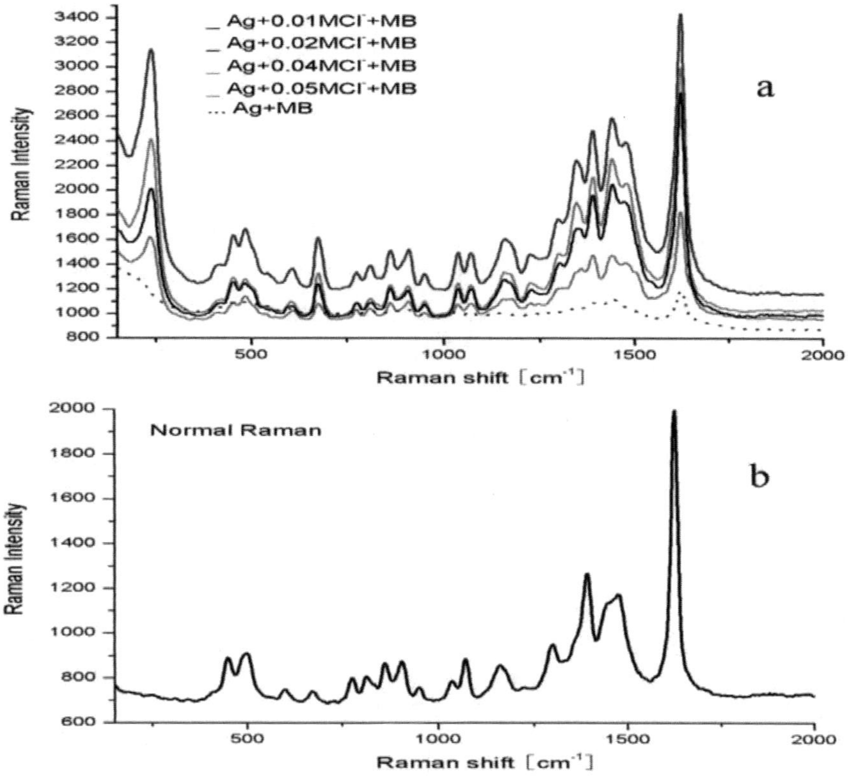


**Figure 2.** a: SERS spectra of MB when adding different concentrations of $Cl^-$ to the colloid. b: Normal Raman spectrum of 100 mM MB aqueous solution.

Figure 3 shows the SERS spectra of MB when adding different concentrations of NaBr to the colloidal system. It is observed that $Br^-$ gives lower enhancement than $Cl^-$ though it has a high affinity for silver surface. The concentrations of $Br^-$ adding to the colloid are 0.02, 0.03, 0.04, and 0.05 mol dm$^{-3}$, respectively. From detecting the SERS spectra of MB, it can be observed that the concentration of 0.02 mol dm$^{-3}$ gives the least enhancement. The bands are observed at 449, 482, 598, 673, 772, 808, 858, 905, 949, 1034, 1072, 1155, 1171, 1221, 1344, 1388, 1438, 1624 cm$^{-1}$. It can be seen that many bands in the SERS spectra have a small shift in position relative to their corresponding normal Raman bands, and some bands split in the SERS spectra. For instance, the bands at 447, 496 cm$^{-1}$ which have been assigned to C-N-C skeletal bending [11] in normal Raman spectrum shift to 449, 482 cm$^{-1}$ in SERS spectra, respectively. The bands at 1393 and 1627 cm$^{-1}$ which have been assigned to C-N stretching and C-C stretching [11] in normal Raman spectrum shift to 1388, 1624 cm$^{-1}$ in SERS spectra. The band at 1163 cm$^{-1}$ in normal Raman spectrum splits up into two bands at 1155 and 1171 cm$^{-1}$. The shift and splits of these Raman bands indicate that MB molecules may be chemisorbed on the silver nanoparticles surface and the chemical effect is the main mechanism responsible for the relative shifts. The intense band at 241 cm$^{-1}$ which is assigned to the Ag-anion bonds doesn't appear. That implies that $Br^-$ adsorbed on the silver particles can't form the strong Ag-Br bonds.

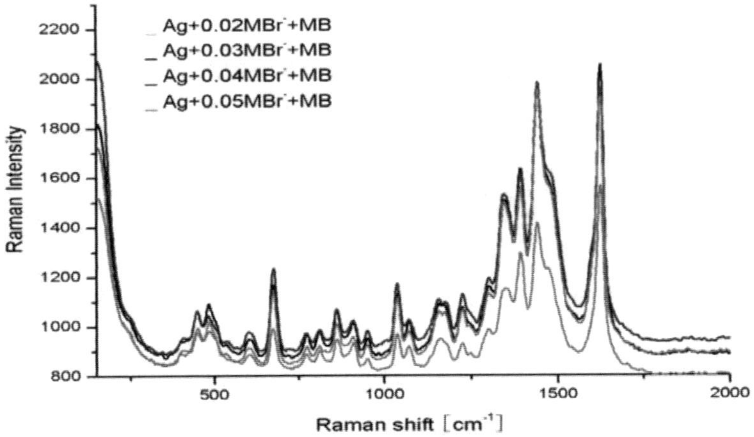


**Figure 3.** SERS spectra of borohydride-reduced silver colloid when adding different concentrations of NaBr.

Figure 4 shows the SERS spectra of MB when adding NaI to the colloid. The bands are observed at 449, 489, 598, 673, 858, 949, 1032, 1068, 1338, 1385, 1435, 1621 $cm^{-1}$. The intensity of SERS signal is greatly reduced, compared with the SERS spectra when adding $Cl^-$ or $Br^-$ as aggregating agents. From these detection it can be calculated that silver colloid with appropriate concentration of anions can enhance the Raman signal several orders of magnitude. The enhancement factors for MB with optimal concentration of $Cl^-$, $Br^-$, and $I^-$ are $3.44\times10^4$, $2.04\times10^4$, and $1.0\times10^4$, respectively. It shows that $I^-$, the most strongly binding anion, gives the least enhancement for SERS. $Cl^-$ gives the greatest enhancement among the series anions. From the numerous investigations on the change of SERS signal from silver colloid on addition of different halideions at various concentrations, it is recognized that there are at least two effects occurring: the extent of aggregation of the colloid and the modification of silver surface chemistry. The differences of the SERS spectra of MB when adding different kinds and concentrations of halideions to the colloid may be attributed to the both effects.

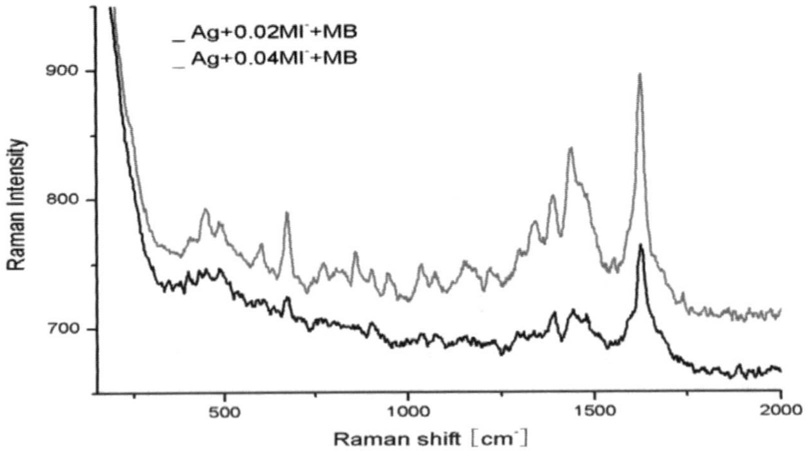


**Figure 4.** SERS spectra of borohydride-reduced silver colloid when adding different concentrations of NaI.

## 4. Conclusion

The intensity of SERS signal of MB was enhanced significantly when adding halideions to the colloid. The species and concentration of the halideions determine the aggregation degree of the silver colloid. The addition of halideions can also modify the silver surface chemistry. Among the three kinds of halideions, chloride gives the greatest enhancement on SERS signal. The addition of NaCl at low concentration mainly causes the modification of the silver surface chemistry but causes little aggregations, aggregation occurs when the concentration of NaCl increases to a higher concentration. From the numerous investigations on the change of SERS signal from silver colloid on addition of different halideions salts at various concentrations, it is recognized that there are at least two effects occurring: the extent of aggregation of the colloid and the modification of silver surface chemistry. The differences of the SERS spectra of MB when adding different kinds and concentrations of halideions to the colloid may be attributed to the both effects. The purpose of this study is to further investigate the effect of halideions on borohydride-reduced silver colloid and to make the experimental conditions suitable for detecting some analytes in high efficiency on rational principles.

## 5. Acknowledgment

This research is supported by the National Natural Science Foundation of China (60678050).

## 6. References

[1] Fleishman M, Hendra P J. and McQuillan A J 1974 *Chem. Phys. Lett.* **26** 163.

[2] Janmaire D L and Van Duyne R P 1977 *J. Electroanal. Chem. Interfacial Electrochem* **84** 1.

[3] Premasiri W R, Moir D T, Klempner M S, Krieger N, Jones G and Ziegler L D 2005 *J. Phys. Chem.* B **109** **(1)** 312.

[4] Peng Y D, Niu Z J, Huang W, Chen S and Li Z L 2005 *J. Phys. Chem.* B **109** **(21)** 10880.

[5] Barhoumi A, Zhang D M, Tam F and Halas N J 2008 *J. Am. Chem. Soc.* **130** **(16)** 5523.

[6] Michaels A M, Nirmal M, Brus L E 1999 *J. Am. Chem. Soc.* **121** **(43)** 9932.

[7] Hering K, Cialla D, Ackermann K, Dörfer T, Möller R, Schneidewind H, Mattheis R, Fritzsche W, Rösch P and Popp J 2008 *Anal. Bioanal. Chem.* **390** 113.

[8] Guingab J D, Lauly B, Smith B W, Omenetto N, Winefordner J D 2007 *Talanta* **74** 271.

[9] Caryn S S, Brittany M G and Russell H G 2009 *J. Phys. Chem.* C **113** **(1)** 74.

[10] Steven E. J. B and Narayana M. S. S 2005 *J. Phys. Chem.* A **109** 7405.

[11] Xiao G N and Man S Q 2007 *Chem. Phys. Lett.* **447** 305.

# Development of a Novel Spectrophotometer for Biochemical Analyzer Based on Volume Holography Transmissive Grating and Linear CCD

**Zhong Ren*[1], Guodong Liu [1], Longmin Dai [2], Zhen Huang [1], Lvming Zeng [1]**
[1]Key Laboratory for Optoelectronics and Communication of Jiangxi Province, Jiangxi Science & Technology Normal University, Nanchang 330013, China
[2]Department of Engineering Technology, Eastern Airlines Co. Ltd, Pudong, Shanghai 201202, China

Email: renzhong0921@163.com

**Abstract.** The classical surface-embossed plane and concave grating are usually used as the diffraction grating in some spectrophotometers. But the minute cracks are produced on the surface of the gratings' grooves, which leads to generate the stray-light and decrease the efficiency of instrument. Therefore, a novel custom-built spectrophotometer for BCA is developed in this paper. Meanwhile, the volum holography transmissive(VHT) grating is used as the diffraction grating in this spectrophotometer. Additionally, a high resolution CCD and data acquisition (DAQ) card with combined the virtual software platform based on LabVIEW are used to design the spectral acquisition and analysis system. Experimental results show that the spectral range and the diffraction efficiency of the spectrophotometer for BCA are greatly increased. The spectral range of the spectrophotometer for BCA can reach 300-1000 nm, its wavelength resolution can reach 1nm. And, it uses the back-splitting-light technology and multi-channel parallel analysis. Compared with other same types, this spectrophotometer has many advantages, such as, higher efficiency, simpler algorithm, higher accuracy, cheaper cost and fewer stray-light and higher imaging quality, etc. Therefore, this spectrophotometer for BCA based on VHT grating will has the greatly potential values in the fields of the biochemical or medical research.

## 1. Introduction

* Corresponding author. Email: renzhong0921@163.com, Phone: +86-0791-3831968.

Biochemical analyzer (BCA) is a kind of instrument based on the theory of Lambert-Beer. It is often used to measure the clinical and chemistry indexes of blood, such as, the glucose, bilirubin, albumin, uric-acid, cholesterol, HGB, etc. Because it can truly and rapidly offer test data for doctor or chemistry identifier, it plays an important role in clinic diagnose and chemistry inspection. Spectrophotometer is the key component in the BCA, the performance of the BCA is connected with the performance of the spectrophotometer[1-2]. From the arising of the spectrophotometer up to now, some classical plane grating splitting-light system, e.g. Czerny-Turner (C-T) system, and the concave grating splitting-light system have been used into some spectrophotometers or spectrometers[3-4]. But these classical surface- embossed gratings will not avoid producing some minute cracks on the surface of the gratings' grooves at the time of manufacture, these cracks will lead to generate the stray-light and decrease the efficiency. Moreover, the conventional reflection gratings work in the Littrow geometry, in which the diffracted light can retro-reflect back toward the source of incident light and the off-axis geometry makes it difficult to correct aberration efficiently. These factors impact the performance of spectrophotometer. However, the VHT grating is a novel grating which is made by the photosensitive polymer material exposed by two periodically-changed beams by virtue of the holography technique, it will not produce the scattered light due to the absence of groove of classical surface-relief gratings. So, its spectral range can be greatly increased and its theoretical diffraction efficiency can approach 100%. The VHT grating has been used into some fields[5-6]. Therefore, a novel custom-built spectrophotometer for BCA is developed in this paper. Meanwhile, the VHT grating is used as the splitting-light apparatus of the spectrophotometer. Experimental results show that some performances of the custom-built spectrophotometer for BCA based on the VHT grating are improved, such as, the wavelength range is widened, the resolution, accuracy and efficiency are all increased, the range of the wavelength can reach 300-1000nm, its resolution can reach 1nm.

## 2. Theory

### *2.1. The theory of Lambert-Beer*

The spectrum analysis theory of the spectrophotometer for BCA is based on the Lambert-Beer theory[7]. That is, when a beam of parallel monochromatic light vertically passes through a homogeneous and non-absorbing or scattering material, the absorbance A is proportional to the concentration of the substance $c$ and the thickness of the absorption layer $b$ .

The Lambert-Beer theory can be expressed as follows:

$$A = Log\left(\frac{I_0}{I_t}\right) = Log\ (I_0) - Log\ (I_t) = \varepsilon bc \tag{1}$$

Where, A is represented to medium absorbance of the incident light; $I_0$ is intensity of the incident light; $I_t$ is intensity of the transmission light; $\varepsilon$ is Moore absorbance parameter of the medium; $c$ is concentration of the medium; $b$ is thickness of the medium.

### *2.2. The theory of the VHT grating*

The mainly manufacturing process of the VHT grating is represented as follows: firstly, a film of photorefractive polymer material is exposed by two beams of coherent light irradiated from a laser

source to take on a holographic fringe pattern. Then, two pieces of glass or silica substrate are merged on two surfaces of the film. To enhance the efficiency of the light, the material of the ultravoliet-enhanced-aluminum coating is coated on the surface of glass or silica substrate.
The structure of the VHT grating is shown in Figure.1.

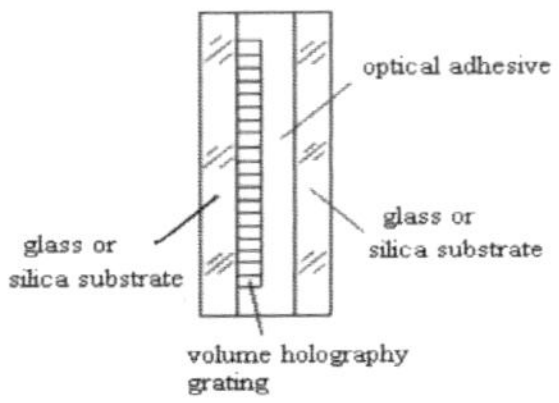


Figure1. The structure of the VHT grating

In order to save the pattern on the grating, the desired refractive modulation can be got by Equation (2)

$$n(x,z) = n + \Delta n \cos(x \cdot 2\pi/\Lambda) \tag{2}$$

Where $n$ is the average value of the modulated refraction, $\Lambda$ is the fringe spacing of the grating equal to 1/v for fringe planes orthogonal to the grating surface (v is the grating frequency) and $\Delta n$ is the refractive modulation.

Supposed that two interference beams irradiate into the photorefractive crystal with the thickness of d and the incident-light with the reference-light lie on the same side of the grating, the VHT grating is formed by the interaction of two interference beams and the generation of the diffraction-light should be satisfied with the Equation (3),

$$2\text{n}\Lambda\sin\theta = \lambda \tag{3}$$

Where $\Lambda$ is the period of the VHT grating, $\theta$ is the angle of the incident light, $\lambda$ is the wavelength of the incident light.

When the angle of incident light $\theta$ is approached to the angle of Bragg, the diffraction efficiency can reach 100%. The Bragg condition can be represented in the Equation (4),

$$m\lambda = \Lambda 2n\sin\theta_0 \tag{4}$$

Where θ0is the Bragg angle in the grating medium.

Kogelnik[8] developed a two-wave, first-order, coupled wave analysis that can be used to approximately estimate the first-order efficiency of a VHT grating. The peak diffraction efficiency $\eta$ of s-polarization light is represented in the Equation (5),

$$\eta_s = \sin^2\left(\frac{\pi\Delta nd}{\lambda\cos\alpha}\right) \tag{5}$$

From Equation.(5) we know that, for the VHT grating, when the angle of incident light equal to the angle of Bragg, the change of the diffraction efficiency η is the square of the function of $\sin(\cdot)$ following the increasing of the thickness $d$ and the refractivity.

## 3. System design of spectrophotometer for BCA based on VHT grating

The spectrophotometer for BCA system mainly consists of the irradiation unit, splitting-light and detection unit, signal processing unit, control unit, etc. where the irradiation unit includes the lamp source, collimation lens, cuvette and objective lens. The splitting-light and detection unit includes the slit, collimation lens, VHT grating, focusing lens, CCD, driver and trigger module of CCD. The signal acquisition and processing unit includes the off-DC amplifier, adjustable-gain amplifier, signal filter, the A/D data acquisition (DAQ) card and the data register. The control unit includes the computer and the software. The basic principle of this system is shown as follows: Firstly, the composite light with certain power is emitted by the lamp source and is collimated by the collimation lens, then castes into the cuvette in which the sample, e.g. blood, is loaded. According the Lambert-Beer theory, parts of light are selective-absorbed by the sample, then the transmitted light is focused by the focalizing lens into the splitting-light unit, the light pass through the slit, collimation lens, VHT grating and objective lens in turn. At this time, the composite light is separated by the VHT grating into the monochromatic light and is focused by the objective lens to become the spectrum-band to cast onto the surface of CCD, The light is transformed into electricity signal by CCD based on the effect of the photo-electricity transform. Then, the electricity signals are amplifier by the off-DC amplifier, the adjustable-gain amplifier and is filtered by filter to remove some noise, the signal is acquired by A/D-DAQ card and transformed into the digital signals, the digital signals are stored by the data register and sent into personal computer through PCI bus to be analyzed or displayed by computer and software. The schematic diagram of the spectrophotometer for BCA is presented in Figure.2. where, the splitting-light unit, CCD detector and CCD driver module are fixed into a sealed and inside-blacked aluminium-alloy box, its size is 10×8×8cm.

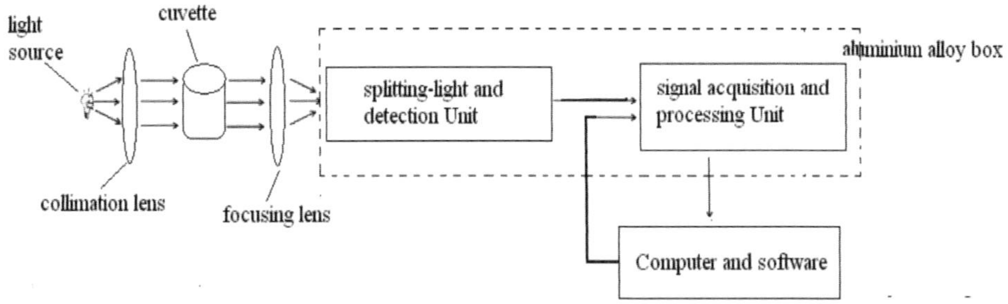


**Figure.2** The schematic diagram of the spectrophotometer for BCA

*3.1. Light source*

According the different demand, there are many light sources can be used into the BCA, such as, the halogen lamp, laser diode, solid laser, etc. In this work, the halogen-tungsten lamp(Philip6605,±12v/20W) is chosen as the light source of the spectrophotometer for BCA, its spectrum range is from 0.3 μ m to 2.5 μ m. It has some characteristics, such as, full-waveband, cheap cost, etc. Its spectrum has the characteristics of continue, no-peaked, the middle of its spectrum is very flat. And its energy is all strong in the range of waveband from 200nm to 1000nm, the waveband is usually used in the bio-chemical analyzer of clinical diagnose or chemical checking.

*3.2. Design of the splitting-light and detection unit based on the VHT grating*

The splitting-light unit is a key part of the spectrophotometer for BCA, its performance is directly association with the performance of the BCA. In past years, the plane and the concave grating have been used in the classical spectrophotometer, As we all know, these conventional gratings are worked by the groove, these grooves are surface-embossed on the surface or inside of the grating at the time of manufacture, So, there has a problem exists in these grating, that is, the embossed mode will generate some minute cracks on the surface of the gratings' groove. These minute cracks will produce the stray-light in the splitting-light system to decrease the performance of the spectrophotometer for BCA. And, this problem is very difficult to avoid in the conventional grating. Therefore, the VHT grating is used in this work. The splitting-light schematic diagram of spectrophotometer for BCA based on the VHT grating is shown in Figure.3.

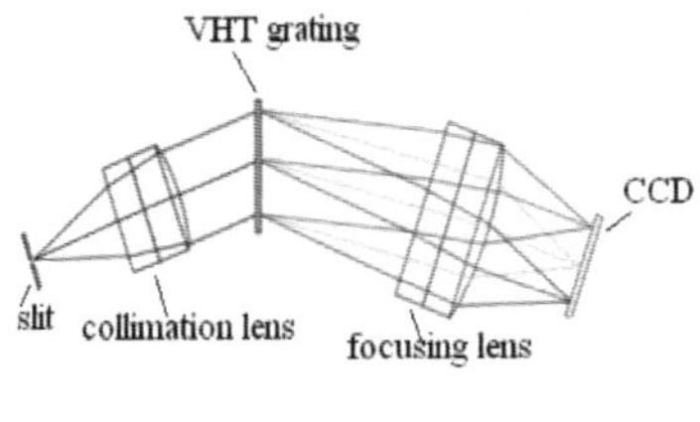


**Figure.3** The splitting-light diagram of spectrophotometer for BCA based on the VHT grating

In the unit of splitting-light, the slit, collimation lens, VHT grating lie on the same optical axis, the VHT grating, focusing lens and CCD lie on another optical axis. The angle of two lines is about 38.4°, and the curvature of the collimation lens and the focusing lens is 26.33mm and 6.24mm respectively, the material of two lenses are fused quartz with the surface ultraviolet-enhanced-aluminum coating. The CCD is custom-built photodiode array, its efficient pixels are 21, the size of each pixel is about 1mm×5mm, the spectral width is about 53mm. The diagram of the custom-built linear CCD is shown in Figure.4. The working frequency of CCD is about 4MHZ, but it can be adjusted by changing the level of trigger in CCD driver module to get more integration time of CCD. In this unit, since the VHT grating is adopted, the efficiency of the spectrophotometer is more than the spectrophotometer based on plane and holography concave grating.

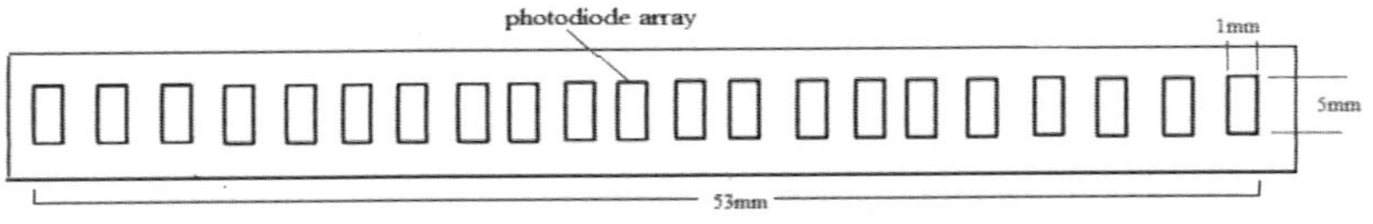


**Figure 4.** The diagram of the custom-built linear CCD

*3.3. Design of signal acquisition and processing Unit*

The biochemical signals acquired by CCD is usually very weak analog signals, it must be amplified by the amplifier. So, the off-current amplifier and the adjustable-gain amplifier are used. In addition, the acquired signals are often polluted by noises, so, the filter must be used. The filtered signals are transformed into the digital signals by the A/D-DAQ card and stored by the data register through the bus trigger unit and the data arrangement unit. Finally, the stored signals are sent into CPU to be processed. The schematic diagram of the signal acquisition and processing unit is shown in Figure.5.

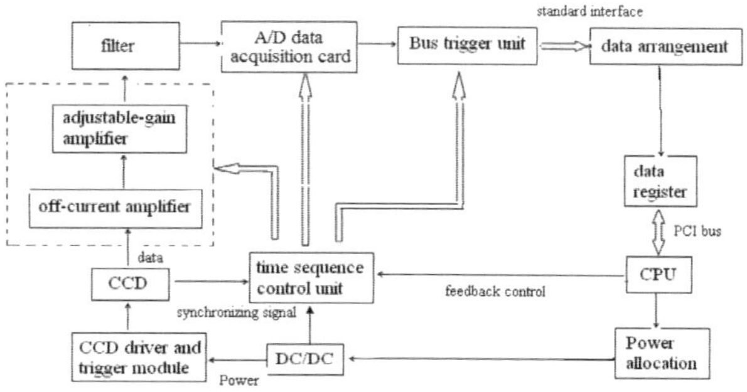


**Figure 5.** The schematic diagram of the signal acquisition and processing unit

*3.4. Design of the virtual-spectrophotometer for BCA based on LabVIEW*

Virtual instrument (VI) is a trend of the future instrument development. In this paper, we use the VI technology based on LabVIEW [9] to acquire and analyze the spectrum. Compared with the traditional spectrophotometer controlled by SCM, DSP chips or other digital circuits, this signal acquisition and processing unit for spectrophotometer based on VI technology have many advantages, such as, higher development efficiency, shorter designing time, cheaper cost and more function, etc. The front panel of the virtual-spectrophotometer for BCA is shown in Figure.6. The virtual- spectrophotometer for BCA has many functions, such as, spectrum acquisition, data storing, threshold estimation, auto-adjustment of the integration time of linear CCD and multi-window displaying, etc.

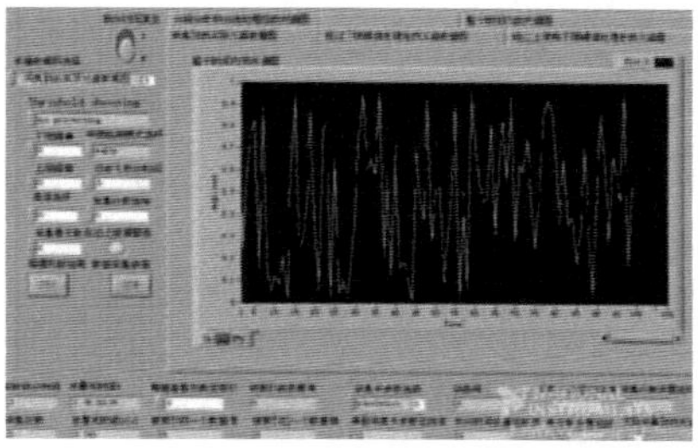

**Figure 6.** The front panel of the virtual-spectrophotometer for BCA

## 4. Experiments and results

*4.1. Materials*

Halogen-tungsten lamp(Philip6605,±12v/20W) is chosen as the lamp source of the spectrophotometer for BCA , its spectrum range is from 0.3μm to 2.5μm. The high-pressure mercury lamp is used to as the calibration light source. The curvature of the collimation lens and the focusing lens is 26.33mm and 6.24mm respectively. Four lenses are made by the fused quartz with surface coated ultraviolet-enhanced-aluminum coating. The quartz colorimetric ware is chosen to as the cuvette of the spectrophotometer, its size is 12.4×12.4×45mm, its optical path is 10mm. The material of the splitting-light unit's shell is aluminum, the width of the slit is 60μm. The VHT grating(1200lines/550nm, incident angle is 19.2°) is made by the synthetic photopolymer with high refractive index modulation, and its encapsulation is the quartz glass with surface coated the ultraviolet-enhanced-aluminum coating, the thickness of the VHT grating is about 10mm, the

efficiency is about 80% at the wavelength of 340nm and 750nm. The focusing lens is F-θ lens, its angle of the imaging is about 35.6°. The CCD (TCD1251UD, TOSHIBA, Japan) is glass capsulated, its size is 41.6×9.65×3.22mm, the length of the photosensitive section is 29.7mm, the efficient pixels is 2700, the size of each pixel is 11×11×11μm. The resolution of the A/D-DAQ card (QY-PCI-S12H, Co.QIYAO, Tianjin, China) is 14bits, its work frequency is about 5MHZ. The control unit is Lenovo Computer(Windows professional XP) and software is LabVIEW(Version.8.0 student edition, NI).
The photography of the custom-built setup for is shown in Figure.7.

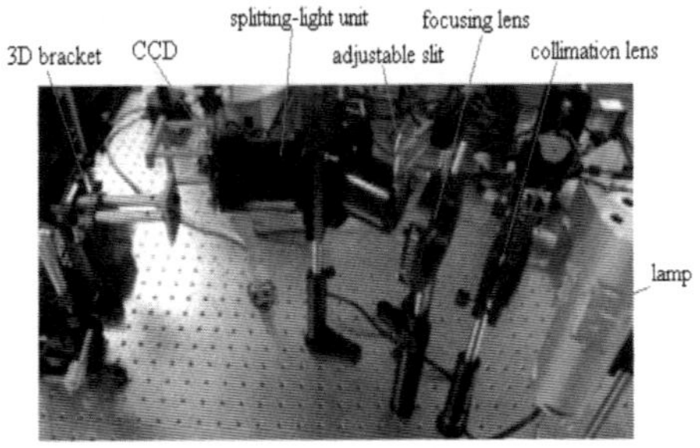


**Figure 7.** The photography of the custom-built setup

*4.2. Calibration of the mercury lamp*

In the experiment, we firstly used the mercury lamp as the calibration light source, the linear CCD with combined the A/D-DAQ card as the spectral detector and the virtual-spectrophotometer for BCA based on LabVIEW to test the resolution of the spectrophotometer for BCA. The characteristic spectrum of the mercury lamp is gotten and shown in the Figure.8. From Figure.8, we can see that the wavelength of 365nm, 404.7nm, 407.8nm, 435.8nm, 546.1nm, 577nm, 579nm can be easily distinguished. So, it is adequately indicate that the spectral resolution of the spectrophotometer for BCA is less than 2nm, wavelength accuracy is satisfactory and the structural design of the spectrophotometer is quite ideal.

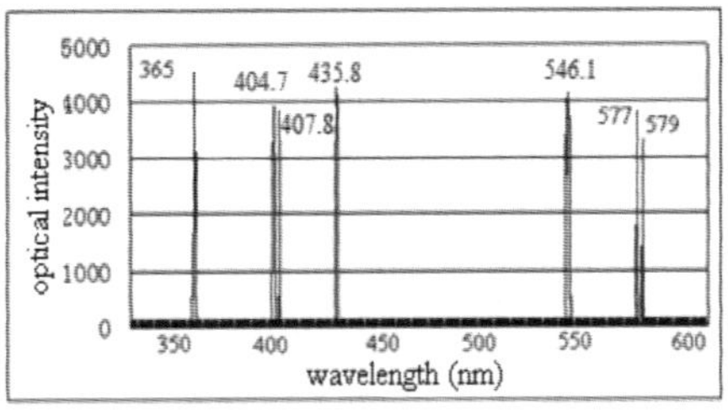


**Figure 8.** The characteristic spectrum of the mercury lamp

*4.3. The experiment of the full wavelength spectrum*

Then, to do the experiment of the full-wavelength, the Halogen-tungsten lamp is used to as lamp source of the spectrophotometer for BCA, the custom-built linear CCD with combined A/D-DAQ card to as the spectral acquisition system. The virtual-spectrophotometer for BCA based on the LabVIEW is used as the software platform to control the spectral acquisition and analysis. The full wavelength spectrum of the spectrophotometer for BCA is got, and the spectrum at the wavelength of 340nm, 380nm, 405nm, 420nm, 450nm, 460nm, 480nm, 505nm, 520nm, 546nm, 578nm, 600nm, 620nm, 650nm, 700nm,

750nm, 800nm, 850nm, 900nm, 950nm and 1000nm can be clearly distinguished.

**Acknowledgements**

This work has been supported by the Chinese National Foundation for Natural Scientific Research grant (60767001). the Science and Technology Pillar Program of Jiangxi Province (2009BSA12700), the Natural Science Foundation of Jiangxi Province (2008GQW0013), and the Scientific Research Foundation of Jiangxi Provincial Education Bureau (GJJ09307,GJJ10243), and the Natural Science Fundation of Jiangxi Science and Technology Normal University (No.KY2009ZY10).

**References**

[1] Massey, P., Strobel, K., Barnes, J. V., & Anderson, E. 1988 Spectrophotometric standards. Astrophysical Journal, Part 1, vol.328, 315-333.

[2] SHI Junfeng, HUI Mei. 2003 Micromation and applications of spectrometers. Optical Technique ,29(1).13-16.

[3] KM Rosfjord, RA Villalaz and TK Gaylord 2000 Constant-Bandwidth Scanning of the Czerny-Turner Monochrometer Applied optics. Vol. 39, Issue 4, 568-572.

[4] Takashi Onaka 1995 Aberration-corrected concave grating for the mid-infrared spectrometer aboard the Infrared Telescope in Space. Applied Optics, Vol. 34, Issue 4,.659-666.

[5] SC Barden, JA Arns, WS Colburn, JB Williams 2000 Volume Phase Holographic Gratings and the Efficiency of Three Simple VPH Gratings. The Publications of the Astronomical Society of the Pacific, Vol. 112, Issue 772, 809-820.

[6] SC Barden, JA Arns 1998 Volume-Phase Holographic Gratings and their Potential for Astronomical Applications. Optical Astronomical Instrumentation. Proc. SPIE Vol.3355, 866-876.

[7] LUO Q Y, DENG Y ZH 1992 A analyze with photometric. Science Press, Beijing. 57-60.

[8] Kogelnik and Herwig 1969 Coupled wave theory for thick hologram gratings. The Bell System Technical Journal, Vol. 48, No. 9, 2909-2947.

[9] Robert H. Bishop 2006 National Instruments, USA.

# Development of a Novel Breast Cancer Detector based on Improved Holography Concave Grating Imaging Spectrometer

**Zhong Ren[1], Guodong Liu, Lvming Zeng, Zhen Huang**
Key Lab of Optic-electronic &Communication, Jiangxi Science &Technology Normal University, Nan Chang, Jiangxi, 330038, China

Email: renzhong0921@163.com

**Abstract.** Breast cancer can be detected by B-mode ultrasonic imaging, X-mammography, CT imaging, and MRI. But some drawbacks existed in these methods, their applications was limited in some certain. So, a novel high resolution breast cancer detector (BCD) is developed in this paper. Meanwhile, an improved holography concave grating imaging spectrometer (HCGIS) is designed. In this HCGIS, the holography concave grating is used as the diffraction grating. Additionally, CCD with combined image acquisition (IAQ) card and the 3D scan platform are used as the spectral image acquisition component. This BCD consists of the light source unit, light-path unit, check cavity, splitting-light unit, spectrum acquisition and imaging unit, signal processing unit, computer and data analysis software unit, etc. Experimental results show that the spectral range of the novel BCD can reach 300-1000 nm, its wavelength resolution can reach 1nm, and this system uses the back-split-light technology and the splitting-light structure of holography concave grating. Compared with the other instruments of breast cancer detection, this BCD has many advantages, such as, compacter volume, simpler algorithm, faster processing speed, higher accuracy, cheaper cost and higher resolution, etc. Therefore, this BCD will have the potential values in the detection of breast disease.

## 1. Introduction

Breast cancer is becoming the major diseases of threatening to the health of women. In some developed Europe countries and USA, the breast cancer has occupied the second place in women's cancer diseases. In some cities of China, such as, Beijing, Tianjin, Wuhan, etc, the incidence of breast cancer is steadily arising. Except for the treatment of the breast cancer, the diagnosis of breast cancer in early stage is most important. In recent decades, some methods based on the ultrasonic, optical and magnetic technology have been used for the detecting of the breast cancer, such as, the conventional hand-touch method, B-mode ultrasonic imaging [1], the X mammography [2], CT imaging [3] and MRI [4], etc. Although these methods have gotten some achievements in some certain, the applications are limited by some drawbacks. For the conventional hand-touch method, this method is limited due to the inconvenience for the patients and the doctors, especially for the different gender. For the B-mode ultrasonic imaging, its spatial resolution is low. For the X mammography, its application is also limited due to the low contrast and spatial resolution, and the X-ray is harmful to the health of the patients. For the CT imaging, although it has the higher resolution and contrast, the algorithm of image reconstruction is very complicated. For the MRI, its cost is very expensive for the patients and hospitals. Since about 70-90% of the breast tissues and gland are composed of the fat, the breast tissue has the characteristic of the light-penetrability [5]. When a beam of the light irradiates

[1] Corresponding author. Email: renzhong0921@163.com. Phone:+86-0791-3831968.

into the breast, the hemoglobin of the blood and the cancer tissues will absorb the light, but the breast fat and the gland are nearly transmissive. Some researches have proved that the optical absorbance of the cancer tissues is 3-5 times more than that of the normal tissues [6]. For the image, the images of the blood vessel and the cancer tissues for the light are the shadow area, the normal tissues are the bright area. For the breast cancer in the early stage, the changes of the tumor cell and the hemoglobin are not obvious. It is difficult to diagnose the cancer tissues simply relies on the transmission image. But, the spectral changes are obvious. And these changes are the important information. Therefore, in order to overcome the drawbacks of these methods above mentioned, a novel breast cancer detector based on the imaging spectrometer is developed in this paper. This detector is the combination of the optics and the image. Not only the image information but the spectral information of the breast tissue can be gotten. So, this novel detector can offer more accurate and more information about the breast diseases. Meanwhile, an improved custom-built holography concave grating imaging spectrometer (HCGIS) also is designed to as the splitting-light unit of the detector. In additional, the virtual instrument (VI) technology based on LabVIEW is used to control the spectral image acquirement of CCD with combined the image acquisition (IAQ) card. Experimental results show that this detector can get not only the clear image of the breast cancer tissue, but also the spectrum of the tissue. It is proved that the method based on the image with combined the optical spectrum is feasible in the breast cancer diagnosis. Compared with others, this detector has many advantages, such as, non-invasive, more accurate, simpler algorithm, more convenient and cheaper cost, etc. And, experimental results show that the spectral range of the imaging spectrometer can reach 300-1200nm, the resolution can reach 1nm. Therefore, this detector has the potential values in the biochemical research and the diagnosis of the breast cancer.

## 2. Theory

From the 20th century up to now, the spectrum analysis technology has been researched by many people in bio-chemical and medical fields [7, 8]. The theory of the breast cancer diagnosis by means of the spectrum analysis is based on the Lambert-Beer theory [9]. That is, we use the absorbed or reflected spectrum to analysis the pathology characteristic of the breast tissues for the patient.
The Lambert-Beer theory can be expressed as follows:

$$A = Log\left(\frac{I_0}{I_t}\right) = Log\ (I_0) - Log\ (I_t) = \varepsilon bc \tag{1}$$

Where, A is represented to medium absorbance of the incident light; $I_0$ is intensity of the incident light; $I_t$ is intensity of the transmission light; $\varepsilon$ is Moore absorbance parameter of the medium; $c$ is concentration of the medium; $b$ is thickness of the medium.

## 3. System design of BCD based on HCGIS

The BCD mainly consists of the lamp source unit, light-path unit, check cavity, splitting-light unit, spectrum acquisition and imaging unit, signal processing unit, control unit, display unit, etc. where, the lamp source unit consists of the lamp, reflecting cup, fan, lamp shade, the lamp switch and the power of the lamp. The light-path unit includes the first collimating lens, filter components and the first focusing lens. The check cavity is used to contain the breast. The splitting-light unit consists of the second collimating lens, the second focusing lens, fiber and HCGIS. The spectrum acquisition and imaging unit consists of the CCD, driver module and IAQ card. The signal processing unit consists of the signal amplifier, signal filter, data buffer and time-sequence synchronous circuit. The control unit includes computer, data analysis software and outside control keypad. The input button to as the outside control unit, the LCD screen to as the signal display unit. The display unit is the LCD screen. The basic principle of this system is shown as follows: Firstly, the composite light is emitted from the lamp and reflected by the reflecting cup to become the parallel light and passes through the lamp shade, then it is collimated by the first collimating lens, is filtered by the filter components, focused by the first focusing lens in turn. The focused light irradiates into the breast tissue. The absorbed and reflected light by the breast tissue penetrate the skin, then is collimated by the second collimating lens and focused by the second focusing lens onto the interface of the fiber, the focused light is coupled into the fiber, then the light transmitted by the fiber pass through the slit to enter into the HCGIS, and irradiate onto the surface of the HCG, the multi-chromatic light is diffracted by the HCG into the monochromatic light and cast onto the surface of the CCD. The light acquired by CCD is transformed into the electricity signals under the driver control of the time-sequence of the CCD, then the electricity signals are acquired by the IAQ card to form the digital image, then the digital image is amplified and filtered by the amplifier and filter in turn, the filtered image signal is stored and sent into the computer to be analyzed by the analysis software. Finally, the image and spectrum of the breast tissues are displayed by the LCD screen, and the diagnosis results are gotten. The integrated system structure of BCD is presented in Figure.1.

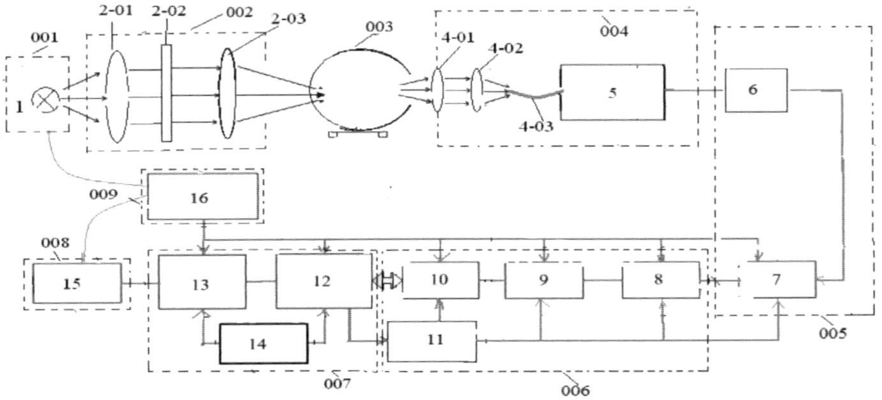


1.Light source, 2-01. The first collimating lens, 2-02 Filter components, 2-03. The first focusing lens, 4-01. The second collimating lens, 4-02. The second focusing lens, 4-03. The fiber, 5. HCGIS, 6. CCD, 7. IAQ card, 8. Amplifier, 9. Digital Filter, 10. Data Buffer, 11. synchronous circuit. 12. Computer, 13. Analysis software, 14. Outside control keyboard, 14. Display screen, 15. Power. 001. The light source unit, 002. Light-path unit, 003. Check cavity, 004. Splitting-light unit, 005. Spectrum acquisition and imaging unit, 006. Signal processing unit, 007 Control unit, 008. Display unit, 009. Power providing unit.

**Figure 1**. The integrated system structure of BCD

### *3.1. Design of the front part unit for BCD*

In this paper, the front part of the BCD includes the light-source unit and the light-path unit. In the light-source unit, the halogen-tungsten lamp is chosen to as the light-source of the BCD because it has the advantages of small volume, high efficiency, stability, low cost, long operating life, etc. And its spectrum range can from 200nm to 2500nm. Its work is control by the lamp switch and power. The reflecting cup is used to reflect the light emitted form the lamp to become parallel light. The material of the reflecting cup and the lamp shade are the glass with high temperature resistance, and inner wall of the reflecting cup is coated the aluminum coating to enhance the reflectivity of the light. The diameter of the export for the reflecting cup and the lamp shade is about 55mm. The fan is used to eliminate the heat of the light-source and prolong the using-life of the lamp. In the light-path unit, the first collimating lens is used to collimate the light reflected from the reflecting cup of the light-source unit. The first focusing lens is used to focus the light and irradiate into the breast tissue, and the focal length of the first focusing lens can be adjustable by the focusing knob, which makes the incidence-depth of the focusing-light adjustable in the breast. To get the different spectral images of the visible and NIR light, the filter component studded 10 pieces of the different wavelength filter is used. The first collimating lens, the filter component and the first focusing lens are fixed in the cylinder of 55mm diameter. The light-path unit and the light-source unit can be connected. The schematic diagram of the front part unit is shown in Figure.2.

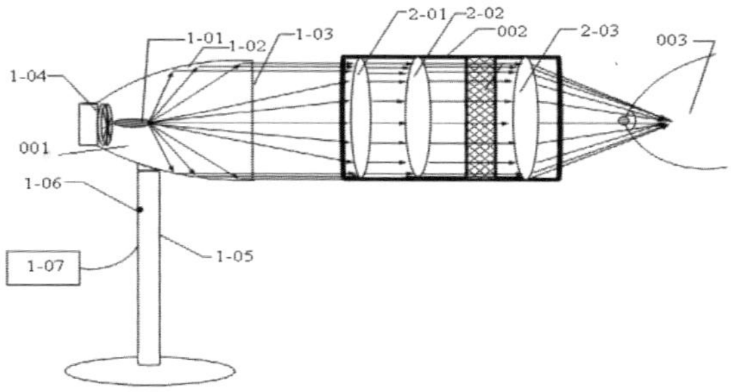


1-01.The lamp, 1-02. The reflecting cup, 1-03. The lamp shade, 1-04.The fan, 1-05.The bracket, 1-06. The lamp switch , 1-07. The power, 2-01.The first collimating lens, 2-02.The filter component, 2-03. The first focusing lens. 001. The light-source unit, 002. The light-path unit, 003. The breast

**Figure2.** The schematic diagram of the front part

*3.2. Design of splitting-light unit based on the HCGIS*

In the system of BCD, the splitting-light unit is the key part. In this splitting-light unit, the improved imaging spectrometer based on the holography concave grating is designed. The splitting-light unit consists of the second collimating lens, second focusing lens, fiber and the HCGIS, where the HCGIS consists of the slit, the plane holographic concave grating and CCD. In this HCGIS, the opening-groove technology is used to eliminate the stray-light. This HCGIS has many advantages, such as, lower cost, compact structure, wider spectrum range, better imaging quality and higher prevision. The light-path structure of HCGIS and the schematic diagram of the splitting-light unit are shown in Figure 3 (a) and (b). In the inner wall of the HCGIS, the depth and width of the groove is 1cm and 0.2cm respectively.

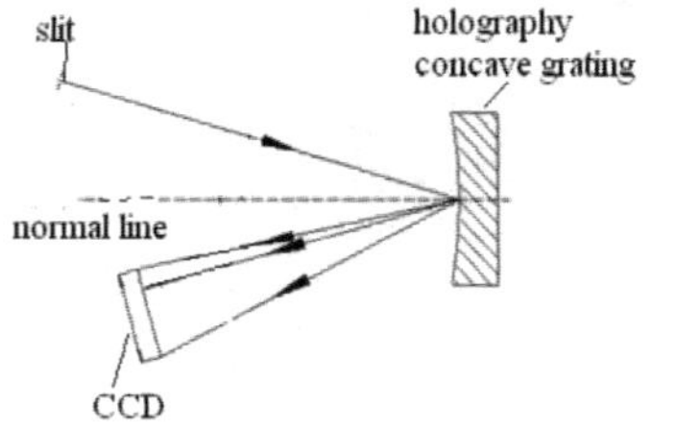


(a)

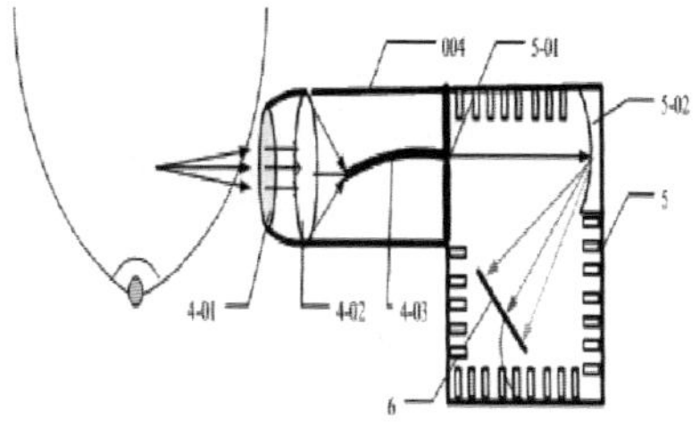


(b)

4-01. The second collimating lens, 4-02. The second focusing lens, 4-03. The filter, 5-01. The slit, 5-02. The holography concave grating, 5. HCGIS, 6.CCD. 004. The splitting-light unit.

**Figure 3**: (a)The light-path structure of the HCGIS, (b) The schematic diagram of splitting-light unit

*3.3. Spectral image acquisition and pre-processing unit of BCD*

In this paper, the spectral image acquisition unit consists of CCD, CCD driver module, high speed IAQ card. The splitting-light unit and the spectrum acquisition imaging unit are merged and fixed on the 3D scan platform. The 3D scan platform is driven by the stepper motor to do movement from up to down, from forth to back and from left to light. The acquired spectral image is amplified by the two-order amplifier, i.e. the first off-current amplifier and the adjustable gain amplifier. Then the amplified spectral image is filtered by the low-pass digital filter. The filter spectrum is send by PCI bus to be stored into data register. The synchronous circuit controls the time-sequence of the CCD driver module, IAQ card, amplifier, filter and data register. Finally, data are read by the computer and are analyzed by the software to get the diagnosis results.

## 4. Experiment and results

*4.1. Materials*

Halogen-tungsten lamp with the reflecting cup(Philip14546,±12v/20W) is chosen as the lamp source of the BCD, its spectrum range is from 340 nm to 2500 nm. The water bottle driven by stepper motor is used as the water-cooling device of the light-source unit. The material of the check cavity is the synthetic resin. The curvature of the first and second collimating lens are all 26.33mm, the curvature of the first and second focusing lens is 6.24mm. Three optic fibers (IBM 39M5696, 100μm) are coupled to transmit the light enter into the HCGIS. The material of the custom-built HCGIS is aluminum-alloy. The HCGIS consists of the slit (50μm), the HCG (600 lines/400 nm) and the camera CCD (ICX404AK, Sony). The length of the input-forearm of the HCGIS is about 104mm, and the output-forearm is about 102mm. The effective pixels of the camera CCD (ICX404AK, Sony) is 510× 492, its package is plastic DIP, 16 pins. The IAQ card (NI PXI-1411, NI) is used to acquire the electricity signals and converted into the digital signal. The amplifier has 4 channels, its input impedance can reach 100MΩ, the range of the gain can reach 10-5000, the frequency wide can reach DC-30KHZ, and its KCMR>100db. The low-pass filter uses the Butterworth low-pass filter, its cut-off frequency is adjustable from 100Hz to 100KHz. In the analysis software, LabVIEW (Version. 8.0 student edition, NI)[7] is used as the software platform to control spectral image acquisition and data analysis.

*4.2. The experiment of the HCGIS for the BCD*

To ensure the precision and feasibility of the BCD based on HCGIS and CCD, the improved custom-built HCGIS for BCD based on the optical platform is firstly established. The photo of this HCGIS is presented in Figure.4.

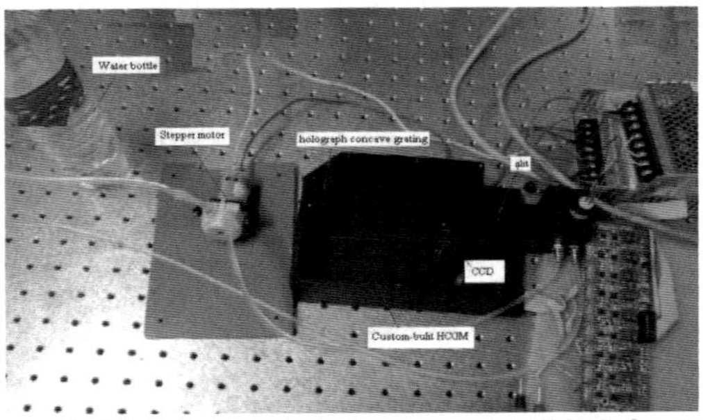


**Figure 4.** The photo of the improved HCGIS for BCD

*4.3. The experimental results of the improved HCGIS for BCD*

The custom-built HCGIS is used in the experiment, the spectrum information are gotten, where the information includes the diffraction, imaging distance and the dispersion of the different wavelength. The results are shown in the Table 1. From table1, we can see that the dispersion of this HCGIS for BCD is improved.

Table 1: The spectrum information of the HCGIS for the BCD

| Wavelength(nm) | Diffraction (angle deg) | Image distance(mm) | Dispersion (nm/mm) |
|---|---|---|---|
| 300 | 10.3700 | 99.3841 | 35.8218 |
| 340 | 11.7709 | 99.8202 | 35.1537 |
| 467 | 16.2899 | 99.4926 | 34.9067 |
| 595 | 20.9159 | 99.8000 | 34.4286 |
| 600 | 21.1000 | 100.2419 | 34.0824 |
| 722 | 26.6897 | 100.8099 | 33.6428 |
| 800 | 28.6851 | 101.8179 | 33.0153 |
| 850 | 30.6635 | 102.6497 | 32.4512 |
| 900 | 32.6830 | 103.1826 | 32.1348 |
| 950 | 34.6625 | 104.0658 | 31.6247 |
| 1000 | 36.6851 | 104.8217 | 31.1563 |

*4.4. The resolution estimation of the HCGIS for BCD*

Spectral resolution is an important standard which is used to estimate the performance of HCGIS for BCD. For the estimation of different wavelength and resolution, Eq.(2) can be used to illustrate their relation. The Eq.(2) is shown as follows:

$$\text{Estimation of the actual resolution} = \frac{\text{integrated width of spectrum}(\Delta l)}{2} \times \frac{1}{\text{line dispersion rate}(\delta l/\delta\lambda)} \quad (2)$$

Where, $\Delta l$ (integrated width of spectrum) $\approx \Delta l_1$ (diffraction width)+ $\Delta l_2$ (the width of slot) + $\Delta l_3$ (aberration width).

The resolution estimations of the wavelength is 300,550, 800, 1000 nm are shown in the Table 2

Table 2: The resolution of the wavelength ($\lambda$) is 300, 550,800 and 1000 nm

| $\lambda$ | $\Delta l_1$ | $\Delta l_2$ | Δl3 | ($\delta l/\delta\lambda$ ) | Resolution |
|---|---|---|---|---|---|
| 300 nm | 10.3700 μm | 50 μm | 6.1 μm | 35.8218 μm /nm | 0.93 nm |
| 550 nm | 18.9050 μm | 50 μm | 3.0 μm | 34.4286 μm /nm | 1.04 nm |
| 800 nm | 28.6851 μm | 50 μm | 2.4 μm | 33.0153 μm /nm | 1.23 nm |
| 1000nm | 33.6851μm | 50 μm | 2.0 μm | 31.1563 μm /nm | 1.38 nm |

From Table 1and 2, we can know that the resolution of the splitting-light of HCGIS for BCD can reach 1.0 nm and the spectrum range can reach 300-1000 nm.

**Acknowledgements**

This work has been supported by the Chinese National Foundation for Natural Scientific Research grant (60767001, F051206). The Science and Technology Pillar Program of Jiangxi Province (2009BSA12700), the Natural Science Foundation of Jiangxi Province (2008GQW0013), and the Scientific Research Foundation of Jiangxi Provincial Education Bureau (GJJ09307, GJJ10243). And, the Natural Science Fundation of Jiangxi Science and Technology Normal University (No.KY2009ZY10).

**References**

[1] Jun Liu, Jie Li, Thomas J Rosol. 2007 Biodegradable nanoparticles for targeted ultrasound imaging of breast cancer cells *in vitro* Phy. Med. Biology. Vol52, No.16.

[2] KC Young, JM Oduko. 2007 Technical Evaluation of the Agfa CR 85-X Mammography System.NHSBSP Equipment Report 0707.

[3] Mitsuaki Tatsumi, Christian Cohade,et.al. 2005 Initial experience with FDG-PET/CT in the evaluation of breast cancer. European J. Nuclear Medicine and Molecular Imaging Vol.33 No.3 254-262.

[4] CD Lehman, C Gatsonis, CK Kuhl,et.al. 2007 MRI Evaluation of the Contralateral Breast in Women With Recently Diagnosed Breast Cancer. Obstetrical & Gynecological Survey Vol. 62 Issue 7. 456-458.

[5] Maillard, Gaston-François M.D; Scheflan, Michael M.D.; Bussien and René M.D. 1997 Ultrasonically Assisted Lipectomy in Aesthetic Breast Surgery. Plastic & Reconstructive Surgery: Vol.100, Issue 1,. 238-241.

[6] A. A. Oraevaky, A. A. Karabutov, and S. V. Solomatin, 2001 Laser optoacoustic imaging of breast cancer in vivo. Proc. SPIE 4256, 352–363.

[7] Robert H. Bishop. 2006 LabVIEW 8 student edition National Instruments, USA.

# Transverse flow-velocity quantification using optical coherence tomography with correlation

**Zhiyuan Shen, Nan Zeng, Yonghong He***
Laboratory of Optical Imaging and Sensing, Graduate School at Shenzhen, Tsinghua University, Shenzhen, 518055, P. R. China

* Corresponding author email: heyh@sz.tsinghua.edu.cn

**Abstract:** We describe a method that uses classic optical coherence tomography to measure the transverse fluid-flow velocity quantitatively without knowing the Doppler angle. An intensity based cross-correlation calculation is taken point-to-pointly between two close cross sections of the scattering fluid to estimate the time delay for scattering particles passing through the two sections which are scanned alternately at a high speed. The transverse velocity distribution of the scattering fluid-flow in the whole section is achieved finally. The experimental results agree well with the preset ones. This method is insensitive to the Doppler angle and provides a variable velocity detection range in different application conditions.

## 1. Introduction

Optical coherence tomography (OCT) [1] is a non-invasive imaging technique capable of providing internal structure information of objects with a micrometer-scale spatial resolution. Doppler optical coherence tomography (DOCT) permits determining location and flow velocity of moving particles in highly scattering media [2-4], which shows a good application prospect, such as in vivo imaging of microvasculature in medical diagnostics [5-6] and flow characteristics measurement in microfluidic devices design [7-8]. However, only the component of the velocity parallel to the incident beam can be measured by conventional DOCT [9]. It is therefore necessary that the angle between the incident beam and the flow vector (the Doppler angle) must be precisely known before an exact value of the fluid flow velocity can be calculated. But in many applications it is difficult to estimate the Doppler angle accurately, flow samples with complex geometry for example, and particularly when the flow is embedded in the highly scattering medium such as the situation of in vivo blood-flow monitoring.

Attempts have been made to reduce the reliance on the Doppler angle. Technique employed with multiple sample beams was developed in several forms [9-10] and often needed OCT systems with relative improvement. The Doppler shift on interference signal and the Doppler spectrum broadening were used combinedly to estimate the values of Doppler angle and average fluid velocity by Proskurin et al [11]. However, like conventional DOCT, this method suffered from detection accuracy decreasing when the transverse velocity was small. Time varying speckle in OCT structural images was studied to carry about tissue structure and flow information [12-13]. Nevertheless, the speckle variance OCT has been used only in qualitative analysis of the fluid-flow velocity up to present.

In this paper, a method which is capable of measuring the transverse flow velocity precisely independent of the Doppler angle is presented. The technique takes a point-to-point cross-correlation calculation between two series of structural images to estimate the time delay for scattering particles passing through two close cross sections, and then achieves the transverse velocity distribution of the scattering fluid-flow in the whole section finally. In our experiments, a classic frequency domain OCT system was used to measure the preset flow velocity. The results agreed well with the preset ones even in the small-velocity condition.

## 2. Materials and Methods

### *2.1. Optical coherence tomography system*

The spectral domain OCT system used in our experiments was shown in figure 1(a). The light source was a super luminescent diode with a center wavelength of 830 nm and a 3-dB bandwidth of 40 nm, corresponding to a coherence length of 7.6 μm in air. The low-coherence light illuminated the source arm of a fiber-based Michelson interferometer. Then the interference signal of the light beams reflected from the reference and sample arms was coupled in the detection arm, which consisted of an achromatic collimating lens (CM: f= 75 mm), a 1200-line/mm transmission grating, an achromatic focusing lens (FL: f= 200 mm), and a line-scan CCD camera (2048 pixels, each 14×14 μm in size). The integration time of the CCD camera was set to be 40 μs and the system was operated at an A-scan frequency of 20 kHz. The output power of the sample arm was 0.5 mW.

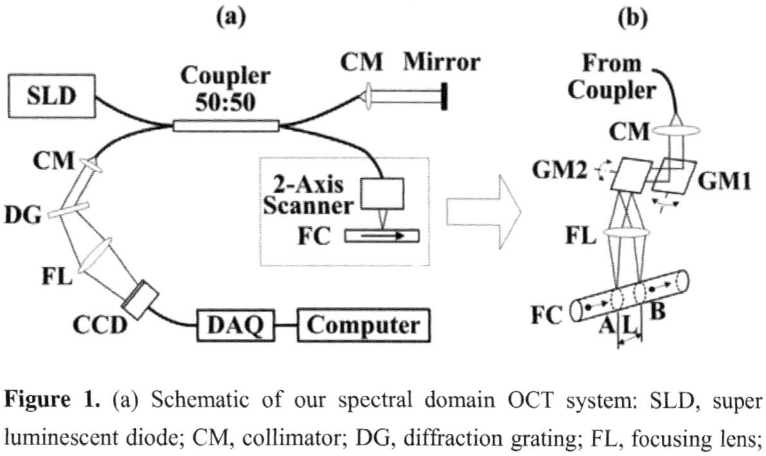


**Figure 1.** (a) Schematic of our spectral domain OCT system: SLD, super luminescent diode; CM, collimator; DG, diffraction grating; FL, focusing lens; DAQ, data acquisition system; FC, flow channel. (b) Sample scanning manner in our experiments. A two-axis scanner with two galvo mirrors was used to scan two close sections (A and B) of the fluid sample alternately at a high speed. Distance L between section A and B can be set flexibly to meet the requirement of the measurement range.

*2.2. Methods*

As for this method, a two-axis scanner with two galvo mirrors was used to scan two cross sections (A and B) of the fluid-flow sample alternately at a high speed (illustrated in figure 1 (b)). One scanning mirror was driven by a sawtooth waveform voltage signal while the other by a square waveform voltage signal. The distance L between section A and section B (along the axis of flow channel) was set to be very small, several ten to several hundred microns for example, so it can be considered that section A and B are parallel to each other. After a certain acquisition time t, we got a series of structural images at each cross section. Then pixel intensity with the same coordinate in the images of one series was extracted and formed an intensity array in the acquisition order, which recorded different scatterers passing through a certain position in the fluid section during the acquisition time. A cross-correlation calculation was taken between two intensity arrays extracted from the corresponding position of the two sections to estimate the time delay $\tau$ for scatterers passing through section A and section B. The average flow velocity v of that position is decided by $v=L/\tau$. By repeating the calculation in the whole fluid section, we achieved the transverse velocity distribution finally.

Exactly speaking, the transverse velocity measured with this method is the velocity along the axis of flow channel, not the general meaning which is in the direction perpendicular to the scanning plan.The velocity measurement range of this method can be calculated with the distance L、 acquisition time t and the section-scan frequency f determined. The period of the section A and B scanning is marked as T, T=2/f. Assuming that we obtain N frame structural image data at each section in the acquisition time t=NT. Theoretically cross-correlation taken between two arrays whose data length are both N can provide a maximum time delay as $\pm$(N-1)T and a minimum time delay as $\pm$T ($\pm$ denotes two opposite flow directions). So the velocity measurement range is $\pm$(L/(N-1)T~L/T). These parameters could be flexibly set to meet variable velocity measurement range.

*2.3. Flow phantom*

In the experiment shown in figure 1 (b), a glass capillary with an inner and outer diameter of 0.3 mm and 1.0 mm was placed on a goniometer stage, and polystyrene microsphere aqueous suspension (mean diameter: 10 μm, volume concentration: 1.5%) was used as working fluid driven by a syringe pump. Then the pump was controlled to generate an average flow rate of 6.29 mm/s in the capillary. And the capillary was tilted by an angle of 85 ° with respect to the incident beam. The distance L between the two sections was adjusted to 0.3 mm. 400 frame structural image data of the two sections were acquired alternately in t = 2 seconds, giving section-scan frequency f =200 Hz. Each frame contained 100 A-Scan. These parameter settings provided a measurement range of $\pm$(0.151 ~ 30 mm/s).

**3. Results and Discussions**

Figure 2(a) shows the OCT intensity image of the capillary filled with microsphere solution. A two-dimensional cross-correlation between two images acquired at section A and section B (figure 1(b)) was calculated to provide a coordinate deviation in order to correct the pixel corresponding relation. Then the transverse velocity distribution of the scattering fluid-flow in the whole section was calculated, filtered and shown in figure 2(b). The velocity profiles before and after smooth filtering along the white line in figure 2(b) were illustrated in figure 2(c). The flow showed a parabolic profile with an average velocity calculated to be 6.27 mm/s, which showed a good agreement with the preset one (6.29 mm/s). A three-dimensional representation of the transverse velocity distribution after smoothed was presented in figure 2(d).

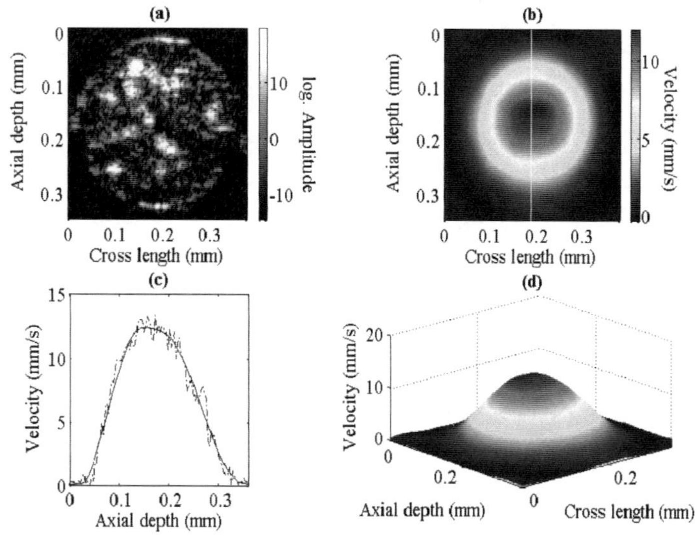


**Figure 2.** (a) Intensity tomogram of the perfused capillary. (b) Transverse velocity distribution of the fluid sample in the whole section. (c) Velocity profiles along the white line in figure 2(b). Dash-dot line: before smooth filtering. Solid line: after smooth filtering. (d) Three-dimensional flow front profile within the capillary.

The principle of this method is based on structural image cross-correlation, and the measurement is performed without the need of knowing the angle between the incident light and the flow velocity field. To examine the influence of the angle of incidence on the velocity measurement, for the second experiment we changed the Doppler angle in steps of 5° , and measured 20 times at the same angle. The averaged results agreed well with three different preset velocities with the angle varied in ±15° , as shown in figure 3. The experiments performed showed that the velocity determined by our method was independent of the orientation of the velocity vector in a large range.

The capability of measuring slow flow is of great importance. However, experimental results of velocity below 0.2 mm/s were seldom reported though some techniques were proved theoretically to have a high velocity sensitivity of the order of 10 μm/s [5,14]. Both DOCT and speckle flow techniques have an inherent tradeoff in flow sensitivity with acquisition time [5,14-15], and the phase-sensitive demodulation based techniques are vulnerable to the background noise or phase instability in slow flow detection mode, especially at high frame rates [14]. To further validate our method in the low-velocity condition, we changed the experiment parameters to L = 0.048 mm, t = 2 seconds and f = 200 Hz, corresponding to a measurement range of 0.024 ~ 4.8 mm/s. We performed a series of measurements at 15 different preset velocities with the same Doppler angle of 85° . Every preset flow was measured 20 times and the averaged results were plotted as a function of preset velocity in figure 4. The average deviation from the set velocity for flows from 0.06 to 0.2 mm/s was 1%. And the correlation coefficient for the data set was calculated to be 0.9999.

The results indicated a high correlation between preset and calculated velocities. This method eliminates the need for phase-sensitive detection or spatial frequency analysis, and is robust to factors confounding both Doppler and speckle analysis techniques, such as non-linear motion of the reference arm mirror and movement of the sample.

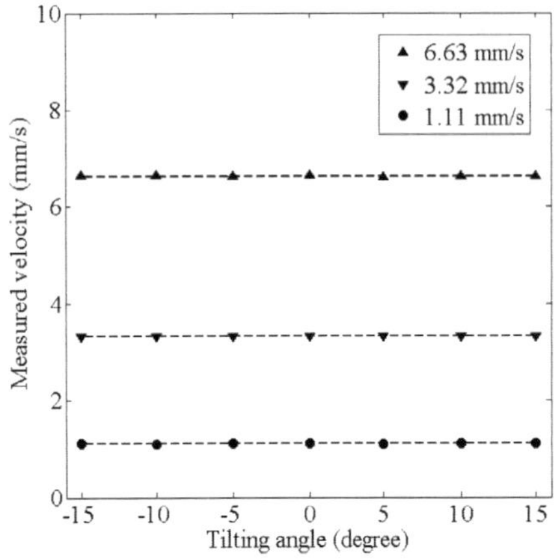


**Figure 3.** Measured velocity for three different preset velocities in different incident angle condition. The incident angle at 0 degree represents normal incidence. Dash lines: preset velocities.

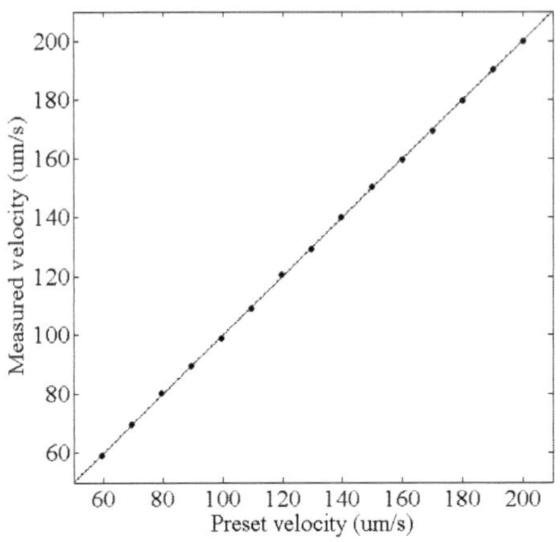


**Figure 4.** Estimation of different preset velocities in slow flow condition. Solid line: line fit. Dot line: unity slope.

In the practical application we found that measurable velocity range by means of this technique was smaller than the theoretical calculation. This could be due to the quasi-synchronous manner of the image acquisition at section A and section B as scatterers recorded at section A may not be totally detected at section B, which also led to measurement error. Increasing the section-scan frequency f and the acquisition time t could enlarge the velocity measurement range and reduce measuring error effectively. Obviously, this method does not work when a flow velocity is parallel to measured cross sections, but it still shows a considerable potential in absolute flow velocity measurement if combined with other techniques, three-dimensional OCT imaging for example. Velocity measurement technique suitable for complex applications will be studied in future research.

## 4. Conclusion

We presented a method to measure the scattering fluid-flow velocity quantitatively using classic OCT system without knowing the Doppler angle. It was demonstrated that one can obtain the transverse velocity distribution of the fluid flow with high accuracy using cross-correlation calculation with intensity from OCT images. This method provides an adaptable measurement in different flow conditions. The good performance in low-velocity detection also indicates a considerable value of this technique in real application.


## Acknowledgements

This research was made possible with the financial support from the 863 project, China (grant 2006AA06Z402), NSFC China (grants 30770592 and 60608019).

## References

[1] Huang D *et al* 1991 *Science* **254** 1178-81

[2] Wang X J, Milner T E and Nelson J S 1995 *Opt. Lett.* **20** 1337-9

[3] Chen Z, Milner T E, Dave D and Nelson J S 1997 *Opt. Lett.* **22** 64-6

[4] Izatt J A, Kulkarni M D, Yazdanfar S, Barton J K and Welch A J 1997 *Opt. Lett.* **22** 1439-41

[5] Zhao Y H, Chen Z, Saxer C, Xiang S H, de Boer J F and Nelson J S 2000a *Opt. Lett.* **25** 114-6

[6] Zhao Y H, Chen Z, Saxer C, Shen Q M, Xiang S H, de Boer J F and Nelson J S 2000b *Opt. Lett.* **25** 1358-60

[7] Wang L, Xu W, Bachman M, Li G P and Chen Z 2004 *Appl. Phys. Lett.* **85** 1855-7

[8] Ahn Y-C, Jung W and Chen Z 2006 *Appl. Phys. Lett.* **89** 064109

[9] Ahn Y-C, Jung W and Chen Z 2007 *Opt. Lett.* **32** 1587-9

[10] Werkmeister R M, Dragostinoff N, Pircher M, Götzinger E, Hitzenberger C K, Leitgeb R A and Schmetterer L 2008 *Opt. Lett.* **33** 2967-9

[11] Proskurin S G, He Y H and Wang R K 2003 *Opt. Lett.* **28** 1227-9

[12] Barton J K and Stromski S 2005 *Opt. Exp.* **13** 5234-9

[13] Mariampillai A et al 2008 *Opt. Lett.* **33** 1530-2

[14] Yang V X D, Gordon M L, Qi B, Pekar J, Lo S, Yue E S, Mok A, Wilson B C and Vitkin I A 2003 *Opt. Exp.* **11** 794-809

[15] Kulkarni M D, van Leeuwen T G, Yazdanfar S and Izatt J A 1998 *Opt. Lett.* **23** 1057-9

# Reduced prefrontal cortex activation in the color-word Stroop task for Chinese dyslexic children: a near-infrared spectroscopy study

**Jinyan Sun[1], Jiahuan Zhai[1], Ranran Song[2], Li Zou[2] and Hui Gong[1]**

[1] Britton Chance Center for Biomedical Photonics, Wuhan National Laboratory for Optoelectronics-Huazhong University of Science and Technology, Wuhan 430074, P.R. China

[2] Department of Child and Adolescent Health and Maternal Care, School of Public Health, Tongji Medical College, Huazhong University of Science and Technology, Wuhan 430074, P.R. China

Email: huigong@mail.hust.edu.cn

**Abstract**. Behavioral studies have investigated the performance of children with developmental dyslexia in conflict resolution, a function connected with the prefrontal cortex (PFC) closely. However, little is known about the prefrontal activation in conflict resolution for dyslexic children. In the present study, the involvement of the PFC in resolving conflict was evaluated for Chinese dyslexic children by means of near-infrared spectroscopy (NIRS). The NIRS instrument is a portable, continuous-wave system and can measure concentration changes of hemodynamic parameters (including oxy-, deoxy-, and total hemoglobin). Considering better sensitivity, the oxy-hemoglobin (oxy-Hb) was chosen to indicate the prefrontal activation. Ten dyslexic children and 11 normal children were recruited to perform the Chinese-character color-word Stroop task, which included the neutral and color (incongruent) tasks. In behavioral performance, both groups showed significant Stroop effect, longer response time or higher error rate for the color task. In particular, the Stroop interference effect was marginally larger for dyslexic children than normal children in response time. What's more, the two groups showed distinct pattern of oxy-Hb activation during the Stroop tasks. The normal group recruited the bilateral PFC to perform the tasks, while the dyslexic group couldn't activate the bilateral PFC in the difficult color task. Moreover, significantly less color Stroop effect was found in the left PFC for the dyslexic group, showing their disability in coping with the Stroop interference. These findings suggest that the PFC is dysfunctional in conflict resolution for Chinese dyslexic children and that NIRS can be an effective tool in neurological research and clinical application.

**Keywords:** Developmental dyslexia, Chinese, Conflict resolution, Stroop, Near-infrared spectroscopy

## 1. Introduction

Developmental dyslexia (DD) is defined as a disorder represented by unexpected reading disabilities in people who otherwise have the normal intelligence and education [1, 2]. Affecting approximately 5-17% of the school-age children, DD is one of the most common problems, and is characterized by difficulties with accurate and fluent word recognition, spelling and decoding. Although DD has been viewed as a specific phonological deficit [3], more and more evidence has pointed out that people with DD also show deficits in other domains which have been demonstrated to contribute to the reading performance for the dyslexic children [4], such as attention [5], visual processing [6] and executive function [4, 7].

In the executive function domain, Although an increasing number of behavioral studies have reported defective performance for dyslexic children in certain functions, such as working memory [7, 8], verbal fluency function [4, 7, 8] and conflict resolution [7, 8], little is known about the neural correlates of executive function in dyslexic children.

Conflict resolution is one major role of executive function, and is critical in keeping adequate performance in the complex situation with significant interference between two or more sources of information. Functional neuroimaging studies have confirmed that the PFC, especially the dorsolateral PFC, is involved in resolving conflict [9, 10]. The PFC reallocates and maintains the attention on the task-relevant stimulus dimensions and processes, while inhibits the task-irrelevant information and responses. The Stroop paradigm is a classical method to study the conflict resolution mechanisms, introduced by John Ridley Stroop in 1935 [11]. The main feature of the Stroop paradigm is selecting to respond to a certain dimension of the stimulus over other dimensions. In the color-word Stroop task, the subject is usually instructed to respond to the color of the stimulus while ignoring to read the word. Since reading the word is a more prepotent response than naming the color, lexical and hue conflict with each other when the presented color of the stimulus is incongruent with its meaning. In addition, the color-word Stroop task has been widely used to investigate the abnormal brain function of conflict resolution in many neurological and psychiatric disorders [12].

Near-infrared spectroscopy (NIRS) is a noninvasive optical functional neuroimaging method. Due to the sensitivity to the oxygenation state of hemoglobin, NIRS can measure concentration changes of oxy-hemoglobin ($\triangle$[oxy-Hb]), deoxy-hemoglobin ($\triangle$[deoxy-Hb]) and tot-hemoglobin ($\triangle$[tot-Hb]), which have been proved to reflect the cortical activation in the activated areas by simultaneous measurement with other imaging methods [13, 14]. Compared with functional magnetic resonance imaging (fMRI) and positron emission tomography (PET), NIRS has advantages in its insensitivity to movements and flexibility. In particular, low susceptibility to movement artifacts makes it available to measure in natural conditions and more suitable in children studies [15] and psychiatric and neurological research, such as depression [16], schizophrenia [16, 17], panic disorder [18], attention-deficit hyperactivity disorder [19], and pervasive developmental disorder [20]. Although with a limited penetration depth by NIRS, an appropriate separation between the source and the detector allows the measurement of hemodynamic activation from the cerebral cortex [13]. What's more, NIRS has been used to measure the brain activation in the PFC during the Stroop task [10, 21].

In this study, we aim to evaluate the PFC function in conflict resolution for Chinese dyslexic children. NIRS was used to measure the activation of the PFC while participates were performing the Chinese-character color-word Stroop task, consisting of the neutral and color (incongruent) tasks. We expected to observe lower prefrontal activation during the Stroop task for dyslexic children than normal children.

## 2. Experimental Procedure

### 2.1. Subjects

Twenty-one Chinese primary school children, grades 4 to 6, participated in this study: 10 dyslexic children (3 girls and 7 boys, age 10-13 years old) and 11 age-matched normal children (1 girl and 10 boys, age 9-12 years old). All participants were right-handed, had normal or corrected-to-normal vision and normal color vision, and were healthy without personal or family history of neurological disorders. According the definition of dyslexia by International Classification of Diseases 10 (IDC-10), children meeting the following criteria were defined to be dyslexic: (1) the score for Pupil Rating Scale Revised Screening for Learning Disabilities (PRS) was equal or less than 65, implying the possibility of learning disability; (2) the score of the Dyslexia Checklist for Chinese Children (DCCC) was equal or greater than 70, suggesting that their reading achievement was two standard deviations or more below the average level; (3) the Chinese achievement fell in the last 10 per cent in the class; (4) the intelligence quotient (IQ) was above 80 evaluated with Wechsler Intelligence Scale for children-Chinese revision (WISC-CR) [22]. The normal children didn't have the PRS, DCCC and IQ tests and were chosen primarily on their Chinese achievement and their teachers' recommendation. The two groups didn't differ in age, education and IQ ($p>0.34$).

After complete depiction about the experiment, informed consent was collected from each subject and their parents before the experiment. This study was approved by the Human Subjects Institutional Review Board at Huazhong University of Science and Technology.

### 2.2. Stimuli and procedure

The Chinese-character color-word Stroop experiment consisted of two kinds of stimuli: neutral (color-irrelevant) characters (贯,奖,球,涂, mean "pass through", "prize", "ball", and "scrawl"), and color characters (红,黄,蓝,绿, mean "red", "yellow", "blue", and "green"). Each character was presented in four colors: red, yellow, blue and green. To reflect the conflict, color characters showed only in the incongruent color were chosen as stimuli. The subjects were instructed to judge the color of the stimulus while ignoring its other dimensions as fast and accurately as possible. If the character was printed in "red", subjects press the D button with the middle finger of the left hand; if in "yellow", they press the F button with the index finger of the left hand; if in "blue", they press the J button with the index finger of the right hand; if in "green", they press the K button with the middle finger of the right hand.

The experiment was conducted in block design [23] and in a single run (figure 1), including two rest blocks, a white cross showing in the centre of the screen for 1 min, and two task blocks, neutral task and color task for neutral and color stimuli respectively. In each task block, 40 stimulus trials were included and were in a pseudorandom order. In each trial, an empty screen remained for 2 s, followed by a stimulus showing for 1 s. The total time of the experiment was 6 min. The order of the task blocks were counterbalanced across subjects. Before the formal experiment, there was a practice block to ensure the subject to understand the task instructions.

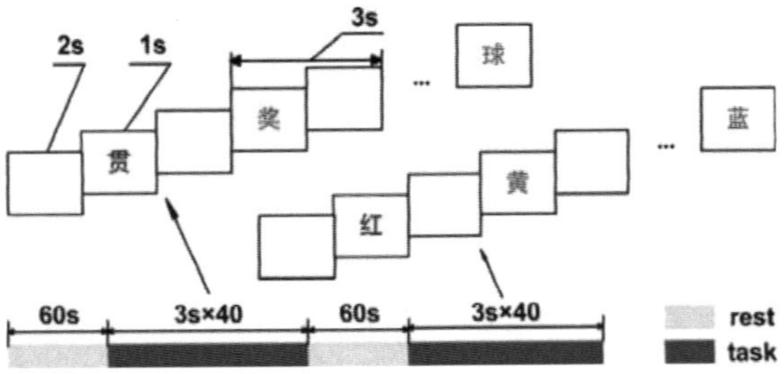


Figure 1. Experimental design of the Stroop task. The stimuli in the neutral task, 贯,奖,球 mean "pass through", "prize", and "ball". The stimuli in the incongruent color task, 红,黄,蓝 mean "red", "yellow", and "blue".

### 2.3. NIRS

A portable, continuous wave NIRS imager with a sampling rate of 3.3 Hz was utilized to detect the concentration changes of oxy-, deoxy-, and tot-hemoglobin. The imager consists of a personal computer, a measuring and controlling module and a probe [24]. The probe includes 4 sources and 8 detectors defining 16 measured areas represented by channels and it covered almost the whole forehead of the subject during the entire experiment, as shown in figure 2. The distance between the source and the detector is 2.45 cm approximately, which allows the measurement of hemodynamic activation from the cerebral cortex [13]. Since $\triangle$[oxy-Hb] has been proved to be more sensitive compared with the other two hemodynamic parameters [15, 25], it was chosen as the indicator of the task activation in this study.

The raw data collected by the NIRS imager were first processed to remove the noise caused by movements, then converted to concentration changes of hemodynamic parameters according to the modified Beer-Lambert Law, and finally bandpass filtered to remove the slow baseline drifts and other physiological noise. The mean signal intensity was calculated for the baseline (30 s before the task) and the vascular response (entire period of the task) for each task and each subject. Difference of the signal intensity between the baseline and the vascular response was taken as the task activation and the task activation was compared against 0 with paired t tests to determine the channels with significant activation for the two tasks. Independent t tests were used to analyze the group differences. The results reached significant level if $p<0.05$.

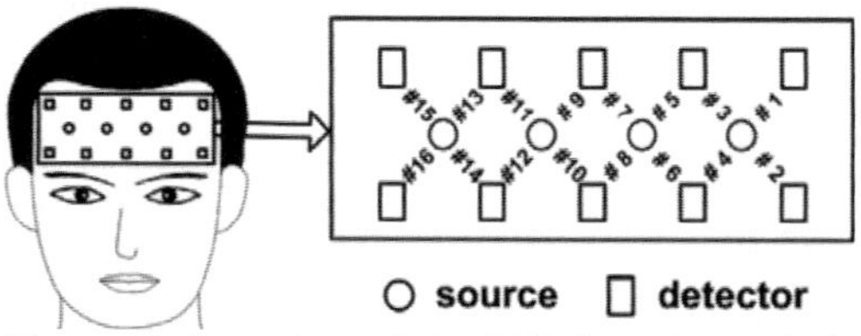


Figure 2. Location of the NIRS probe and the 16 measured areas represented by channels. "#"denotes channel.

## 3. Results

### 3.1. Behavioral data

The behavioral results are displayed in figure 3. Concerning the response time, a repeated measure task (neutral, color) × group (normal, dyslexic) analysis of variance (ANOVA) revealed a significant main effect for task ($F(1,19)=13.946$, $P=0.001$) and marginally significant task × group interaction ($F(1,19)=4.13$, $P=0.056$), whereas there were no main effect for group. The response time was evidently longer for the color task than the neutral task in both groups (normal: $t=2.323$, $p=0.043$; dyslexic: $t=4.136$, $p=0.003$; paired t test). In particular, there was a statistical trend towards a lager Stroop interference effect in dyslexic group ($t=2.032$, $p=0.056$). For error rate, the repeated ANOVA demonstrated a significant main effect for task ($F(1,19)=11.347$, $P=0.003$), without the main effect of group or an interaction. The error rate of the color task was significantly higher than that of the neutral task for the dyslexic group ($t=2.61$, $p=0.028$, paired t test), but not for the normal group. Although the dyslexic group had more errors, they did not differ from the normal group in error rate for the two tasks or the Stroop effect significantly.

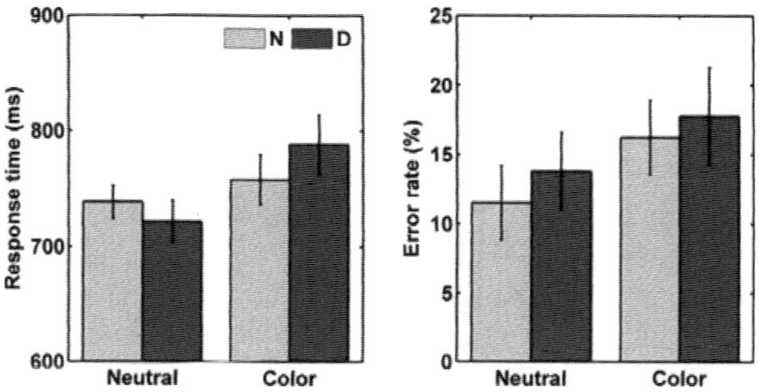


Figure 3. The response time and error rate of the neutral and color tasks for the two groups. N for the normal group, and D for the dyslexic group. Mean ± SEM (standard error of measurement).

### 3.2. NIRS data

For the neutral task, the paired t tests demonstrated evident oxy-Hb increase in three channels (channels 1, 2, 15) for the normal group, and in four channels (channels 1, 3, 5 14)for the dyslexic group. No statistically significant difference between groups was found. For the color task, the normal group activated nearly all the channels (except for channel 14), while the dyslexic group couldn't effectively activate any channel. What's more, independent t tests revealed that the normal group showed larger oxy-Hb activation in channel 3 (t=2.601, p=0.018, Cohen'd=1.14), channel 7 (t=2.23, p=0.038, Cohen'd=0.97) and channel 11 (t=2.177, p=0.042, Cohen'd=0.95) than that of the dyslexic group.

With regard to the color Stroop interference effect, activation difference between the color task and neutral task, it was significant in channel5 (t=2.319, p=0.043) and channel 6 (t=2.627, p=0.025) for the normal group, whereas it was not evident in any channel for the dyslexic group. Comparison between groups showed greater color Stroop interference effect in channel 5 (t=2.273, p=0.035, Cohen'd=0.92) and channel 7 (t=2.199, p=0.04, Cohen'd=0.9) for the normal group than dyslexic group (figure 4).

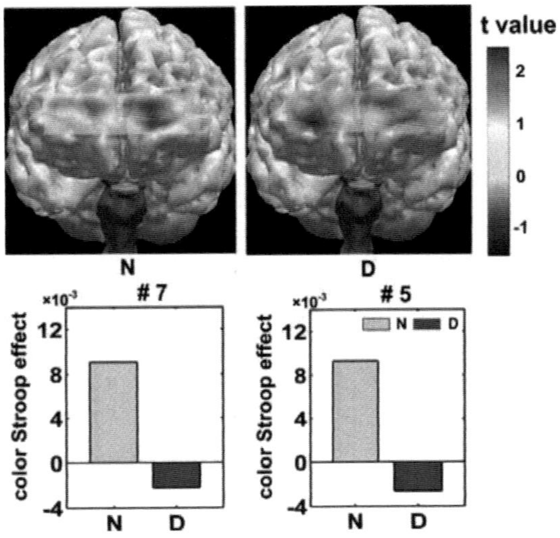


Figure 4. Stroop effect in the normal group (N) and dyslexic group (D). T maps of the Stroop effect in two groups (top) and significant difference of the Stroop effect in channel 5 and channel 7 between groups (bottom). "#" means channel.

## 4. Discussion

In the present study, NIRS was utilized to evaluate the PFC activation in conflict resolution during the Stroop task for Chinese dyslexic children. In behavior, there was a statistical trend towards a lager Stroop interference effect for dyslexic group, though without significant difference in response time or error rate. The result consists with previous studies showing that the Stroop interference is negatively associated with reading abilities [26, 27]. The larger Stroop interference suggested the poor cognitive control for dyslexic children. In addition, NIRS data revealed distinct brain activation pattern for the two groups, the reduced prefrontal activation showing the dysfunction of the PFC in conflict resolution for dyslexic children and conforming to our initial assumption.

The normal children recruited the bilateral PFC to perform the Stroop task, and the activation level increased and the activation field enlarged from the neutral to the color task. Further, when comparing the color task with the neutral task, only activation in the left PFC (Channels 5, 6) reserved, which proved the involvement of the left PFC in conflict resolution for children. The results are concordant with the well-established notion that the dorsolateral PFC plays a top-down role in the conflict resolution by maintaining the attention demands on the current task while suppressing the distracting information [9, 28, 29]. Our finding is also well in line with the recent functional imaging studies on children in different kinds of conflict tasks. Schroeter et al. used a Stroop paradigm with event-related fMRI and reported that children utilized the left lateral PFC to copy with the Stroop-related interference [30]. Another study by Konrad and colleagues, using the flanker task, found that children activated the left middle frontal gyrus (MFG) in executive control of the conflict [31]. What's more, our results concur with some studies on adults [9, 10].

Compared with the controls, the PFC was not significantly activated in the color task for dyslexic children. What's more, there was a significantly smaller color Stroop effect in the left PFC (Channels

5, 7) for the dyslexic children. The decreased activation in the PFC suggested that this area couldn't be recruited in conflict resolution for the dyslexic group, which might contribute to their larger Stroop interference in behavior. The weaker activation in the PFC was also reported in other cognitive task, such as working memory task, for dyslexic children [32]. Siok et al. have reported reduced gray matter volume in the MFG for Chinese dyslexic children [33] and it might be possible that the structural abnormality caused the PFC dysfunction and made it less sensitive to the increasing attention demands and task difficulty and more active in undemanding task.

There were some limitations in the present study. Firstly, the NIRS probe used in this study covered just part of the PFC. Hence the involvement of other brain regions in conflict resolution could not be assessed, such as posterior parietal cortex. Further studies measuring more brain regions can demonstrate this issue more comprehensive. Secondly, the sample size was small. However, it should be pointed out that the activation difference between groups was significant with large effect size.

Taken together, NIRS can effectively measure reduced prefrontal activation during the Stroop task for Chinese dyslexic children, suggesting the PFC dysfunction in conflict resolution for dyslexic children. This study contributes to our knowledge about the neural correlates of executive function in DD, and suggests that the executive function should be considered in the diagnosis and treatment of the dyslexia. Our findings elucidate that NIRS can be an effective tool in neurological research and clinical application.

**Acknowledgments**

This research was supported by the Program for Changjiang Scholars and Innovative Research Team in University, Ph.D. Programs Foundation of Ministry of Education of China (Grant No. 20070487058). We thank all subjects for their participation.

**References**

[1] Eden G F and Moats L 2002 *Nat. Neurosci.* **5** 1080-1084

[2] Temple E, Deutsch G K, Poldrack R A, Miller S L, Tallal P, Merzenich M M and Gabrieli J D 2003 *Proc. Natl. Acad. Sci. U. S. A.* **100** 2860-2865

[3] Ramus F 2004 *Trends Neurosci.* **27** 720-726

[4] Menghini D, Finzi A, Benassi M, Bolzani R, Facoetti A, Giovagnoli S, Ruffino M and Vicari S 2009 *Neuropsychologia* **48** 863-872

[5] Valdois S, Bosse M L and Tainturier M J 2004 *Dyslexia* **10** 339-363

[6] Siok W T, Spinks J A, Jin Z and Tan L H 2009 *Curr. Biol.* **19** R890-R892

[7] Brosnan M, Demetre J, Hamill S, Robson K, Shepherd H and Cody G 2002 *Neuropsychologia* **40** 2144-2155

[8] Reiter A, Tucha O and Lange K W 2005 *Dyslexia* **11** 116-131

[9] MacDonald A W, Cohen J D, Stenger V A and Carter C S 2000 *Science* **288** 1835-1838

[10] Zhai J H, Li T, Zhang Z X and Gong H 2009 *J. Biomed. Opt.* **14** 054022

[11] Stroop J R 1935 *J. Exp. Psychol.* **18** 643-662

[12] Melcher T, Falkai P and Gruber O 2008 *Brain Res. Rev.* **59** 96-124

[13] Toronov V, Webb A, Choi J H, Wolf M, Michalos A, Gratton E and Hueber D 2001 *Med. Phys.* **28** 521-527

[14] Kleinschmidt A, Obrig H, Requardt M, Merboldt K D, Dirnagl U, Villringer A and Frahm J 1996 *J. Cereb. Blood Flow Metab.* **16** 817-826

[15] Hoshi Y 2007 *J. Biomed. Opt.* **12** 062106

[16] Suto T, Fukuda M, Ito M, Uehara T and Mikuni M 2004 *Biol. Psychiatry* **55** 501-511

[17] Ehlis A C, Herrmann M J, Plichta M M and Fallgatter A J 2007 *Psychiatry Res.* **156** 1-13

[18] Nishimura Y, Tanii H, Hara N, Inoue K, Kaiya H, Nishida A, Okada M and Okazaki Y 2009 *Psychiatry Res. Neuroimaging* **172** 168-172

[19] Moser S J, Cutini S, Weber P and Schroeter M L 2009 *Psychiatry Res. Neuroimaging* **173** 190-195

[20] Kuwabara H, Kasai K, Takizawa R, Kawakubo Y, Yamasue H, Rogers M A, Ishijima M, Watanabe K and Kato N 2006 *Behav. Brain Res.* **172** 272-277

[21] Schroeter M L, Zysset S, Kupka T, Kruggel F and von Cramon D Y 2002 *Hum. Brain Mapp.* **17** 61-71

[22] Song R and Wu H 2006 *Chin. J. Clin. Rehabil.* **10** 175-177

[23] Ehlis A C, Herrmann M J, Wagener A and Fallgatter A J 2005 *Biol Psychol* **69** 315-31

[24] Lv X, Zheng Y, Li T, Zhang Z and Gong H 2008 *Front. Optoelectron. China* **1** 279-284

[25] Hoshi Y 2003 *Psychophysiology* **40** 511-520

[26] Guo T, Peng D and Liu Y 2005 *Cognition* **98** B21-34

[27] Protopapas A, Archonti A and Skaloumbakas C 2007 *Cogn. Psychol.* **54** 251-282

[28] Banich M T, et al. 2000 *Cogn. Brain Res.* **10** 1-9

[29] Milham M P, Erickson K I, Banich M T, Kramer A F, Webb A, Wszalek T and Cohen N J 2002 *Brain Cogn.* **49** 277-296

[30] Schroeter M L, Zysset S, Wahl M and von Cramon D Y 2004 *Neuroimage* **23** 1317-1325

[31] Konrad K, Neufang S, Thiel C M, Specht K, Hanisch C, Fan J, Herpertz-Dahlmann B and Fink G R 2005 *Neuroimage* **28** 429-439

[32] Beneventi H, Tonnessen F E, Ersland L and Hugdahl K 2010 *Int. J. Neurosci.* **120** 51-59

[33] Siok W T, Niu Z, Jin Z, Perfetti C A and Tan L H 2008 *Proc. Natl. Acad. Sci. U. S. A.* **105** 5561-6

# Reflective mesoscopic spectroscopy for noninvasive detection of reflective index alternations at nano-scale

**Yuanhao Tao, Zhihua Ding**[1]

State Key Lab of Modern Optical Instrumentation, Zhejiang University, 38 Zheda Rd., Hangzhou 310027, P.R.China

E-mail:zh_ding@zju.edu.cn

**Abstract**. Cancer has been one of the most serious threats to human life. However, there is no substantial improvement in overall treatment of cancer patients. One of the key reasons is the unavailability of convenient method to detect cellular alterations in ultra-early stage of carcinogenesis processes, where genetic aberrations at nano-scale have not yet resulted in histological changes. In this paper, we described an optical method based on reflective mesoscopic spectroscopy for ultra-early cancer detection. According to mesoscopic light transport theory, photons propagating in one dimension (1D) within a weakly disordered medium have the non-self-averaging effect. Reflected signal after 1D propagating is sensitive to any length scale of refractive index fluctuations due to multiple interferences of light waves travelling along 1D trajectory. The principle of mesoscopic spectroscopy for perceiving reflective index fluctuations at length scale of nanometers is introduced. A system for the measurement of reflective mesoscopic spectroscopy based on spatial-incoherence broadband source and spectrometer is established. Simulations on light propagation in cell-emulating model with controlled refractive index distribution are done by finite-difference time-domain (FDTD) approach.

## 1. Introduction

Cancer has been one of the most serious threats to human life. Although a series of new therapy methods were developed, including immunotherapy, endocrine therapy, gene therapy and so on, however, there is no substantial improvement in overall treatment of cancer patients. One of the key reasons is the unavailability of convenient method to detect cellular alterations in ultra-early stage of carcinogenesis processes.

Most current cancer detection methods rely on medical imaging, tumor-specific serum markers, and tissue biopsy. The sensitivity and resolution of medical imaging methods such as ultrasound, computed tomography, magnetic resonance imaging and positron emission tomography have serious limitations. Tissue biopsy is the most effective method of tumor diagnosis. Overall, limited by optical resolution, only large-scale structural abnormality information can be detected. There is rare study on the relationship between cell's nano-scale structures changes and cell's carcinogenesis process. One possible way is to use transmission electron microscopy [1, 2], which can achieve nano-resolution imaging for thin tissue. But the flaw is thin tissue don't have the complete information of living cells under natural status. New optical techniques, such as optical coherence tomography, confocal optical

[1] Corresponding Author: zh_ding@zju.edu.cn; phone 86-0571-87951194

microscopy, optical diffusion tomography, and imaging technique based on the elastic scattering, fluorescence or Raman spectroscopy [3-8], compared to conventional medical imaging techniques have higher resolution and sensitivity, but they are still limited by the far-field optical resolution(at best, 200nm). So it's impossible to recognize the earlier stages of carcinogenesis that precede histological changes up to now. Ultra-high resolution optical imaging techniques including near-field and far-field [9-11] have been developed to break through the optical diffraction limitation, but limited by subwavelength contact area, or require additional exogenous nano tags, even for certain specific sample which has nonlinear absorption properties, all of these techniques can not be widely used in non-invasive observation of cells under natural status.

Reflective mesoscopic spectroscopy method is introduced in this paper, which is based on one dimensional light transport property in weakly disorder and weakly scattering medium such as biological cells and it's sensitivity to nano-scale refractive index fluctuations. It can be used to describe the subtle changes of optical properties in the ultra-early stage of cell's carcinogenesis process. The proposed method might be useful in probing statistical structural changes at nano-scale which are unperceivable up to now, and hence avoiding the limitation due to optical diffraction in far field.

## 2. Method

### 2.1. Theoretical basis

Spectroscopy of elastic scattering has been widely used to probe tissue morphology. However, the traditional scattered light detecting methods are generally record the information after photos propagation in 3D space, due to the Fourier-transform relationship between the scattering signal and the scattering potential [20], the sensitivity to sub-wavelength length scales decreases fast with size. However, according to mesoscopic light transport theory, while photos propagating in 1D are recorded, situations will be completely different [19].

According to 1D mesoscopic light transport theory [24, 25], photons propagating in a one-dimensional weakly disordered refractive-index variable medium have the non-self-averaging effect [21-23]. Reflected signal after 1D propagating is sensitive to any length scale of refractive index fluctuations due to multiple interferences of light waves travelling along 1D trajectory. In other words, the reflected signal is sensitive to any length scale of refractive index fluctuations including those below the wavelength.

The previous research results have confirmed that nano-scale structural abnormalities are the biology basis of ultra-early detection of carcinogenesis cell [12-14], while they are accompanied by refractive index changes which is the most significant changes. In order to describe the disorder degree of refractive index fluctuations in a 1D channel, the statistical parameter disorder strength can be introduced [18, 19]:

$$L_d = \langle \Delta n^2 \rangle \cdot l_c \tag{1}$$

$\langle \Delta n^2 \rangle$ is the variance of refractive index fluctuations in the channel and $l_c$ is the spatial correlation length of these refractive index fluctuations. The disorder strength quantifies the spatial variability of refractive index and, thus, the local concentration of intracellular material. At a given 1D channel, $\langle \Delta n^2 \rangle$ is proportional to the local concentration of intracellular solids, whereas $l_c$ can be viewed as the characteristic size of the intracellular ultra-structure of a cell. Although the disorder strength do not directly related to structural changes within the channel at certain specific location, increased disorder strength represents 1D channel toward a more disordered state, so it is feasible to regard disorder strength as optical fingerprint of cell's carcinogenesis.

The next step is to analysis the association between disorder strength and reflective mesoscopic spectrum from a given 1D channel. Because fluctuating part of reflective mesoscopic spectrum $R(\lambda)$ is formed primarily by 1D back-scattered photos, $R(\lambda)$ can be analyzed by means of 1D mesoscopic light transport theory in disorder medium [24,25]. If $kl_c < 1$ and $4k^2L_dL/n_0^2 << 1$, where $k$ is the center wave number, $L$ is channel's length, $n_0$ is the mean refractive index in the channel (These three parameters are all known or can be estimated by reference to typical values), with experimentally acquired spectrum $R(\lambda)$, disorder strength $L_d$ can be calculated as [15, 16, 18, 19]:

$$L_d = A\left(n_0^2/2k^2\right) < R > (l/L) \qquad (2)$$

where A is the calibration constant, $< R >$ means the root mean square of fluctuation part of 1D reflective spectrum. The equation (2) shows that the disorder strength is linearly depend on the root mean square of $R(\lambda)$.

2.2. System description

The design of the reflective mesoscopic spectroscopy instrument is shown in figure 1[17]. The broadband white light from a Xe lamp is collimated by a 4f system (L1-A1-L2) and is focused onto a sample by a low-numerical-aperture objective (NA of objective=0.4, NA of illumination=0.2, NA of collection=0.4, Nikon CFI Plan). The spatial coherence length of the illumination beam is in the order of the diffraction limitation, ≈700 nm. The illumination beam diameter is 120μm (the diameter of sample is ≈10μm), which ensured that plane-wave illumination can be assumed for analyzing light scattering signatures. The backscattered light is collected by the same objective and focused by a tube lens to form a magnified image. The resulting backscattered image is projected with a 40× magnification onto the slit of an imaging spectrograph (20μm slit width, PI SP 2156) coupled with a CCD (PI PIXIS 400BR) camera. The flipper mirror directed the image onto a digital camera for quick visualization of the image while adjusting the distance between the objective lens and the sample. The sample is mounted on a scanning stage (KOHZU).

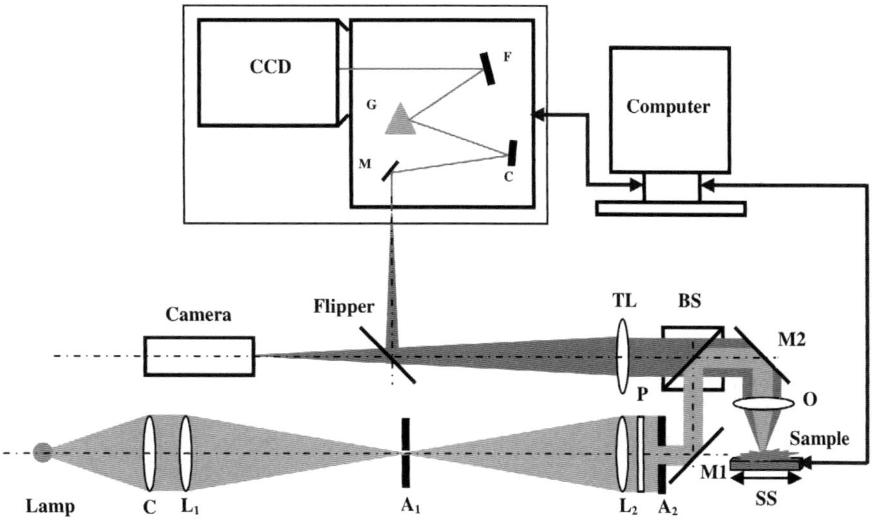


**Figure 1.** Schematic of the reflective mesoscopic spectroscopy system. C, condenser; L1, L2, lenses (f =150mm); A1, A2, apertures D(A1) = 1.8 mm, D(A2)=4 mm;BS, beam splitter; TL, tube lens, f = 400 mm; F, flipper; SS, scanning stage

Assume that sample such as a single cell was placed flat on a slide. In the sample space, the spatial coherence length is in the order of the diffraction limited lateral resolution(≈700nm in this paper). As shown in figure 2, the whole cell (Figure 2 (a)) was virtually divided into several parallel 1D channels (Figure 2 (b)), each with a diffraction limited lateral size. Compared to the typical diameter of biological cells (such as 10μm), refractive index distribution in the transverse direction within a single channel can be considered constant. So the photos can be approximately regarded as one-dimensional transmission along a channel while there is only one dimension of refractive index fluctuations.

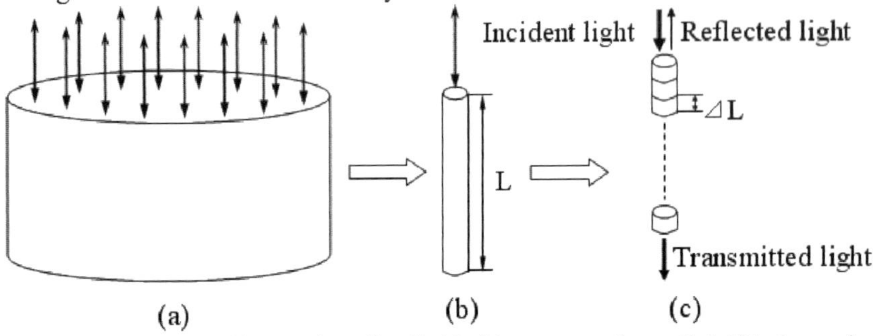


**Figure 2.** Single cell was virtually divided into several parallel 1D channels

This instrument can simultaneously obtain a diffraction-limited resolution microscopic image of sample and record the spectrum from each pixel (each 1D channel). The back-scattering image is acquired by linearly scanning the sample with a 0.5μm step. The size of a pixel in the image plane (image pixel) is 20μm×20μm, while the size of a pixel in the object plane (sample pixel) is 500nm×500nm. At each scanning step x, the CCD camera records a matrix with one axis corresponding to $\lambda$ and the other to the spatial position of the image y, resulting in a data cube $I(x,y;\lambda)$. This also means that each spectrum $I(x,y;\lambda)$ is from a particular one-dimensional channel. So the spectrum of the back-scattered light intensity $I(x,y;\lambda)$ which is formed by interference of photons reflected from refractive index fluctuations within a channel is recorded. Experiments on nano-structured models and biological samples are on schedule.

## 3. Simulation

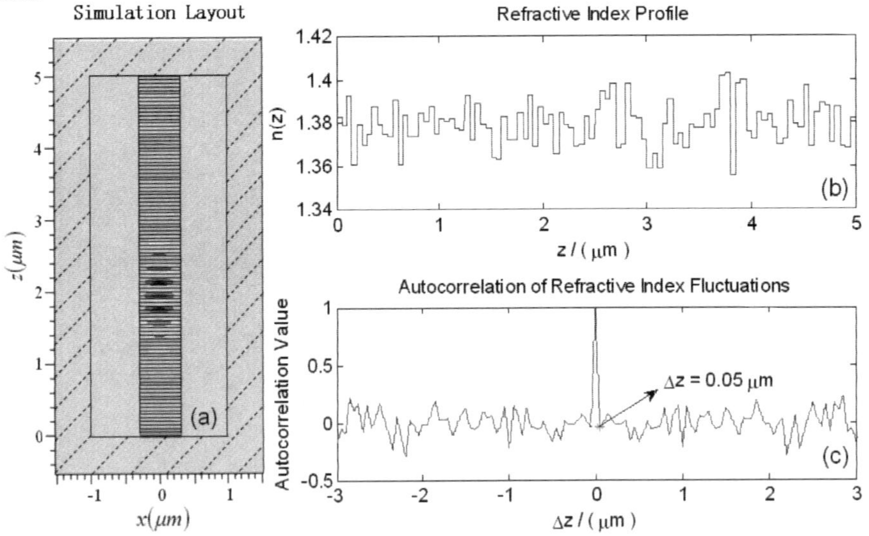


**Figure 3.** Simulation layout, refractive index profile and its autocorrelation function

We have confirmed the dependence of $L_d$ on nano-scale refractive index fluctuations by conducting rigorous numerical experiments using finite-difference time-domain (FDTD) approach which provides an accurate solution for the propagation of light in essentially arbitrary media. All the simulation parameters settings were refer to reported conclusions [15-19].

As shown in figure 3 (a), we used 1D slab model to simulate the 1D channel in real sample. The lateral size (x axis) was set to 600nm which was in the order of the diffraction limited lateral resolution. The simulated channel's length was set to 5μm which was the height when a cell was placed flat on a slide. Refractive index fluctuations exist in only one direction (z axis) and obey the gaussian white noise distribution. The spatial distribution was ladder-like with a uniform spatial step (50nm in Figure 3). Figure 3 (b) is the refractive index profile of 1D channel as shown in Figure 3 (a), the variance of refractive index fluctuations is $<\Delta n^2>=1\times 10^{-4}$ , while Figure 3(c) is the autocorrelation function of refractive index fluctuations. The spatial correlation length of refractive index fluctuations was approximately equal to ladder step. Both standard deviation and spatial correlation length of the refractive index fluctuations along the one-dimensional channel were under precise control.

In the FDTD simulation layout, the grid was terminated using a Berenger perfectly matched layer (PML) absorbing boundary condition, the shadow area in Figure 3 (a). A modulated Gaussian pulse as the time-domain source waveform which accommodates the complete frequency range of visible light was employed to source an x-polarized plane wave propagating in the FDTD grid. The back-scattered frequency response is extracted via a discrete Fourier transform (DFT) of the time-domain data recorded on the backward surface and normalized by the spectrum of the source pulse. Then the wave number was converted to wavelength, the final reflective mesoscopic spectrum along 1D channel was acquired.

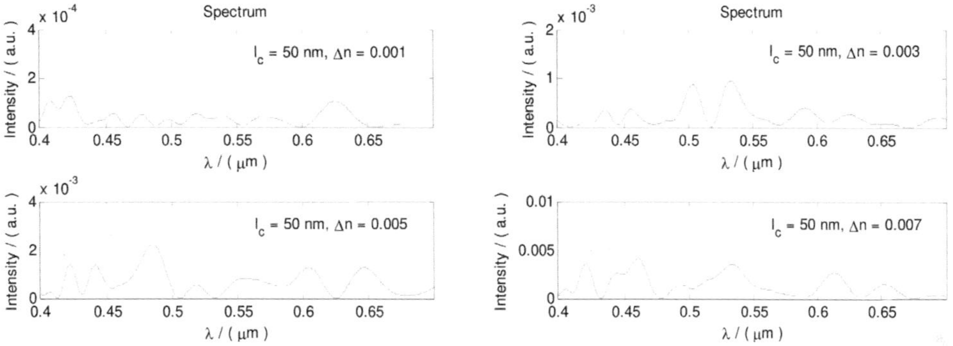


**Figure 4.** Simulated 1D mesoscopic reflective spectrum versus standard deviation

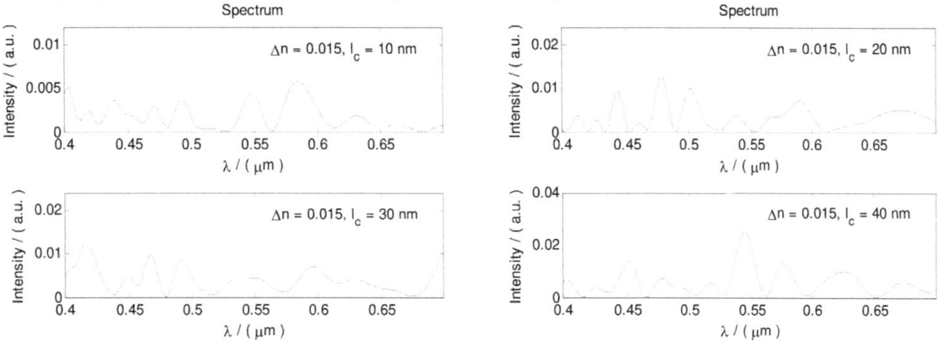


**Figure 5.** Simulated 1D mesoscopic reflective spectrum versus spatial correlation length

If the spatial correlation length was constant ($l_c = 50nm$), while the standard deviation of the refractive index fluctuations was varied, simulated 1D mesoscopic reflective spectrums were shown as figure 4. In turn, if the standard deviation of the refractive index fluctuations was constant ($\Delta n \equiv 0.015$), while the spatial correlation length was varied, simulated 1D mesoscopic reflective spectrums were shown as figure 5. Known from figure 4 and figure 5, the reflective spectrums were highly sensitive to subtle changes of standard deviation $\Delta n$ and spatial correlation length $l_c$ of refractive index fluctuations. In other word, the 1D reflective mesoscopic spectrums were also highly sensitive to subtle changes of disorder strength.

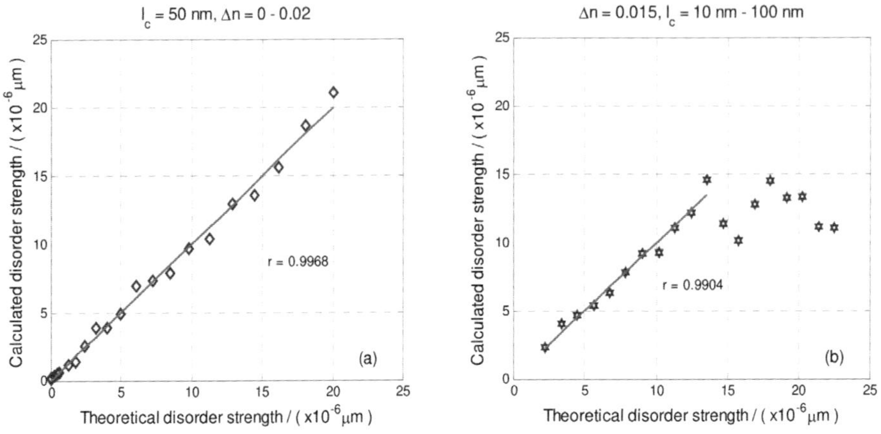


**Figure 6.** Comparison of theoretical disorder strength and calculated disorder strength.

Through precise control of the standard deviation and the spatial correlation length of refractive index fluctuations along a one-dimensional channel, calculated disorder strength (calculated by equation (2)) and theoretical disorder strength (calculated by equation (1)) were quantitatively analyzed. The simulation results were illustrated in figure 6.

Figure 6 (a) showed situations of the spatial correlation length was constant ( $l_c = 50nm$ ), while the standard deviation of refractive index fluctuations $\Delta n$ varied from 0 to 0.02. Figure 6 (b) showed situations of the standard deviation of refractive index fluctuations was constant ( $\Delta n \equiv 0.015$ ), while the spatial correlation length $l_c$ varied from 10nm to 100nm. In the areas of marked with red solid lines in both figure 6 (a) and figure 6 (b), the linear correlation of theoretical disorder strength and calculated disorder strength were as high as r = 0.9968 and r = 0.9904 respectively. So calculated disorder strength was essentially equal to theoretical disorder strength. However, in the area of without marked with red solid lines in figure 6 (b), calculated disorder strength was generally smaller than theoretical disorder strength. The main reason was that the approximation condition $kl_c < 1$ which was used in deriving calculated disorder strength formula was not fully met.

## 4. Conclusion

The mesoscopic theory was applied to analysis the properties and optical information of the photon transmission in 1D medium which has nanometer scale refractive index fluctuations. One-dimensional photonic transmission channel model of weakly disordered and weakly scattering biological cell was established. The root mean square of the fluctuating part of reflective spectrum from 1D channel was linearly depends on the disorder strength. Although the disorder strength do not directly related to structural changes within the channel at certain specific location, increased disorder strength represents 1D channel toward a more disordered state, so regarding disordered strength as optical fingerprint of cell's carcinogenesis state is feasible. The FDTD simulation results show that reflective spectrum is highly sensitive to disorder strength. Under certain approximating conditions, disorder strength is proportionally enlarged with increasing variance and spatial correlation length of the refractive index fluctuations. An experimental instrument has been established, which can simultaneously obtain a diffraction-limited resolution microscopic image of sample and record the spectrum from each pixel (each 1D channel). Experiments on nano-structured models and biological samples are on schedule.

## 5. Acknowledgement

This work was supported by the National Natural Science Foundation of China (Grant No. 60978037, No. 60878057).

**References**


[1] Bozzola J J and Russell L D 1999 Electron Microscopy: Principles and Techniques for Biologists(Second Edition)

[2] Bartek M, Wang X, Wells W, Paulsen K D and Pogue B W 2006 J. Biomed. Opt. **11** 064007

[3] Wang T D, Van Dam J 2004 Clinical Gastroenterology and Hepatology **2** 744

[4] Sung K B, Rebecca R K, Follen M, Malpica A, Chen L, and Descour M 2003 Opt. Exp. **11** 3171

[5] Gurjar R S, Backman V, Perelman L T, Georgakoudi I, Badizadegan K, Itzkan I, Dasari R R and Feld M S 2001 Nature Medicine **7** 1245

[6] Georgakoudi I, Jacobson B C, Van Dam J, Backman V, Wallace M B, Muller M G, Zhang Q G, Badizadegan K, Sun D, Thomas G, Perelman L T and Feld M S 2001 Gastroenterology **120** 1620

[7] Wang K and Ding Z H 2008 Chin. Opt. Lett. **6** 902

[8] Chalut K J, Kresty L A, Pyhtila J W, Nines R, Baird M, Steele V E, Wax A 2007 Cancer Epidemiology Biomarkers & Prevention **16** 223

[9] Courjon D 2003 *Near field microscopy and near field optics* (London, Imperial College Press)

[10] Hell S W 2003 Nat. Biotechnol **21** 1347

[11] Hell S W, Willig K I, Dyba M, Jakobs S, Kastrup L and Westphal V 2006 *Nanoscale Resolution with focused light: STED andother RESOLFT microscopy concepts, Handbook of biological confocal microscopy*, third edition (New York, Springer Science Business Media) pp571-579

[12] Roy H K, Liu Y, Wali R K, Kim Y L, Kromin A K, Goldberg M J, Backman V 2004 Gastroenterology **126** 1071

[13] Liu Y, Brand R E, Turzhitsky V, Kim Y L, Roy H K, Hasabou N, Sturgis C, Shah D, Hall C and Backman V 2007 Clin Cancer Res **13** 4392

[14] Wax A, Ynag C H, Backman V, Badizadegan K, Boone C W, Dasari R R and Feld M S 2002 Biophys J **82** 2256

[15] Liu Y, Pradhan P, Li X, Kim Y L, Wali R K, Roy H and Backman V 2006 Biomedical Topical Meeting (BIO) paper: TuD1

[16] Subramanian H, Pradhan P, Kunte D, Deep N, Roy H and Backman V 2008 Biomedical Optics (BIOMED) paper: BTuC5

[17] Liu Y, Li X, Kim Y L and Backman V 2005 Opt. Lett. **30** 2445

[18] Vadim Backman, Yang Liu, Young Kim, Hemant Roy, Michael Goldberg, Randall Brand, Prabhakar Pradhan and Hariharan Subramanian, *Method for identifying refractive-index fluctuations of a target*, US 2008/0278713 A1

[19] Subramanian H, Pradhan P, Liu Y, Capoglu I R, Rogers J D, Roy H K, Brand R E and Backman V 2009 Opt. Lett. **34** 518

[20] Born M and Wolf E 1999 *Principles of Optics: Electromagnetic Theory of Propagation, Interference and Diffraction of Light* (Cambridge,UK, Cambridge Univ Press) pp554-633

[21] Anderson P W 1958 Phys. Rev. **109** 1492

[22] Anderson P W, Thouless D J, Abrahams E and Fisher DS 1980 Phys. Rev. B **22** 3519

[23] Kumar N 1985 Phys. Rev. B **31** 5513

[24] Haley S B and Erdos P 1992 Phys. Rev. B **45** 8572

[25] Rammal R and Doucot B 1987 J. Phys. France **48** 509

# Simulation on scattering features of biological tissue based on generated refractive-index model

**Baoyong Wang, Zhihua Ding***

State Key Lab of Modern Optical Instrumentation, Zhejiang University 38 Zheda Rd., Hangzhou 310027, P.R.China

*Corresponding author: zh_ding@zju.edu.cn

**Abstract**. Important information on morphology of biological tissue can be deduced from elastic scattering spectra, and their analyses are based on the known refractive-index model of tissue. In this paper, a new numerical refractive-index model is put forward, and its scattering properties are intensively studied. Spectral decomposition[1] is a widely used method to generate random medium in geology, but it is never used in biology. Biological tissue is different from geology in the sense of random medium. Autocorrelation function describe almost all of features in geology, but biological tissue is not as random as geology, its structure is regular in the sense of fractal geometry[2], and fractal dimension can be used to describe its regularity under random. Firstly scattering theories of this fractal media are reviewed. Secondly the detailed generation process of refractive-index is presented. Finally the scattering features are simulated in FDTD (Finite Difference Time Domain) Solutions software. From the simulation results, we find that autocorrelation length and fractal dimension controls scattering feature of biological tissue.

## 1. Introduction

Since J. M. Schmitt and G. Kumar[2] discuss the turbulent nature of biological tissue index, the fractal structure of biological tissue has been intensively studied[3,4,5]. Tissue images obtained by various imaging technology and tissue slice[6,7,8,9] are the most usually used and direct way to study biological tissue structures. And their results indeed shows that tissue structures ,such as vasoganglion under skin, membranes of organelles, texture of nucleus, are fractals[6,8,9]. Unfortunately the interaction between light and fractal structures is not studied enough.

J. M. Schmitt and G. Kumar[2] are pioneers of this topic. Based on their study of the mouse liver tissue image under phase-contrast microscopy, they found that the spectrum of tissue index variations possesses a power-law behavior for spatial frequencies over a range of scales. Inspired by the similar structure function of the classical Kolmogorov model of turbulence, they consider biological tissue index variation as a kind of turbulence too, and take it as the fundamental feature of biological tissue. Since then biological tissue can be called fractal random media.

M. Xu and R. R. Alfano[10] develop this theory. They start their study from the postulate that the biological tissue is a kind of continuous random media. Based on potential scattering theory and Born approximation, light scattering by continuous random media is modeled and studied. Their model shows that biological tissue is a complex of random media with different autocorrelation lengths of fractal distribution. In their theory, the scattering properties derived from this model, e.g. reduced scattering coefficient, the anisotropy factor, the phase function are studied.

Because of the complexity of biological tissue, all the research groups above use approximates in their work, and one thing more they ignore one nature of lights that lights are vectors with phases and polarization states in their theory. FDTD (Finite Difference Time Domain)[11] is a numerical method solving Maxwell's equations, and a powerful tool to study scattering problems of structures. Applying this method, there are two advantages. One is unlimited structures that can be studied, another is full vector scattering field you can get. Therefore, numerical simulation approach provides a novel and powerful way to study scattering problems.

## 2. Principle

The exact theory of the scattering by fractal structures is not available yet. But when the index variations of tissue is very weak, e.g. relative index $m \approx 1.01 \sim 1.1$, Born approximate and Rytov approximate[12] etc. are valid, and for most cases in biological tissue, Born approximate is a reasonable assumption. So under these approximates a lot of useful deductions can be achieved, and these deductions are foundations of other researches.

### 2.1. Scattering theory of fractal medium

Potential scattering theory[13] declares that under Born approximate the scattering amplitude is derived from a special Fourier component of potential scattering of tissue. Martin Hunter[14] shows that the potential scattering of random media is the autocorrelation function of the media. For the simplification in math, biological tissue can be treated as an isotropy stationary random media, and its autocorrelation function can be the following form,

$$C(r) = C_0 \times \exp(\frac{r}{l_c}) \tag{1}$$

Where $l_c$ is the autocorrelation length, $C_0$ is a constant, $r$ is the distances between any two points. And $l_c$ follows a fractal distribution[10],

$$f(l_c) = cons\tan t \times l_c^{\,3-Df} \tag{2}$$

Where $Df$ is fractal dimension. According to Wiener-Khinchin theorem, the power spectrum density of random media is the Fourier transformation of autocorrelation function. So, we get the power spectrum density[10],

$$\hat{R}(k) = \int_0^{l\max} \frac{\varepsilon^2 l^3}{\pi^2 (1 + k^2 l^2)^2} f(l) dl \tag{3}$$

And the scattering intensity[10] is,

$$|S(\theta)|^2 = \int_0^{k*l\max} \frac{\varepsilon^2 \times cons\tan t \times k^{Df-1} x^{6-Df}}{2\pi [1 + 2(1-\mu) x^2]^2} dx \tag{4}$$

Where the $\theta$ is the scattering angle , $\mu = \cos(\theta)$ , $\varepsilon^2 = 4 n_0^{\,4} (m-1)^2$ , $k$ is the wave vector, $l_{\max}$ is the maximum autocorrelation length. From equation (4), we can get scattering intensity distribution as a function of either scattering angle $\theta$ , or wave vector $k$ .Based on this, other scattering coefficients can be derived.

### 2.2. Spectral decomposition method

The FDTD method demands a exact refractive index distribution as an input, so we have to generate a special random media with a controlled statistic. Spectral decomposition method is widely used to generate random medium in geology[1], and this method has universality in generating a random media, so it can be used in biological. In general, spectral decomposition is based on Wiener-Khinchin theorem too, the only difference is that we multiply the power spectrum density by random phases. The process is demonstrated in the flow chart below,

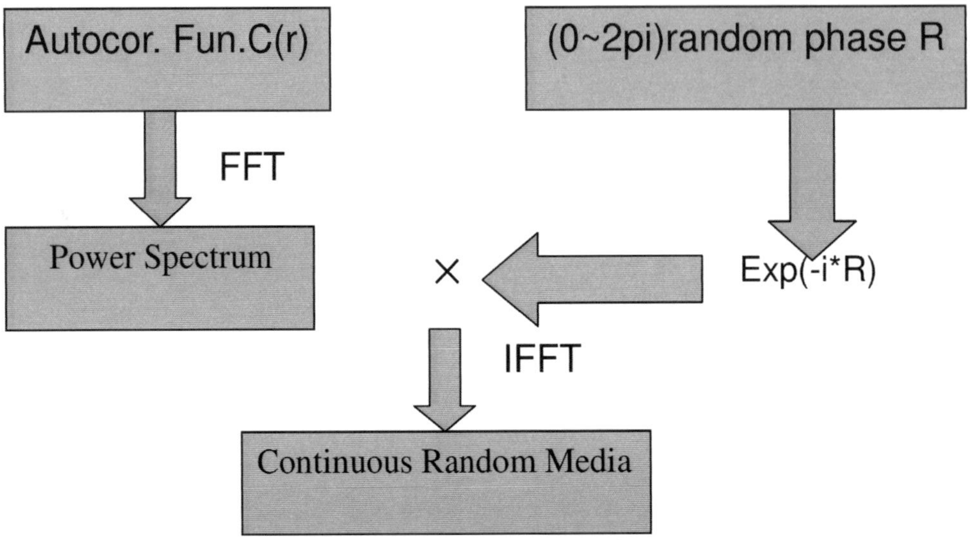


**Figure 1.** Flow chart of spectral decomposition method

Applying this method we get an exact refractive index distribution show below.

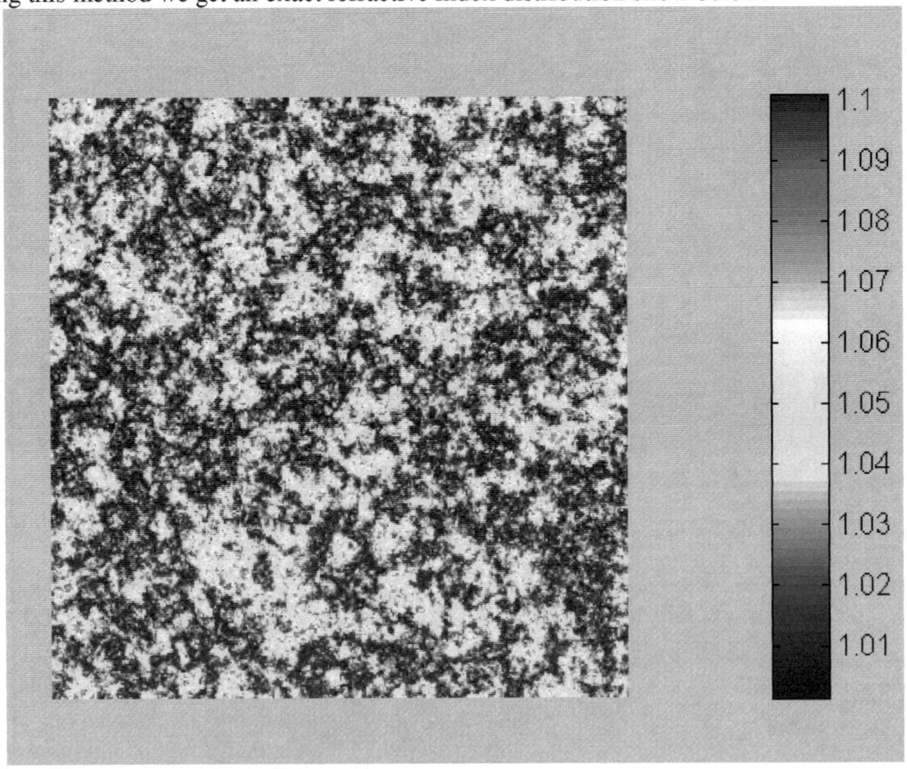


**Figure 2.** A 2D random medium.

This 2D random medium with a mean refractive index value 1.03, *Df* 6.32, $l_{max}$ 8.5um, standard deviation 0.01, and 100um×100um in area.

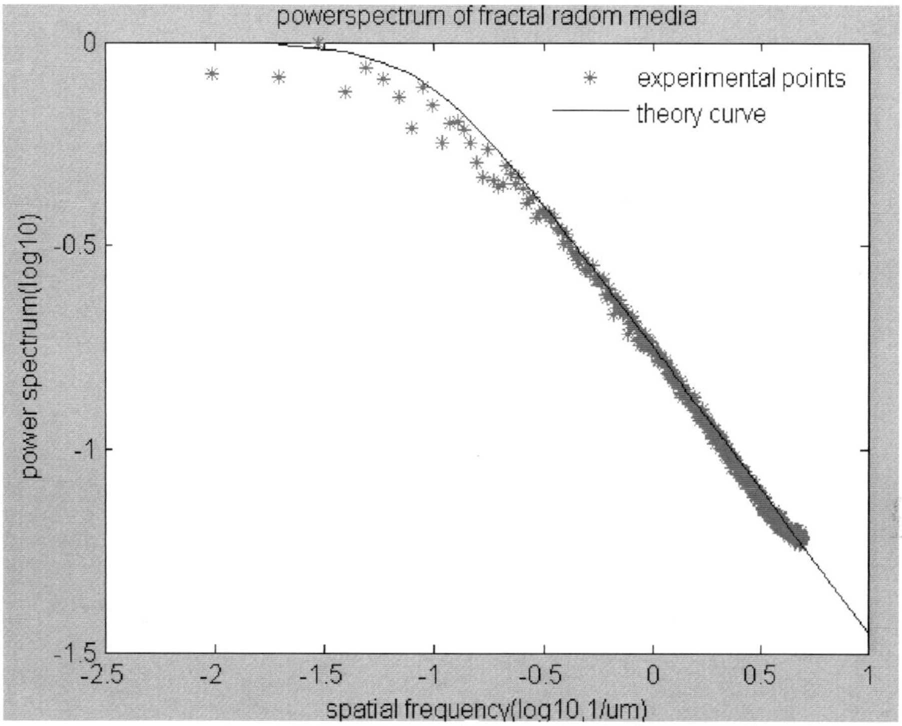


**Figure 3.** Power spectrum of refractive index variations

Figure 3 show the power spectrum of generated fractal random media. The power spectrum of refractive index variations is calculated using the algorithm mentioned in Ref.2. Applying least square fit to the experimental data, we get a $Df$ 6.32, $l_{max}$ 8.2um, this fit our initial setup exactly.

## 3. FDTD Simulations

$Df$ and $l_{max}$, together with the mean value of refractive index $m_0$ and standard deviation $\sigma$ (these two combined as $\varepsilon^2$), we get four controllable input parameters. In most application, tissue is semi-infinite, and a limited range of scattering angle can be detected. So in this simulation, only backscattering properties are studied. We use multi-waves in the simulation, so we will get scattering amplitude as the function of scattering wavelength.

### 3.1. Simulation Setup

Using the method mentioned above, We generate a 3 dimension random media. It has a $Df$ 5.2, $l_{max}$ 0.5um, mean refractive index 1.05, standard deviation 0.01, and 5um×5um×5um in volume, 0.05um in step size. Then we put the generated media in the software FDTD Solutions 6.5.11 (lumerical) to run the simulation. In my simulation, we use the wavelength range 400nm~700nm, and set the periodic boundary conditions in the direction normal to wave vector and a absorbing boundary condition on other boundaries.

### 3.2. Simulations of bulk scattering

Figure 4 shows the parallel polarized part of scattering intensity as the function of wavelength. Based on the simulated data, we apply a least square fit in the log-log coordinates, and we get $Df$ 5.4, $l_{max}$ 0.6um, these results fit our initial setting roughly. Because of limited simulation area and accuracy, there is a little error between the two.

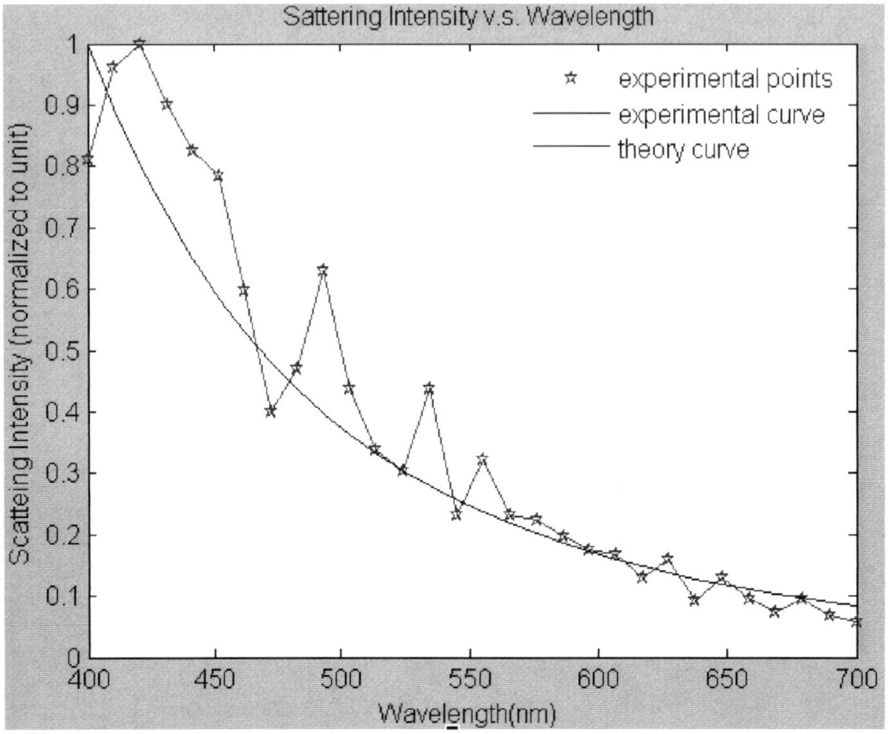


**Figure 4.** parallel polarized scattering intensity vs wavelength

### 3.3. *Df* and $l_{max}$

From equation 3, we know that *Df* and $l_{max}$ control the shape of power spectrum density, and as the figure 2 shows, the curve of power spectrum density follows a line-curve-line pattern. So which parameter controls the position of the curve, which parameter controls the slope of line? We do the simulation below.

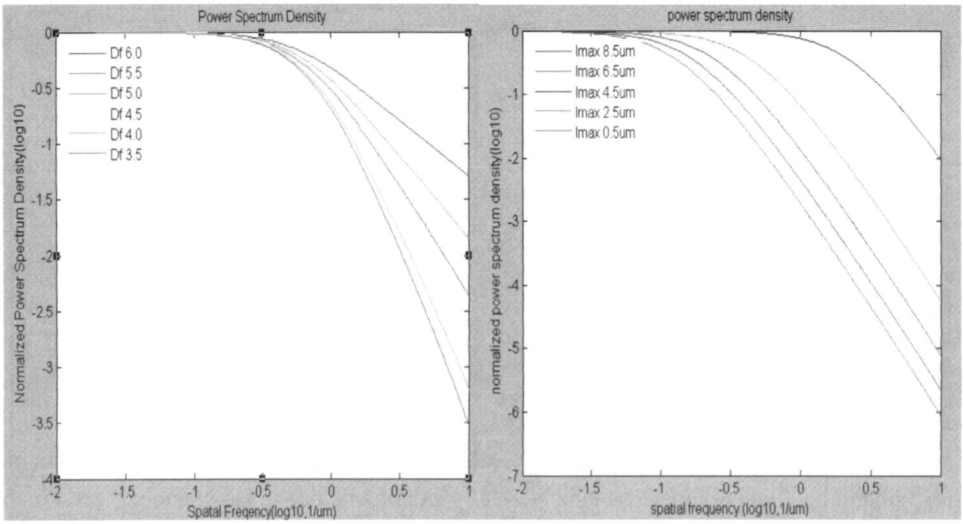


**Figure 5.** Scan *Df* or $l_{max}$ while keep the other unchanged

This figure shows the power spectrum changes while $Df$ or $l_{max}$ changes. The left one is the case that keep $l_{max}$ 1.5um,change $Df$ every time, and this shows that $Df$ changes the slope of oblique line only. The right one is the case that keep $Df$ 3.5, change $l_{max}$ every time, and this shows that $l_{max}$ change the position of curve only.

Ground on figure 5, we know that $l_{max}$ controls the position of curve, while $Df$ only affects the slope of the line. Further more, the slope of the oblique line changes when lower limit of autocorrelation length $l_{min}$ changes. All in all, $Df$ is more sensitive to small scale structures. Size-testing shows that the sensitive size range is from 1um to 0.01um. FDTD simulations support this proposition too.

## 4. Discussion

Based on my simulation, $Df$ and $l_{max}$ can be derived from the backscattering intensity spectrum. But as the equation 4 shows, $m_0$ and $\sigma$ are constants for either wavelength or scattering angle. So $m_0$ and $\sigma$ variations don't change the shape of the scattering curve, and if you want to get them, you have to get the absolute quantity of scattering intensity. This is difficult for both experiments and practical applications. Fortunately tissues are more complex than fractal random media, as the nucleuses are the most import scattering centers for most kinds of surface tissues, large particles make a great contribution to the scattering pattern in small scattering angles, and $m_0$ coupled with $\sigma$ are parameters of particle scattering. So make use of nucleus scattering is a good way to solve this problem[15].

The backscattering spectrum of figure 3 obtains a fluctuation deflected from theory curve. An initial guess is that this fluctuation is caused by randomness of the input media, but when applying 35 times average, this fluctuation still exits, only smaller a little. So this fluctuation is an inherent property of scattering field, and the relation between the scattering field fluctuation and refractive index variation will tell us more information[12].

## 5. Conclusions

The numerical model of tissue generated by the spectral decomposition method performs good in the backscattering direction. Based on this model, we can derive $Df$ and $l_{max}$ from the backscattering spectrum detected in experiments or practical applications. while $Df$ means the overall complexity of refractive index variations, particularly $Df$ is sensitive to small structures ranging from 1um to 0.01um, and $l_{max}$ stands for maximum scale in biological tissue which is an important characteristic in tissue morphology.

## 6. Acknowledgement

This work was supported by National Natural Science Foundation of China (60878057, 60978037).

## References

[1] L.T.Ikelle,S.K.Yung etc. 2D Random media with ellipsoidal autocorrelation functions,1993 *Geop.*, **Vol. 58, NO. 9**.

[2] J. M. Schmitt and G. Kumar, Turbulent nature of refractive-index variations in biological tissue 1996 *Opt. Lett.*, **Vol. 21, No. 16**.

[3] P. Waliszewski, J. Konarski, Tissue as a self-organizing system with fractal dynamics,2001 *A. Spa. Resea*, **Vol. 28. No. 4**, pp. 545-548.

[4] Gabriele A. Losa,etc. F*ractals in biology and medicine*,2005,pp.153-209.

[5] Carla M. Rosetti, Bruno Maggio,etc. The self-organization of lipids and proteins of myelin at the membrane interface Molecular factors underlying the microheterogeneity of domain segregation,2008, *Bio. et Biophysica Acta* **1778**,pp.1665–1675.

[6] Wolfgang Bauer,etc. Cancer Detection via Determination of Fractal Cell Dimension,1995 *arXiv:patt-sol* **vl.9506003**.

[7] Nicola Dioguardi, Fabio Grizzi, etc. Liver fibrosis and tissue architectural change measurement using fractal-rectified metrics and Hurst's exponent,2006 *World J. Gastr.* **12(14)**: 2187-2194.

[8] Valcinir V. Morariu, etc. A Fractal and Long-range Correlation Analysis of Plant Nucleus Ultrastructure,2006 *Romanian J. Biophys.*, **Vol. 16, No. 4**, P. 243–252.

[9] V.Bedin,etc. Fractal dimension of chromatin is an independent prognostic factor for survival in melanoma 2010, *BMC Cancer*, **10:**260.

[10] M. Xu and R. R. Alfano, Fractal mechanisms of light scattering in biological tissue and cells, 2005 *Opt. Lett.*, **Vol. 30, No. 22.**

[11] Allen Taflove and Susan C. Hagness, *Computational Electrodynamics: The Finite – Difference Time-Domain Method, 3rd ed*, Artech House Publishers. ISBN 1-58053-832-0.

[12] Akira Ishimaru, *Wave Propagation and Scattering in Random Media*,Academic Press,1978,346-351,402-406.

[13] Born and wolf, *Principles of Optics*, 1999. Publishing House of Electronics Industry 580-586.

[14] M. Hunter, etc. Tissue Self-Affinity and Polarized Light Scattering in the Born Approximation: A New Model for Precancer Detection,2006 *PRL* **97, 138102**.

[15] M.Xu, etc. Unified Mie and fractal scattering by cell and experimental study on application in optical characterization of cellular and subcellular structures,2008 *JBO* **13(2), 024015**.

# Simulation of medical Q-switch flash-pumped Er:YAG laser

**WANG -Yan-lin, HUANG-Chuyun*, YAO-Yucheng and Zou Xiaolin**

Physics school, Hubei University of Technology, Wuhan, China 430068

Wangyanlin0@126.com chuyunh@163.com yyuch@soho.com zouxiaol@126.com

**Abstract:** Er: YAG laser, the wavelength is 2940nm, can be absorbed strongly by water. The absorption coefficient is as high as 13000 $cm^{-1}$. As the water strong absorption, Erbium laser can bring shallow penetration depth and smaller surrounding tissue injury in most soft tissue and hard tissue. At the same time, the interaction between 2940nm radiation and biological tissue saturated with water is equivalent to instantaneous heating within limited volume, thus resulting in the phenomenon of micro-explosion to removal organization. Different parameters can be set up to cut enamel, dentin, caries and soft tissue. For the development and optimization of laser system, it is a practical choice to use laser modeling to predict the influence of various parameters for laser performance. Aim at the status of low Erbium laser output power, flash-pumped Er: YAG laser performance was simulated to obtain optical output in theory. the rate equation model was obtained and used to predict the change of population densities in various manifolds and use the technology of Q-switch the simulate laser output for different design parameters and results showed that Er: YAG laser output energy can achieve the maximum average output power of 9.8W under the given parameters. The model can be used to find the potential laser systems that meet application requirements.

Keyword: Er:YAG laser; rate equation; Q-switch

## 1. INTRODUCTION

The wavelength of Er(50at.%):YAG laser radiation($\lambda$=2.940μm) matches a pronounced absorption band of water.So the small penetration depth of the radiation (little more than 1μm in water) and the high intensity of the Er:YAG laser pulses promote effective ablation of soft and hard human tissues (the latter group is of special interest for dentistry) with minimum inevitable destruction of the area closed to the treated location.However the output power of Erbium laser is low,so aim at this flash-pumped Er: YAG laser performance was simulated to obtain optical output in theory and we also used Q-switch technology to short pules[1],all of this based on the rate equation.The result show that Er: YAG laser output energy can achieve the maximum average output power of 9.8W under the given parameters.

## 2. CRYSTAL CHARACTER

Zhavikov E V have firstly reported the output of 2.94μm Er(30at.%):YAG in 1967;Li-Mengyu ,in 1987, have also obtained the 2.94μm laser output by use Er(50at.%):YAG.

In 1990,Li-Yuanqi use optical crystal with 2mm deep gain fluorescence emission spectrum of $Er^{3+}$,show in Fig 1 and Fig 2 [2].

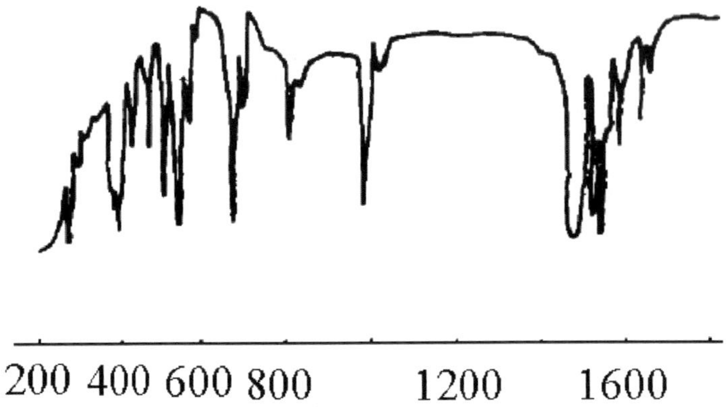


**Figure 1.** Absorption spectrum of $Er^{3+}$

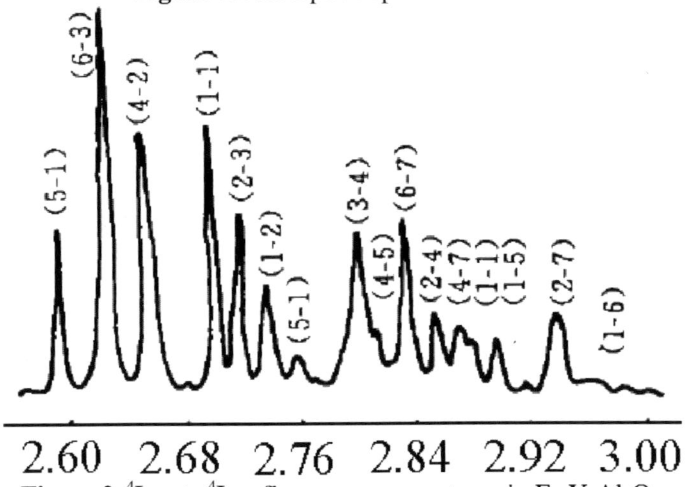


**Figure 2.** $^4I_{11/2}$ to $^4I_{15/2}$ fluorescence spectrum in $Er{:}Y_3Al_5O_{12}$

## 3. SPECIAL APLICATION

The interaction between body and laser, Whether through photochemical effect makes organizing molecular fracture, or by thermal effects to tissue vaporization cells,all of this require cell to fully absorb the laser energy. In biological cells and tissues, the water content of more than 70% usually (some up to 90%)[2], due to the laser wavelength of 2.94 μm is consistent with the vibration frequency of water molecules, it was very strong in the water absorption, absorption coefficient to $\mu_a$=13000 $cm^{-1}$, can use less energy to achieve greater the effect. The superiority is unmatched by other wavelengths, using it can be very precise cutting and soft tissue surgery. In addition, the interaction of water saturated tissue and 2.94μm laser focus, equivalent to a limited volume of transient heating, increased pressure within the volume, with the substance (eg blood) from the incision emission, the phenomenon does not produce blood clotting. This feature determines the Er: YAG lasers in the medical field wide application, Figure 3 shows the absorption spectra of water molecules and the commonly used $CO^2$, Nd: YAG, Er: YAG laser output wavelength of the location , can be seen, for laser energy is absorbed by tissue, select the output wavelength of 2.94 μm of Er: YAG laser is the most ideal outcome[3].

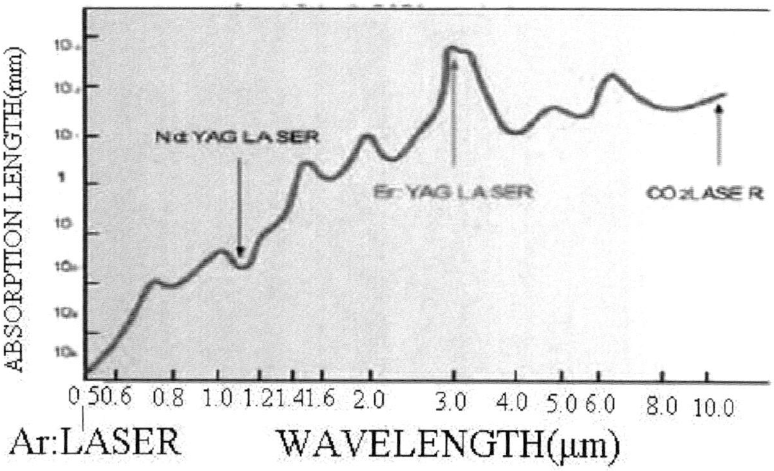


**Figure 3.**Absorption spectrum of water molecules

## 4. RATE EQUATION FOUNDATION

Two pumping channels are effectively present in Er:YAGlasers . The so-called direct-pumping channel consists of the $Er^{3+}$ ground-state absorption with transition to states $^4I_{11/2}$, $^4I_{9/2}$ and $^4F_{9/2}$ and a nonradiative relaxation of the latter two to the upper laser level $^4I_{11/2}$. The indirectpumping channel, on the other hand, starts with a shorterwavelength excitation to states $^4S_{3/2}$, $^2H_{11/2}$ and $^4F_{7/2}$,and a nonradiative relaxation of the later two to the high-lying metastable level $^4S_{3/2}$, Through inter-ionic cross relaxation of this level, both the upper and the lower laser level ($^4I_{13/2}$)are excited to the same extent[4].

Despite a shorter lifetime of the emitting level as compared with the terminal one, efficient long-pulse or even CW 2.94μm memission was obtained at room temperature in concentrated erbium systems[5]. Excited state absorption (ESA) from the terminal laser level and/or energy transfer processes as a cooperative upconversion inside the system of $Er^{3+}$ ions were invoked to explain the success of the 2.94μm erbium laser. The presence of efficient laser emission in concentrated systems favors the cooperative upconversion model. According to this model, the driving mechanism for 2.94μm erbium lasers is the upconversion process($^4I_{13/2} \rightarrow {}^4I_{15/2}$) +($^4I_{13/2} \rightarrow {}^4I_{9/2}$)followed by the rapid multiphonon transition to $^4I_{11/2}$.This mechanism depopulates the terminal laser level and repopulates the initial one. In order to understand the peculiarities and the possibilities of the 2.94μm Er lasers working in the Q–switch regime, we constructed and used rate equatio erbium systems is given in (1).

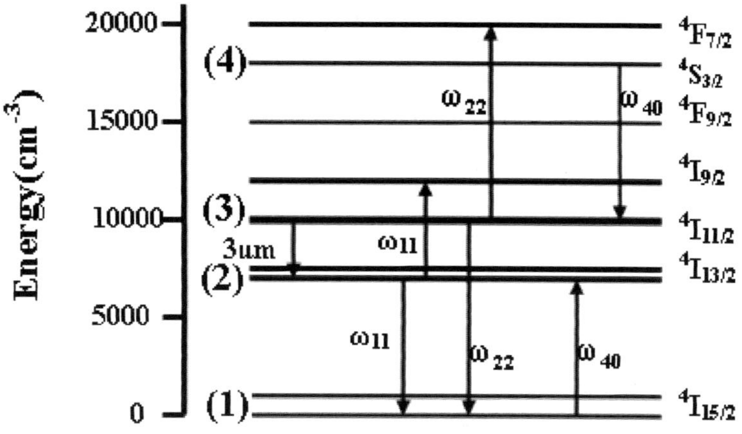


**Figure 4.** Energy level scheme of $Er^{3+}$ in YAG

$$\frac{dN_4}{dt} = R_1 N_0 - \frac{N_4}{\tau_4} + \omega_{22} N_3^2$$

$$\frac{dN_3}{dt} = R_2 N_0 + \omega_{40} N_1 N_4 + \omega_{11} N_1^2 - 2\omega_{22} N_2^2 - \frac{N_3}{\tau_3} + \frac{N_4}{\tau_4}$$

$$\frac{dN_2}{dt} = -\frac{N_2}{\tau_2} + \frac{N_3}{\tau_3} - 2\omega_{11} N_1^2 - c\sigma(\alpha N_2 - \beta N_1)\phi$$

$$N_1 = N_{er} - \sum_{i=2}^{4} N_i$$

$$\frac{d\phi}{dt} = \frac{c\sigma}{\left(n_r - 1 + L/l\right)}\left[\sigma(\alpha N_2 - \beta N_1) - \rho\right] \quad (1)$$

Here, in addition to differential equations for the populations $N_2$ and $N_3$ of the lower ($^4I_{13/2}$) and the upper ($^4I_{11/2}$)laser level, this model accounts explicitly for the population $N_4$ of the state $^4S_{3/2}$, and $\phi$ is the photos desity,while the relaxations of other electronic states are assumed to be instantaneous. The energy transfer processes are represented in (1) by $\omega_{11}$,the rate of upconversion from $^4I_{13/2}$, $\omega_{22}$ the rate of upconversion from $^4I_{11/2}$ and $\omega_{40}$ and the rate of the cross relaxation from $^4S_{3/2}$ and $\alpha$, $\beta$ are the Boltzmann population coefficients for the Stark sublevels involved in the laser transition. $c$ is the speed of light.The parameter $l$ and $L$ are crystal length and optical length in cavity,the connection of them can expressed as this:

$$L = L_0 + (n_r - 1)l$$

in this formulary: $L_0$ is the length of cavity; $n_r$ is the refractive index of the crystal.

The total loss per round trip $\rho$ is expressed as[5]:

$$\rho = -\ln\frac{R_1 R_2 T_Q^2}{2l} + \rho_0$$

where $R_1, R_2$ are the reflectivity of the total (rear) and partial (coupling) laser mirrors, $T_Q$ is the one-way optical transmission of the Q-switch, and $\rho_0$ is the total passive loss.

The Q-switch operation is simulated in (1) by changing the optical transmission of the Q-switch. This can be done by assuming various temporal dependencies of transmission. we found that the change of the -switch transmission can be quite realistically represented by the expression:

$$T_Q(t) = T_Q^{low} + (T_Q^{low} - T_Q^{high}).\{[\cos(\frac{2\pi}{\tau_Q}(t - t_1))]^m - 1\} \quad (2)$$

where $T_Q^{low}$ and $T_Q^{high}$ are, respectively, the values of the transmission in the closed and open state of the Q-switch and $t_1$ is the Q-switching moment. With two parameters, $\tau_Q$ and $m$ (where $m$ is an even integer), we can simulate various frontsand duration of the transmission interval of the Q-switch.

**5. SIMULATION**

Firstly,we simulated the Q-switch transmission (2) under those parameters:

$T_Q^{low} = 0.3$ $\quad T_Q^{high} = 0.95$ $\quad \tau = 16\mu s$

$m = 12$ $\quad t_Q \approx 1.5\mu s$

The picture we obtained (as up given parameter) as figure 5:

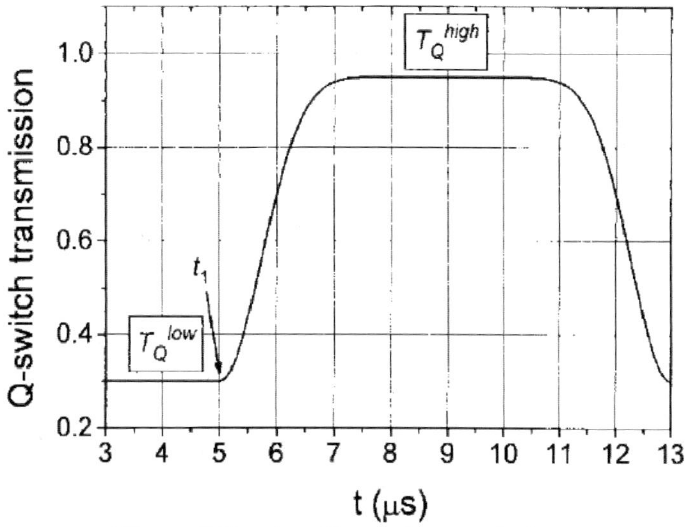


**Figure 5.**Simulate Q-switch transmission

Then we simulated the rate equation ,we chose suitably parameter as follows:

$\sigma = 2.6\times10^{-24} m^3$ $\omega_{22} = 3.7\times10^{-21} m^3 s^{-1}$ $\omega_{11} = 1.3\times10^{-21} m^3 s^{-1}$

$\omega_{40} = 1.06\times10^{-21} m^3 s^{-1}$ $h = 6.626\times10^{-34}$ $c = 3.0\times10^8 ms^{-1}$

$n_r = 1.838$, $R_1 = 0.6$ $R_2 = 0.4$ $\tau_3 = 0.1\times10^{-3} s$

$\tau_2 = 2.0\times10^{-3} s$ $\tau_4 = 1.6\times10^{-5} s$ $T_{out} = 0.80$ $T=1$,

then we made use of matlab to simulate and result have showed as follows:

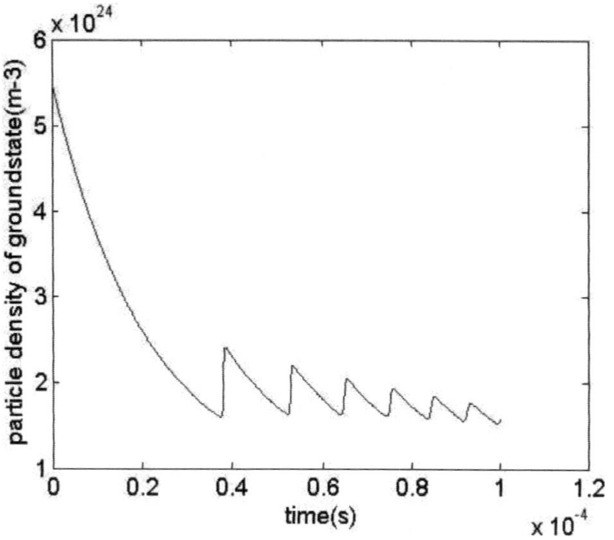

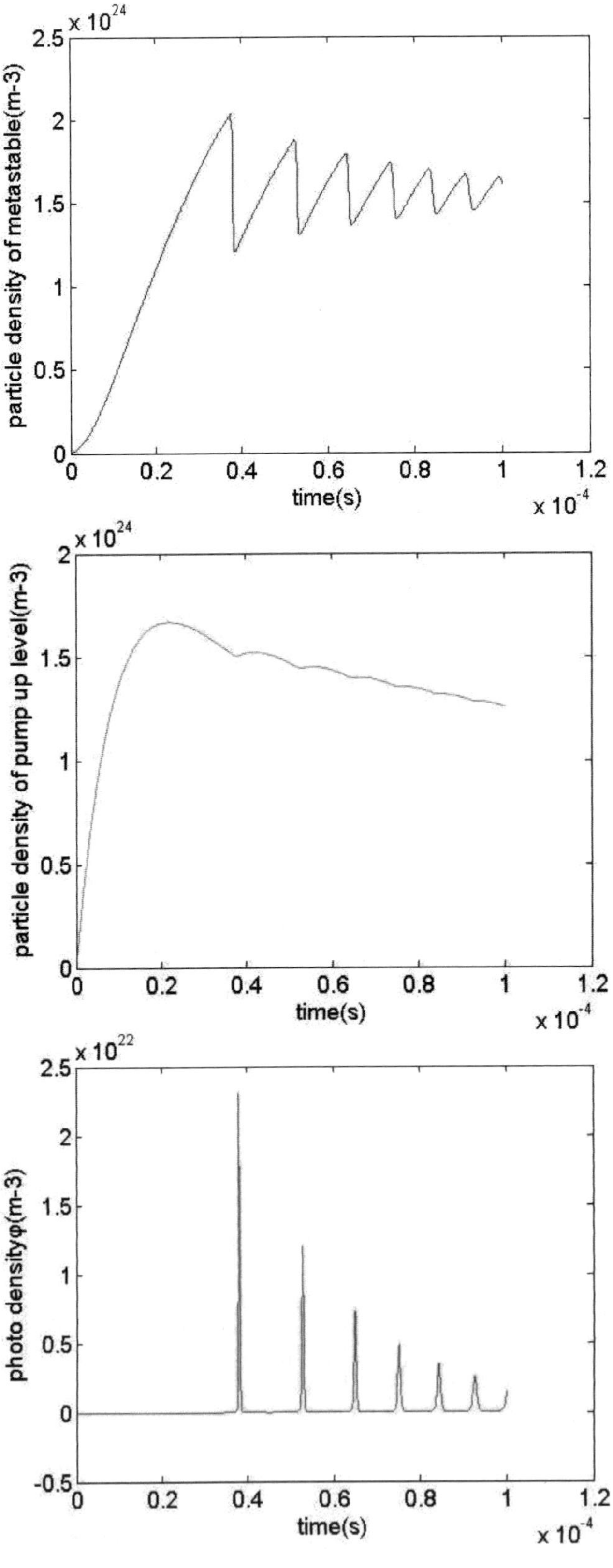

x 10^24
particle density of metastable(m-3)
time(s)
x 10^-4
x 10^24
particle density of pump up level(m-3)
time(s)
x 10^-4
x 10^22
photo densityφ(m-3)
time(s)
x 10^-4

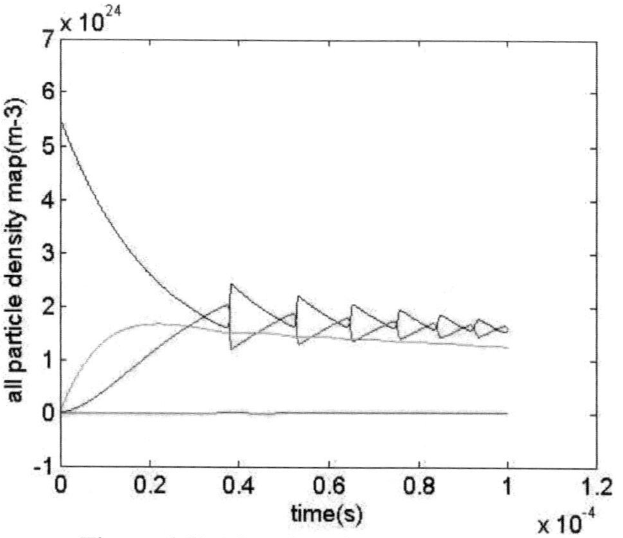


**Figure 6.**Simulate Q-switch rate equation

Long pulse Q-switch laser is similar to CW laser,so the output power can expressed by[7]:

$$P_{out} = \phi.\sigma.h\frac{c^2}{\lambda}$$

based on the simulation ,the output power about 10W.

## 6. CONCLUSION

We have simulated the Q-switch flash-pumped Er:YAG laser and calculated the output power about 10W in theory,this model based on rate equation can afford some suggestions to manufacture and application of Er:YAG, but what we done just in theroy so it existence error .Actually a lot of parameters affect the output of the laser,such as bulk of model,total efficiency of pump in different condition.So what we did in this article,compare to practice,is just a guid for the law.

## REFERENCES

[1] Maris Ozolinsh, Karl Stock, Raimund Hibst, and Rudolf Steiner,"Q-Switching of Er:YAG (2.9μm) Solid-State Laser by PLZT Electrooptic Modulator",IEEE J. Quantum Electron, vol.33, no.10,pp:1846-1849,1997

[2] Bao Liang-bi, Ye-Bin, Yuan Zi-jun, Cheng- Ping, Liu Fu-xia,"Study of Er□YAG laser and its application", School of Scienc,vol.24,no.1,pp:23-27,2001

[3] Liang Jian-wei,Jin-Tao,Jiang Shao-ji,"The Study and Fabrication of 2.94μm Er:YAG Laser System for Medical Use", Chinese Journal of Medical Physics,vol.26,no.4,pp:1345-1348,2009

[4] Boris Majaron,Tanja Rupnik and Matjai Lukac, "Temperature and gain dynamics in flashlamp-pumped Er:YAG", IEEE J.Quantum Electron,vol.32,no.9, pp:1636-1644,1996

[5] Serban Georgescu and Voicu Lupei,"Q-Switch Regime of 3-μm Er:YAG Lasers", IEEE J. Quantum Electron,vol.34,no.6,pp:1301-1040,1998

[6] Wang Mo-ran,"Matlab Application",Electron industry publish,no.2,pp;271,2003

[7] Zhou Bin-kun,"Laser Principle",defence industry publish,no.5,2007

# The bromination Hypocrellin A and the spectra character of the reaction of gelatin with the derivative

**Wei Liu, Jucheng Zhang, Yong Min, Rui Chen, Aiping Fan, Zelin Yang**

[a] Key Laboratory of Natural Pharmaceutical & Chemical Biology of Yunnan Province, Honghe University, Mengzi, Yunnan Province, P. R. China; b school of sicence, Honghe University, Mengzi, Yunnan Province, P. R. China.

Correspondence author Email: hhusci@163.com

**Abstract**. The bromination derivates of hypocrellin A (Br-HA) was prepared in this paper, and the photodynamic activity also was appraised. UV-Vis spectroscopy and Fluorescence spectrophotometer were employed to detect the reaction of Br-HA with gelatin. The result shown the Br-HA can react with gelatin, and the HA-Br also photodegraded the pyridinoline cross-link formation in gelatin.

## 1. Intraduction

Hypocrellin A (HA), one of the nature photosensitizer, was extracted from the fungus of *Hypocrella bambusae*. For several decades, the photochemical, photophysical and photobiological properties were extensively studied. Hypocrellins are well known for its excellent PDT activity and super low dark-toxicity [1-6]. Some water-soluble particles were successfully prepared for drug-delivery purpose [7]. Recently, hypocrellins, as the PDT reagent, were proved to have the curative in some vascular diseases [8-10]. But the water soluble is poor, so the dosage form is the difficulty.

In this paper, we researched the bromination reaction of HA, prepared the derivate (Br-HA), detected the UV spectroscopy of HA and Br-HA. The solution ability of Br-HA product in water and the reaction of Br-HA with gelatin also were investigated by UV spectroscopy and fluorescence spectrometer.

## 2. Materials and methods

### 2.1. The bromination of HA

1.2 ml $Br_2$ was been dissolved in 5.0ml $CHCl_3$ and 20mg HA was dissolved in 20 ml $CHCl_3$ at round flask in room temperature. The $Br_2/CHCl_3$ solution was dropped into HA with stirring. The bromination derivative was purified by thin-layer chromatography after the reaction finished.

### 2.2. The solubility of Br-HA in water

Token 3 mL distilled water in a 1 cm×1 cm quartz cuvette, A concentrated solution of Br-HA in dimethylsulfoxide (DMSO) (58.3 mmol/L) was added to the water by microlite quantity and mixed quickly by shaking in dark.

### 2.3. The dark reaction of Br-HA with gelatin

16.8μL Br-HA solution was injected into 3mL 12.5mg/mL gelatin solution (pH=7.4) and mixed quickly by shaking in dark. The spectrum characters of the reaction system were detected by UV spectroscopy and Fluorescence spectrometer in different time points.

## 3. Results

3.1. The bromination derivative of hypocrellin A

The Br-HA was prepared by bromination, the equation shown in Scheme 1. The UV-Vis spectrum of Br-HA and HA was detected, and shown in figure 1

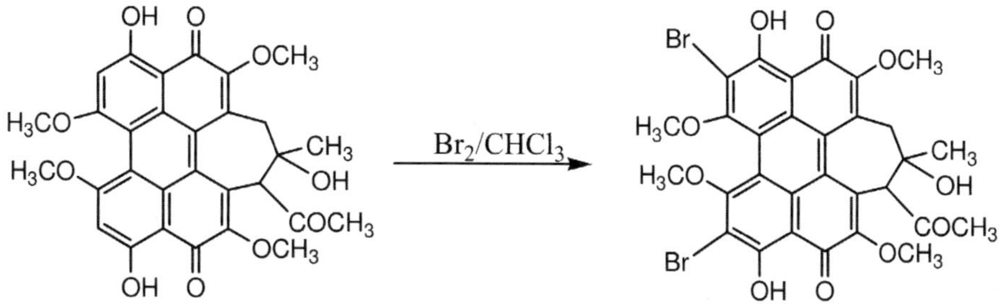


Scheme 1 the bromination reaction of HA

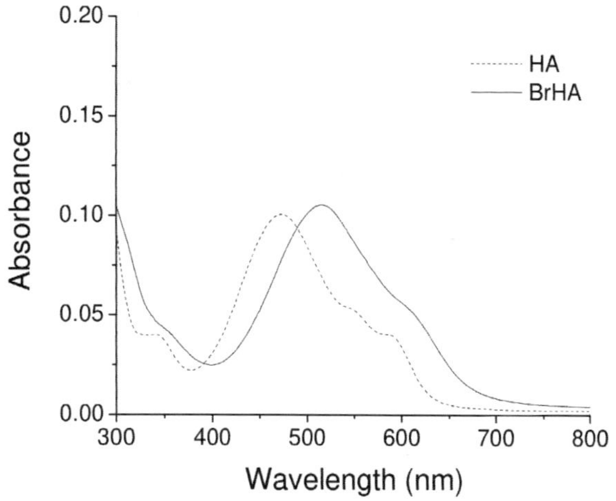


Fig. 1 the UV spectrum of HA and Br-HA

The Br-HA was purified by thin-layer chromatography (TLC), and Br-HA was crystal in chloroform before use. The UV spectrum of Br-HA shown the max absorbance peak of bromine derivative was red shift. This mean the Br-HA can absorb the more length wavelength light. The color of Br-HA is different from the HA, its rose.

3.2. The solubility of Br-HA in water

Hypocrellins is fat-soluble so it ease gather in water solution. The solubility of Br-HA in water was been investigated. The Br-HA solution was injected into water with shaking in dark. The absorption of the aqueous solution was detected, the figure 2 shown the plot of the 476 nm absorption to the concentration.

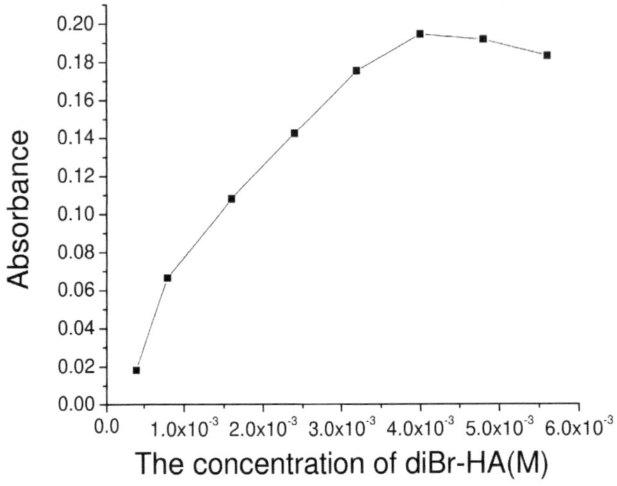


Fig. 2 the solubility curve of Br-HA in water

In figure 2, the Br-HA would gather when the concentration of Br-HA reach $4.0\times10^{-3}$ mol.L$^{-1}$. Therefore, the concentration of Br-HA in aqueous solution should less than $4.0\times10^{-3}$ mol.L$^{-1}$. The volume of Br-HA was 16.8μL in this paper.

3.3. The reaction of Br-HA with gelatin

16.8μL Br-HA solution was injected into 3mL 12.5mg/mL gelatin solution (pH=7.4) and mixed quickly by shaking in dark. The spectrum characters of the reaction system were detected by UV spectroscopy and Fluorescence spectrometer in different time points.

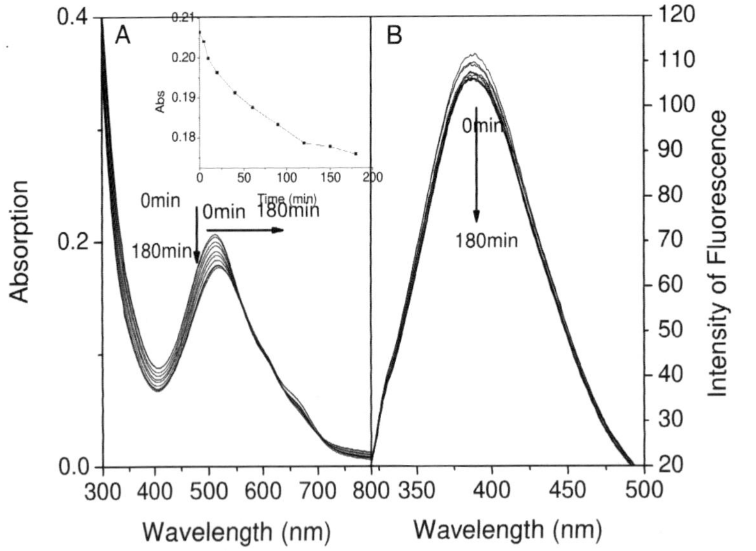


Fig. 3 the spectrum of the reaction of Br-HA with gelatin in water solution. A is the absorbance spectra. B is the fluorescence spectra.

The absorbance spectra shown the absorption of Br-HA was decrease with the reaction process. But the isobestic point not existed in this reaction. The fluorescence spectrum shown the fluorescence 295nm/295nm of gelatin declined, this suggested the fluorescence of gelatin was quenched by Br-HA. The reaction of Br-HA with gelatin wasn't adsorption phenomenon. This also could prove by extraction experiment, which using CHCl3 to extract the Br-HA from the complex, but the Br-HA couldn't be extracted from the system. So we think the Br-HA combined with gelatin, it didn't the adsorption.

## 4. Conclusions

In this paper the bromide derivative of HA was been prepared, and the spectrum character of Br-HA with gelatin was investigated. The UV spectra shown the Br-HA combined with gelatin, but the reaction didn't have the isobestic point. And the UV spectra are different with the spectra of HB with gelatin.

**Acknowledgement**: This research was supported by the Key Discipline Foundation of Yunnan province and the Scientific Research Project of Honghe University (Project XJ1Y0902).

## References

[1] Jiang, L. J. (1990) Chin. Sci. Bull. 21, 1608–1616.
[2] Y.Z. Hu, J.Y. An, L. Qin, L.J. Jiang, J. Photochem.Photobiol., A Chem. 78 (1994) 247–251.
[3] M.J. Fehr, S.L. Carpenter, Y. Wannemuehler, J.W. Petrich, Biochimie 34 (1995) 15845– 15848.
[4] J. Moan, K. Berk, Photochem. Photobiol. 55 (1992) 931–948.
[5] Trojanowska M, LeRoy E C, Krieg B E T. J Mol Med, 1998, 76: 266—274.
[6] Dennis A H, David R E. J Biol Chem, 1996, 271: 26508—26516.
[7] Zhao B Z, Xie J, Zhao J Q. BBA-Gen Subjects, 2004, 1670: 113—120.
[8] Huang N Y, Liu F G, Gu Y. Acta Laser Biol Sin, 2007, 16, 359—363.
[9] Liu F G, Gu Y, Yu C Q, et al. Chin J Laser Med Surg, 2002, 11: 137—140.
[10] Wang K, Gu Y, Liu F G, et al. Chin J Laser Med Surg, 2002, 11: 190—192.

# Fast photoacoustic tomography by use of acoustic lens

**Yadong Wei[1,2], Zhilie Tang[2†], Xian Chen[2], Yongheng He[2], and Haifeng Liu[2]**

1. School of Electronic and Engineering, Dongguan University of technology, 523808, Dongguan, People's Republic of China

2. School of Physics and Telecom Engineering, Laboratory of Photonic Information Technology, South China Normal University, 510006, Guangzhou, People's Republic of China

E-mail: wyd@dgut.edu.cn; tangzhl@scnu.edu.cn

**Abstract**. Photoacoustic tomography has the potential to image the configuration or function of animal or human organs with simultaneous high contrast and high spatial resolution. This article provides an overview of photoacoustic tomography with an acoustic lens. And a new fast photoacoustic tomography technique is introduced. An acoustic lens is able to map the initial photoacoustic pressure distribution in an object in real time without the need of complicated computational reconstruction. Further more, this method produces no artifacts in reconstructed images. Finally, a new fast photoacoustic tomography system based on acoustic is discussed in detail in this paper.

## 1. Introduction

When electromagnetic (EM) energy, such as optical or radio-frequency wave, is absorbed by a medium, its temperature will rise rapidly, and then a Photoacoustic (PA) signal will be generated as a result of thermoelastic expansion. That is called PA effect which was discovered by Bell in 1880.[1] Readers are referred to earlier reviews, books and conference proceedings, and original study papers.[2-6]

Because a PA pressure is directly proportional to the optical absorption in a sample, it is possible to reconstruct the optical absorption distribution in the sample by measuring PA pressures at all directions. Another method to map the optical absorption distribution is using acoustic lens to transform the pressure distribution on the image plane. Although several other techniques such as ultrasound imaging, x-ray photography, computerized tomography (CT), magnetic resonance imaging (MRI), and positron emission tomography (PET), have appeared, they are all limited more or less when applied in practice. For example, ultrasound imaging is limited by the relatively poor echogenicity of blood vessels;[7] x-ray photography, which is based on the mass density of biological tissue, cannot tell whether the fracture is beginning to recover or not;[8] the sensitivity of MRI in many instances is not very impressive, and not all types of breast tissues can be differentiated by MRI, what is more, it is too expensive. The significance of photoacoustic tomography (PAT) imaging is that it overcomes the above problems and yields images of high optical contrast at high ultrasonic

† To whom any correspondence should be addressed.

resolution in relatively large volumes of biological tissues. And with the advantages of high optical contrast and high acoustic penetration depth, PAT imaging is widely used in medical imaging technology.[9-12] A number of PA imaging techniques have been proposed and studied. The biomedical applications of PA imaging are comprehensive, including tomographic imaging of the skin and other superficial organs by laser-induced PA microscopy, breast cancer detection by near-infrared light or radio-frequency–wave induced PA imaging, and small animal imaging by laser induced PA imaging. Because an acoustic lens is able to map the initial PA pressure distribution in an object in real time without the need of complicated computational reconstruction, PAT based acoustic lens had became an common method.

This article provides an overview of the acoustic lens based PA imaging method. And then a new fast PA imaging system based on acoustic lens is introduced. With the new fast PAT imaging system, we can obtain multilayer PA images simultaneously. The experimental results from phantom samples show that the reconstructed PA images are in agreement with original samples. The most importance is that we can really achieve the multilayer tomographic images simultaneously, and improve the PAT imaging system efficiency greatly.

## 2. Photoacoustic imaging techniques

PA imaging is a noninvasive imaging technique for visualizing both structural and functional information of biological tissues. This method has become an active research area in recent years. A number of PA imaging methods have been proposed, such as computed tomography with unfocused transducers,[13-19] depth profiling in layered media,[20,21] scanning tomography with focused ultrasonic transducers,[22-25] optoacoustic imaging using Fabry-Perot etalon[26-29] and with other optical detection methods,[30,31] and imaging with an acoustic lens[32-39]. Each of these methods has its own benefit in some aspects. The acoustic lens based PAT may produce nearly real-time images and eliminate the artifacts associated with the restruction based method.

The acoustic lens based PA imaging was first reported in Xu's paper.[33] The paper put forward the PA imaging based on acoustic lens, and described the sound field distribution by using complex amplitude. And then the theory of PAT imaging is reported in the next paper.[34] The acoustic lens was made of polymethylmethacrylate in the two papers. In 2004, J. J. Niederhauser demonstrated a real-time PA imaging with a 4f acoustic lens system.[35] In the paper, an optical dark-field stereo imaging system using a 30ns flash illumination light was used to capture a snapshot of the pressure-induced refraction index changes in a water container at a predetermined time after the original laser pulse.[36] The result showed that the images from different layers agree well with the original samples. the theoretical results indicates that a 4f acoustic lens is able to image the PA pressure distribution and guarantees axial and lateral unit magnification of image and in 3D PAT imaging, the OTI imaging system may provide the PA images with the same magnification on axial and lateral, and the 3D structure of the object can be imaging really.[37] Zhang reported a two dimensional PA imaging system based on an acoustic lens, and provided a new method to produce a two dimensional (2D) PA image using the peak-hold technology.[38] Wan investigated the theory of PAT imaging using an acoustic lens imaging system and presented the PA Fourier imaging property of an acoustic lens.[39] Wei presented a new high-contrast photoacoustic tomography (PAT) system using a 4f acoustic lens and a 64-element linear transducer array by use of peak-hold technology.[40] Without the integration limit of the boxcar, this system has the potential advantage to form real-time images and acquire them more rapidly. However, they only acquire a peak value one time during once experiment. In order to solve this problem, we developed a fast photoacoustic imaging system based on a high speed data acquisition system. With the new experiment setup, we can get multilayer PAT images simultaneously without the need of the Boxcar and peak-hold module. The experiment results from phantom samples show that the reconstructed results agree well with the original samples.

## 3. Fast photoacoustic tomography imaging

According to the Fourier imaging theory, the PA signals from different object planes in the range of the focal length could precisely image on the same imaging plane at the different time because of the long focal depth of the acoustic lens. So with the time-resolved technique, the PA signals from different object planes can be distinguished. In the previous paper, the proper delay time must be adjusted to catch the peak values of the PA signal.[36-40] The system with BOXCAR should adjust the delay time in BOXCAR, and the system with peak-hold module should adjust the delay time in the module for acquiring the maximum value in the proper time gate. Since the peak-hold module has the potential advantages of forming real-time images and acquiring them more rapidly without any complex algorithms, it only can reconstruct image for one layer at a time. If there were several layers in the sample, corresponding times experiment must be done. This will waste more time and bring error which is caused by scanning.

To simplify the experimental system and improve the imaging speed of multilayer PAT images, we used one high speed data acquisition card (ADC, Model: PCI-4712AS1) to replace the card employed in the previous system (ADC, Model: Advantech PCL-818HG). The new card has a 50Msps 12bit ADC and an independent programmable gain channel composed of high speed precise operational amplifier and precise attenuation filter network are integrated to achieve an voltage signal acquisition range of ±1V or ±10V. Gain errors and zero-bias of each channel can be eliminated independently by data acquisition controller's fine-tuning, which facilitates this system with high measurement accuracy and phase consistency. No manually adjustable component ensures the high reliability and stability of the system. With this high speed data acquisition card, we can catch the complete PA signals for further process. Figure 1 shows the experimental setup of the fast PAT imaging system.

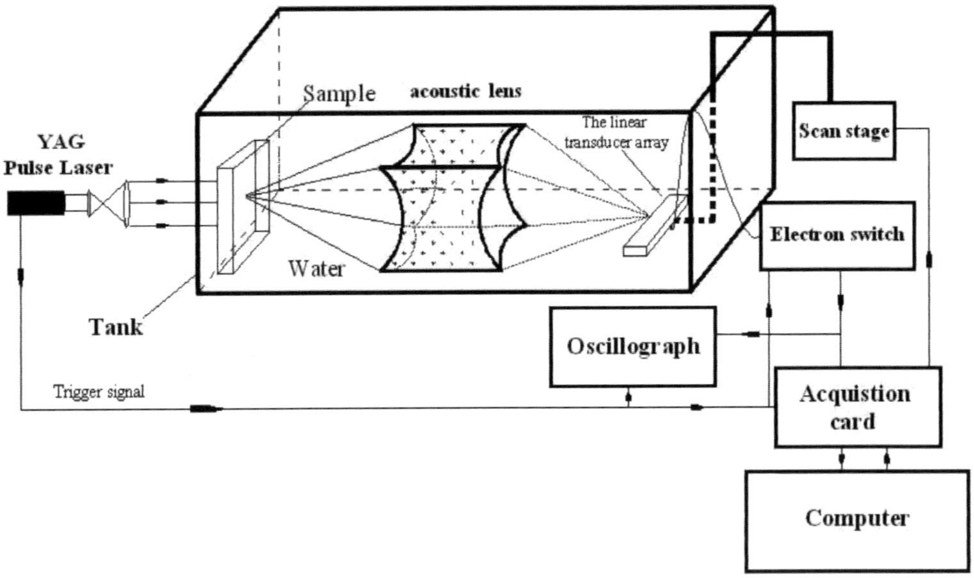


**Figure 1**.Experiment setup of the fast PAT imaging system

In a 60cm×20cm×15cm water tank, the sample, acoustic lens and the linear transducer array were placed in line. The distance between the sample and acoustic lens was about 10cm and the distance between the acoustic lens and the transducer was about 12cm. In order to simulate the scattering surrounding, we poured 250ml milk with a concentration of 10% into the water tank. Owing to the scattering effect of milk, the samples became invisible. The samples were irradiated by a Q-switched ND: YAG laser (model: PRO-230, Spectra Physics, USA) operating at 532nm with a 7 ns pulse width and 10mJ optical pulse. The repetition rate of the laser was 30Hz. A beam expander collimator and spatial filter were used to expand the laser output beam to a diameter in excess of 8 mm. PA signal generated from the samples were focused on the imaging plane by a concave acoustic lens made of polymethylmethacrylate. The aperture of the acoustic lens was 40mm, with about 50mm focal length in water. A multi-element linear array transducer with 1MHz frequency response was specially produced to collect PA signals and change them into the homologous electric signals. The transducer consists of 64 0.22mm × 1.0mm PVDF detectors, and the distance between two neighboring detectors was 1.5mm. In order to change the parallel signals acquired by the transducer into a series, we designed a 64 electronic switch. The output signals from the electronic switch were amplified by a low-noise amplifier with about tenfold gain and then transmitted to the high speed data acquisition card (ADC, Model: PCI-4712AS1) for DAQ. The scan stage was controlled by the same DAQ card. All of the operations were assigned automatically by the LabVIEW program in the computer. The oscillograph was used to scout the PA signals for comparing with the signals which were shown on the computer's monitor.

For testing the fast PAT imaging system, the PA imaging of one single layer was first performed. Figure 2 shows the experiment result. An advantage of the fast PAT imaging system is that it can yield the multilayer images of the sample simultaneously.

a b

**Figure 2**.Sample consisting of two black adhesive tape dots adhered to a piece of polymethylmethacrylate submerged in milk (a) and the reconstructed PA image using the new system (b).

Figure 3 shows the picture of a sample with a three-layer structure: (1) two black adhesive tape dots stuck to the front (shown as figure 5. O1), (2) three black adhesive tape dots stuck to the middle (shown as figure 5. O2) and (3) one black adhesive tape circle stuck to the back (shown as figure 5. O3) of two pieces of polymethylmethacrylate.

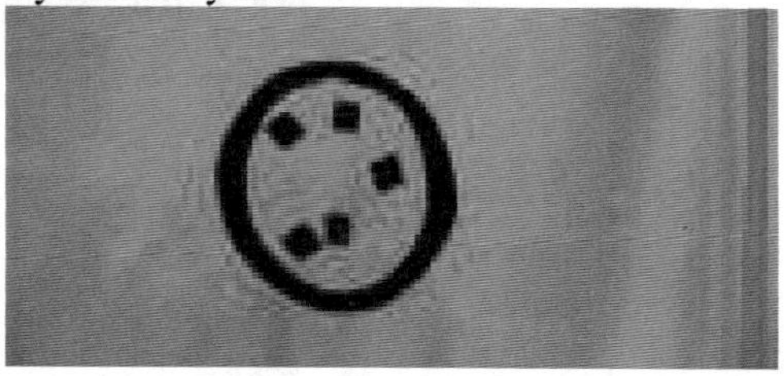

**Figure 3**. The planform photo of the three-layer sample

The distance between the front layer and middle layer is 5mm, and that between the middle layer and back layer is 8mm. The PA signal from the three-layer sample was monitored by an oscillograph (Model: TDS1002). As figure 4 shown, the separate time slot is $2\mu s$ and $3\mu s$ in the PA signal.

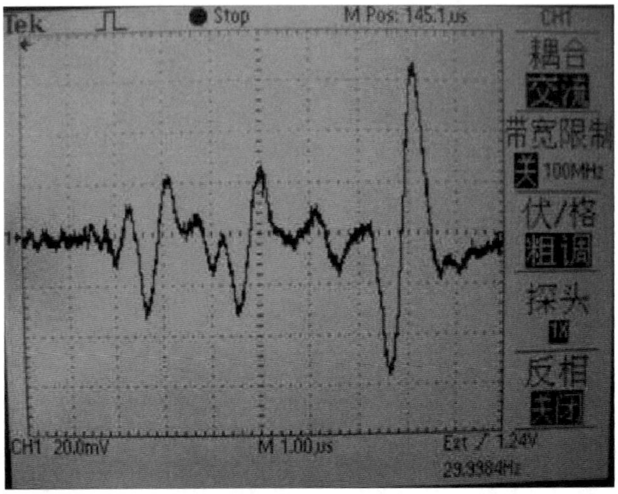


**Figure 4**. The photoacoustic signal of the three-layer sample

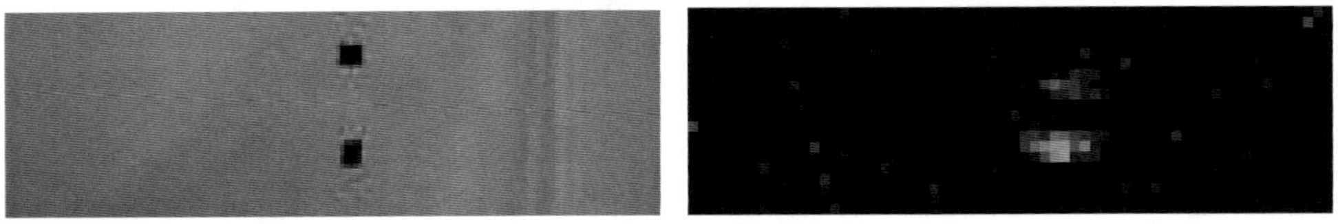

O1 I1

O2 I2

O3 I3

Figure 5. The objects at the different layer O1、O2、O3 and the reconstructed PA images I1、I2、I3

The complete system is controlled with LabVIEW. The sample rate and sample length are 20MHZ and 256, respectively. With the 20M sample rate, the sampling interval is 50ns. The duration of the PA signal is about 1 $\mu s$ . According the Nyquist Sampling Theorem, this data acquisition is valid. With the 256 sample length, the totaled sampling time is 0.05*256=12.8 $\mu s$ . The PA signal of the front layer is at about 2 $\mu s$ , and the PA signal of the back layer is at about 10 $\mu s$ . Thus, all of them are in the sampling range of the system. And the complete data of the PAT signals can be acquired and saved in a computer. If we wanted to get the multilayer tomographic images simultaneously, we need only set the proper rows before the program running, the images of different layers would be reconstructed at the same time. For example, if we wanted to reconstruct the images of the front layer we need only set 40 to 70 rows at the front panel because the PA signal of the front layer was at 2-3.5 $\mu s$ ; if we wanted to reconstruct the images of the middle layer we need only set 84 to 110 rows at the front panel because the PA signal of the front layer was at 4.2-5.5 $\mu s$ ; and so on. Figure 5 shows the pictures of the three-layers of the sample and the corresponding reconstructed results.

The experimental results indicate that the fast PAT imaging system may produce multilayer PAT images simultaneously save them for further process.

## 4. Summary and discussion

The key advantages of the PAT imaging system with an acoustic lens are (1) the initial photoacoustic pressure distribution can be imaged onto an image space in real time without the need of complicated computational reconstruction; (2) No artifacts appear in the reconstructed images. The fast PAT imaging system presented in this paper is simplified and the imaging efficiency of multilayer PAT is much improved by using the high speed data acquisition card. It can acquire the complete PA signals of multi-layers quickly and save them in a computer. The images of multilayer can be reconstructed simultaneously. Further study must put the emphases on the two dimension PA imaging at the same time to reveal the advantage of the PAT imaging system with an acoustic lens.

## 5. Acknowledgement

This work is supported by the National 863 Program Project of China (Grant No. 2006AA02Z4B4), National Natural Science Foundation of China (Grant No.60877068), National Natural Science Foundation of China (Grant No.60377009), and the Natural Science Foundation of Guangdong Province, China (Grant No.05005926).

## References

[1] Bell A.G., On the production and reproduction of sound by light. *Am.J.Sci.*,1880,20, 305-324.

[2] Patel C. K. N. and Tam A. C., Pulsed optoacoustic spectroscopy of condensed matter, *Rev. Mod. Phys.* 1981, 53, 517-550.

[3] Tam A. C., Application of photoacoustic sensing techniques, *Rev. Mod. Phys.* 1986, 58, 381-431.

[4] M. H. Xu and L.H. Wang, Photoacoustic imaging in biomedicine, *Rev. Sci. Instrum.* 2006, 77, 041101

[5] Principles and Perspectives of Photothermal and Photoacoustic Phenomena, edited by A. Mandelis (Elsevier Science, New York, 1992)

[6] V. E. Gusev and A. A. Karabutov, Laser Optoacoustics (AIP, New York, 1993)

[7] Cox B. A., Kelly K. M., Ko P., Hertzog L. and Stain S. C., Ultrasound characteristics of breast carcinoma., *Am. Surg.* 1998,64934–8

[8] Esenaliev R. O., Tittel F. K., Thomsen S. L., Fornage B., Stelling C., Karabutov A. A. and Oraevsky A. A., Laser optoacoustic imaging for breast cancer diagnostics: limit of detection and comparison with x-ray andultrasound imaging.,1997, Proc. *SPIE297971–82*

[9] Hoelen C. G. A., de Mul F. F. M., Pongers R. and Dekker A., Three-dimensional photoacoustic imaging of blood vessels intissue., *Opt. Lett.* 1998,23648–50

[10] Hao F. Zhang, Konstantin Maslov, Meng-Lin Li, George Stoica, Lihong V. Wang,In vivo volumetric imaging of subcutaneous microvasculature by photoacoustic microscopy, *Opt. Express*, 2006, 14, 9317.

[11] K. H. Song, G. Stoica, and L. V. Wang, In vivo threedimensional photoacoustic tomography of a whole mouse head, *Opt. Lett.* 2006, 31, 2453-2455

[12] X. D. Wang, X. Y. Xie, G. N. Ku, and L. H. V. Wang, Noninvasive imaging of hemoglobin concentration and oxygenation in the rat brain using high-resolution photoacoustic tomography, *J. Biomed. Opt.* 2006, 11, 024015.

[13] Kruger R, Liu P, Fang R, Photoacoustic ultrasound (PAUS) reconstruction tomography, *Med. Phys.*, 1995, 22(10), 1605-1609.

[14] Hoelen C.G.A., de Mul F.F.M., Pongers R., and dekker A., Three-dimensional photoacoustic imaging of blood vessels in tissue, *Opt. Lett.* 1998, 23, 648-650.

[15] Köstli K. P., Frenz M., Weber H. P., et al., Optoacoustic tomography: time-gated measurement of pressure distributions and image reconstruction, *Appl. Opt.*, 2001, 40(22), 3800-3809.

[16] Xu M. H., Wang L. H., Time-Domain Reconstruction for Thermoacoustic Tomography in a Spherical Geometry, *IEEE Trans. Med. Imaging,* 2002, 21,814-822.

[17] Minghua Xu and Lihong V. Wang, Universal back-projection algorithm for photoacoustic computed tomography, *Phys. Rev. E.*, 2005, 71, 016706.

[18] D. W.Yang, D Xing., H. M.Gu, Y. Tan, and L. M. Zeng, Fast multielement phase-controlled photoacoustic imaging based on limited-field-filtered back-projection algorithm, *Appl. Phys. Lett.* 2005, 87, 194101.

[19] D. W. Yang, D. Xing, Y. Tan, H. Gu, and S. H. Yang, Integrative prototype B-scan photoacoustic tomography system based on a novel hybridized scanning head, [J] *Appl. Phys. Lett.* 2006, 88, 174101.

[20] A. A. Oraevsky, S. L Jacques, and F. K. Tittel, Measurement of tissue optical properties by time-resolved detection of laser-induced transient stress, *Appl. Opt.* 1997, 36, 402.

[21] Köstli K.P., Frenz M., Weber H.P., et.al, optoacoustic infrared spectroscopy of soft tissue, *J. Appl. Phys.* 2000, 88, 1632-1637.

[22] Kruger R. A., Liu P. Photoacoustic ultrasound: pulse production and detection in 0.5% liposyn. *Med. Phys.* 1994; 21:1179-1184

[23] G.Ku and L. H. Wang, Scanning thermoacoustic tomography in biological tissue, *Med. Phys.* 2000, 27(5), 1195–1202.

[24] G. Ku and L.H. Wang, Scanning microwave-induced thermoacoustic tomography: Signal, resolution, and contrast, *Med. Phys.* 2001, 28, 4.

[25] A. A. Oraevsky and A. A. Karabutov, in Biomedical Photonics Handbook, edited by T. Vo-Dinh (CRC, Boca Raton, FL, 2003), Chap. 34.

[26] P. C. Beard, and T. N. Mills. Extrinsic optical-fiber ultrasound sensor using a thin polymer film as a low-finesse Fabry-Perot interferometer, *Appl. Opt.* 1996, 35, 663-675

[27] P. C. Beard, and T. N. Mills. 2D line scan photoacoustic imaging of absorbers in a scattering tissue phantom, *Porc. SPIE, 4256,* 2001,34-42

[28] Shai Ashkenazi, Yang Hou, Takashi Buma, and Matthew O'Donnell, Optoacoustic imaging using thin polymer étalon, *Appl. Phys. Lett.* 2005, 86, 134102.

[29] P. C. Beard, 2D ultrasound receive array using an angle-tuned Fabry Perot polymer film sensor for transducer field characterisation and transmission ultrasound imaging, *IEEE Transactions on Ultrasonics, Ferroelectrics and Frequency control*, 2005, 52(6), 1002-1012.

[30] Paltauf G, Schmidt-Kloiber H, Guss H. Light distribution measurements in absorbing materials by optical detection of laser-induced stress waves, *Appl. Phys. Lett.*, 1996, 69: 1526

[31] G. Paltauf, H. Schmidt-Kloiber, KP Köstli, and M. Frenz, Optical method for two-dimensional ultrasonic detection, *Appl. Phys. Lett.* 1999, 75, 1048–1050

[32] X. F. Xu, Z. L. Tang, G. S. Chen, H. J. Pei, and X. G. Yu, Studies on acoustic lens imaging of photoacoustic signal, *J. of South China Normal University (Natural Science Edition).* 2003, 1, 62

[33] X. F. Xu, Z. L. Tang, J. Wang, and G. S. Chen, Studies on acoustic lens imaging of photoacoustic signal, *Acta Opt . Sin.* 2003, 23, 1103

[34] J. J. Niederhauser, M. Jaeger, and M. Frenz, Real-time three-dimensional optoacoustic imaging using an acoustic lens system, *Appl. Phys. Lett.* 2004, 85, 846-848.

[35] Y. H. He, Z. L. Tang, Z. X. Chen, W. Wan and J.H. Li, A novel photoacoustic tomography based on a time-resolved technique and an acoustic lens imaging system, *Phys. Med. Bio.* 2006, 51, 2671-2680.

[36] Z. X. Chen, Z. L. Tang and W. Wan, Photoacoustic tomography imaging based on a 4f acoustic lens imaging system, *Opt. Express*, 2007, 15(8), 4966-4976

[37] H. C. Zhang, Z. L. Tang, Y. H. He, and L. N. Guo, Fast Photoacoustic System Based on an Acoustic Lens and the Peak Detection Technology, *J. Eng. Mater.,* 2007, 364-366, 1157-

1161.
[38] W. Wan, R. S. Liang, Z. L. Tang, Z. X. Chen, H. C. Zhang and Y. H. He, The Imaging Property of Photoacoustic Fourier Imaging and Tomography Using an Acoustic Lens Imaging System, *J. Appl. Phys.* 2007, 101, 063103.
[39] Y. D. Wei, Z. L. Tang, H. C. Zhang, Y. H. He and H. F. Liu, Photoacoustic tomography imaging using a 4f acoustic lens and peak-hold technology, *Opt. Express*, 2008, 16(8), 5314-5319.

# Simulation on Temperature Field of Radiofrequency Lesions System Based on Finite Element Method

**D Xiao[1], L Qian[2], Z Qian[1*] and W Li[1]**
[1] Department of Biomedical Engineering, Nanjing University of Aeronautics and Astronautics, 29 Yudao Street, Nanjing 210016, China
[2] College of Telecommunications and Information Engineering, Nanjing University of Posts and Telecommunications, Nanjing, 210046
[*]Corresponding author: Zhiyu Qian Email: zhiyu@nuaa.edu.cn Tel: 086-025-84891938-602

**Abstract.** This paper mainly describes the way to get the volume model of damaged region according to the simulation on temperature field of radiofrequency ablation lesion system in curing Parkinson's disease based on finite element method. This volume model reflects, to some degree, the shape and size of the damaged tissue during the treatment with all tendencies in different time or core temperature. By using Pennes equation as heat conduction equation of radiofrequency ablation of biological tissue, the author obtains the temperature distribution field of biological tissue in the method of finite element for solving equations. In order to establish damage models at temperature points of 60℃, 65℃, 70℃, 75℃, 80℃, 85℃ and 90 ℃ while the time points are 30s, 60s, 90s and 120s, Parkinson's disease model of nuclei is reduced to uniform, infinite model with RF pin at the origin. Theoretical simulations of these models are displayed, focusing on a variety of conditions about the effective lesion size on horizontal and vertical. The results show the binary complete quadratic non-linear joint temperature–time models of the maximum damage diameter and maximum height. The models can comprehensively reflect the degeneration of target tissue caused by radio frequency temperature and duration. This lay the foundation for accurately monitor of clinical RF treatment of Parkinson's disease in the future.

## 1. Introduction

Parkinson's disease (PD) [1] is one of the most common degenerative diseases in the general population, especially in the old people. This disease affects the patient's motor function and life skills seriously in China at present as the result of an increasing trend of incidence rate year by year. According to the literatures that published during the last two years, drug control method and neuronal ablation [2, 3] method are widely used for the treatment of Parkinson's disease at present. The method, which is widely used for neuronal ablation today, of removing the ablated brain section [4] from living body in order to measure the size of damaged region, is insufficiently studied on the information of damaged region and affection of surrounded brain tissue.

This paper mainly describes the way to get the volume model of damaged region according to the simulation on temperature field of radiofrequency ablation lesion system in curing Parkinson's disease based on finite element method. This volume model reflects, to some degree, the shape and size of the damaged tissue during the treatment with all tendencies in different time or core temperature.

## 2. RF ablation Technology

Radiofrequency (RF) [5] with highly concentrated current is a kind of electromagnetic wave at the frequency between 50 kHz and 2000 kHz. The friction caused by the high frequency among the ions heats up the tissue in organisms to make it dissolved or gasified. RF ablation means ablating the target tissue with RF instruments. The RF generator generates high frequency current which arrives at the damage probe to form a current loop with another probe parked on the surface of skin. Powerfully and highly thermal, current from the sharp point of probe kills nerve cells and fibers by heating the tissue while the tissue gives a feedback thermal to the probe. We get the temperature of sharp point of probe with RF generator in order to deduce the temperature distribution of damaged area. Probe diameter, length, tissue parameters and other circumstance affect the range of damaged area, especially the duration and temperature of operation.

RF ablation, which has been broadly applied in clinical medicine as a result of rapid development of electronics, is a rising technology during the last ten years. High frequency bistouries, RF ablation instruments, heat treatment instruments for prostate and so on are advanced medical installations in recent years. As is repeatable, lightly hurt, easily controlled, short time treatment, little side-effect, little syndrome, RF ablation is more and more widely used in clinical medicine, such as treatment of cardiac arrhythmia, pain, prostatoplasia, tumor of brain, lung, liver, kidney, pancreas, galactophore and matrix[6, 7].

RF ablation heat treatment instruments used domestic mainly contains Leksell of temperature-controlled, Radionics RFG-3C, ASA-601[8] produced in Anke limited company of high technique in Shenzhen.

## 3. Finite element method

The first step of Finite Element Method (FEM) [9] is to divide the whole solve field into limited small conjoint fields that are defined as finite elements. We get an imprecisely solution by respectively getting the solution of each finite element. The solution is imprecisely because of the predigestion of the question. FEM become an effective method to analytics as a result of highly precision, adaptability to different shapes.

Radically, FEM distinguishes from other methods of solving boundary problems by confining the similarity in much smaller fields. Dr. Clough, who firstly gave a definition to FEM early in 1960s, vividly described that "FEM = Rayleigh Ritz + partial function", that is FEM is localized by Rayleigh Ritz [10]. Unlike Rayleigh Ritz, which solves the question in whole field, FEM defines the function in ordinary shapes which is much more advanced than other methods.

## 4. Heat transfer model in biological tissue

Pennes equation [11] is the most widely used thermal conductive model in biomedical circle. This model simplifies the process of thermal conduction in organism tissues with the formulation following:

$$\rho c\frac{\partial T}{\partial t}=\nabla\cdot\left(k\nabla T\right)+W_bC_b\left(T_a-T\right)+q_m+q_r \tag{1}$$

Table 1 listed the parameters and their physical significance. In this simulation, $q_m$ and $q_r$ are ignored while normal temperature in vivo tissue is selected.

**Table 1.** Parameters and their physical significance

| Parameter | Physical significance |
|---|---|
| $T$ | normal temperature of tissue |
| $\rho$ | density of tissue |
| $c$ | specific heat of tissue |
| $k$ | heat exchange rate of tissue |
| $W_b$ | filling rate of blood |
| $C_b$ | specific heat of blood |
| $T_a$ | normal temperature of arterial blood |

| | |
|---|---|
| $q_m$ | febriferous rata of metabolize |
| $q_r$ | external heat energy |

The assumptions of Pennes are listed as follows:

(1) Tissue where capillary vessels (the diameter is between 0.005 mm and 0.015 mm) are existed is the major situation that exchange between vessels and tissue occurs;

(2)Blood at the temperature of $T_a$ will achieve a balance to tissue around vessel immediately while feeding in the capillary vessel bed and then feeds back to the vein at the temperature of $T$;

(3)The heat exchange between blood and tissue can be simulated by a scalar heat quantity, of which the size has direct ratio with filling rate of blood and temperature margin of arterial blood and tissue.

## 5. Simulation of temperature distribution in RF Ablation

### *5.1 Tissue model*

The RF system selected to damage tissue is Leksell RF instrument produced in Sweden, which is an intact system specially used in nerve exciting, solid beamed damage, ache treatment and bipolar electro coagulating. The frequency of Leksell, which is higher than any other congeners, insures that the increasing speed of temperature can be 5℃/s. This could cut down the pain and operation time.

As is shown is figure 1, simulation of temperature distribution [12] in RF ablation setting out in this paper supposes the PD′s model to be an even, infinite sphere with the probe tip (0.5mm, -1mm) around the centre (0, 0). The diameter of the sphere is 20mm while the probe is 1mm with 2mm exposed.

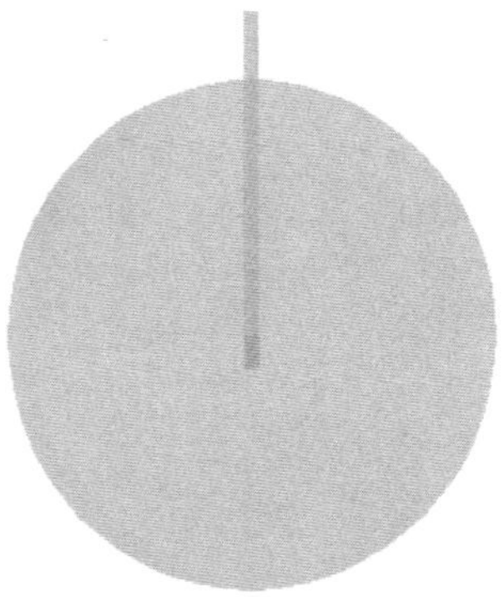

**Figure 1.** Model of idealized tissue

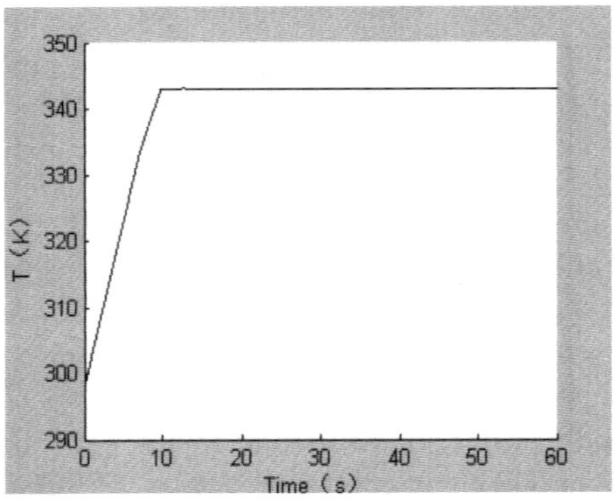


**Figure 2.** Temperature curve of the stylus

The fuscous part in the centre in figure 1 is the exposed part of the probe, of which the temperature curve is showed in figure 2. The temperature of the probe increase at the speed of 5℃/s from normal temperature(25℃) to damage temperature(70℃) linearly and then keep constant at the latter one.

Equation (1) is chosen to be the simulate formula while mice tissue is chosen to be the organism. Thermal parameters [13, 14] are listed in table 2.

**Table 2.** Thermal parameters of mice tissue

| | $\rho(kg/m^3)$ | c(J/kg · ℃) | k(w/m · ℃) | T(℃) | h(w/m2 · ℃) |
|---|---|---|---|---|---|
| tissue | 1090 | 3400 | 0.54 | 25 | 4.33 |

### *5.2 Simulation results*

Figure 3 shows the result of finite element simulation.

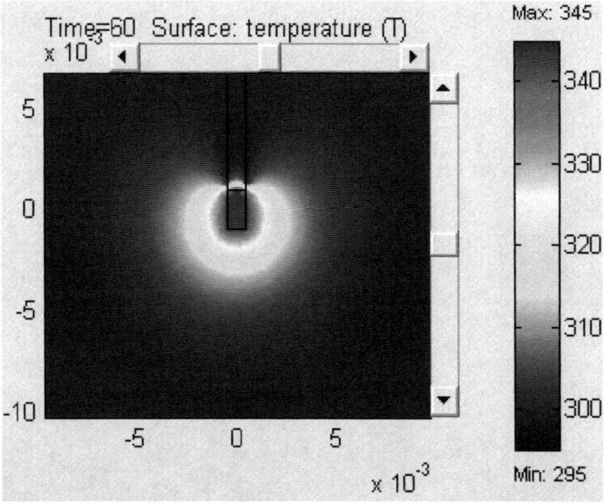


**Figure 3.** Result of finite element simulation

Dark rectangular portion in figure 3 is the exposed part of the probe. Probe temperature rises as figure 2 describes. A tissue temperature distribution is formed in consequence of the rise in temperature. Temperature distribution of RF ablation presented to half of the elliptical according to the effective picture. The unit of colour bar is Kelvin.

*5.2.1 Temperature distribution at constant temperature*

The simulation establishes damage models at temperature points of 60℃, 65℃, 70℃, 75℃, 80℃, 85 ℃ and 90℃ while the time points are 30s, 60s, 90s and 120s. Take the case of the simulation results at four different time points while the temperature is constant, for example, 70℃.

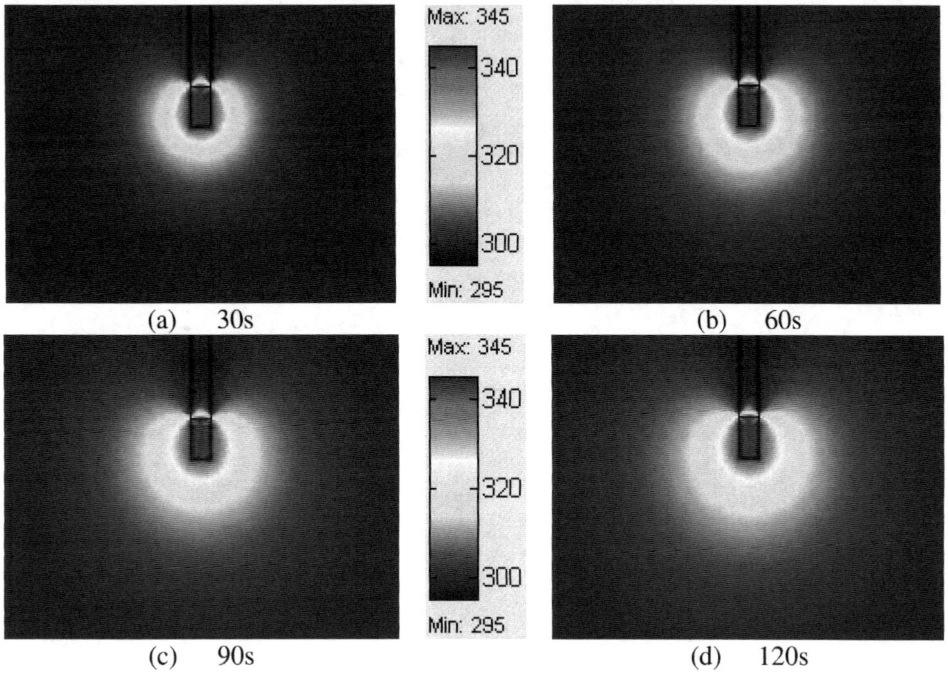

**Figure 4.** Results of temperature distribution at 70 degree centigrade

While heating the probe to 70℃ for 20 seconds, a small oval is formed surrounding as the temperature distribution. This oval becomes larger while the damage time increases at constant temperature. We can deduce that the volume of damage area is increased with damage time while the temperature is constant.

*5.2.2 Temperature distribution at constant time*

Figure 5 shows different distributions at different temperatures(30℃, 60℃, 90℃, 120℃) while time is constant(60s).

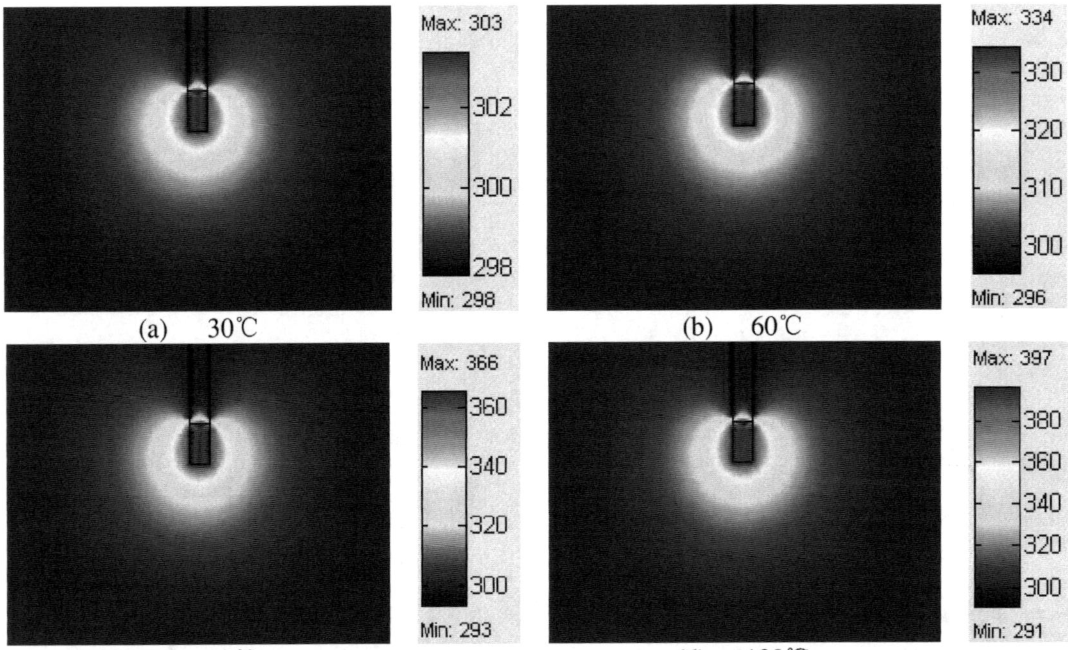


(a) 30℃ (b) 60℃

(c) 90℃ (d) 120℃

**Figure 5.** Results of temperature distribution at 60 degree centigrade

While heating the probe to 60℃ for 60 seconds, a small oval is appeared as the temperature distribution. This oval becomes larger while the damage temperature increases in constant time. We can draw a conclusion that the volume of damage area is increased with damage temperature while the time is constant.

*5.3 Analysis*

Through data analysis of maximum radial direction length and maximum longitudinal length, we could draw the temperature curves in figure 6:

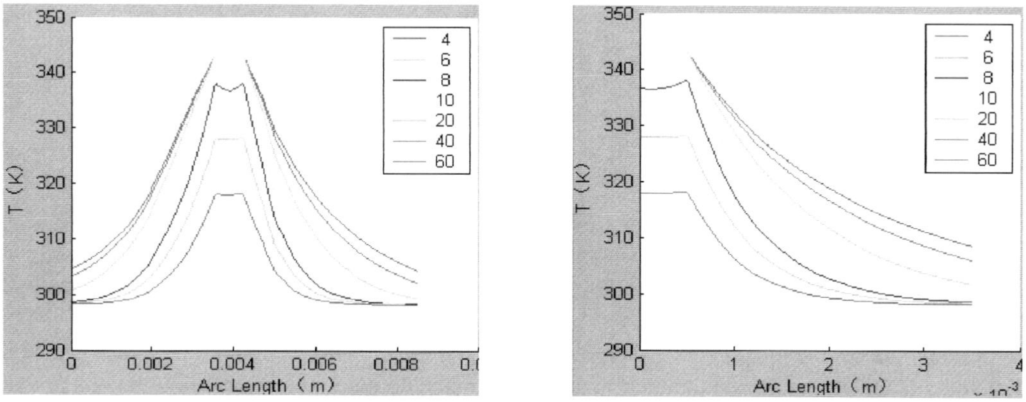


(a) Temperature curves of maximum diameter (b) Temperature curves of maximum height

**Figure 6.** Temperature curves

The maximum radial direction temperature holding time of curves shown in figure 6(a) are respectively 4 seconds, 6 seconds, 8 seconds, 10 seconds, 20 seconds, 40 seconds, 60 seconds while the maximum height of curves is shown in figure 6(b) with the same temperature holding time. According to the irreversible change of clinical RF ablation while temperature rises up to 45℃ [15], we are mainly concerned about the irreversible data. Through analyzing the simulated data, we could get the volume of damaged area. Figure 7 shows the volume data comparison of simulation and mice radio frequency experiment. Chart (a) is the trend comparison when the time maintains at 30 seconds and the temperature changes from 60 centigrade to 90 centigrade. Similarly, chart (b) is the same comparison when the time maintains at 60 seconds with the same range of temperature while the maintained time of chart (c), (d) is 90 second and 120 second.

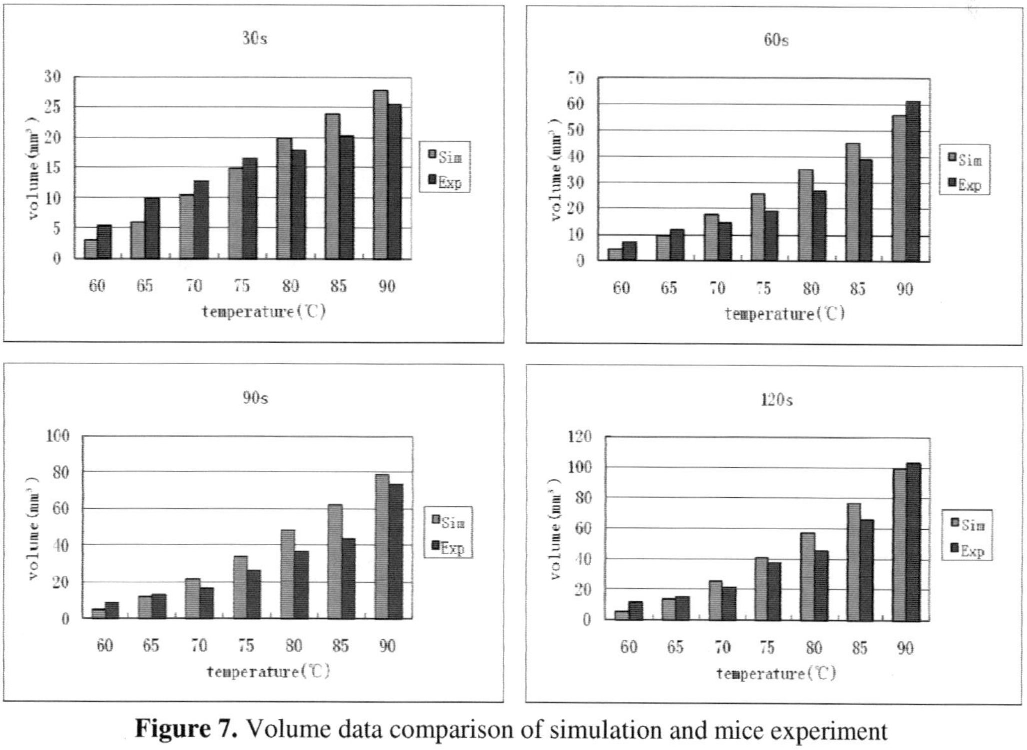


**Figure 7.** Volume data comparison of simulation and mice experiment

Figure 8(a) shows the three dimensional plot of the simulated volume result at different time and different temperature while figure 8 (b) shows the finally model according to fitting algorithm. Figure 8(c) shows the three dimensional plot of the volume result in mice experiment at different time and different temperature while figure 8 (d) shows the finally model according to fitting algorithm.

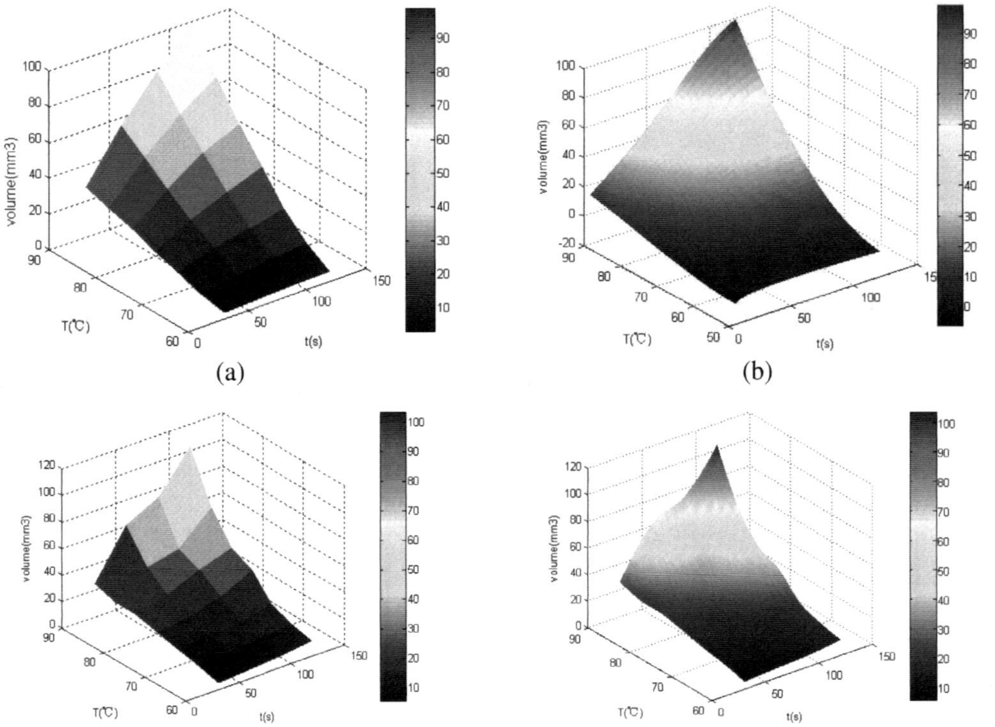


(a) (b)

(c) (d)

**Figure 8.** Time-temperature distribution of effective damage volume

The top value of color bar means the maximum value of volume in the simulation while the bottom means the minimum.

We can draw a conclusion that the simulation on temperature field of RF lesions system based on FEM has the same effect with mice experiment and could comprehensively reflect the denaturation of damaged tissue.

## 6. Conclusions

As an easy, fast and effective method, FEM referenced in this paper is widely used to simulate the temperature distribution in biological tissue optics domain. During the research, the author draws several conclusions as followings:

- The damage area presented to be a smaller oval
- While the centre temperature is constant, the volume of damaged area is increased with holding time.
- While the holding time is constant, the volume of damaged area is increased with centre temperature.
- Simulation on temperature field of RF lesions system based on FEM has the same effect with mice experiment and could comprehensively reflect the denaturation of damaged tissue.

The nonlinear model, which can synthetically reflect denaturation of target brain tissue, is developed to describe foregoing conclusions that can make sense to the dosage theory.

**Acknowledgments**
This work was supported by Bio-photonics Laboratory of Nanjing University of Aeronautics and Astronautics and the National High Technology Research and Development Program ("863" Program) in China (2008AA02Z438).

**References**

[1] Arash S, Heike R, François J. G. V, Catherine D, Yves B, Pierre R. B and Kamiar A 2004 *IEEE J. Biomed. Engineering.* **51** 1434
[2] Hu X, Zhou X 2002 *Chin. J. Min. Inv. Surg.* **2** 266
[3] Gielen, F.L.H. 2003 *Proc. IEEE Int. Conf. on Neural Engineering* (Netherlands: Maastricht) pp 489-491
[4] Chen L, Liu J, Yang D, Lin X and Gao L 2003 *Chin. J. Exp. Surg.* **20** 807
[5] Yang H, Wang D and Ma Q 2001 *J. Med. Information.* **14** 364
[6] Zhang H, Filip B, Stella M, Enrique C N, Hernan A and Kevin C 2007 J. *Proc. of SPIE.* **6509** 650922-1
[7] Jiang D 1994 *Research on Practical Neurosurgery Department Surgery* (China: Shanghai)
[8] Niu C, Wang Y, Ling Z, Jiang X, Ling S, Ding M and Fu X 2003 *Chin J. Stereotact Funct Neurosurg* **16** 11
[9] Clough RW 1990 *J. Finite Elem. Anal. Desig.* **7** 89
[10] MacDonald J.K.L 1993 *J. Phys. Rev.* **43** 830
[11] Pennes H. H. 1948 *J. Appl. Physiol.* **1** 93
[12] Dai L, Wang H and Qian Z 2007 *J. Nanjing Univ. Aeron. Astron.* **39** 176
[13] Surya C. G, Chen Y, Wu F, Kenneth E. B, Jerry W. R, James P.W, Liu H and Wei R. Chen 2007 *J. Proc. of SPIE* **6438** 64380D-1
[14] Pennes H. H., 1948 *J. Applied Phys.* **1** 93
[15] Mirza AN, Fornage B and Sneige N 2001 *Cancer J.* **7** 95
[16] Athi N. N, Urmi D and Victor M 2007 *J. Am. Chem. Soc.* **129** 5673

# Near-infrared spectroscopy technique to evaluate the effects of drugs in treating traumatic brain edema

**J Xie[1], Z Qian[1*], T Yang[2], W Li[1] and G Hu[1]**
[1] Department of Biomedical Engineering, Nanjing University of Aeronautics and Astronautics, 29 Yudao Street, Nanjing 210016, China
[2] School of Clinical medicine, Southeast University, 87 Dingjiaqiao Road, Nanjing 210009, China

[*]Corresponding author: Zhiyu Qian Email: zhiyu@nuaa.edu.cn Tel: 086-025-84891938-602

**Abstract.** The aim of this study was to evaluate the effects of several drugs in treating traumatic brain edema (TBE) following traumatic brain injury (TBI) using near-infrared spectroscopy (NIRs) technology. Rats with TBE models were given hypertonic saline (HS), mannitol and mannitol+HS respectively for different groups. Light scattering properties of rat's local cortex was measured by NIRs within the wavelength range from 700 to 850 nm. TBE models were built in rats' left brains. The scattering properties of the right and left target corresponding to the position of normal and TBE tissue were measured and recorded *in vivo* and real-time by a bifurcated needle probe. The brain water contents (BWC) were measured by the wet and dry weight method after injury and treatment hours 1, 6, 24, 72 and 120. A marked linear relationship was observed between reduced scattering coefficient ($\mu_s'$) and BWC. By recording $\mu_s'$ of rats' brains, the entire progressions of effects of several drugs were observed. The result may suggest that the NIRs techniques have a potential for assessing effects *in vivo* and real-time on treatment of the brain injury.

## 1. Introduction

Trauma is the commonest cause of death in young people (aged <45 years) worldwide, and up to half the deaths are caused by traumatic brain injury (TBI) in the United States [1]. Traumatic brain edema (TBE) following TBI is defined as increased liquid content in brain substance. TBE leading to an expansion of brain volume has a crucial impact on morbidity and mortality as it increases intracranial pressure (ICP), impairs cerebral perfusion and oxygenation, and contributes to additional ischemic injuries. Dehydration therapy is the major way to the TBE treatment. Mannitol, a preferred dehydration agent, has been extensively used in clinical therapy. However, numerous studies on the mannitol agent have recently revealed some adverse effects after repeating doses, including renal failure [2-8] and edema rebound phenomena[9-12]. Instead of mannitol, hypertonic saline (HS) solutions have been increasingly used as an attractive alternative in basic science and clinical studies [13-21]. Different administration routes, concentrations (2% to 23.4% NaCl) and doses were investigated in these studies. Therefore, it is necessary to establish an *in vivo* monitoring method to determine the therapeutic efficacy, and thus, to guide the treatment protocols in clinic.

Near infrared technique features some advantages in that it provides a relatively high temporal resolution, it is cheap to build, and is easily maintained. It has been used widely in the differentiation tissue types in recent years [22, 23]. It is known that near-infrared (NIR) light in the 700-850 nm wavelength range is special to tissues as light scattering is more prominent than light absorption. There is a significant difference in light-scattering properties between normal tissues and injured tissues due to their differences in anatomical substructure [24, 25]. In our previous work, a linear correlation between $\mu_s'$ and water content in rat brain tissues was found [26], which implied that $\mu_s'$ may serve as an indicator for the brain water content. In this study, an effort was made to develop NIR technology as an alternate method for monitor of brain trauma.

The aim of this study is to establish an innovative method to real time evaluate the therapeutic efficacy of dehydration agents treatments. The functional near infrared spectroscopy (fNIRs) was used to real time monitor the changes of reduced scattering coefficient $\mu_s'$ in rat brain tissues. Acute regional brain trauma was induced in Sprague-Dawley rats (SD) by Feeney's apparatus. The effect of mannitol+ hypertonic saline (HS) treatment monitored by fNIRs was compared to that of commonly used mannitol agent and HS agent.

## 2. Materials and methods

### *2.1. Instruments and Methods*

The fNIRs experimental system[26, 27] for optical parameter measurement consisted of a halogen light source（HL-2000,Ocean Optics Inc. Dunedin, FL）, a bifurcated needle probe(600 $\mu$m o.d. with two single mode fibers of 100 $\mu$m), a USB optical spectrometer(USB2000,Ocean Optics Inc.) with wavelength grating from 340-1020nm, a rat brain stereotactic instrument (Jiangwan type1), a stepper power driver (MID-7604 National Instruments) , a step motor and controller (PCI-7344) and a computer. Absorption coefficient ($\mu_a$), reducing scattering coefficient ($\mu_s'$), oxygen saturation($SO_2$）, oxygenated hemoglobin ($HbO_2$), deoxygenated hemoglobin (Hb), oxygenated hemoglobin total (HbTotal), cerebral blood flow (CBF) and cerebral blood volume (CBV) of tissue are continuously recorded with this fNIRs equipment [28, 29]. In brief, the wide band light was delivered by the deliver fiber inside the probe to the surface of the tissue, and the reflected spectrum was collected by another fiber inside the probe. The slope of the spectrum within 700~850nm was obtained by a fitting method. The reduced scattering coefficient ($\mu_s'$) was calculated based on the empirical equations, which were deduced by extensive simulation experiment of the NIRs system [29]. Comparing the $\mu_s'$ measurement of NIRs to that of Oximeter, the $\mu_s'$ on 690nm wavelength by NIRs for the specific probe (0.2mm diameter) was proven to be more accurate. The empirical equation of $\mu_s'$ on 690nm is displayed as the following, which was used in this study.

$$\mu'_{s690} = 4.5203\exp(-0.3241 * \text{slope}) \qquad (1)$$

*2.2. Experimental animals*

One hundred Sprague-Dawley rats (half male and half female, weight about 230-280g, the Medical Animal Center of Southeast University, Nanjing, P.R.China) were randomly divided into 5 groups (*n*=20 for each group). Control group was normal rats for control experiment. Traumatic group was traumatic model without treatment. Treatment group (TGroup) 1, 2 and 3 were traumatic models with different treatment of 2g/kg mannitol for TGroup 1, 7.5% HS for TGroup 2 and 1g/kg mannitol+7.5% HS for Tgroup 3 after trauma, respectively, as listed in Table 1. Each group was divided into five subgroups (n=4) for brain water contents measurement at different time points of 1 hour (h), 6h, 24h, 72h and 120h, as shown in Table 1. The traumatic model rats were achieved by hitting the left epidural in a preset windows by Feeney's free-falling body against devices [30] (30g heavy falling from the height of 20cm above the rat brain to produce an impact of 600g • cm), causing a focal cerebral contusion in the left parietal lobe. Apnea for a few seconds indicated the success of inducing a brain injury model. On right sides, a burr hole were performed with 1 mm dental drill, but were not traumatic. Rats in the control group only had a bone window drilled at the same position (left and right sides), but were not injured. All rats were kept in their habitual environment until the day of the experiment. All experiments were carried out in compliance with the Animal Management Rules of the Ministry of Health of the People's Republic of China (document no. 55, 2001), and the guidelines for the Care and Use of Laboratory Animals of Southeast University Medical Center.

**Table 1.** Protocol of animal model (n =100).

| Groups | Subgroups | | | | |
|---|---|---|---|---|---|
| Time | 1h | 6h | 24h | 72h | 120h |
| Control group | Control, not injured | | | | |
| Traumatic group | Traumatic model without treatment | | | | |
| Treatment group 1 (TGroup 1) | Traumatic model with 2g/kg mannitol treatment | | | | |
| Treatment group 2 (TGroup2) | Traumatic model with 7.5% HS treatment | | | | |
| Treatment group 3 (TGroup 3) | Traumatic model with 1g/kg mannitol +7.5% HS treatment | | | | |

*2.3. Near-infrared Spectroscopy (NIRS) monitoring in real-time*

All rats were kept in their habitual environment until the day of the experiment. The subject rats were anesthetized with 1% Nembutal (40mg/kg) and mounted into the stereotactic frame in a flat-skull position. After the skull was exposed, the dual-fiber probe was installed on stereotaxic frame [28, 29]. The step motor drove the fiber probe deep into the skull at an interval of 0.7mm until it reached the surface of the dura mater. The reduced scattering coefficient ($\mu_s'$) of the cortex was continuously monitored and recorded (2Hz in frequency, 10min) by fNIRs system. After 10min of data acquisition, a pause was given for tail vein injection, viz. 2g/kg mannitol for TGroup 1, 7.5% HS for TGroup 2, 1g/kg mannitol +7.5% HS for TGroup 3 and saline for the control group. Then measurements were resumed until the end of the dehydration process (about 4-5 hours). Data on both the injured (left) and contralateral (right) sides were obtained. In addition, rats in each subgroup were monitored by fNIRs at different time points (1h, 6h, 24h, 72h and 120h). Control group was monitored following the same procedures on the ipsilateral and contralateral sides, respectively.

### *2.4. Measurement of brain water content*

At the end of the designated monitoring time points (1h, 6h, 24h, 72h and 120h), all rats were sacrificed by decapitation under deep pentobarbital anesthesia. And the brains were removed immediately. Cortex tissue (about 1 mm3 in volume) was removed around the injury focus and the corresponding position in the contralateral side; The dissected cortex tissue was firstly placed on the electronic analytical balance to measure the wet weight (WW) and then placed in an oven at 100°C for 72h until the weight was constant (a difference of less than 0.2mg in twice weighting) to measure the dry weight (DW). Brain water content was calculated according to the following equation:

$$[(WW-DW) / WW] \times 100\%.$$

### *2.5. Statistical analysis*

All statistical analysis on the data of μs' and brain water contents were analyzed by SPSS statistical software package (SPSS 13.0 for Windows) and a t-test was performed. Group data were presented as mean ± S.D. A value of $P < 0.05$ was considered statistically significant. The correlation of μs' and brain water contents were analyzed by linear forward regression.

## 3. Results

### *3.1. Rat cortex reduced scattering coefficient $\mu_s'$ ($cm^{-1}$)at different time points*

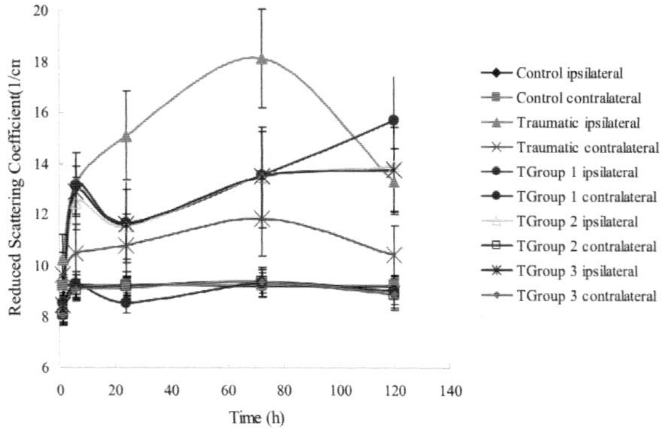


**Figure 1.** $\mu_s'$–versus–time curves.

Rats reduced scattering coefficient $\mu_s'$ values were recorded at different time points. The time profiles were illustrated in Figure 1. For the rats in traumatic group ( with traumatic model, but without treatment), rat cortex reduced scattering coefficient $\mu_s'$ values in the traumatic side increased at about 60 minutes after injury and significant difference was observed at about 6 hours after injury compared to control group ($p<0.01$), the $\mu_s'$ values were 3.95 $cm^{-1}$ higher than that at 72 hours after injury, the $\mu_s'$ values achieved a peak value of 18.12cm$^{-1}$, which was higher than control group ($p$<0.01); 120 hours after injury, the $\mu_s'$ values dropped slightly but were higher than control group($p$<0.01).

The cortex $\mu_s'$ values of each treatment group at 1 hour, 24 hours, 72 hours after injury were significantly lower than traumatic group ($p$ <0.01) at 1 hour, 24 hours, 72 hours after injury.

$\mu_s'$ values of traumatic group and all treatment groups were higher on traumatic side than that on contralateral side, as shown clearly in Figure 1.

### *3.2. The brain water content*

Table 2 shows the changes of brain water content (BWC) in different groups at different time points. BWC around the injury focus of traumatic group and all treatment groups increased compared with that of control group at about 60 minutes after injury. But no statistically significant differences were observed. BWC of traumatic group and all treatment groups was significantly higher than that of control group at about 6 hours after injury ($p$ <0.01). Then BWC of traumatic group increased rapidly and reached peak edema (82.43±0.81%) at 72 hours after injury ($p$ <0.01). In traumatic group, edema at about 120 hours after injury was lower than that at 72 hours after injury. BWC of traumatic group was still significantly higher than that of control group ($p$ <0.01) at about 120 hours after injury. BWC of contralateral side in traumatic group was also slightly increased from 1 to 120 hours after injury. But only at the peak of edema (24 hours to 72 hours after injury), the significant differences ($p$<0.01) compared with control group were observed. The increased rate of BWC on contralateral side of traumatic group was far less than that on traumatic side. BWC on contralateral side of traumatic group returned to normal at 120 hours after injury.

**Table 2.** Rat brain water content at different time in different groups (mean ± S.D., n =5).

| | | Brain water content (%) | | | | |
|---|---|---|---|---|---|---|
| | Times | 1h | 6h | 24h | 72h | 120h |
| Control | Ipsilateral side | 77.51±0.37 | 77.49±0.41 | 77.56±0.47 | 77.39±0.53 | 77.46±0.40 |
| | Contralateral side | 77.60±0.32 | 77.62±0.39 | 77.43±0.36 | 77.47±0.41 | 77.52±0.36 |
| Traumatic | Traumatic side | 78.08±0.53 | 79.25±0.56$^{**f}$ | 82.10±0.59$^{**ff}$ | 82.43±0.81$^{**ff}$ | 79.42±0.51$^{**ff}$ |
| | Contralateral side | 77.71±0.49 | 78.10±0.58 | 78.42±0.43$^{**}$ | 79.20±0.66$^{**}$ | 77.92±0.57 |
| TGroup 1 | Traumatic side | 77.16±0.44 | 78.61±0.53$^{**}$ | 80.12±0.61$^{**\$}$ | 80.26±0.79$^{**\$}$ | 81.06±0.83$^{**\$}$ |
| | Contralateral side | 76.47±0.31$^{**\$}$ | 77.68±0.39 | 76.92±0.47$^{\$}$ | 77.59±0.56$^{\$}$ | 76.68±0.53$^{\$}$ |
| TGroup 2 | Traumatic side | 76.92±0.40$^{\$}$ | 78.85±0.41$^{**}$ | 79.99±0.63$^{**\$}$ | 78.83±0.60$^{**\$\#}$ | 79.60±0.66$^{**\#}$ |
| | Contralateral side | 76.42±0.31$^{**\$}$ | 77.58±0.42 | 77.30±0.40$^{\$}$ | 77.56±0.58$^{\$}$ | 76.63±0.60$^{\$}$ |
| TGroup 3 | Traumatic side | 77.22±0.36 | 78.76±0.51$^{**}$ | 79.89±0.61$^{**\$}$ | 79.2±0.83$^{**\$}$ | 79.56±0.70$^{**\wedge}$ |
| | Contralateral side | 76.50±0.32$^{**\$}$ | 77.62±0.41 | 77.15±0.42$^{\$}$ | 77.60±0.52$^{\$}$ | 76.7±0.51$^{\$}$ |

* $p$<0.05, ** $p$<0.01: traumatic group and TGroup 1,2,3 *vs*. control group; $ $p$<0.01: TGroup 1,2,3 *vs*. traumatic group; *f* $p$<0.05, *ff* $p$<0.01 traumatic side *vs*. contralateral side; # $p$<0.01: TGroup 1 *vs*. TGroup 2; ^ $p$<0.01 TGroup 3 *vs*. TGroup 1or 2

There was no significant difference on traumatic side between BWC of TGroup 1, 3 and BWC of traumatic group at 1 hour after injury. BWC on traumatic side of TGroup 2 were significantly lower than that of traumatic group ($p$<0.01). There was no difference on traumatic side between BWC of all treatment groups and traumatic group at 6 hours after injury. At the peak of edema (24 hours to 72 hours after injury), traumatic side BWC of all treatment groups was significantly lower than that of traumatic group ($p$<0.01). BWC of TGroup 1 was significantly different (higher) from traumatic group ($p$ <0.01) at 120 hours after injury. BWC of TGroup 2 was higher than that of traumatic group at 120 hours after injury. But no statistically significant differences were observed. For the rats of TGroup 3, no statistically significant differences were observed compared with TGroup 1, 2 at any time point except 120 hours after injury.

Contralateral side BWC of all treatment groups was significantly lower than that of traumatic group at 1h, 24h, 72h and 120h after injury.

### *3.3. Linear regression analysis*

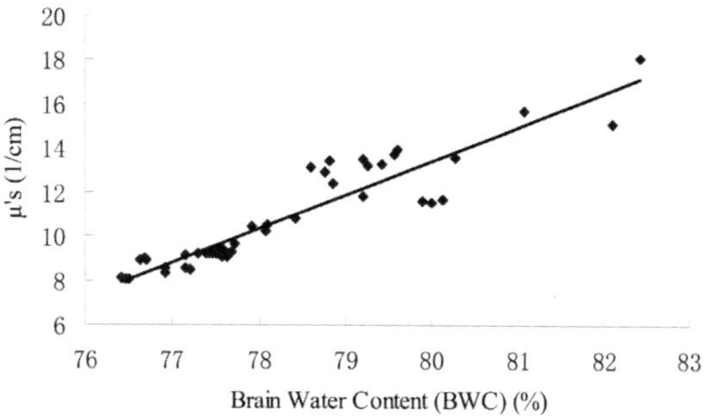


◆: $\mu_s'$ *vs.* BWC ; —: Regression curve of $\mu_s'$ and BWC

**Figure 2.** Regression curve of $\mu_s$' and BWC.

As shown in figure 2, two parameters, the average reduced scattering coefficient ($\mu_s'$) of rats' local cerebral cortex and BWC of this region at different observation points, were analyzed by linear regression. We obtained the fundamental equations of functions as following,

$$\mu_s' = 1.5494 \times (\text{BWC} \times 100) - 110.54 \quad (2)$$

$R^2 = 0.8709$ , $p<0.001$, a marked linear relationship was observed between the two parameters, which is similar to previous literature [27].

### *3.4. The reduced scattering coefficient ($\mu_s'$) measured by fNIRs*

In order to compare the treatment efficacy in different treatment groups, the $\mu_s'$ values on traumatic sides on the representative rats in TGroup 1,2 and 3 were plotted, respectively, as shown in Figure 3. Curves 1, 2 and 3 represented the $\mu_s'$ value of TGroup 1 rat , TGroup 2 rat and TGroup 3 rat on traumatic side, respectively. A noticeable downward trend were observed on Curves 1, 2 and 3 just 20, 14 and 30 minutes after the dehydrating agent injection ($p<0.05$); the $\mu_s'$ value on both the treatment groups reached a relative steady phase at 60 minutes, 40 minutes and 70 minutes after injection of mannitol, HS and mannitol + HS, respectively. The steady state maintained about 2.5 hours. Finally, Curves 1, 2 and 3 rose back to pre-injection levels at about 5 hours post-injection, as shown in Figure 3. The recovery rate for TGroup 2 (HS treatment) is slower than that of mannitol and mannitol + HS. The time profile formed a "U-shape" curve.

The time profiles of $\mu_s'$ on treatment groups and control group were compared in Figure 4. Rat brain cortex $\mu_s'$ values of control group had no significant change before and after the administration of saline (curve 1). The $\mu_s'$ values in TGroup 1, 2 and 3 decreased after administration of different dehydrating agent. Compared to control group, the significant differences of $\mu_s'$ values occurred at about 20 minutes for TGroup 1, at about 14 minutes for TGroup 2 and about 30 minutes for TGroup 3, respectively ($p<0.05$). At

60 minutes after administration of mannitol, $\mu_s'$ reached a relative steady status with a maximum reduction of 2.89% (curve 2). At 36 minutes after administration of HS, the $\mu_s'$ reached a relative steady phase with maximum decline of 3.43% (curve 3). At 70 minutes after administration of mannitol + HS they reached the relative steady phase with the maximum decline of 2.78% (curve 4).

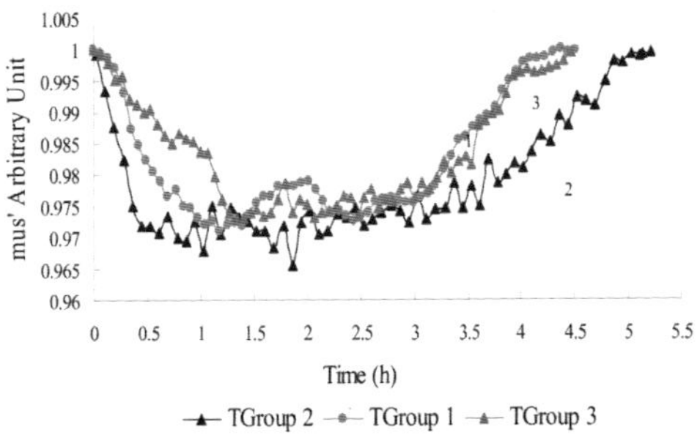


**Figure 3.** *In vivo* and real-time $\mu_s'$ Determined by NIR system on the rats of treatment groups.

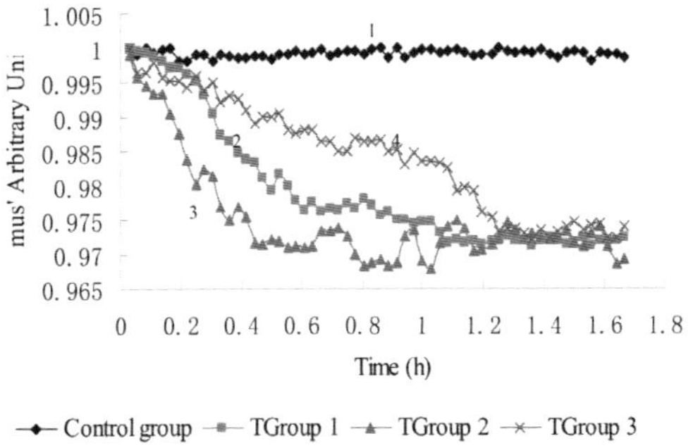


**Figure 4**. Comparison of $\mu_s$' changes on different group rats of control group and TGroup1,2,3.

## 4. Discussion

In previous studies on brain trauma, typical means of evaluating edema following brain injury include the dry/wet weight method and the special proportion calculation method [31]. These methods displayed many disadvantages, such as sacrifice of animals, error in weight measurement and so on. To avoid these shortcoming, a special bifurcated needle-shaped functional near infrared probe was used in this study to *in vivo* and real-time measure the reduced scattering coefficient( $\mu_s'$) in rats' local cerebral cortex.

The $\mu_s'$ values in traumatic side of TGroup1, 2 and 3 were significantly lower than that of traumatic group at about 1 hour after injury, ($p$ <0.01), whereas no significant difference in BWC was observed during the same period. This result indicated that $\mu_s'$ values were more sensitive to change of brain edema than BWC. Brain edema could be predicted earlier by using $\mu_s'$ values.

Usually, the edema is observed to reach the peak 24~72 hours after injury. For rats in TGroups 1,2 and 3, the maximum decrease of $\mu_s'$ values in traumatic side occured at 72 hours after injury (about 25.2% ~25.7%) . And the $\mu_s'$ values in contralateral side of TGroups 1, 2, 3 rats showed the same decrease percent (about 20%). The result may suggest that HS and mannitol with equal osmotic gradient had the same capability of dehydration on both traumatic and contralateral side. There was not the synergistic effect between mannitol and HS. During the same period, the decrease percent of BWC was much smaller than that of $\mu_s'$ values, indicating that $\mu_s'$ values are more sensitive to edema than the BWC.

The regression equation, the equation (2), showed a marked linear relationship between $\mu_s'$ values and brain water content. The good correlation between $\mu_s'$ and BWC implied that $\mu_s'$ values may serve as an indicator for edema assessment. The dynamic changes of $\mu_s'$ value after the administration of dehydrating agent could be *in vivo* continuously monitored and recorded which avoid the complicated procedures of BWC measurement.

After analyzing the $\mu_s'$ data, we found that the therapeutic efficacy of different dehydration treatments on brain edema can be assessed from there parameters, including the onset time, the maximum reduction percent of $\mu_s'$ values, the time required to achieve the maximum effectiveness and efficacy, as shown in Table 3. The onset time of TGroup 3 was longer than that of TGroup 1 and 2. The maximum reduction percent of $\mu_s'$ values of TGroup 3 was 2.78%, which is similar to TGroup 1 and lower than that of TGroup 2. The time required to achieve the maximum effectiveness and efficacy in TGroup 3 was 70 minutes, which was substantially larger than that in TGroup 2 (36 minutes) and larger than TGroup 1 (60 minutes). The results may indicate that the therapeutic efficacy of mannitol + HS was not superior to that of mannitol or HS alone.

**Table 3.** $\mu_s$' values of treatment groups.

| Group | The onset time (min) | The time required to achieve the relative steady phase (min) | The maximum reduction percent of $\mu_s'$ (%) |
|---|---|---|---|
| TGroup 1 | 20 | 60 | 2.89 |
| TGroup 2 | 14 | 36 | 3.44 |
| TGroup 3 | 30 | 70 | 2.78 |

## 5. Conclusion

This study demonstrated that fNIRs has the ability to *in vivo* and in real-time monitor the development of traumatic brain edema following injury and the treatment effect of dehydrating agents. Rats' cortical $\mu_s'$ values measured by fNIRs are a reliable parameter for assessing brain edema and drug treatment. The occurrence and development of cerebral edema could be detected earlier with the sensitive indicator $\mu_s'$ values.

By recording the changes of $\mu_s'$ value, the entire progress of dehydration therapy was disclosed. According to the experimental data, the following conclusions can be obtained: there was not the synergistic effect between mannitol and HS; the therapeutic efficacy of mannitol + HS was not superior to that of mannitol or HS alone.

The result demonstrated that the fNIRs technique has a potential for *in vivo* and real-time assessing the efficacy of treatment on the brain injury. Based on this study, developing an *in vivo* non-invasive optical monitoring method on traumatic brain edema will be our future direction.

## Acknowledgments

This work was supported Biophotonics Laboratory at the University of Aeronautics and Astronautics at Nanjing, the National High Technology Research and Development Program ("863" Program) of China (2008AA02Z438) and the Natural Science Foundation of Jiangsu Province of China (BK2009371).

## References

[1] Adamides A A, Cooper D J, Rosenfeldt F L, Bailey M J, Pratt N, Tippett N, Vallance S and Rosenfeld J V 2009 *Acta Neurochir.* 151 1399

[2] Pérez-Pérez A J, Pazos B, Sobrado J, Gonzalez L and Gándara A 2002 *Am. J. Nephrol.* 22 573

[3] Luo R and Su Y Y 2008 *Adverse Drug Reactions J.* (*ADRJ*) 10 312

[4] Dziedzic T, Szczudlik A, Klimkowicz A, Rog T M and Slowik A 2003 *Clin. Neurol. Neurosurge.* 105 87

[5] Fang Y and Wu X C 2008 *Chin. J. Intergrat. Med. Cardio-/Cerebrovascular Dis.* 6 1511

[6] Dorman H R, Sondheimer J H and Cadnapaphornchai P 1990 *Medicine* (*Baltimore*) 69 153

[7] Whelan T V, Bacon M E, Madden M, Patel T G and Handy R 1984 *Arch. Intern. Med.* 144 2053

[8] Gondim F A A, Aiyagari V, Shackleford A and Diringer M N 2005 *J. Neurosurg.* 103 444

[9] Donato T, Shapira Y, Artru A and Powers K 1994 *Anesth. Analg.* 78 58

[10] Hartwell R C and Sutton L N 1993 *Neruosurg.* 32 440

[11] Zhu G, Yao J, Qin Z and Sun X 1997 *Chin. J. Nerv. Ment. Dis.* 23 286

[12] Kobayashi H, Ide H, Kabuto M, Handa Y, Kubota T and Ishii Y 1995 *Surg. Neurol.* 44 55

[13] Oddo M, Levine J M, Frangos S, Carrera E, Maloney-Wilensky E, Pascual J L, Kofke W A, Mayer S A and LeRoux P D 2009 *J. Neurol. Neurosurg. Psychiatry* 80 916

[14] Rockswold G L, Solid C A, Paredes-Andrade E, Rockswold S B, Jancik J T and Quickel R R 2009 *Neurosurg.* 65 1035

[15] Yada Langui MM, Anjos Valotta EA, Sannomiya P, et al. 2004 *Exp. Biol. Med.* 229 684

[16] Hirsh M, Dyugovskaya L, Bashenko Y and Krausz M M 2002 *Crit. Care. Med.* 30 861

[17] Deitch E A, Shi H P, Feketeova E, Hauser C J and Xu D Z 2003 *Shock* 19 328

[18] Kolsen-Petersen J A 2004 *Acta Anaesthesio.l Scand.* 48 667

[19] Qureshi A I, Suarez J I 2000 *Crit. Care Med.* 28 3301

[20] White H, Cook D and Venkatesh B 2006 *Anesth. Analg.*102 1836

[21] Ogden A T, Mayer S A and Connolly E S 2005 *Neurosurg.* 57 207

[22] Johns M, Giller C, and Liu H 1998 *J. Biomed. Opt.* 3 437

[23] Mourant J R, Bigio I J, Boyer J, Conn R L, Johnson T and Shimada T 1995 *Lasers Surg. Med.* 17 350

[24] Cheong F W, Prahl S A, and Welch A J 1990 *IEEE J. Quantum Electron.* 26 2166

[25] Bonthan J, Minet O, Helfmann J, Herrig M and Mülle G 1996 *Phys. Med. Biol.* 41 369

[26] Xie J, Qian Z, He L and Yang T 2009 *Spectr. Spectr. Anal.* 29 3254

[27] Qian Z, Chen R, Gu Y, Giller C A and Liu H 2004 *J. Nanjing Univ. Aeron. Astron.* 36 369

[28] Qian Z, Gu Y and Liu H 2005 *Chin. J. Med. Phys.* 22 463

[29] Dai L, Qian Z, Li K, Yang T and Wang H 2008 *J. Biomed. Opt.* 13 044003

[30] Feeney D M, Boyeson M G, Linn R T, Murry H M and Dail W G 1981 *Brain Res.* 211 67

[31] Okiyama K, Smith D H, Thomas M J and Mcintosh T K 1992 *J. Neurosurg.* 77 607

# Relationship between muscle oxygenation by NIRS and blood lactate

[1]Guodong Xu, [2]Zongzhen Mao, [2]Yanjie Ye, [2]Kunru Lv

1. School of Physical Education, Jianghan University, Hubei Wuhan 430056, P. R. China

2. School of Health Sciences, Wuhan Institute of Physical Education, Hubei Wuhan 430079, P. R. China

E-mail: xguodong@wipe.edu.cn

**Abstract**: The aim of the study was to investigate the relationship of muscle oxygenation in term of oxy-hemoglobin concentration change ($\triangle HbO_2$) by NIRS and blood lactate (BLA) in local skeletal muscle and evaluate the capability of NIRS in the research of exercise physiology. Twenty-three athlete in the national fin-swimming team took the increasing load training on the power bicycle while their $\triangle HbO_2$ and BLA were simultaneously recorded. The initial powers used in the training were set as 100 w for males and 40 w for females. During the experiment, the power kept constant for 3 min before each abrupt increment of 30 w until the limit of the athlete's capability. Statistical analysis and data visualization were performed. Following the increasing load training, $\triangle HbO_2$ step-likely increased in the phase of aerobic metabolism but linearly decreased in the phase of anaerobic metabolism. The variation tendency of BLA was the same as $\triangle HbO_2$ and the concurrency of crucial turning points between $\triangle HbO_2$ and BLA was revealed. This relationship between $\triangle HbO_2$ and BLA presented in the increasing load training suggested that $\triangle HbO_2$ might be capable for taking the place of the invasively measured parameter BLA. Considering that $\triangle HbO_2$ can be noninvasively measured by NIRS, $\triangle HbO_2$ has the potential in the evaluation of athletes' physiological function and training effect on the athletes and accordingly NIRS can be well used in this field.

**Key words**: near-infrared spectroscopy (NIRS), increasing load training, oxy-hemoglobin concentration change ($\triangle HbO_2$), blood lactate (BLA), exercise physiology

## 1. Introduction

As a traditional method, Blood lactate (BLA) played a critical role in sports training. Generally, BLA would first slowly increase and then intensely increase in incremental exercise. The turning point of such procedure has been named as lactate threshold (LAT), at which time point the body changed into anaerobic status from aerobic status. LAT is widely used to evaluate the sports capacity of the sportsmen and study the body mechanism in exercise. However, the current method to measure BLA is invasive and non real-time.

Near-infrared spectroscopy (NIRS) can provide noninvasive measurements of the concentration changes of oxy-hemoglobin ($\triangle HbO2$), deoxy-hemoglobin (Hb) and total hemoglobin (blood volume, BV) simultaneously and in real-time [1-3]. Since Jobsis reported that NIR light can penetrate to the muscle and the reflected light intensity signal was quantitatively related with the concentration of those $O_2$-dependent materials, quite a lot of researchers devoted to the research of NIRS [4-9]. Some of them had successfully tested the feasibility of NIRS in measuring the muscle $O_2$ consumption ($VO_2$) [10]. Some had used NIRS to measure the blood flow [11, 12] and found NIRS is super in providing the measurement of local muscle compared to plethysmography [13-16].

Among the parameters measured by NIRS, $\Delta HbO_2$ reflects the oxygen metabolism especially the

oxygen supply in the local skeletal muscle. Accordingly, some researchers began to apply NIRS in sports training in use of $\triangle HBO_2$. Finally, the reliability and the validity of this application had been approved [1, 17, 18]. Recently, the application research of NIRS in specific ports training was highlighted, although it is just at the beginning.

Considering that BLA is the product of glucolysis and that glucolysis is closely related to the supply and consumption of oxygen, $\triangle HbO_2$ might be related with BLA. This study attempted to study the relationship of BLA and $\triangle HbO_2$ and further discuss the feasibility and superiority of replacing BLA with $\triangle HbO_2$ in sports training. In the study, we also use other technique to record physiologic indices, such as the respiration exchange ratio (RER) which has been commonly used for monitoring sports training. Of note, the measurement of these indices is used as supplemental evidence in our research.

## 2. Materials and Method

### Subjects

Twenty-three athletes (13 females and 10 males) in the national fin-swimming team were recruited in the experimental research. The age of female subjects were 19.5±2.4 years and the age of male subjects were 20.0±1.5 years. The heights of female subjects were 172.5±3.7cm and those of males were 181±2.8cm. The weights of female subjects were 66.6±5 kg and those of males were 82±6.3 kg.

### Equipments

The equipments used in this research included a portable three-wavelength NIRS which was developed by Huazhong University of Science and Technology. a MAX-II cardio-pulmonary function monitoring system made in the U.S, a MONARK（829）power bicycle made in Sweden, a P-Lar heart rate watch made in Finand and a BLA monitor made in Japan.

### Experimental paradigm

Incremental exercise on power bicycle was adopted in this research. The initial power of the exercise for female subjects was 40 w and that for male subjects was 100 w. During the exercise, the power of exercise increased by 30 w per 3 min until the exercise limit of the subjects [19].

### Data collection

The probe of NIRS was tied to the surface skin on the right quadriceps femora's muscle. The bottom edge of the probe is 10 cm upper to the crevice of the knee-joint. The separation between light source and detectors was 3 cm. the simplified demonstration of detection principle of NIRS was shown by the light transport in layered tissue with muscle (see Fig. 1). The variation of △HBO2 was recorded by NIRS during the experiment. The MAX-II cardio-pulmonary function monitoring system was used to record oxygen uptake, $VO_2$ and carbon dioxide output ($VCO_2$), which were also related with oxygen metabolism in the body and thus would be used in supplemental analysis to the relationship between △HBO2 and BLA. During the rest phase right before the exercise, BLA was recorded as a contrast. Then, when starting the exercise, increasing the exercise load and ending the exercise, at 1.5 min, 3.0 min, 6 min, 9 min just after the exercise, the values of BLA were recorded accordingly.

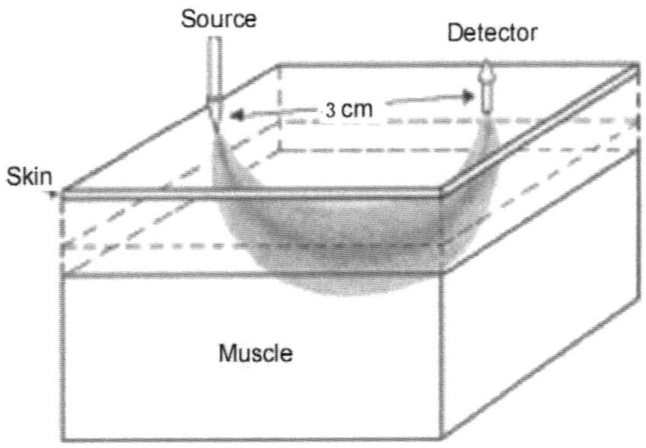


Fig. 1 Schematic of NIR light transport in the tissue with muscle.

**Data analysis**

Except NIRS, other used equipment belonged to commercial systems which provided the measurement of correspondent parameters without need of additional computing. The prototype NIRS instrument provided the measurements of the reflected light intensities of all used wavelengths during the experiment. A translation algorithm, which was based on the modified Beer-Lambert law, was used initially to change the data on the reflected light intensities to the data on the relative concentration change of $\triangle HBO_2$, △Hb and blood volume. The algorithm had been validated by a serial of published reports. In this research, we mainly analyze the data of $\triangle HBO_2$.

Time course analysis was carried out to compare the variations of $\triangle HBO_2$ and BLA, with intense attention on both turning points, that were, the BLT and the turning point of $\triangle HBO_2$ which were called as oxygen dissociation point.

## 3. Results

Fig. 2 showed typical primary result curves from a single subject. Compared among △HBO2, BLA, $VO_2$ and $VCO_2$, we can easily see that the turning point of $\triangle HbO_2$ occurred concurrently with BLT at about 900 s and that the crossing point between $VO_2$ and $VCO_2$ occurred at the same time to BLT and oxygen dissociation point. This finding kept consistent in other subjects' results. The crossing point between $VO_2$ and $VCO_2$ were generally known as respiratory exchange ratio turning point (RERT). The concurrence of the oxygen dissociation point by $\triangle HbO_2$, BLT and RERT marked the conversion of the sportsmen body from the status of aerobic metabolism to the status of anaerobic metabolism.

During the incremental exercise, $\triangle HbO_2$ decreased slowly and step-shapely before the time point of BLT and then decreased intensely and linearly. On the contrary, BLA increased slowly before the occurrence of BLT and then increase sharply and linearly. Except for the tendency of increase or decrease, there is another difference between $\triangle HbO_2$ and BLA, that is, the step-shaped variation of $\triangle HbO_2$ before BLT but no obvious step-shaped variation of BLA. Considering that it was hard to measure BLA in real time and thus it was almost unrealistic to improve the sampling rate of BLA, it is unreasonable to expect that the curve of BLA before BLT might form the similar step-shape of $\triangle HbO_2$.

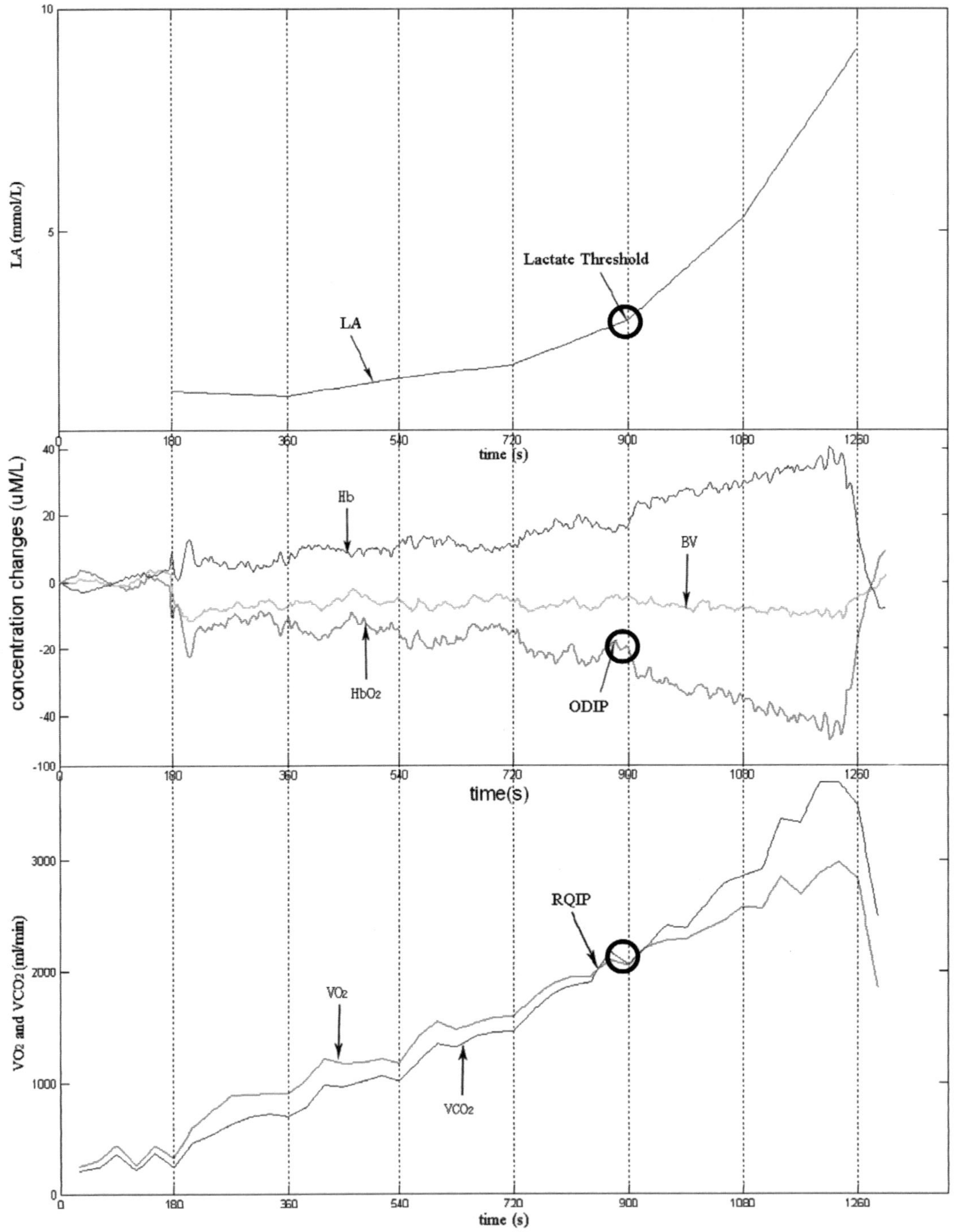


Fig. 2. Time courses of BLA, $\triangle HBO_2$, and $VO_2/VCO_2$. The occurring time point of BLT, the turning point of $\triangle HBO_2$, and the crossing between curves $VO_2$ and $VCO_2$ were marked with circles.

In order to test the above expectation and further analyzing the relationship between $\triangle$HbO2 and

BLA, we increased the sampling rate and redraw the variation curve of BLA, which was shown in blue color in Fig. 3. It was quite clear that the step-shape of $\triangle HbO_2$ was almost repeated by the new curve of BLA. In addition, it was found that the earlier occurred step shape of $\triangle HbO_2$ showed a faster and shorter increase phase and a slower and longer decrease phase. However, following with the increment of the exercise load, the $\triangle HbO_2$ trended to be flat, which was obvious during the fourth increment of the exercise. This variation could not be significantly observed by the data of BLA, but we could found the time points linking the increasing phase and the decreasing or flat phase under all increment of the exercise loads kept consistent between $\triangle HbO_2$ and BLA. Moreover, it should be emphasized that, the expectation on the step-shape of BLA variation in the aerobic status of metabolism could also be supported by the theory analysis.

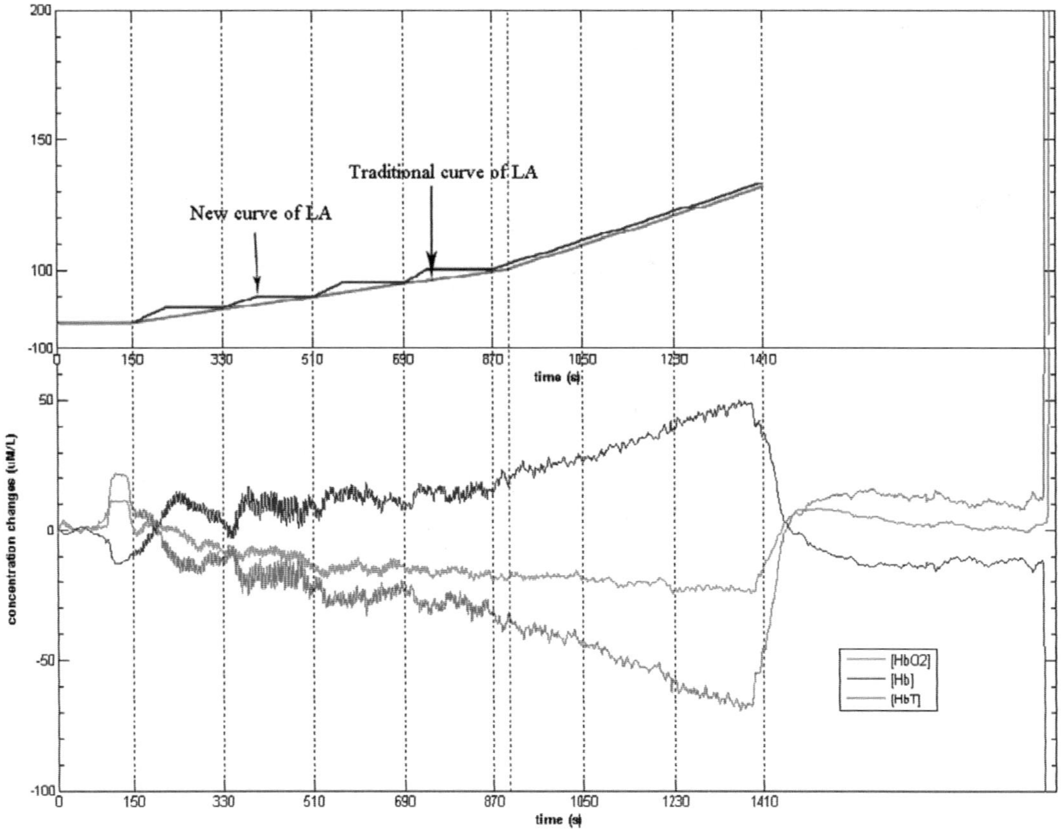


Fig. 3. Comparison between BLA and hemodynamic changes on each step-shape segment corresponding to the increment of the exercise loads. The blue curve for BLA was redrawed by increasing the sampling rate of the primary curve of BLA.

In order to compare the total tendency of $\triangle HbO_2$ and BLA during the whole experiment, we decreased the sampling rate of $\triangle HbO_2$ to be the same as BLA. After the operation of $\triangle HbO_2$ data, the curve of $\triangle HbO_2$ was updated as shown in black color in Fig.4. It was clear that the opposite $\triangle HbO_2$ curve would be same to BLA in variation in the complete duration of the experiment. Specifically, in the status of aerobic metabolism, $\triangle HbO_2$ presented the similar slower increasing variation to the BLA; in the status of anaerobic metabolism, $\triangle HbO_2$ presented nearly the same sharp increasing variation to the BLA. In detail, there is only a small difference between the reduced curve of $\triangle HbO_2$ and the primary curve of BLA, which is that, $\triangle HbO_2$ in the status of aerobic metabolism showed a nonlinear increment but BLA showed linear increment.

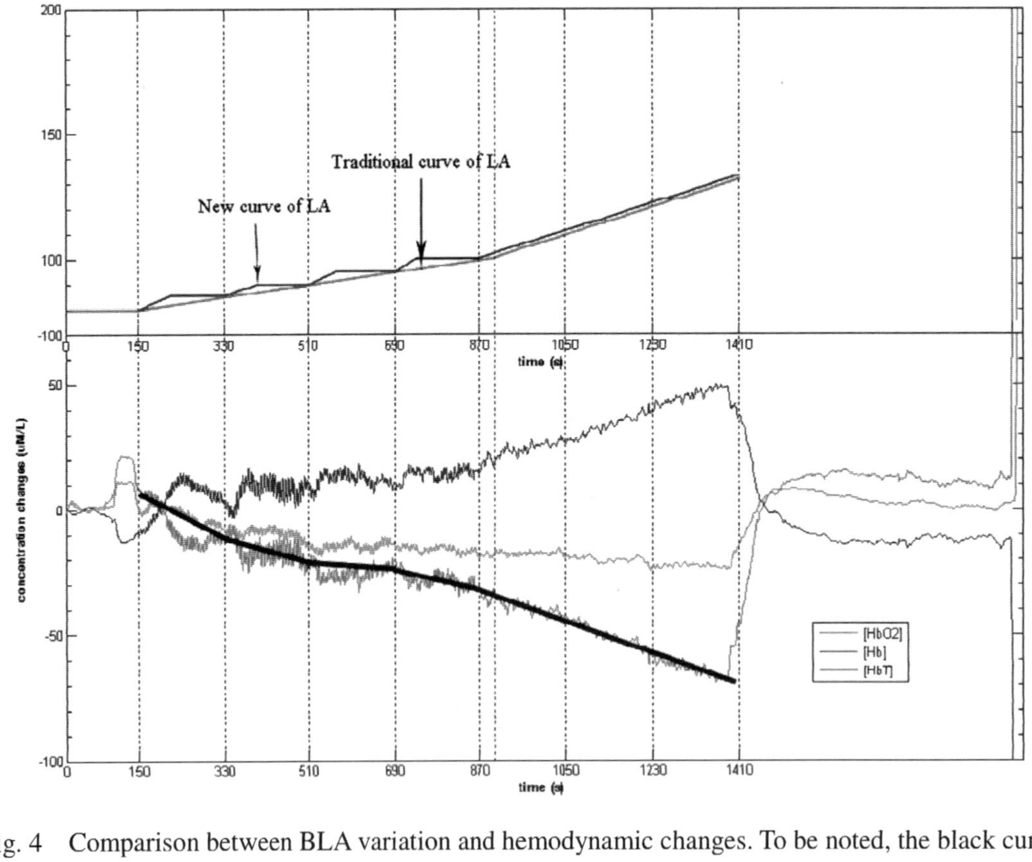


Fig. 4 Comparison between BLA variation and hemodynamic changes. To be noted, the black curve displayed the variation of $\triangle HbO_2$ at the same sampling rate with BLA.

It is worth noting thta, the parameter $\triangle Hb$ also showed the nearly the same turning point and the variation tendency to BLA. To mention, the curve of $\triangle Hb$ was almost parallel to the updated curve of

BLA when it was smoothed in Figs. 2 - 4, no matter in the status of aerobic metabolism or in the status of anaerobic metabolism. On the contrary, the blood volume did not exhibit obvious consistence with BLA.

Last but not least, the oxygen dissociation point is not the precise time point but a small time segment, which was the same either in BLT or in the crossing point between the curves of $VCO_2$ and $VO_2$. In theory, it should take a while for the body turned into the anaerobic status of metabolism from the aerobic status of metabolism. In the practical test, it was common to observe that the curve of RER fluctuated around the crossing point between $VCO_2$ and $VO_2$. In the measurement of BLA, we found the time point of BLT but considering that sampling rate of BLT was quite low, the actual BLT should be within the sampling interval around BLT and we could not conclude BLT was a point but not a segment. As for the whole body, all skeletal muscles could not turn into the anaerobic status of metabolism simultaneously. Actually, the transferring time of each muscle between those statuses was dependent on its sustainability to the exercise load. Consistent with the physiology theory, the curve of $\triangle HbO_2$ also showed a fluctuation period around the analyzed oxygen dissociation point. Rightly in this fluctuation period, the metabolism status of the body changed. Among all the subjects, the time range (870 ~ 900 s) of the oxygen dissociation period (See Fig. 4) was quite small and showed coincidence with BLT and RERT.

## 4. Discussion and Conclusion

This study investigated the relationship of muscle oxygenation in term of oxy-hemoglobin concentration change ($\triangle HbO_2$) and blood lactate (BLA) in local skeletal muscle. Twenty-three athletes in the national fin-swimming team were recruited to take the increasing load training on the power bicycle. The variation of $\triangle HbO_2$, BLA, and other oxygen metabolism related physiological parameters ($VO_2/VCO_2$) were simultaneously recorded. The time courses among hemodynamic changes (especially $\triangle HbO_2$), BLA, and $VO_2/VCO_2$ were compared and analyzed. The consistence between the variation of $\triangle HbO_2$ and BLA were revealed and were supported by the observation on $VO_2/VCO_2$.Our study showed that the traditional important parameter used in sports training, BLA, which owned the drawbacks including invasive approaches and non real-time measurement, could be possibly predicted by △$HbO_2$ and △Hb by NIRS, which is noninvasive and real-time in measure.

During the incremental exercise, all the sportsmen showed the consistence between the variation of hemodynamic changes mainly revealed by △$HbO_2$ and BLA, which specifically included below. Firstly, the oxygen dissociation point revealed by △HbO2 and △Hb appeared concurrently with BLT and RER, which is the crossing point between $VCO_2$ and $VO_2$. This point means for the body's transition from the status of aerobic metabolism to the status of anaerobic metabolism. More importantly, the oxygen dissociation points revealed by △$HbO_2$ and △Hb in all subjects were actually a time segment ranged from 870 s to 900 s in practical test, which was also in coincidence with BLT and RER crossing point. Secondly, the complete variation tendency of △$HbO_2$ and △Hb were consistent to BLA. During the aerobic status of metabolism, they both exhibited a slow increasing phase; while during the anaerobic status of metabolism, they both exhibited a fast and linear increasing phase. In addition, they both showed step-shaped increase during the aerobic status of metabolism and interestingly, the time point connecting the sharp increasing phase and the slow decreasing or flat phase in each step-shaped segment showed up nearly at the same time among △ $HbO_2$, △Hb and BLA. Accordingly, although the value of each point in the BLA curve could not be predicted by △HbO2 or △Hb, the variation of BLA in the whole time sequence and some critical point for further physical analysis could be likely to predict by △HbO2 or △Hb. There seems to be a proportional ratio between the measured values of △HbO2/△Hb and BLA. In this situation, the invasive and non real-time approach to measure BLA is potentially substituted by noninvasive and real-time method, NIRS.

Here, it might be questioned that why △$HbO_2$ changed as BLA in a similar linear way? A possible explanation would be as follows: the association and dissociation of △Hb and $O_2$ could be affected by the PH value. As demonstrated by the Bohr effect [20, 21], when the PH value in the blood decreased, the association between △Hb and $O_2$ would be suppressed and the dissociation between △Hb and $O_2$ would increase [22]. In this case, the concentration of △$HbO_2$ would decrease and the concentration of △Hb would increase. During the initial state of increasing the exercise load, strong glycolysis activity exploded which caused the in vivo accumulation of blood lactic acid and accordingly decreased the PH value in the blood. Therefore, the concentration of △$HbO_2$ and △Hb would change correspondingly respond to the variation of BLA. Such process benefited the supply of $O_2$ to the body in exercise. However, further study was required to test this explanation and explore the quantitative relationship between $\triangle HbO_2/\triangle Hb$ and BLA, by which we could predict the value of BLA with NIRS measured $\triangle HbO_2/\triangle Hb$.

In summary, our study attempted to apply NIRS in sports training and found that the concentration change of $\triangle HbO_2$ and $\triangle Hb$ measured by NIRS showed the same variation tendency and critical turning point as BLA. In addition, the oxygen dissociation point revealed by both $\triangle$HbO2 and $\triangle$Hb was proved by the crossing point between $VCO_2$ and $VO_2$. This study presented the strong capability of NIRS in sports training and proposed the strong potential of predicting BLA variation tendency with the data of $\triangle HbO_2$ or $\triangle Hb$ by NIRS. Considering that the BLA was one of the crucial parameter used in sports training and its measuring approach is invasive and non real-time and that NIRS is noninvasive and real-time, our finding on the relationship between $\triangle HbO_2/\triangle Hb$ and BLA would be likely to enlighten a novel way to measure BLA.

## 5. Acknowledgement

This work was supported by the National Nature Science Foundation of China (grant No. 30770554).

## References

[1] Delpy, D.T., and Cope, M.. Quantification in tissue near-infrared spectroscopy. Philos. Trans. R. Soc. Lond. B. Biol. Sci. 1997, 352: 649-659.

[2] Rolfe, P.. In vivo near-infrared spectroscopy. Annu. Rev. Biomed. Eng. 2000, 2: 715-754.

[3] Strangman, G, Boas, D.A., and Sutton, J.P.. Non-invasive neuroimaging using near-infrared light. Biol. Psychiatry. 2002, 52: 679-693.

[4] Boushel, R., and Piantadosi, C.A.. Near-infrared spectroscopy for monitoring muscle oxygenation. Acta Physiol. Scand. 2000, 168: 615-622.

[5] Boushel, R., Langberg, H., Olesen, J., Gonzales-Alonzo, J., Bulow, J., and Kjaer, M.. Monitoring tissue oxygen availability with near infrared spectroscopy (NIRS) in health and disease. Scand. J. Med.

Sci. Sports. 2001, 11: 213-222.

[6] Ferrari, M., Binzoni, T., and Quaresima, V.. Oxidative metabolism in muscle. Philos. Trans. R. Soc. Lond. B. Biol. Sci. 1997, 352: 677-683.

[7] Madsen, P.L., and Secher, N.H.. Near-infrared oximetry of the brain. Prog. Neurobiol. 1999, 58: 541-560.

[8] McCully, K.K., and Hamaoka, T.. Near-infrared spectroscopy: What can it tell us about oxygen saturation in skeletal muscle? Exerc. Sport Sci. Rev. 2000, 28: 123-127.

[9] Owen-Reece, H., Smith, M., Elwell, C.E., and Goldstone, J.C.. Near infrared spec-troscopy. Br. J. Anaesth. 1999, 82: 418-426.

[10] De Blasi, R.A., Cope, M., Elwell, C., Safoue, F., and Ferrari, M.. Noninvasive mea-surement of human forearm oxygen consumption by near infrared spectroscopy. Eur. J. Appl. Physiol. 1993, 67: 20-25.

[11] Hamaoka, T., Iwane, H., Shimomitsu, T., Katsumura, T., Murase, N., Nishio, S., Osada, T., Kurosawa, Y., and Chance, B.. Noninvasive measures of oxidative metabo-lism on working human muscles by near-infrared spectroscopy. J. Appl. Physiol. 1996, 81: 1410-1417.

[12] De Blasi, R.A., Ferrari, M., Natali, A., Conti, G., Mega, A., and Gasparetto, A.. Noninvasive measurement of forearm blood flow and oxygen consumption by near-infrared spectroscopy. J. Appl. Physiol. 1994, 76: 1388-1393.

[13] Homma, S., Eda, H., Ogasawara, S., and Kagaya, A.. Near-infrared estimation of $O_2$ supply and consumption in forearm muscles working at varying intensity. J. Appl. Physiol. 1996, 80: 1279-1284.

[14] Bangsbo, J., Krustrup, P., Gonzalez-Alonso, J., Boushel, R., and Saltin, B.. Muscle oxygen kinetics at onset of intense dynamic exercise in humans. Am. J. Physiol. 2000, 279: 899-906.

[15] Boushel, R., Langberg, H., Green, S., Skovgaard, D., Bulow, J., and Kjaer, M.. Blood flow and oxygenation in peritendinous tissue and calf muscle during dynamic exercise in humans. J. Physiol. 2000, 524: 305-313.

[16] Boushel, R., Langberg, H., Olesen, J., Nowak, M., Simonsen, L., Bulow, J., and Kjaer, M.. Regional blood flow during exercise in humans measured by near-infrared spectroscopy and indocyanine green. J. Appl. Physiol. 2000, 89: 1868-1878.

[17] Ferrari, M., Wei, Q., Carraresi, L., De Blasi, R.A., and Zaccanti, G.. Time-resolved spectroscopy of the human forearm. J. Photochem. Photobiol. B. 1992, 16: 141-153.

[18] Zhao, H., Tanikawa, Y., Gao, F., Onodera, Y., Sassaroli, A., Tanaka, K., and Yamada, Y.. Maps of optical differential pathlength factor of human adult forehead, so-matosensory motor and occipital regions at multi-wavelengths in NIR. Phys. Med. Biol. 2002, 47: 2075-2093.

[19] van Beekvelt, M.C., Borghuis, M.S., van Engelen, B.G., Wevers, R.A., and Colier, W.N.. Adipose tissue thickness affects in vivo quantitative near-IR spectroscopy in human skeletal muscle. Clin. Sci. 2001, 101: 21-28.

[20] Riggs A. Molecular adaptation in haemoglobins: nature of the Bohr effect. Nature. 1959 Apr 11;183(4667):1037-1038.

[21] Rossl L, Marchi S, Milla E. The Bohr effect in relation to the state of polymerization of the hemoglobin molecule. Boll Soc Ital Biol Sper. 1959 Dec 31;35:2214-2216.

[22] McKay BR, Paterson DH, Kowalchuk JM. Effect of short-term high-intensity interval training vs. continuous training on O2 uptake kinetics, muscle deoxygenation, and exercise performance. J Appl Physiol. 2009 Jul;107(1):128-138

# Research on the Photography Measurement of the Optical Parameters of the Vegetable Tissues

**Yang Li[1,2], Jiang Zhun-fei[1], Sun Jian-hua[1]**

[1]Science College, Civil Aviation University of China, Tianjin 300300, China;

[2]Key Laboratory of Opto-electronic Information Science and Technology, Ministry of Education, Tianjin University, Tianjin 300072, China)

e-mail address：y_lyang@yahoo.com.cn

**Abstract**. To measure the optical parameters of biological tissue, the laser imaging system of the diffusing reflection and the transmission measurements based on CCD was designed. The absorption coefficient ( $\mu_a$ ) and the reduced scattering coefficient ( $\mu_s'$ ) of the biological tissue simulating liquid ( Intralipid-20% diluents ) were derived from nonlinear regression of the diffusion equation with spatially resolved steady-state diffuse reflectance which was measured by CCD. At the same time, the reduced scattering coefficient ( $\mu_s'$ ) of the biological tissue simulating liquid ( Intralipid-20% diluents ) was also derived from linear regression of the Beer-Lambert equation with transmission light intensity of the biological tissue simulating liquid ( Intralipid-20% diluents ) which was measured by CCD for the different titer, which is used as the calibration of the non-invasively method. The optical parameters of the same sample measured by CCD diffusing reflection and transmission measurements were compared. The results of the two methods were approximately consistent. Furthermore, the optical parameters of Eggplant and Murphy were measured by CCD diffusing reflection measurements. So the technique of laser scattering imaging analysis is useful for measuring the optical properties of the vegetable tissues.

## 1. Introduction

Optical transmission in the biological tissue is one of the research focus on tissue optics[1-3]. By measuring the spatially resolved diffuse light of tissue surface, which is the result of the interaction of the incident light and the effect of absorption and scattering by tissue, we can study the optical properties of the tissue.

The mainly methods for measuring the spatially resolved diffuse reflectance of tissue surface are point by point scanning by optical fiber and CCD video camera measurement. The traditional method of point by point scanning by optical fiber has the following problems: Long measurement time, as a result, optical parameters of biological tissue and the distance between the optical fiber and tissue surface may change in every measurement; the light distribution of the region near source can not be measured; as the integrated optical fibers array will touch the tissue when measuring, the physiological state of the pressurized tissue may change. Measurement of light distribution of tissue diffuse reflectance or transmission with the CCD is new technology appeared in the 90's of the last century [4]. Compared with the traditional method of point by point scanning by optical fiber, the technology of

CCD video camera measurement has greatly improved accuracy and efficiency traditional optical scanning point by point comparison, CCD video camera technology, accuracy and efficiency has been greatly improved and can measure the spatial distribution of light intensity near light source. R. A. Bolt[5] measured the radiance distribution at the surface of the tissue phantoms with CCD and got the Optical Properties with absolute measurement method. Alwin Kienle[6] measured the radiance distribution at the surface of the tissue with the CCD non-contact method and the wavelength of 633nm light.

In this study, the laser imaging system of the diffusing reflection and the transmission measurements based on CCD was designed. The absorption coefficient and the reduced scattering coefficient of the biological tissue simulating liquid ( Intralipid-20% diluents ) were derived from nonlinear regression of the diffusion equation with spatially resolved steady-state diffuse reflectance which was measured by CCD. At the same time, the reduced scattering coefficient of the biological tissue simulating liquid ( Intralipid-20% diluents ) was also derived from linear regression of the Beer-Lambert equation with transmission light intensity of the biological tissue simulating liquid ( Intralipid-20% diluents ) which was measured by CCD for different titer. The absorption coefficient and the reduced scattering coefficient of Murphy and eggplant were measured by the CCD diffusing reflection measurements.

## 2. Method and Materials

2.1 Measuring $\mu_a$ and $\mu_s'$ by diffuse reflection

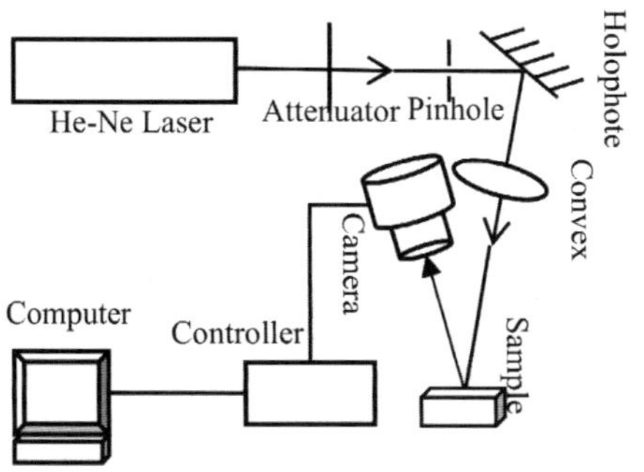


**Figure. 1** Optical path of experiment setup of CCD diffusing reflection_measurement

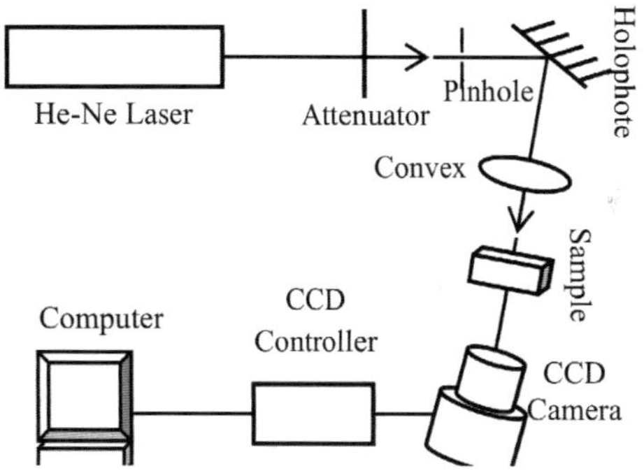


**Figure.2** Optical path of experiment setup of CCD transmission measurement

The liquid tissue phantom for this experiment is the Intralipid-20% diluents, which is denoted as sample I10. Configuration proportion of the sample I10: 1000ml sample I10 is made by 100ml Intralipid-20% diluted by adding 900ml distilled water. The Experimental device is shown in Fig.1.
The laser pass through the attenuator, pinhole ($\phi = 0.1mm$) and the convex lens, at last, it is projected onto the surface of the sample with a smaller incidence angle ($5^o \sim 10^o$). The CCD in this measurement is the 150CLM cooled CCD used for digital microscope, pixel resolution of which is 1392×1040, the sensitivity is set to 200, the exposure time is set to 1/10sec. By adjusting the CCD focal length and the amount of light into CCD we can obtain image of diffuse light on the experimental sample surface, the data of which would send to CCD controller for store and simple calculation, then send to computer for further image analysis and processing. Whole experiment is carried out in the darkroom.

Simple-PCI image analysis software developed by Hamamatsu can analyse the CCD image, accurately read the relative intensity value of any point in the image coordinates; it also can directly export relationship between the relative intensity values the image of and spatial pixels to Excel, therefore, in a short period of time we can obtain the analysis data of CCD images.

The method of calibration is that putting a millimeter-scaled ruler at the location at the same focal plane with the experimental samples, and take the photo of the ruler by CCD, then from the scale of the ruler image we can determine the ratio between the length and the number of pixels. Use this constant ratio, we can transform the relationship between relatively light intensity and pixels to the relationship between relative intensity and spatial coordinates.

Light transmission in tissue can be described by the radiative transfer equation[7]. As the size tissue sample is much larger than the actual transport distance of photons, the tissue sample can be considered as the semi-infinite medium. When unlimitedly thin collimated beam vertically project to the tissue surface, under the diffusion approximation [8] condition and the extended boundary condition, and define the diffusion coefficient D as:

$$D = \frac{1}{3(\mu_a + \mu_s')} \quad (1)$$

The solution of the diffusion approximation is

:

$$R(\rho) = \frac{1}{4\pi}\left[ z_0\left(\mu_{eff} + \frac{1}{r_1}\right)\frac{\exp(-\mu_{eff} r_1)}{r_1^2} + (z_0 + 2z_b)\left(\mu_{eff} + \frac{1}{r_2}\right)\frac{\exp(-\mu_{eff} r_2)}{r_2^2}\right] \quad (2)$$

where: $z_0$=3D ; $\mu_{eff} = (\mu_a/D)^{1/2}$ ; $r_1^2 = z_o^2 + \rho^2$ ; $r_2^2 = (z_o + 2z_b)^2 + \rho^2$ ; $z_b = \frac{1+R_{eff}}{1-R_{eff}} \cdot 2D$ ;

$R_{eff}$=0.493[9]. $\rho$ is the vertical distance at tissue surface from the study point to the incident light beam.

It can be made the non-linear fitting to R (ρ) that is measured using CCD and equation (2) non-linear fitting by MATLAB program, and the absorption coefficient $\mu_a$ and the reduced scattering coefficient $\mu_s'$ of biological tissue can be obtained.

2.2 Measuring scattering coefficient $\mu_s$ and $\mu_s'$ by transmission method

The Experimental device is shown in figure.2. It is used to measure the scattering coefficient of liquid tissue phantom I10 and the reduced scattering coefficient, to verify the accuracy of the diffuse reflection method. Laser beam (632.8nm) pass through the attenuator, pinhole （$\phi = 0.1mm$）, and the convex lens, then, it is vertically projected in the sample I10 inside the quartz slot. The CCD is in the other side of the slot out of 130cm from the sample to receive the transmitted light intensity. The principle of transmission method which is used to measure the scattering coefficient is based on Beer-Lambert Law：

$$E_c = E_i \cdot \exp[-(\mu_a + \mu_s)\cdot cd] \quad (3)$$

in the equation (3)， $E_c$ is the approximately direct transmission light intensity； $E_i$ is the light intensity which is vertically projected in the sample slot; *c* is the concentration of sample I10； *d* is the thickness of the sample in the sample slot. Adopt logarithm on both ends of equation (4)， then between $\ln(E_c)$ and (*cd*) exists a linear relationship， so the negative of its slope is $\mu_a + \mu_s$. As the intralipid solution is a highly scattering medium with low absorption[10]， $\mu_s >> \mu_a$， therefore $\mu_a + \mu_s \approx \mu_s$. During the experiment, under the fixing condition of the incidence light intensity, the location of sample pool, as well as the CCD focal length and the amount of light which goes through the CCD, then we can obtain the relative transmission light intensity with corresponding solutions of different concentrations of I10, by changing the concentration of sample I10 in sample pool with the distilled water. By making the linear fitting between $\ln(E_c)$ and (*cd*) using the MATLAB program and calculating the negative of its slope, $\mu_s$. could be acquire. As under the approximate condition of diffuse reflection[9], there exists the relation as follows:

$$\mu_s' = \mu_s \cdot (1 - g) \qquad (4)$$

The anisotropic factor g of the biological tissue liquid phantom of I10 which used in the experiment is 0.75[5], put the $\mu_s$ which obtaind by the transmission method in the equation (4), we could obtain an reduced scattering coefficient ( $\mu_{s-t}'$ ).

## 3. Results and analysis

3.1 Measurement results on samples I10

Diffuse reflection measurement is shown in figure 3. Figure 4 is the nonlinear fitting curve of the measurement data with the equation (2), thus there are: $\mu_a = 0.0442mm^{-1}$ ; $\mu_s' = 2.987mm^{-1}$.

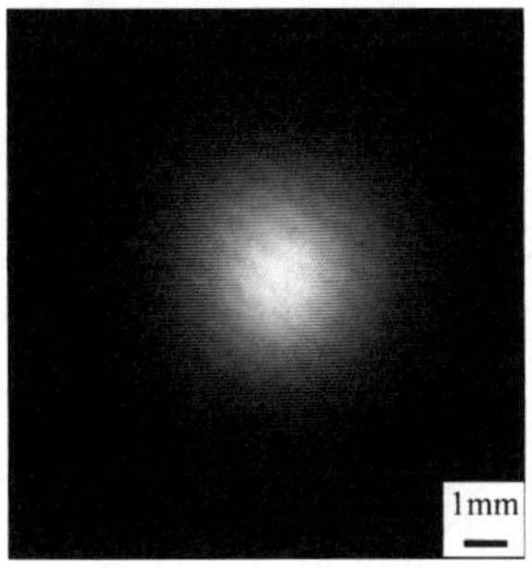


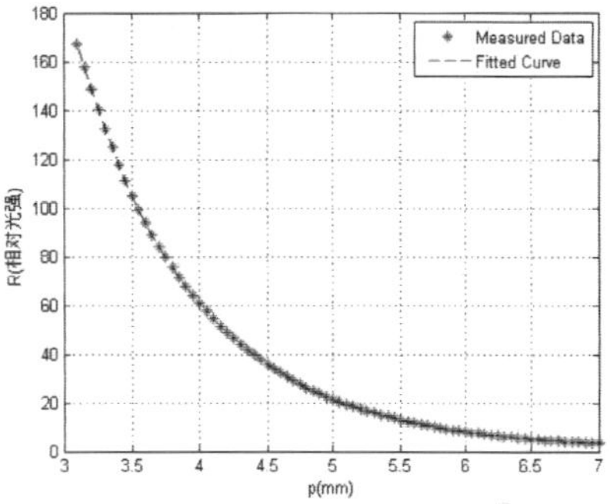


**Figure.3** Diffusing reflection distributing of the sample's surface

**Figure.4** Nonlinear fitting curve of diffusing reflection measurement

According to figure 2, we can measure the transmission intensity of sample I10 in different concentration. After the analysis of Simple-PCI software, we can get the relative transmitted light intensity ($E_c$) corresponding to sample I10 of different concentration. Taking the linear fitting between $\ln(E_c)$ and ($cd$) by MATLAB software, the slope is $-11.353$, i.e. the scattering coefficient of sample I10 is $\mu_s = 11.353$. Putting $\mu_s$ into the equation (4), we could obtain the reduced scattering coefficient : $\mu_{s-t}' = 2.838mm^{-1}$. Figure

Comparing the reduced scattering coefficients $\mu_s'$ and $\mu_{s-t}'$, which obtained by the methods of diffuse reflection and transmission respectively, the error is: $\Delta\mu_{s-t}' / \mu_s' = 5.25\%$ . Taking into account that the different measurement methods and mathematical models used in the experiment may cause some differences to the final outcomes, therefore, the above error is reasonable.

Table 1. The comparison of data between this paper and R. A. Bolt, et al

| Sample | Method | $\mu_s'/mm^{-1}$ | $\mu_a/mm^{-1}$ | Error |
|---|---|---|---|---|
| I10 | this paper | 2.987 | 0.0442 | $\Delta\mu_s'/\mu_s'$ :4.73% $\Delta\mu_a/\mu_a$ :9.79% |
| | R. A. Bolt, et al[4] | 2.852 | 0.049 | |

Table 1 is the Comparison in this paper's absorption coefficient $\mu_a$ and reduced scattering coefficient $\mu_s'$ to literature [4]. From Table 1, It can be seen the optical parameters obtained in this paper is very close to the reported data from RA Bolt, the error of reduced scattering coefficient is 4.73%, the error of absorption coefficient is just about 9.79%, the results which obtained in this paper are allowed within the system error [11]. The reasons for these differences may include the following aspects: the use of Intralipid-20% solution are from different manufacturers, the particle's size and uniformity will differ in the inside of liquid; CCD's own dark current as well as the tiny existence of the environment noise source; the error caused by the algorithm of measured data fitting for the optical parameters.

By comparison, we can see that both the devices which designed basing on CCD's diffuse reflection method to measure the optical properties of biological tissue parameters and the dispose methods all have a high accuracy

3.2 Measurement results on fresh vegetable tissue

The basis of the above-mentioned study, we also measure optical properties parameters of fresh vegetable tissue such as potatoes, eggplant and so on. The transportation of photon in the vegetable tissues, like other biological tissue, is compliance with the Boltzman transport equation. If the distribution of the light on the surface of the tissue was given, we could use the analytical expression to replay the optical parameters easily, so with the analytical expression of light distribution on the surface and light distribution in the vegetable tissue surface, we could replay the vegetable tissue optical parameters of vegetable tissue. Before the experiment, keep fresh potatoes and eggplants ready, use the surface diffuse reflection optical image to measure them when they are each in full and slice situation, as is shown in Figure 5.

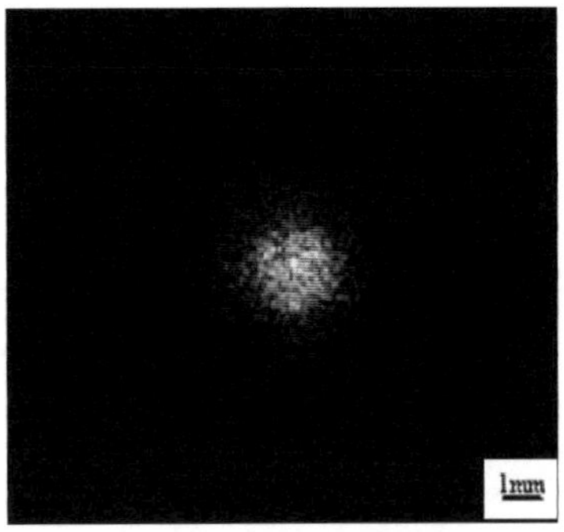


(*a*) Complete potatoes

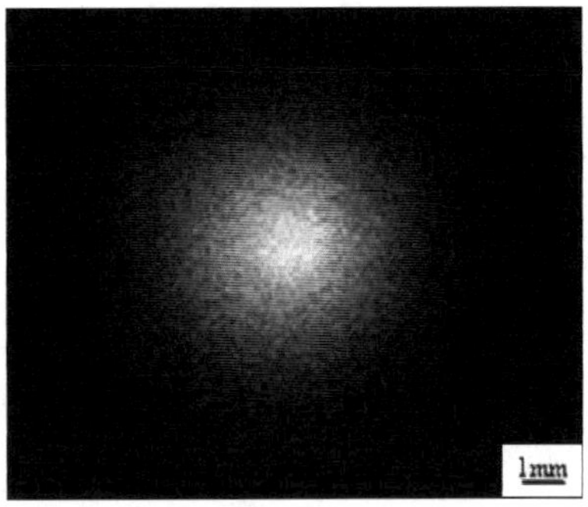


(b) Sliced potatoes

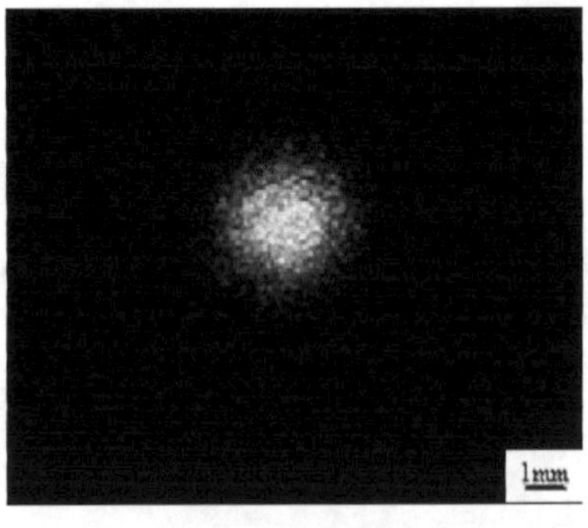


(*c*) Sliced eggplants

**Figure 5**, The diffuse reflection light distribution on the surface of different vegetable tissue

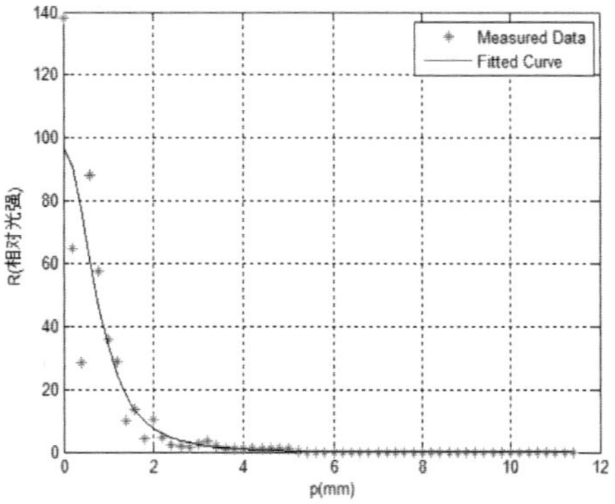


(*a*) Complete potatoes

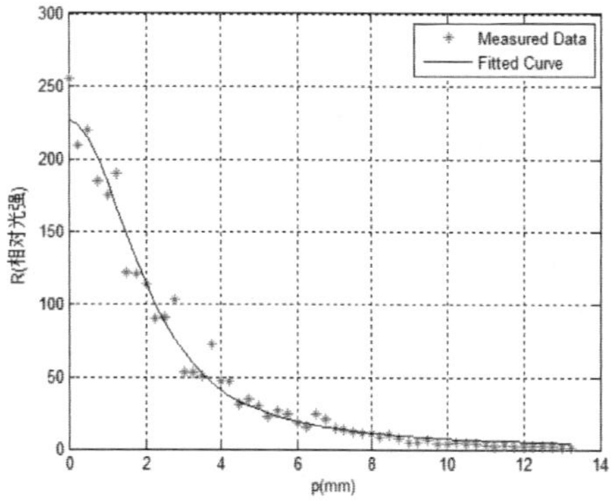


(b) Sliced potatoes

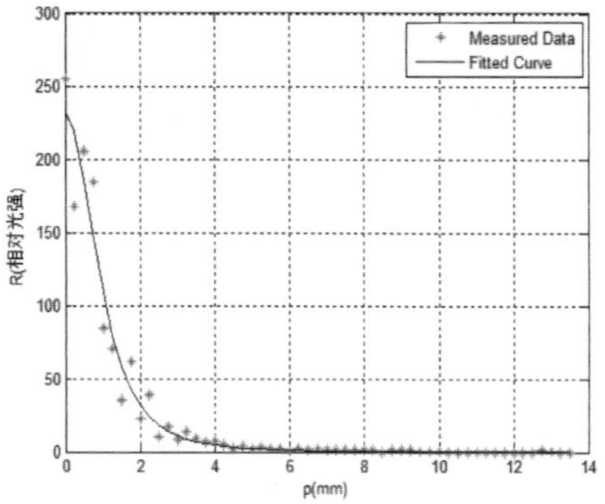


(*c*) Sliced eggplants

**Figure 6,** The non-linear fitting curve of different vegetables tissue

According to diffuse optical image data, the analysis results are shown in figure 6. Potatoes and eggplants' fitted optical parameters are in Table 2.

Table 2 the absorption and reduced scattering coefficient of potato and eggplant

| project | Potato | Eggplant |
|---|---|---|
| complete | $\mu_a$=0.1011 $mm^{-1}$<br>$\mu_s'$=0.9607 $mm^{-1}$ | |
| slice | $\mu_a$=0.0127 $mm^{-1}$<br>$\mu_s'$=0.2977 $mm^{-1}$ | $\mu_a$=0.0714 $mm^{-1}$<br>$\mu_s'$=0.7809 $mm^{-1}$ |

According to the figures 5, there is a big difference in the diffusion images of different types and under different conditions, they are related to the surface texture of vegetable tissue (including: color, moisture, sugar content, smoothness and other conditions). As the transportation of the light in the vegetable tissue decided by the organization optical transmission parameters, therefore, diffuse light images of different types, under different conditions is indicative of different optical parameters

## 3. Summary

In short, we have designed that laser image measurement device by CCD to determine the optical parameters of biological tissue. The absorption coefficient ( $\mu_a$ ) and the reduced scattering coefficient ( $\mu_s'$ ) of the biological tissue simulating liquid ( Intralipid-20% diluents ) were derived from nonlinear regression of the diffusion equation with spatially resolved steady-state diffuse reflectance which was measured by CCD. At the same time, the reduced scattering coefficient ( $\mu_s'$ ) of the biological tissue simulating liquid ( Intralipid-20% diluents ) was also derived from linear regression of the Beer-Lambert equation with transmission light intensity of the biological tissue simulating liquid ( Intralipid-20% diluents ) which was measured by CCD for the different titer, which is used as the calibration of the non-invasively method. The optical parameters of the same sample measured by CCD diffusing reflection and transmission measurements were compared. The results of the two methods were approximately consistent. Furthermore, the optical parameters of Eggplant and Murphy were measured by CCD diffusing reflection measurements. So the technique of laser scattering imaging analysis is useful for measuring the optical properties of the vegetable tissues.


### Acknowledgements

This work was supported by the Basic Research Operating Expenses of Central University, the Ministry of Education of the People's Republic of China (No. ZXH2010C003)


### References


[1] Patterson M.S. Wilson B. C. and Wyman Y. D. R. 1991 *Lasers in Medical Science*, 6 (4) 379

[2] Zhongming L. Xiaohua L. 2002 *Optoelectronics Laser,* 13 (8) 860 (in Chinese)

[3] Jiancheng L Zhenhua L. Chunyong W. *et al.* 2007 *Laser Technology* 31(3) 265. (in Chinese)

[4] Splinter R. Nnaney G A. Littman L. *et al.* 1994 *Lasers in the life Scienc* 6(1) 15

[5] Bolt R A. Bosch J J. 1993 *Applied Optics* 32 (24) 4641

[6] Kienle A. Lilge L. and Patterson M S. *et al.* 1996 *Applied Optics* 35 (13) 2304

[7] Kienle A. and Patterson M.S. 1997 *Journal of the Optical Society of America* 14 (1) 246

[8] Ishimaru A. Dudley D.G. 1978 *New York: Academic Press*: 142-157.

[9] Haskell R.C. Svaasand L.O. and Tsongtsen T. *et al.* 1994 Optical Society of America 11 (10) 2727.

[10] Van Staveren H.J. Moes Christian J.M. and Van M.J. *et al.* 1991 *Applied Optical* 30 (31) 4507.

[11] Heping X. and Patterson M.S. 2006 Optics Express 14 (14) 6485

# Plasmonic-enhanced fluorescence emission using D-shape microstructured optical fiber

**D Yong,** [1,2], **X Yu**[1,6], **S Zhang** [1,3], **C C Chan**[2], **Y Zhang**[1], **H P Ho**[4] **and D Liu**[5]

[1] Singapore Institute of Manufacturing Technology, 71 Nanyang Drive, S638075, Singapore

[2] School of Chemical and Biomedical Engineering, Nanyang Technological University, Block N1.3, 70 Nanyang Drive, S637457, Singapore

[3] School of Electrical and Electronic Engineering, Nanyang Technological University, Block S2, 50 Nanyang Avenue, S639798, Singapore

[4] Department of Electronic Engineering, The Chinese University of Hong Kong, Ho Sin Hang Engineering Building, CUHK Sh`atin, N.T. Hong Kong

[5] Department of Optical Communication and Optical Network Engineering, Huazhong University of Science and Technology, 1037 Luoyang Road, Wuhan, China

Email: xyu@simtech.a-star.edu.sg

**Abstract.** Highly sensitive side-polished D-shaped optical fiber sensors have been fabricated based on surface plasmon resonance (SPR) technology. Current techniques in plasmonic-enhanced total internal reflection microscopy (TIRM) and evanescent wave microscopy, although advantageous, require cumbersome set-ups and encompass large coupling losses. A gold coated D-shaped optical fiber was demonstrated to provide fluorescence enhanced spectroscopy. Comparison was made between a gold coated and uncoated D-shaped microstructured optical fiber (MOF) with respect to excitation of Rhodamine B (Rh B). Results highlighted improved fluorescence emission intensity and heightened sensitivity in fluorescence spectroscopy in the gold coated device, indicating potential in enhanced bio-imaging applications.

## 1. Introduction

SPR has been vastly explored for optical sensing applications [1]. Optical fibers, notably, have provided a means of miniaturization for SPR sensors. In particular, a side-polishing method, where the cladding of the optical fiber is removed locally, rendered the evanescent field of the optical fiber's guided mode accessible. The side-polished D-shaped optical fiber was further coated with a thin gold film allowing its support of surface plasmon waves (SPW) [2-3]. Additionally, since the refractive index of the adjacent medium (analyte) drastically alters the SPW's propagation constant, coupling

[6] To whom any correspondence should be addressed.

between the fiber mode and SPW is highly dependent on the analyte, enabling the fabrication of highly sensitive side-polished D-shaped optical fiber sensors [4-5].

Apart from applications in optical sensing, SPR has enabled heightened spatial resolution and background suppression of fluorescent labels in plasmonic-enhanced TIRM [6] as compared to conventional TIRM [7]. The numerous lenses and filters for guidance and focusing of light, however, render the set-up bulky and inflexible. Furthermore, background auto-fluorescence was inevitably elicited due to propagation of excitation light through the glass objective. Nevertheless, despite further modifications made in auto-fluorescence mitigation, via avoidance of propagating excitation light through the objective [8], the set-up remains equally cumbersome. Additionally, planar waveguides have also been employed in evanescent wave fluorescence microscopy. Unlike TIRM, excitation is confined at the sample surface in contact with the illumination platform, enabling study of cell surface events and morphology [9]. Coupling of the excitation source to the planar waveguide via manual alignment with optical fibers, however, elicits huge losses.

D-shaped optical fibers, on the contrary, provide a less cumbersome alternative to the illumination platform due to the flexibility and minute size of optical fibers. Furthermore, coupling of excitation sources directly into optical fibers diminishes losses. In this paper, a novel fiber device for plasmonic-enhanced fluorescence spectroscopy, structured upon a D-shaped optical fiber, is proposed and demonstrated. Comparison was made between a gold coated and uncoated D-shaped optical fiber, where different forms of fluorescence excitation modes are offered, namely SPWs and evanescent waves correspondingly. The presence of the gold coating demonstrated an improvement in fluorescence emission intensity and heightened sensitivity in fluorescence spectroscopy. In addition, results were indicative of increased efficiencies in energy coupling from SPWs to fluorescence molecules as compared to that by evanescent waves. Furthermore, the gold coated D-shaped optical fiber showed higher accuracy in data projection.

## 2. Methodology

An organic fluorescence, Rh B, was separately excited by evanescent waves and SPWs via a gold coated and an uncoated D-shaped optical fiber respectively. The resultant emissions were analysed and compared. The optical fiber, an all solid microstructured optical fiber (MOF) described in [10], was embedded in a grooved silica block and subsequently polished till its photonic crystal core. The resultant D-shaped MOF is illustrated in Fig 1(a). An identical D-shaped MOF was further coated with a thin gold layer of an approximate 35 nm thickness, and is shown in Fig 1(b).

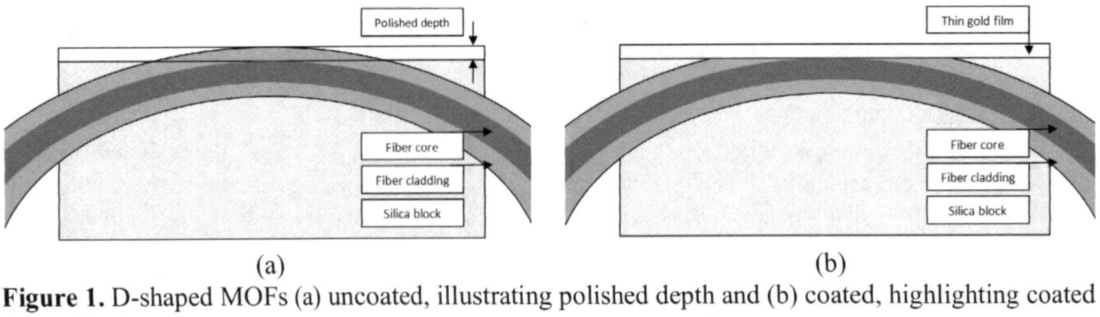


**Figure 1.** D-shaped MOFs (a) uncoated, illustrating polished depth and (b) coated, highlighting coated thin gold film.

The experimental set-up, as depicted in Fig. 2, was utilised in the detection of fluorescence emission for both the coated and uncoated D-shaped MOFs. A laser of wavelength, $\lambda = 532$ nm was butt-coupled into the D-shaped MOF via a translation stage. Collected signal was guided through a fiber optic cable to a PC-linked spectrometer (Ocean Optics, Maya 2000) for spectral analysis. Position of the objective lens was maintained for the remaining collection of emission data. A droplet of Rh B, overlapping the D-shaped region, was dripped onto the surface of the silica block, and its emission was subsequently collected. Respective emission spectra were obtained for Rh B of concentrations 10 μM, 35 μM, 50 μM, 75 μM, 100 μM, 350 μM and 500 μM.

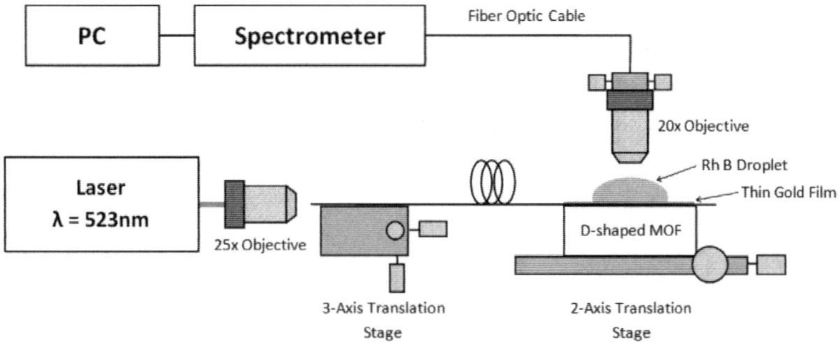


**Figure 2.** Experimental set-up for detection of fluorescence emission from gold coated D-shaped MOF. An identical set-up was utilised for the uncoated D-shaped MOF.

## 3. Results and Discussion

Fluorescence emission spectra for both gold coated and uncoated D-shaped MOFs were studied and further contrasted. Comparisons were made on the emission spectra, the peak emission wavelengths as well as the peak emission intensities. The heightened fluorescence emission intensity and fluorescence spectroscopy sensitivity under surface plasmon excitation were consequently highlighted.

### 3.1. Comparison of fluorescence emission spectra

Fluorescence emission spectra resided between 550 nm and 770 nm for both gold coated and uncoated D-shaped MOFs and are shown in Fig. 3(a) and 3(b) correspondingly. Average peak emission wavelengths of 597 nm and 587 nm were further calculated respectively and indicated a red shift in the presence of a thin gold film. Emission at a longer wavelength, for the gold coated D-shaped MOF, signifies a lower emission energy therefore implying a lower required excitation energy. This demonstrated reduced energy losses to the surrounding and thus increased efficiency in energy coupling from SPWs, elicited by the thin gold film, to Rh B molecules.

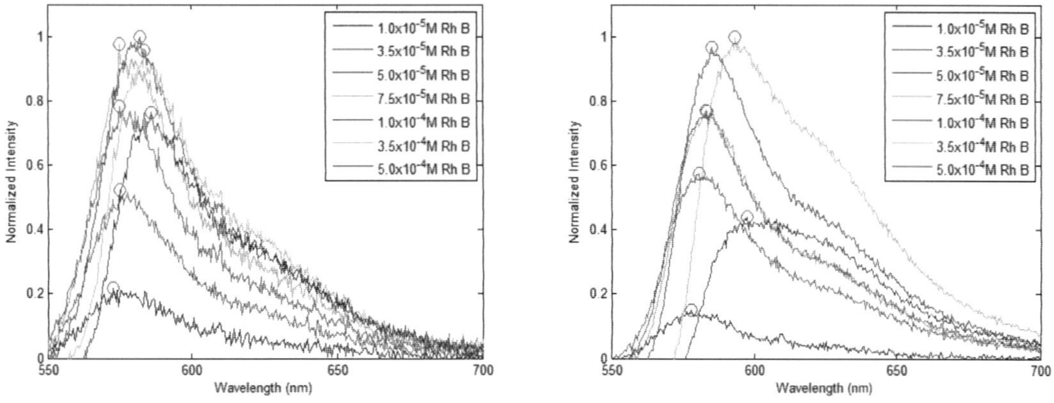


(a) (b)

**Figure 3.** Fluorescence emission spectra of (a) uncoated and (b) gold coated D-shaped MOFs.

3.2. Comparison of peak emission wavelength

A relationship between peak emission wavelength shifts with respect to changing Rh B concentrations was drawn for both gold coated and uncoated D-shaped MOFs to provide a lucid comparison.

As illustrated in Fig 4, it is evident in both cases that the peak emission shifts to higher wavelengths with increasing Rh B concentrations. This phenomenon is educed due to re-absorption of emissions. Since excitation and emission follows Stokes shifting, the emission energy is lower as compared to that of excitation. Emission energy post re-absorption, therefore, was constantly being reduced after each excitation-emission step [11]. The lower emission energy consequently resulted in a longer emission wavelength. In short, an increase in Rh B concentration raised the probability for emission reabsorption, essentially reducing the energy of emission and hence resulted in a shift of peak emission to higher wavelengths.

Conversely, a distinctive difference was observed between the gradients of the two plots. A sharper rise in wavelength with respect to the increasing Rh B concentration, as indicated by the gradient, was shown by the gold coated D-shaped MOF. The steeper gradient demonstrates a higher sensitivity in fluorescence spectroscopy, indicating improved coupling of core energy to the evanescent field by SPWs. In addition, the larger R-square value of 0.9926, for the gold coated D-shaped MOF, was indicative of its higher accuracy in projection of data.

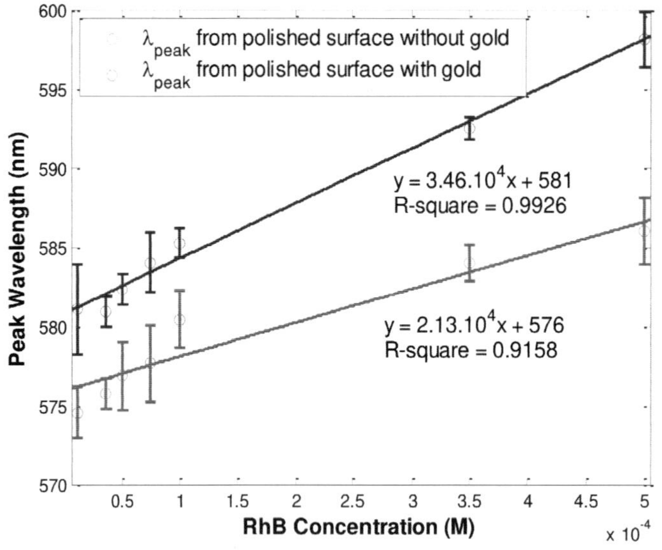


**Figure 4.** Comparison of shifts peak emission wavelengths with change in Rh B concentration, between gold coated (blue) and uncoated (red) D-shaped MOFs.

3.3. Comparison of peak emission intensity

Further study was made on the peak emission intensities with respect to Rh B concentration. Fig. 5 portrays a similar trend for both gold coated and uncoated D-shaped MOFs, where initial increment of Rh B concentration (10 μM to 100 μM) led to a rise in peak emission intensity whereas further increment resulted in a drop. During the initial increment, increasing amounts of Rh B molecules became available for excitation; the amount of Rh B molecules was thus the limiting factor. Upon further increment, two phenomena occur concurrently. First, a maximum rate of excitation was attained, eliciting a saturation of Rh B emission; here the excitation source was the limiting factor. Second, the further increment in Rh B concentration caused a larger proportion of Rh B molecule not being in contact with the SPWs or evanescent waves, amplifying the probability of collisions between excited and non-excited Rh B molecules. This resulted in higher efficient quenching of excited Rh B molecules via non-radiative energy transfer to non-excited Rh B molecules. Hence, the likelihood of actual emission from an excited Rh B molecule is diminished, reducing the overall emission. The collective effect of the two described phenomena, thus, grounds the drop in peak emission intensity [12].

Emission intensities, on the contrary, were distinctively different. Notably, emission intensity was easily influenced by experimental factors, namely the size of the fluorescence droplet as well as the alignment at the butt-coupled input. Experimental data were therefore collected and averaged over three data sets to minimise the influnce of such factors. It is hence evident that the 2 fold higher overall fluorescence emission intensity, for the gold coated D-shaped MOF, was contributed by plasmonic enhancements. In addition, fluorescence excitation at the D-shaped surface with emission collected at the fiber end face was also conducted on the uncoated D-shaped MOF. Peak emission intensity, in this case, was several orders lower.

Furthermore, Rh B concentration for optimum peak emission was higher for the gold coated D-shaped MOF at 1.90 M as compared to 1.61 M. This indicates saturation of exciation at higher Rh B concentrations, implying a higher maximum rate of excitation as provided by the SPWs. A higher efficiency of energy transfer from SPWs to Rh B molecules is again demonstrated.

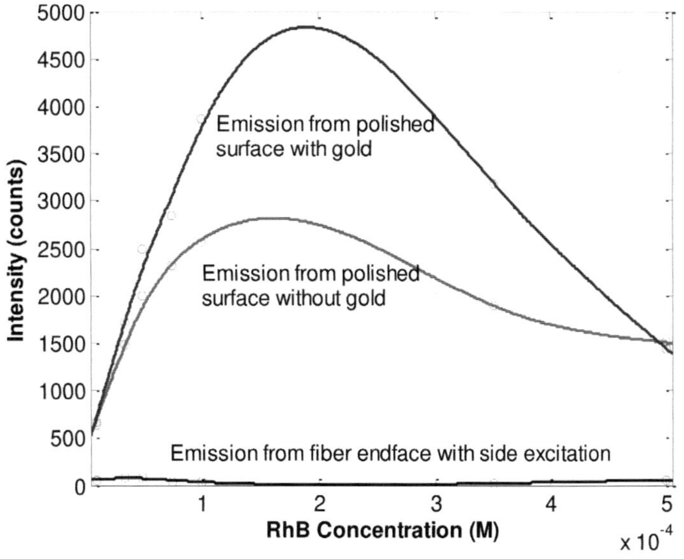


**Figure 5.** Comparison of peak emission intensity with change in Rh B concentration, between gold coated (blue) and uncoated (red) D-shaped MOFs as well as excitation at D-shaped surface (black).

## 4. Conclusion

Plasmonic-enhanced fluorescence spectroscopy has been demonstrated on a gold coated D-shaped MOF. Results highlight the improved fluorescence emission intensity as well as heightened sensitivity in fluorescence spectroscopy due to plasmonic enhancements. Results also indicated higher efficiencies in energy coupling for SPWs to fluorescence molecules as well as showed higher data projection accuracy in the plasmonic-enhanced device.

These exhibit the potential of plasmonic-enhanced D-shaped MOFs as compact illumination platforms with high coupling efficiencies, for applications in improved bio-imaging with superior histogram output.

**References**


[1] J. Homola, *et al.*, "Surface plasmon resonance sensors: review," *Sensors and Actuators B: Chemical,* vol. 54, pp. 3-15, 1999.

[2] J. Homola and R. Slavik, "Fibre-optic sensor based on surface plasmon resonance," *Electronics Letters,* vol. 32, p. 480, 1996.

[3] H.-Y. Lin, *et al.*, "Side-polished multimode fiber biosensor based on surface plasmon resonance with halogen light," *Appl. Opt.,* vol. 46, pp. 800-806, 2007.

[4] R. Slavík, *et al.*, "Single-mode optical fiber surface plasmon resonance sensor," *Sensors and Actuators B: Chemical,* vol. 54, pp. 74-79, 1999.

[5] R. Slavík, *et al.*, "Novel spectral fiber optic sensor based on surface plasmon resonance," *Sensors and Actuators B: Chemical,* vol. 74, pp. 106-111, 2001.

[6] T. P. Burghardt, *et al.*, "In Situ Fluorescent Protein Imaging with Metal Film-Enhanced Total Internal Reflection Microscopy," *Biophysical Journal,* vol. 90, pp. 4662-4671, 2006.

[7] D. Axelrod, "[1] Total internal reflection fluorescence microscopy in cell biology," in *Methods in Enzymology*. vol. Volume 361, M. Gerard and P. Ian, Eds., ed: Academic Press, 2003, pp. 1-33.

[8] T. P. Burghardt, *et al.*, "Around-the-objective total internal reflection fluorescence microscopy," *Appl. Opt.,* vol. 48, pp. 6120-6131, 2009.

[9] B. Agnarsson, *et al.*, "Evanescent-wave fluorescence microscopy using symmetric planar waveguides," *Opt. Express,* vol. 17, pp. 5075-5082, 2009.

[10] X. Yu, *et al.*, "Silica-Based Nanostructure Core Fiber," *Photonics Technology Letters, IEEE,* vol. 20, pp. 162-164, 2008.

[11] R. F. Kubin and A. N. Fletcher, "Fluorescence quantum yields of some rhodamine dyes," *Journal of Luminescence,* vol. 27, pp. 455-462, 1982.

[12] I. D. Johnson and M. W. Davidson. 19/07/2010). *Jablonski Energy Diagram - Java Tutorial.* Available: http://www.olympusmicro.com/primer/java/jablonski/jabintro/index.html

# Improved electroporation parameters of delivering silver nanoparticles into living C666 cells for surface-enhanced Raman scattering

**Yun Yu[1], Juqiang Lin[1], Zufang Huang[1], Gangqin Xi[1], Duo Lin[1], Yongjian Chen[1], Rong Chen[1]* and Haishan Zeng[2]**

1. Key Laboratory of Optoelectronic Science and Technology for Medicine, Ministry of Education, Fujian Normal University, Fuzhou 350007, China

2. Cancer Imaging Department, British Columbia Cancer Research Centre, Vancouver, B.C., V5Z 1L3, Canada

E-mail: chenr@fjnu.edu.cn

**Abstract**. Electroporation assisted metallic nanoparticle delivery has been shown by our previous work to significantly reduce the time of sample preparation for surface-enhanced Raman spectroscopy (SERS) measurements of biological cells. In this research note, we report our experimental work to optimize the electroporation parameters, including adjustment of the pulse pattern and operation temperature, for fastest delivery of silver nanoparticles into living C666 cells (a human nasopharyngeal carcinoma cell line). The delivery efficiency was evaluated by the integrated intensity of whole cell SERS spectrum. Our work concluded that the silver nanoparticle delivery rate is best under the electroporation condition of using 4 consecutive 350 V (875 V/cm) rectangular electric pulses of 1 ms, 10 ms, 10 ms, and 1 ms durations respectively. Low temperature (0~4°C) is necessary for improving the delivery efficiency of silver nanoparticles.

## 1. Introduction

With advantages in detection sensitivity, selectivity, and specificity, surface-enhanced Raman scattering (SERS) spectroscopy is an analytical method with rapid expanding applications in chemical and biochemical analysis and detection. One such new application is intracellular SERS measurement, in which the metallic nanoparticles such as colloidal silver or gold are delivered into living cells to serve as the enhancing agent. At present, "endocytosis" is the dominant technique for delivering metal nanoparticles into living cells[1-3]. Besides this technique, biosynthesis of nanoparticles has also been tried in intracellular SERS study[4-5]. However, both the endocytosis technique and the biosynthesis technique are time-consuming and silver or gold nanoparticles have to be incubated with living cells at 37°C for more than several to twenties hours prior to the SERS experiments. For many biomedical applications, such as cancer screening, keeping the cells alive for such a long time and delivering nanoparticles in such a slow fashion bring extra procedures and increase the cost.

---

* To whom any correspondence should be addressed.

To overcome the disadvantages of the two existing delivery techniques, we have recently developed an electroporation based method that can rapidly deliver silver nanoparticles into living cells within one minute and successfully carried out intracellular SERS measurements[6]. Electroporation involves the application of controlled electrical pulses to induce transient and reversible pores on the cell membrane. For a moment, the cell membrane becomes highly permeable to exogenous substances in the surrounding media[7-8]. Although electroporation is a powerful delivery technique for many molecules[8], silver nanoparticles are not so easy to be delivered into living cells, because, electroporation efficiency is influenced by many factors, such as voltage, pulse width, cell condition, operation temperature, serum concentration in electrical buffer[9-11]. Therefore, the primary objective of this study is to investigate the optimal electroporation parameters for obtaining the greatest delivery efficiency and at the same time ensuring cell viability. This optimization is important for practical applications of electroporation assisted intracellular SERS spectroscopy.

## 2. Materials and Methods

### 2.1. Preparation of silver nanoparticles

Colloidal silver was prepared by the aqueous reduction of silver nitrate with hydroxylamine hydrochloride[12]. First 4.5 mL sodium hydroxide (0.1 mol/L) was added to 5 ml hydroxylamine hydrochloride ($6\times10^{-2}$ mol/L) and then the mixtures were added rapidly to 90 mL silver nitrate ($1.11\times10^{-3}$ mol/L) with vigorous shaking. The silver colloids were kept in dark at room temperate. The absorption maximum of the silver colloid was located at 419 nm, very close to the value from literature of 416 nm[12].

### 2.2. Cell culture

Human nasopharyngeal carcinoma cells (C666 cell line) were obtained from the Fujian Medical University Union Hospital. C666 cells grow in the complete RPMI 1640 medium (supplemented with 100 IU/mL penicillin/streptomycin and 10% fetal calf serum, hereafter referred to as RPMI 1640) at 37°C and 5% $CO_2$. Before electroporation, C666 cells were harvested from monolayer cultures, then the cells were washed three times by phosphate buffered saline (PBS) to remove any residual medium. After the cells were centrifuged, the cells were suspended in electroporation buffer with a density of $10^6$ cells/mL, finally 800 μL cell suspension and 200 μL silver colloids were mixed thoroughly and added into an electrode cuvette. The silver colloid concentration used in this experiment was reduced by 10 times compared to our previous work[6] to avoid the excessive accumulation of silver nanoparticles surrounding the cells, which is hard to wash off.

### 2.3. Stimulation parameters and operation conditions for cell electroporation

Based on our previous study[6], we used a constant voltage of 350 volts (corresponding to an field strength of 875 V/cm) for all the experiments. Other parameters including different series of electronic pulses and operation temperature were varied to identify optimum settings for electroporation delivery of silver nanoparticles into living C666 cells. Thus two optimization experiments were performed as outlined below. (1) Influence of different pulse patterns. The electroporation voltage was kept at 350 V (field strength of 875 V/cm), while four different pulse patterns were tested for best silver nanopartilce delivery efficiency: ① 1 ms pulse width pulses for two times, ② three consecutive pulses with pulse width of 1, 10, and 1 ms respectively, ③ four consecutive pulses with pulse width of 1, 10, 10, and 1 ms respectively, ④ five consecutive pulses with pulse width of 1, 10, 10, 10, and 1 ms respectively. For each experiment, the cell/silver nanoparticle mixture was first added into an electrode cuvette and incubated for 10 min at 4°C on ice and then transferred to an electroporation chamber to carry out the electroporation. The electrical buffer used was the complete RPMI 1640 medium. After each electroporation, the cuvette was incubated again for 10 min at 4°C on ice. We also had a cell/silver nanoparticle mixture without electroporation as the control experiment. After the electroporation procedure, the C666 cells were washed twice with PBS to remove any residual buffer medium. After that, the cells were measured under the Renishaw Micro-Raman spectrometer to obtain the SERS spectrum. In the control experiment, the C666 cells were also washed twice with PBS before the SERS spectral measurements. (2) Influence of operation temperature. For the electroporation experiments with 350 V voltage and four consecutive pulses of pulse width 1, 10, 10, 1 ms and in a complete RPMI 1640 medium buffer, two different operation temperatures were tested: cold (0~4°C on ice) and ambient room temperature (20~25°C).

2.4. SERS measurements

A Renishaw Raman microscope (InVia system) with a ~2 spatial resolution, a 20 mW, 785 nm, semiconductor laser as excitation source, was used for the collection of SERS spectra. The microscope was operated under a 50× objective, which focused the laser beam onto a spot on the sample and the illumination pinhole was adjusted for the laser spot to cover the whole cell. The typical spectral accumulation time in our study was 10 seconds. Raman spectra were scanned over a wavenumber range of 400 ~ 1800 $cm^{-1}$ with respect to the excitation frequency.

## 3. Results and discussion

After electroporation, the cell morphology was observed by 50×microscopy. Figure 1 shows three example images. All the cells treated by pulse pattern ① (1 ms pulse width pulses for two times) and pattern ② (three consecutive pulses with pulse width of 1, 10, and 1 ms) and most of the cells treated by pulse pattern ③ (four consecutive pulses with pulse width of 1, 10, 10, and 1 ms) have normal looking morphology (Figure 1(a)). A significant portion of the cells treated by pattern ④ (five consecutive pulses with pulse width of 1, 10, 10, 10, and 1 ms) are broken (Figure 1(b)); some are fused together (Figure 1(c)). The darker part in Figure 1(b) shows the accumulation silver nanoparticles, probably originated from the silver nanoparticle rushing-in after the cell membrane was irreversibly broken.

We measured the SERS spectra of the control cell and cells at different electroporation treatment experiments. The signal of spectrum measured from the control cell (without electroporation) is very noisy; no reliable Raman peaks are seen. The signal of SERS spectrum measured from individual living C666 cells treated by electric pulse pattern ① (1 ms pulse width pulses for two times) are no reliable Raman peaks; SERS spectrum measured from individual living C666 cells treated by electric pulse pattern ② (three consecutive pulses with pulse width of 1, 10, and 1 ms) are visible, but not good enough S/N. Good quality SERS spectra were obtained from individual living C666 cells treated by electric pulse pattern ③ (four consecutive pulses with pulse width of 1, 10, 10, and 1 ms), containing a wealth of intracellular surface enhanced Raman signals. This is the optimal electroporation parameter for efficient silver nanopartical delivery into living C666 cells. The quality of example SERS spectrum measured from individual living C666 cells (normal looking morphology) treated by electric pulse pattern ④ (five consecutive pulses with pulse width of 1, 10, 10, 10, and 1 ms) is as good as treated by electric pulse pattern ③. However, a significant portion of the cells treated by pattern ④ are broken, making this pulse pattern unusable. Moreover, SERS spectra from the broken cells are different from each other.

The delivery efficiency is evaluated quantitatively by the integrated SERS intensity. For each electroporation experiment, SERS measurements were performed on fifteen cells and two spectra were acquired from each cell by repeated measurements. Therefore, in total thirty SERS spectra were obtained during each experiment. While for every spectrum, we calculated the total integrated area under the curve from 400 $cm^{-1}$ to 1800 $cm^{-1}$. Then the integrated intensity was divided by the area of the cell to generate a standardized "integrated SERS intensity", representing the total amount of silver nanoparticles delivered into the cell. With these data processing applied to the four different electroporation experiments and the control experiment, we obtained a comparison of the integrated SERS intensity. For the control experiment and the electroporation experiment with electric pulse pattern ① (1 ms pulse width pulses for two times), the integrated SERS intensities are minimal, indicating that very little silver nanoparticles were delivered into the cells. Compared with this, the cells treated by the electric pulse pattern ② (three consecutive pulses with pulse width of 1, 10, and 1 ms) achieved considerable integrated SERS intensity. This suggested that considerable amount of silver nanoparticles were delivered into living C666 cells. The cells treated by the electric pulse pattern ③ (four consecutive pulses with pulse width of 1, 10, 10, and 1 ms) showed strong integrated SERS intensity. This result demonstrated that 875 V/cm electric field strength and four pulses of 1, 10, 10, 1 ms width led to a high delivery efficiency of silver nanoparticles into living C666 cells. The integrated SERS intensity of cells treated by the electric pulse pattern ④ (five consecutive pulses with pulse width of 1, 10, 10, 10, and 1 ms) is even higher, however, this is not useful because a significant portion of cells are dead and corresponding Raman signals are very different from the cells that are still alive after the electroporation. In the electric pulse pattern ③ treatment, the silver nanoparticles are effectively delivered into living cells and the cell membrane also recovered its integrity and functions normally, maintained the viability of the cells.

The temperature at the electroporation process might play a role in the efficiency of delivery[13-15]. The electroporation procedure was performed at (0~4ºC) or at room temperature (20~25ºC) to determine the optimal electroporation temperature with the same electrical parameters (875 V/cm field strength and four pulses of 1, 10, 10, 1 ms pulse width). We also used the integrated SERS intensity to compare the efficiency of silver nanoparticles delivery in these two experiments. Compared with low operation temperature, the integrated SERS intensity of cell electroporation in room temperature decreased dramatically. One explanation for this result is that keeping the cells at low temperature allows the electroporation-induced pores on the membrane to remain open longer and more silver nanoparticles in the medium will get into the cell. Therefore, it is importance to maintain the cells at low temperature condition for the electroporation operation for optimal silver nanoparticle delivery efficiency.

## 4. Conclusion

In conclusion, we optimized the electroporation parameters for delivering silver nanoparticles into living C666 cells for intracellular SERS spectroscopy. The optimum delivery is achieved under the electroporation condition of using 4 consecutive 350 V (875 V/cm) rectangular electric pulses of 1 ms, 10 ms, 10 ms, and 1 ms durations respectively. Low temperature (0~4°C) is necessary for improving the delivery efficiency of silver nanoparticles. With the optimized electroperation procedure, intracellular SERS analysis can be performed within minutes, offering great promises for biomedical applications such as high-throughput cancer cell screening. However, there are still several questions to be solved and further study on electroporation for intracellular SERS spectroscopy measurements will be given in following research note.

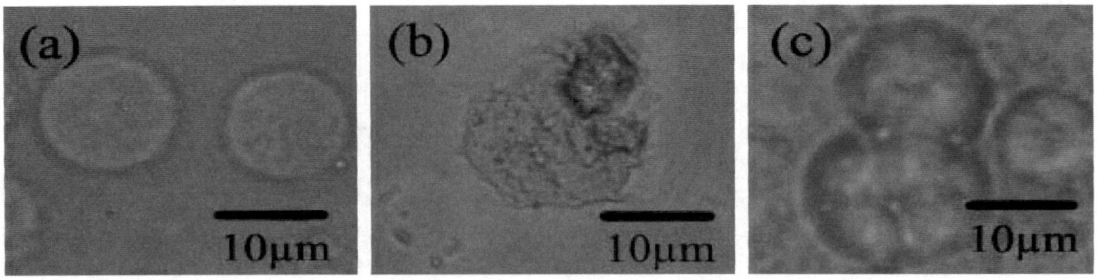


**Figure 1**. Optical microscopic image of C666 cells after electroporation. (a) normal looking cell; (b) broken cell; (c) cell fusion

## 5. Acknowledgements

This work was supported by the National Natural Science Foundation of China (Nos. 60778046 and 60910106016), the Science and Technology Project of Fujian Province (Nos. 2008J0016), the Project of Science Foundation of Ministry of Health and United Fujian Provincial Health and Education Project for Tackling the Key Research (No.WKJ2008-2-046).

**References**

[1] Kneipp K, Haka A S, Kneipp H, Badizadegan K, Yoshizawa N, Boone C, Shafer-Peltier K E, Motz J T, Dasari R R and Feld M S 2002 Surface-Enhanced Raman Spectroscopy in Single Living Cells Using Gold Nanoparticles *Appl. Spectrosc.* **56** 150-4

[2] Kneipp J, Kneipp H, Rice W L and Kneipp K 2005 Optical Probes for Biological Applications Based on Surface-Enhanced Raman Scattering from Indocyanine Green on Gold Nanoparticles *Analytical Chemistry.* **77** 2381-5

[3] Chourpa I, Lei F H, Dubois P, Manfait M and Sockalingum G D 2008 Intracellular applications of analytical SERS spectroscopy and multispectral imaging *Chemical Society Reviews.* **37** 993-1000

[4] Anshup, Venkataraman J S, Subramaniam C, Kumar R R, Priya S, Kumar T R S, Omkumar R V, John A and Pradeep T 2005 Growth of gold nanoparticles in human cells *Langmuir.* **21** 11562-7

[5] Shamsaie A, Jonczyk M, Sturgis J, Robinson J P and Irudayaraj J 2007 Intracellularly grown gold nanoparticles as potential surface-enhanced Raman scattering probes *Journal of Biomedical Optics.* **12** 020502

[6] Lin J, Chen R, Feng S, Li Y, Huang Z, Xie S, Yu Y, Cheng M and Zeng H 2009 Rapid delivery of silver nanoparticles into living cells by electroporation for surface-enhanced Raman spectroscopy *Biosens Bioelectron.* **25** 388-94

[7] Heiser W C 2000 Optimizing electroporation conditions for the transformation of mammalian cells *Methods Mol Biol.* **130** 117-34

[8] Kandušer M and Miklav i D 2009 Electrotechnologies for Extraction from Food Plants and Biomaterials *Electroporation in Biological Cell and Tissue: An Overview (Food Engineering Series)* ed G V Barbosa-Cánovas (New York: Springer) pp 1-37

[9] Hui S W 2008 Electroporation Protocols *Overview of Drug Delivery and Alternative Methods to Electroporation (Methods in Molecular Biology* Vol 423*)* ed J M Walker (Humana Press) pp 91-107

[10] Hui S W 1995 Animal Cell Electroporation and Electrofusion Protocols *Effects of Pulse Length and Strength on Electroporation Efficiency (Methods in Molecular Biology* Vol 48*)* ed J M Walker (Humana Press) pp 29-40

[11] Gehl J 2003 Electroporation: theory and methods, perspectives for drug delivery, gene therapy and research *Acta Physiol Scand.* **177** 437-47

[12] Leopold N and Lendl B 2003 A New Method for Fast Preparation of Highly Surface-Enhanced Raman Scattering (SERS) Active Silver Colloids at Room Temperature by Reduction of Silver Nitrate with Hydroxylamine Hydrochloride *The Journal of Physical Chemistry B.* **107** 5723-7

[13] Canatella P J, Karr J F, Petros J A and Prausnitz M R 2001 Quantitative Study of Electroporation-Mediated Molecular Uptake and Cell Viability *Biophysical Journal.* **80** 755-64

[14] Davalos R V, Rubinsky B and Mir L M 2003 Theoretical analysis of the thermal effects during in vivo tissue electroporation *Bioelectrochemistry.* **61** 99-107

[15] Davalos R V and Rubinsky B 2008 Temperature considerations during irreversible electroporation *International Journal of Heat and Mass Transfer.* **51** 5617-22

# Second harmonic generation imaging of dermal collagen component in human keloid tissue

**H B Yu*, S Chen*, X Q Zhu, H Q Yang and J X Chen**
Institute of Laser and Optoelectronics Technology, Fujian Provincial Key Laboratory for Photonics Technology, Key Laboratory of OptoElectronic Science and Technology for Medicine of Ministry of Education, Fujian Normal University, Fuzhou 350007, P. R. China

E-mail: chenjianxin@fjnu.edu.cn

**Abstract.** In this paper, we report second harmonic generation (SHG) imaging of human keloid tissue. High resolution SHG images of collagen component were obtained in the superficial, medial and deep dermis of human keloid tissue, respectively. Our results show that this method has a capability to observe the structure of collagen component in human keloid tissue, which will help to better understand the formation process of human keloid scar at the molecular level.

## 1. Introduction

The excessive collagen deposition and broadened bundles were prominently observed in keloid skin, and different depths of dermis were visualized by difference of image in keloid [1]. Second harmonic generation microscopy based on SHG signals is a nonlinear optical microscopy, which allows to image microstructure of the noncentrosymmetric molecular organization such as collagen and is used to monitor changes associated with disease progression [2]. In this study, the morphological variation of collagen among the whole dermis were obtained using SHG microscopy in order to better understanding the formation process of human keloid scar and a parameter was developed to quantify these differences.

## 2. Materials and Methods

Keloid specimen was obtained from a Chinese female patient (age 36 years) undergoing reconstructive surgical procedures. Immediately after being excised from patient, tissue sample was snap-frozen and

* Equally contributed to this work

kept in liquid nitrogen (–196°C) until use. The tissue sections for SHG microscopic imaging were cut into 120 μm thickness and sandwiched between the microscope slide and cover glass. The sample was imaged with the cover glass facing the microscopic objective.

SHG microscope used in this study has been described previously [3]. The excitation light source is a Ti: sapphire laser (Coherent Mira 900-F). The excitation wavelength of 810nm was used in this work. A commercial inverted microscope (Zeiss LSM 510 META) was used to image. The backward SHG signals were detected with the META detector. In this study, all images have a 12-bit pixel depth. The images were obtained at 2.56 μs per pixel. The META detector has eight independent-channels and each channel can selectively be set to detect emission signals within the random range from 377 to 716 nm to achieve imaging. In this work, a channel was selected to obtain the images of collagen with the wavelength range of 398 - 409 nm, when the excitation wavelength $\lambda_{ex}$ = 810 nm were used. The Fast Fourier Transform (FFT) of the SHG images was implemented using ImageJ (http://rsb.info.nih.gov/ij). The FFT method can calculate the collagen orientation index (COI) which is defined as

$$COI = 1 - \frac{W}{H} \qquad (1)$$

where W and H represent the width and height of the ellipse or circle in the power plots of SHG images.

**3. Results and Discussion**

Figure 1 displays the 2-D SHG image of collagen from keloid specimen, which was used to characterize the microstructural differences of collagen in the dermis. Figure 1(a) shows the microstructure of collagen in the superficial dermis of keloids. In superficial dermis, the collagen bundles are organized in a parallel manner. On the contrary, in the medial dermis (shown in Fig. 1(b)), the collagen bundles are randomly oriented and loosely arranged in the pattern of fine mesh. Compared with the others, Fig. 1(c) indicates that most of the collagen bundles are largely oriented in the same direction. Furthermore, a part of collagen bundles are disrupted (marked in the Fig. 1(c)), which is consistent with the histological result [4-6].

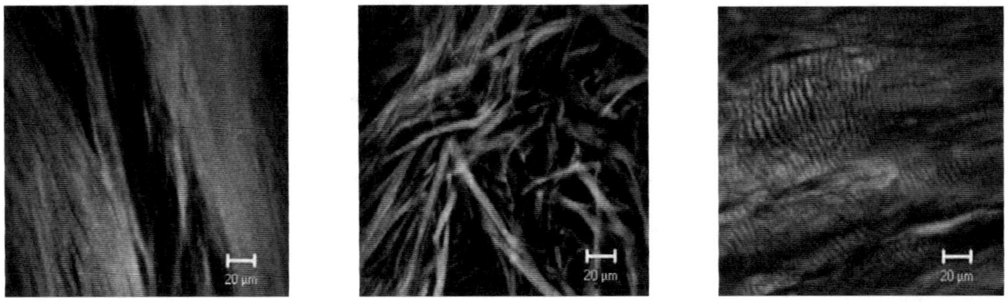


(a) (b) (c)

**Figure 1.** SHG images of collagen in (a) the superficial dermis (b) the medial dermis (c) the deep dermis of keloid originated from a patient aged 36 years olds

To better quantify the orientation of the collagen bundles, the FFT of collagen SHG images was implemented. According to the definition (1), the COI of perfectly randomly oriented collagen bundles is 0. The Fourier analysis of an image with parallel collagen orientation yields an elongated power spectrum that leads to a higher orientation index (maximum is 1). The power plots showed significant differences within the whole dermis of the keloid. The circle in Fig. 2(b) indicated collagen bundles arranged randomly in medial dermis whereas the ellipse in Fig. 2(a), 2(c) indicated a higher orientation of collagen bundles in superficial and deep dermis. In order to better differentiate the whole dermis of the keloid, The COI was developed, which was defined in (1). In our measurements, five square areas with 512×512 pixels are selected in each specimen to calculate the index. The results showed that the COI is (0.7880±0.0317) (the level superficial), (0.2739±0.0318) (the level medial) and (0.7467±0.0126) (the deep), respectively. The distance (D) between the centers of gravity of the first order maxima is used to calculate the distance between the centers of the collagen bundles (λ) by using the formula [7]:

$$\lambda = 794\,\mu m \times \frac{1}{0.5 \times D} \tag{2}$$

The distance λ is called the bundle distance (BD). BD depends on the collagen bundle thickness, the collagen bundle density and the space between the adjacent collagen bundles. In order to obtain D, five square areas with 512x512 pixels are selected in the specimen. According to the formula (2), the BD is (24.58±18.89) in the superficial dermis and (40.65±25.21) in the deep dermis.

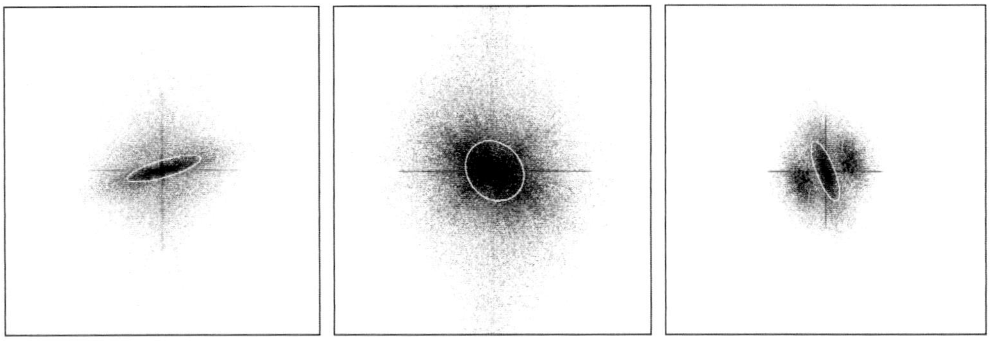

(a) (b) (c)

**Figure 2.** The representative power plot of Fourier Transform (FFT) in (a) the superficial dermis of keloid (b) the medial dermis of keloid (c) the deep dermis of keloid

### 4. Conclusion

In conclusion, our results demonstrated that SHG microscopy has ability to discriminate the whole dermis of keloid based on the SHG images of collagen. The developed parameter COI can quantitatively identify the three parts of dermis in keloid. With the advent of clinical portable SHG microscope, this technique has the potential to be a powerful tool for diagnosing keloid and assessing the response of the therapy *in vivo*.


### Acknowledgements

The work was supported by the Natural Science Funds for Distinguished Young Scholar in Fujian Province (2009J06031), and the Research Fund for the Fujian Province University (JK2009005).


### References

[1] Atiyeh B S, Costagliola M and Hayek S N 2005 Keloid or Hypertrophic Scar *Ann. Plas. Surg.* **54** 676-80

[2] Zipfel W R, Williams R M, Christie R, Nikitin A Y, Hyman B T, Webb W W 2003 Live tissue intrinsic emission microscopy using multiphoton-excited native fluorescence and second harmonic generation *Proc. Natl. Acad. Sci. USA.* **100** 7075–80

[3] Zhuo S M, Chen J X, Luo T S and Zou D S 2006 Multimode nonlinear optical imaging of the dermis in ex vivo human skin based on the combination of multichannel mode and Lambda mode *Opt. Express* **14** 7810–20

[4] Ehrlich H P, Desmoulière A, Diegelmann R F, Cohen IK, Compton C C, Garner W L, Kapanci Y and Gabbiani G 1994 Morphological and immunochemical differences between keloid and hypertrophic scar *Am. J. Pathol.* **145** 105–13

[5] Lee J Y, Yang C C, Chao S C and Wong T W 2004 Histopathological differential diagnosis of keloid and hypertrophic scar *Am. J. Dermatopathol.* **26** 379–84

[6] Santucci M, Borgognoni L, Reali U M and Gabbiani G 2001 Keloids and hypertrophic scars of Caucasians show distinctive morphologic and immunophenotypic profiles *Virchows Arch.* **438** 457–63

[7] Verhaegen P D, van Zuijlen P P, Pennings N M, van Marle J, Niessen F B, van der Horst C M and Middelkoop E 2009 Differences in collagen architecture between keloid,hypertrophic scar, normotrophic scar, and normal skin: An objective histopathological analysis *Wound Rep. Reg.* **17** 649–56

# *In vivo* photoacoustic imaging of osteosarcoma on animal model

**Menglei Yu[1], Fei Ye[2], and Jun Hu[1*]**

[1] Department of Orthopaedics, The First Affiliated Hospital, Shantou University Medical College, Shantou 515041 , China

[2] MOE Key Laboratory of Laser Life Science & Institute of Laser Life Science, College of Biophotonics, South China Normal University, Guangzhou 510631, China

Email: hjzkm@yahoo.com.cn

**Abstract.** Osteosarcoma is the commonest primary malignant tumor of bone, and the second highest cause of cancer-related death in the paediatric age group. Although there are several methods for osteosarcoma detection, e.g. X-ray, CT, MRI and bone scan, they are not satisfied methods because they can hardly detect osteosarcoma in early stage. Photoacoustic imaging (PAI) is an emerging hybrid imaging modality that is noninvasive, nonionizing, with high sensitivity, satisfactory imaging depth and good temporal and spatial resolution. In order to explore this new method to detect osteosarcoma, we established SD rat models with osteosarcoma and utilized PAI to reconstruct the osteosarcoma image *in vivo*. This is the first time detecting osteosarcoma *in vivo* using PAI, and the results suggested that PAI has potential clinical application for detecting osteosarcoma in the early stage.

## 1. Introduction

Osteosarcoma is the prominent primary bone cancer in humans, excluding hemopoietic malignancies. It is usually highly aggressive and eventually lethal, and mainly affects children and adolescents. More than 80% of these tumors tend to occur in the long bones of the appendicular skeleton which are undergoing rapid growth[1]. The age at presentation ranges from 10 to 25 years. Several years ago, all patients with osteosarcoma were treated by amputation but the cure rate was under 10% and almost all patients died within a year from diagnosis. Today, for localized osteosarcoma at onset (80% of cases) treated in specialized bone tumor centres with pre- and postoperative chemotherapy associated with surgery, the percentage of patients cured varies between 60% and 70%. However, in patients with metastasis at onset cure rate is about 30% [2]. Obviously, this great therapeutic effect progress is based on the confirmed diagnosis in the early stage and the tumor is localized osteosarcoma which has not ever migrated to other organs. Early detection and treatment are critical for managing patients with osteosarcoma. Although there are several methods for osteosarcoma detection in clinic, e.g. plain radiographs, computed tomography, magnetic resonance imaging, angiography, dynamic bone scintigraphy and, if necessary, the type of reconstruction, there is not a satisfied enough method can detect osteosarcoma in the early stage [2-5]. Therefore, it is very necessary to find a new way to detect osteosarcoma, especially at the early stage.

PAI, also referred as photoacoustic tomography (PAT) or optoacoustic tomography, is a new imaging technology. It is a potential candidate for osteosarcoma detection because this hybrid technique combines the high intrinsic optical contrast with the high spatial ultrasound resolution. What's more, it is non-invasive and non-destructive and does not rely on the use of ionizing radiation [6]. PAI is based on the detection of acoustic waves generated by absorption of pulsed light in tissue. Owing to the initial ultrasonic pressure is proportional to the local optical energy deposition in the tissue, the induced acoustic signals exhibit heterogeneity of optical absorption in tissue and the PAI reveals the optical absorption distribution of tissue. PAI has been applied to in vivo imaging of vasculature in small animals and humans[7-11], and early cancer detection[12-14]. The PA technique has also been used for non-invasive monitoring of blood oxygenation (especially cerebral oxygenation), cerebrovascular activities in small animals [15-17] and monitoring of skin abnormalities [18].

In this study, osteosarcoma models in rats was established. The rat leg was imaged in vivo with PAI. As an essential step toward our ultimate goal, the feasibility of PAI in imaging osteosarcoma in vivo was verified through the study on rats osteosarcoma models.

## 2. Materials and methods

### *2.1 Establishment of osteosarcoma rat model*

Male Sprague-Dawley rat, 4 weeks old, was purchased from the laboratory animal center of Sun Yat-sen University. The rat was maintained in microisolator cage under specific pathogen-free conditions, in accordance with the institutional guidelines of the South China Normal University Ethics Committee and under the supervision of authorized investigators. The UMR 106 osteosarcoma cell line syngeneic to SD rat was purchased from the American Type Culture Collection (ATCC; Manassas, VA) and cultured in RPMI-10% FBS (Sigma) supplemented with L-glutamine (2 mM), penicillin (100 U/ml) and streptomycin (100 μg/ml) confirmed to be mycoplasma-free by routine testing. The number of the cells were adjusted to about $2\times10^7$ /ml and the rats were anesthetized intraperitoneally with 10% chloral hydrate at a dosage of 0.1 ml/30 g live weight, then the cells were injected directly into the proximal part of the tibia shaft surface of anesthetized SD rat with a syringe and 25 gauge needle. After implantation, the rat was returned to cage and fed for 15 days as before. The rat was treated with cyclosporin A (10 mg/kg) every day until one week after the implantation and the image reconstruction was performed 15 days following the methods as described in detail below [19,20].

### *2.2 PAI principle and image reconstruction*

PA signals are caused by optical energy deposition in target tissue. The strength of the PA signal is proportional to the local absorbed energy density. When the laser pulse is short enough, thermal diffusion can be ignored, and the resulting PA pressure $p(r,t)$ that reaches a detector at position $r$ and time t can be expressed as [21,22] ,

$$p(r,t) = \frac{I_0\beta}{4\pi C_p}\iiint \frac{d^3(r')}{|r-r'|}A(r')T(t')$$

Where $\beta$ is the isobaric volume expansion coefficient, $C_p$ is the specific heat, $I_0$ is a factor proportional to the incident optical energy density, $T(t')$ denotes the temporal profile of the light with units of 1/s, which can be regarded as a Dirac delta function δ(t) for a short laser pulse, and $A(r')$ is the absorbed optical energy deposition per unit volume of soft tissue at position $r'$.

The PA signal propagates from its originating region can be detected by an ultrasonic transducer or an array detector. Through a reconstruction algorithm, the distribution of absorbed optical energy $A(r')$ in a target tissue can be reconstructed from a set of measured PA signals $p(r,t)$. In the current study, a modified filtered back-projection algorithm developed by our early work is used to reconstruct the distribution of absorbed optical energy density[23].

*2.3 Experimental setup*

A PAI prototype system was employed in this study, as shown in Fig. 1 A Q-Switched Nd: YAG laser (LOTIS TII Ltd, Minsk, Belarus) is used as excitation source, which operates at 1064 nm with pulse duration of 10 ns and pulse repetition rate of 15 Hz The laser beam was expanded by a concave lens and then illuminated into the rat leg (diameter of the beam is approximately 1.3 cm). The incident energy density on the leg surface was set below 20 mJ/cm$^2$. PA signals are detected by a 64-element linear transducer array. Each element of the array has a width of 0.3 mm, a height of 4 mm, and a center frequency of 7.5 MHz with a 70% bandwidth. A custom-built water tank coupled the PA signals between the transducer array and the mouse leg. The mouse was fixed in a box, and the box was controlled by a precise stepper motor to scan circularly around its axial in the horizontal plane X–Y plane. To cover a 360° receiving angle for imaging the mouse leg, a total of 20 steps with a constant 18° interval were taken. The PA signals acquired from the transducer array were transmitted by a custom 64-ch acquisition system to the personal computer, and then processed by the reconstructed program. The ultrasound speed was assumed to be exactly 1500 m/s for every PA reconstructed image.

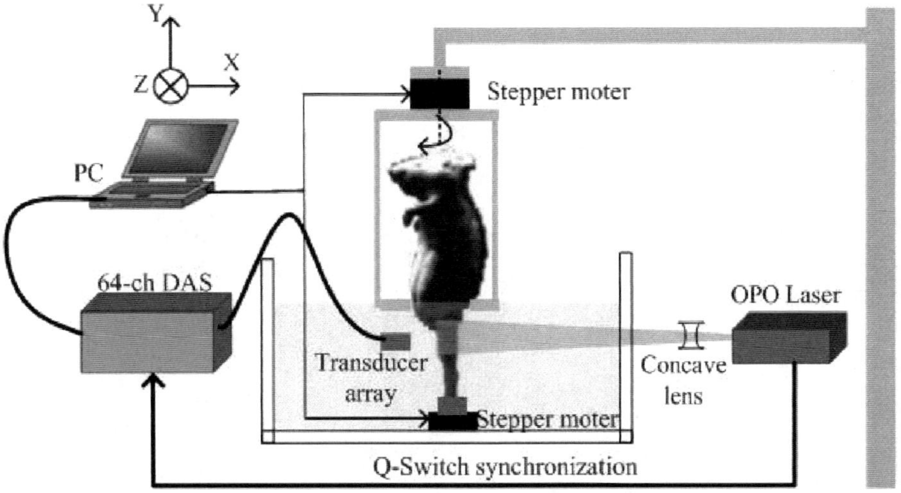


**Figure 1.** Schematic of the PAI system.

*2.4 PAI of rat leg in vivo*

The rat was anesthetized using the same method as previously mentioned and fixed on the experiment instrument as shown in the Figure 1. The PAI image of horizontal cross section of rat leg with osteosarcoma, at the proximal end of tibia, was reconstructed using the same PAI methods at 15 days after the rat was injected the osteosarcoma cells.

*2.5 Histopathological examination*

After the PAI image was reconstructed, the rat leg with tumor was cut down and saved in 10% buffered formalin for 2 days. The leg was then decalcified with formic acid for 4 days and monitored with a Faxitron MX-20 x-ray machine. Once decalcification was completed, the specimen was dehydrated with graded alcohol, embedded in paraffin, cut into blocks, and sectioned to 7-μm thickness with a Reichert-Jung 20/30 metal knife. Hematoxylin and eosin staining of specimen sections on glassslides was conducted. Last, histological pictures of specimen section was taken with a 10×magnification.

## 3. Results

*3.1 PAI of sick rat leg in vivo*

The PAI of the sick rat leg 15 days after injected UMR-106 osteosarcoma cells, is shown in Fig. 2(a). Compared with its transverse section anatomical photograph Figure 2(b), the tumor was unequivocally detected with PAI and matched well both in position and shape. The images suggest that PAI not only can detect the osteosarcoma, but also can distinguish the range of the tumor. The images from PAI provide the information on tumor changes in addition to osteosarcoma tissue in the leg.

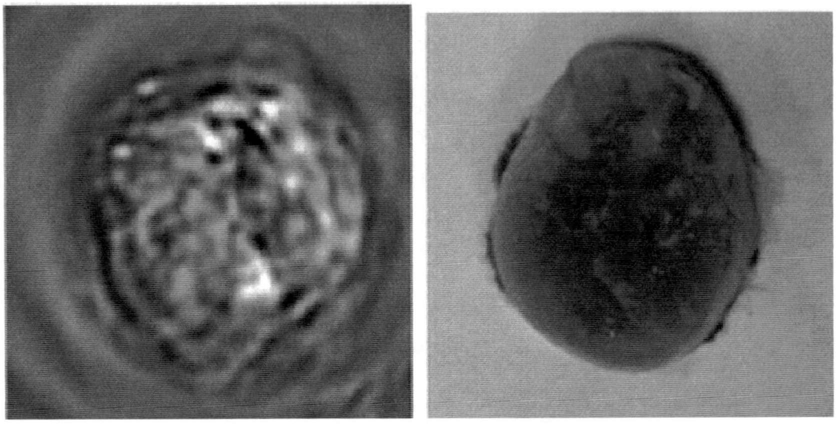

**Figure 2.** Noninvasive PAI images of sick rat leg in vivo. (a) PAI image on 15 days after injected tumor cells; (b) Histological photograph of a similar cross section in the sick rat leg on 15 days after PAI was finished.

*3.2 Histopathological evaluation*

The rat was submitted to necropsy after experiment as described before. (Fig.2) Cut the tissue corresponded the region with increased optical absorption. The specimen was analysis by routine pathology and definited the tissue was osteosarcoma. (Figure 3)

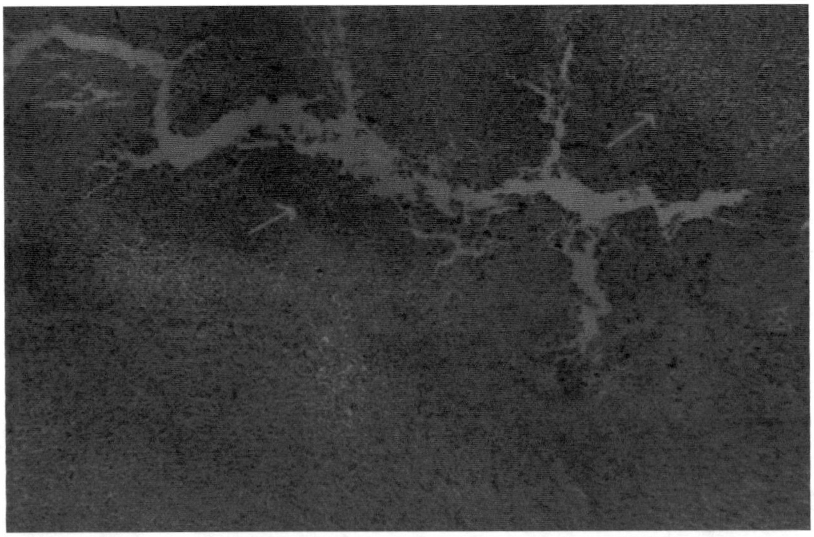

**Figure 3.** Histopathological test with hematoxylin-eosin staining demonstrated an implanted osteosarcoma ( × 100). Implanted tumor was characterized by fusiform or triangle cells (arrow) along nutrient vessels with nuclear pleomorphism, mitoses, bone invasion and necrosis.

## 4.Discussion and conclusion

In the study, the PA image of horizontal cross section of the sick rat leg was achieved in vivo, and the feasibility of osteosarcoma detection with PAI technique has been demonstrated. With the high resolution, PAI can be expanded to help definite the tumor site, evaluate the stage and the developmental trend of the tumor by series of PA images. Therefore, PAI provides a new method for early detection of osteosarcoma.

In our study, the diameter of the rat leg is about 12 mm, and its cross section can be visualized. According to the experiments reported, 30 mm depth PA image can be achieved [24]. While for the osteosarcoma diagnosis on human, 30 mm is not deep enough for most patients. To get a deeper imaging depth, the contrast should be employed in the imaging. With the development of NIR contrast, a new deeper imaging depth can be reached.

To get a full view data, the PA signals of 50 steps with each step of 3.6° were recorded, the scanning time is about 15 minutes. To get a volume data, more time must be expensed, which will hinder the PAI application in osteosarcoma diagnosis in clinic. In the system, the data acquisition time was limited by the scanning process and the repetition of laser. Therefore, transducer plane should be developed for reducing the scanning time. With the development of high repetition of the laser, the data acquisition time can be reduced greatly.

Though some details of cross section of the rat leg can be visualized in the experiment, the resolution is still not good enough for early detection of osteosarcoma. The resolution of the system is limited to the bandwidth of the transducer array. To get a higher resolution, wider bandwidth transducer array should be considered to employ.

In summary, we have first used the noninvasive PAI technology in osteosarcoma diagnosis for imaging osteosarcoma at the early stage and monitoring the osteosarcoma developing. The in vivo imaging results using a rat model suggest that PAI have the feasibility for mapping osteosarcoma and monitoring osteosarcoma developing. Therefore, the PAI technology has the potential to be applied in osteosarcoma diagnosis, especially at the early stage, and the osteosarcoma patients will have a better therapeutic effect with help of PAI technology.

## 5.Acknowledgments

This research is supported by the National Basic Research Program of China (2010CB732602), the Program for Changjiang Scholars and Innovative Research Team in University (IRT0829), the National Natural Science Foundation of China (30627003;30870676;30570513), and the Natural Science Foundation of Guangdong Province (7117865;7008220).

## References

[1] Dorfman H D and Czerniak B 1995 *Cancer* **75** 203–210
[2] Piero Picci 2007 Eur. J. Cancer **5** 227-234
[3] Kurth A A and Müller R 2001 Skeletal Radiol. **30** 94-98
[4] Dickson G R, Geddis C and Fazzalari N 2006 J. Orthop. Res. **26** 729-736
[5] Chen F, Sun X and Keyzer F 2006 Radiology **239** 554-562

[6] Wang Y, Xing D and Zeng Y G 2004 Phys. Med. Biol. **49** 3117-3124
[7] Wang X D, Pang Y J and Ku G 2003 Opt. Lett. **28** 1739–1741
[8] Yang D W, Xing D and Gu H M. 2005 Appl. Phys. Lett. **87** 194101
[9] Yang D W, Xing D and Tan Y 2006 Appl. Phys. Lett. **88** 174101
[10] Zhang H F, Maslov K and Stoica G 2006 Nat. Biotechnol. **24** 848–851
[11] Kolkman R, Steenbergen W and van Leeuwen T G 2003 Lasers Med. Sci. **21** 134–139
[12] Siphanto R I, Thumma K K and van Leewen T G 2005 Opt. Express **13** 89–95
[13] Oraevsky A A, Savateeva E V and Solomatin S V 2002 Proc. SPIE **4618** 81–94
[14] Ermilov S A, Coniusteau A and Mehta K 2006 Proc. SPIE **6086** 608609
[15] Zhang H F, Maslov K and Sivaramakrishnan M 2007 Appl. Phys. Lett. **90** 053901
[16] Yang S H, Xing D and Lao Y Q 2007 Appl. Phys. Lett. **90** 243902
[17] Yang S H, Xing D and Zhou Q 2007 Med. Phys. **34** 294–3302
[18] Yamazaki M, Sato S and Ashida H 2005 J. Biomed. Opt. **10** 064011
[19] Cherrier B, Gouin F and Heymann M F 2005 Tumour Biol. **26** 121–130
[20] Yu Z, Sun H H and Fan Q Y 2009 Oncol. Rep. **21** 1175–1180
[21] Xu M H and Wang L V 2006 Rev. Sci. Instrum. **77** 041101–041123
[22] Kruger R A, Kiser W L and Miller K D 2003 Mol. Imag. **2** 113–123
[23] Wang Y, Xing D and Chen Q 2004 Phys. Med. Biol. **49** 3117–3124
[24] Kim C, Favazza C and Wang L V 2010 Chem. Rev. **110** 2756–2782

# Numerical Simulation of the Micro-explosion during Ho:YAG laser lithotripsy

**Yao Yucheng[1,2], Huang Chuyun[1], Xu Guowang[1], Yan Xudong[1], Wang Yanlin[1]**

[1]Physics school, Hubei University of Technology, HuBei Wuhan
[2]Institute of Optoelectronic Science and Engineering, HUST, HuBei Wuhan

E-mail: yyuch18@sohu.com

**Abstract:** The micro-explosion during Ho:YAG laser lithotripsy may cause calculus fragmentation and migration. It plays an important role to the surgery. A numerical simulation of the micro-explosion during Ho:YAG laser lithotripsy has been developed. The explosion problem in water environment was solved by the Euler algorithm and the piecewise parabolic method (PPM) was selected in the calculation. This simulation investigated the explosion dynamics evolution in the lithotripsy area. The pressure and intensity of the calculus surface were calculated for different laser pulse energy and different distance between calculus and fiber tip. The calculation results indicate that the micro-explosion's properties are determined by the pulse energy, pulse duration and the water distance. Though Short pulse duration and large pulse energy cause high ablation efficiency, it mains more calculus retropulsion at the same time. The ideal surgery results need property laser parameters.

**Key words:** laser lithotripsy, numerical simulation, micro-explosion, Ho:YAG laser

## 1. Introduction

Laser lithotripsy is considered to be a most convenient, economical and less painful technique for destruction of calculus formed in human body. The Ho:YAG pulse laser energy is delivered to the calculus by a thin fiber and ablated it in water environment[1, 2]. The water between calculus and fiber tip high absorbs the pulse infrared laser energy and induced micro-explosion threshold[3]. The micro-explosion dynamics may give help to the ablation ability and make calculus migrates away from the laser delivery fiber tip on one hand, on the other had it cause the calculus retropulsion during laser lithotripsy. This retropulsive calculus movement will prolong operative time and affect the surgery precision, so the investigation of micro-explosion evolution in the lithotripsy is important to the operation. The micro-explosion evolution was simulated for different laser pulse energy and different distance between calculus and fiber tip here. The pressure and intensity of the calculus were calculated and the retropulsion would be predicted. It will give help to the operation.

## 2. Theoretical model

The pulse laser is delivered to calculus by a thin fiber and had a small spot during laser lithotripsy. The micro-explosion effect area on the calculus is very small[2], so the calculus surface will be assumed to be plan and the laser beam irradiates perpendicularly in our numerical model. As shown in fig.1, we will investigate micro-explosion process from the axis-symmetric two-dimensional (2D) model.

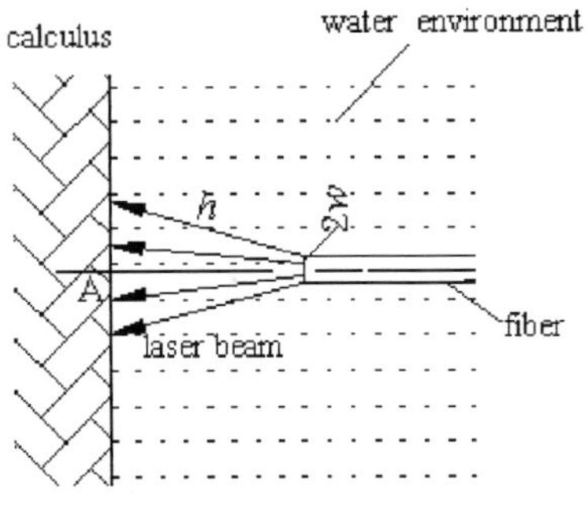


Figure1. Scheme of the numerical model

The experimental investigations indicate that the explosion process of water for free infrared laser irradiation is sudden vaporization and expansion[4,5]. No plasma is observed in the explosion process, so the explosion may be considered from thermodynamic viewpoint[6-9]. The explosion dynamics problem may be solved from the model with the fluid equations for the conservation of mass, momentum and energy. The Euler algorithm is applied to simulate the explosion problem in the calculation field. The piecewise parabolic method (PPM) is selected in the calculation[10, 11].

During the explosion the water absorbs great energy suddenly and transits to vapor rapidly. The heat conduction can be ignored because of the explosion duration is so short and the water heat conductivity is very small relatively. The inviscid fluid 2D Euler equations (continuity, momentum and energy) for axis-symmetric geometries are as follows:

$$\frac{\partial \rho}{\partial t}+\frac{\partial}{\partial z}(\rho u)+\frac{1}{r}\frac{\partial}{\partial r}(r\rho v)=0 \tag{1}$$

$$\rho(\frac{\partial u}{\partial t}+u\frac{\partial u}{\partial z}+v\frac{\partial u}{\partial r})=-\frac{\partial p}{\partial z} \tag{2}$$

$$\rho(\frac{\partial v}{\partial t}+u\frac{\partial v}{\partial z}+v\frac{\partial v}{\partial r})=-\frac{\partial p}{\partial r} \tag{3}$$

$$\rho(\frac{\partial e}{\partial t}+u\frac{\partial e}{\partial z}+v\frac{\partial e}{\partial r})=-p[\frac{\partial u}{\partial z}+\frac{1}{r}\frac{\partial}{\partial r}(rv)]+S \tag{4}$$

Where the $u$ and $v$ is the velocity vector in respective axial (z) and radial (r) co-ordinates; $\rho$ is the mass density; $p$ is the pressure and $e$ is the specific internal energy. The source term, $S$, represents the laser energy absorbed by water.

In this study the reflection and the influence to the absorption by the water temperature and phase transition is not been considered. According to the Lambert-Beer theory, the intensity of the laser beam inside water at every instant is given by:[12, 13]

$$I(r,z)=I_0\exp(-k_0\rho z)\exp(-2r^2/w^2) \tag{5}$$

Where $z$ is the laser transmit depth in the water; $I_0$ is the radiation intensity of the central point on the fiber tip. It may be calculated from the pulse energy and pulse duration; $k_0$ is the absorption constant of water and be calculated from the absorption coefficient listed in table 1[14], $w$ is the beam waist and $r$ is the radial distance from the center of the laser spot. It should be noted that the intensity of the laser beam is constant over time. The local heat deposition, S, per unit area and time over a slice of material with thickness $z$ is given by:

$$S(r,z) = -\frac{\partial I(r,z)}{\partial z} = k_0 \rho I(r,z) \qquad (6)$$

The absorption of radiation by the water was simulated by generating heat in certain elements. The heat generating value was considered constant over each element and calculated according to Eqs.(5,6) by taking the coordinates of one of the element's nodes.

Table1. laser absorption coefficient of water

| Laser | Laser Wavelength ($\mu m$) | Absorption Coefficient of $H_2O$ ($cm^{-1}$) |
|---|---|---|
| Ho:YAG | 2.1 | 25 |

The system of Eqs.(1-4) with definitions by Eqs.(5,6) is closed by means of an equation of state (EOS). During the explosion process the water will be in liquid, vapor or liquid vapor coexisting phase. Precisely describing the water thermodynamic properties during the phase evolution needs water thermodynamic state line or water thermodynamic state database. In this study the water thermodynamic properties will be simply described by bender equation.[15,16]

$$p = \rho T[R + B\rho + C\rho^2 + D\rho^3 + E\rho^4 + F\rho^5 + (G + H\rho^2)\rho^2 \exp(-a_{20}\rho^2)] \qquad (7)$$

Where $T$ is the water temperature; $R$ is the constants. $B$, $C$, $D$, $E$, $F$, $G$ and $H$ is the function of $T$ and $\rho$. The expression of them and the parameters $a_1$--$a_{20}$ for water have been given in the literature[15]. The water temperature $T$ is given by[17]:

$$e = e_\rho + c_v(T - T_0) \qquad (8)$$

Where $c_v$ is the specific heat, here using the average value; $e_\rho$ and $T_0$ are the specific internal energy and temperature determined at the water melting point.

Modeling the micro-explosion formal approach to the PPM programming, firstly, the fluid dynamic equations are solved in Lagrangian co-ordinates for the Riemann problem and get the average velocities and energy of the every unit. Secondly, the results remapped on to the original finite difference Eulerian grid.[11]

In our calculation model, a boundary condition of rigid wall is used at the surface of the hard target and the symmetry axis. The conditions at the outer boundary can be taken as free boundary condition. The water parameters are taken as at normal conditions.

### 3. Calculation results and discussion

According to the general operation, we sat the fiber diameter 0.4mm and pulse duration 400μs in our calculation. The numerical model provides all of the details of the laser induced explosion process. Fig.2 shows the pressure evolution in the calculation field during the explosion process ($E$=2J,

$h$=0.5mm). The water absorbs great laser energy and transits to vapor rapidly. The rapid phase transition leads to strong expansion and the explosion takes place. The pressure of the water between calculus and fiber tip is increasing rapidly after the beginning of the laser pulse. The vapor plume and water droplets formed during the explosion are accelerated by the rapid expansion and eject around. The great shock wave caused high pressure on the calculus surface. The vapor pushed aside the water and a vapor hole came into being. These properties of the numerical model accord well with the experiment results in the literature.[2]. Fig.3 shows the vapor and water density evolution during the explosion. The vapor and water density in the explosion area decreasing rapidly and the Ho:YAG laser may transmit to the calculus throng the vapor hole at that time.

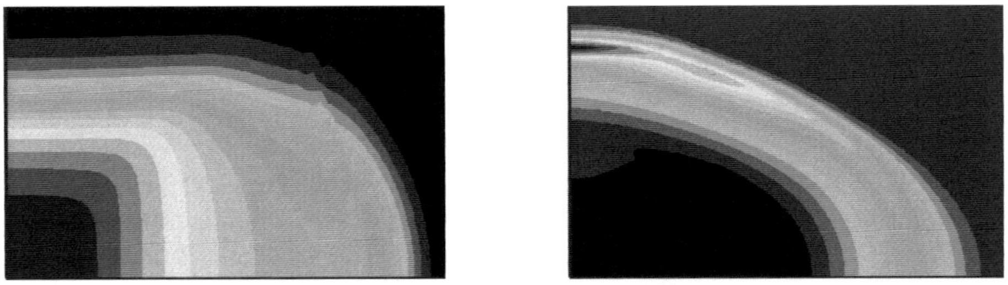

(t=5μs) (t=20μs)

Figure2. Pressure distribution evolution during the explosion ($E$=2J, $h$=0.5mm)

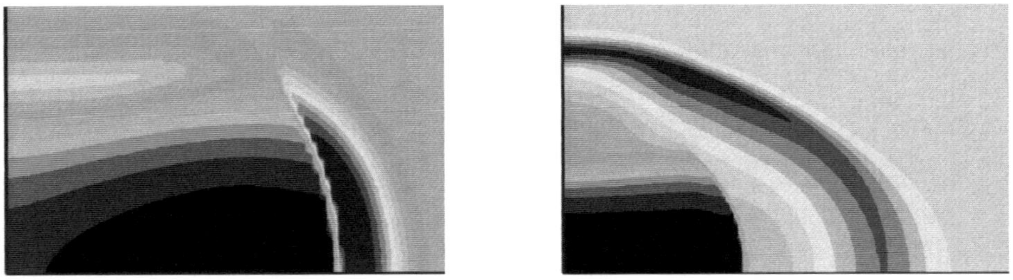

(t=5μs) (t=20μs)

Figure3. Water vapor density distribution evolution during the explosion ($E$=2J, $h$=0.5mm)

Fig.4 gives the pressure distribution evolution induced by the explosion shock wave on the calculus surface. It indicates that the pressure on the calculus doesn't have a steady increase during the explosion. There is a particular duration for the special parameters. The high-pressure area centralizes under the laser spot and the peak pressure takes place near the symmetry central point A (see Fig.1). From the figure we should find that the pressure evolution of point A approximately represents the pressure evolution process of the whole calculus surface. The dynamics characters for the special parameters will be got from the pressure of point A. Fig.5 shows the pressure evolution of point A for different laser pulse energy. It indicates that the pressure increases rapidly and has a high peak pressure value for high laser pulse energy. The peak pressure value for the laser pulse energy 2J is about 70MPa and the value for the laser pulse energy 0.3J is only about 15MPa. The micro-explosion caused pressure on the calculus surface increases rapidly with the laser pulse energy increase.

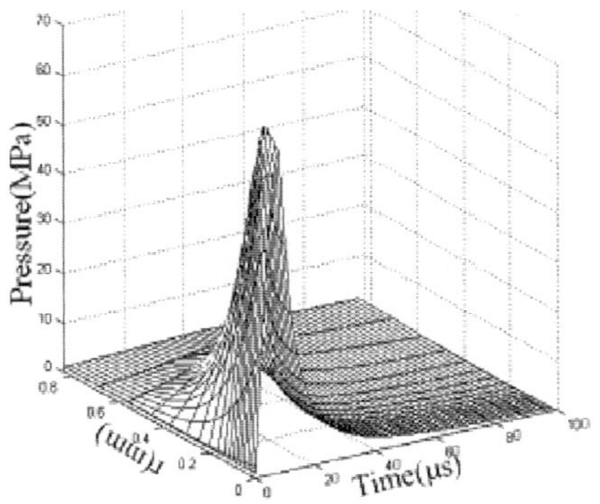


Figure4. Pressure distribution evolution on the calculus surface ($E$=2J, $h$=0.5mm)

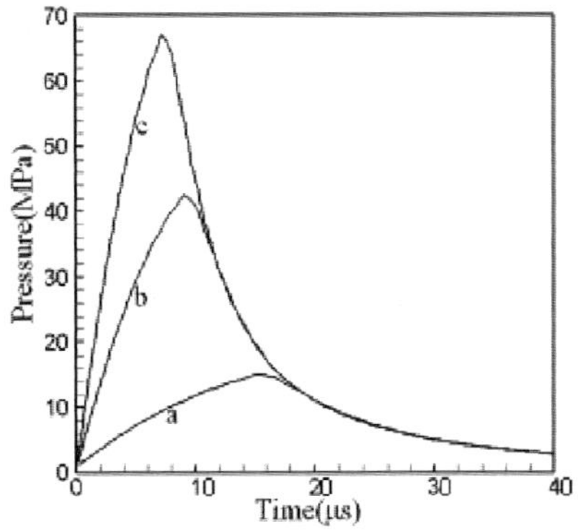


Figure5. Pressure evolution of the point A for different pulse energy (a. 2J, b. 0.8J, c. 0.3J, $h$=0.5mm)

Fig.6 gives the comparison of the pressure evolution in different distance between calculus and fiber tip. It indicates that the high pressure duration increases with the water distance increase. From fig.7 we can see that the intensity of impulse increases rapidly with the distance between calculus and fiber tip increase. It will cause more calculus retropulsion in the laser lithotripsy.

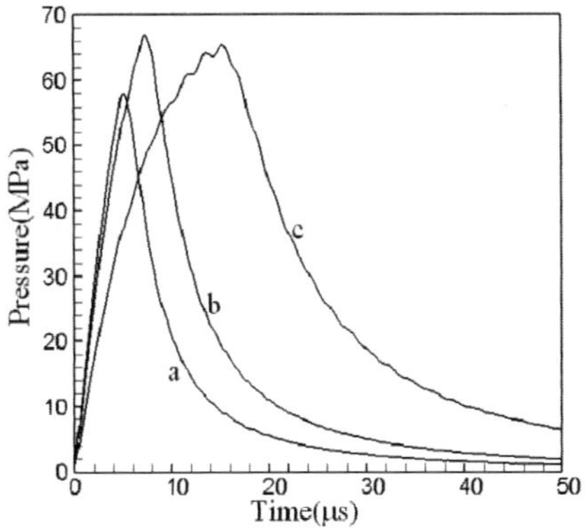


Figure6. Pressure evolution of point A for different water distance (a. 0.2mm, b.0.5mm, c. 1.5mm, $E$=2J)

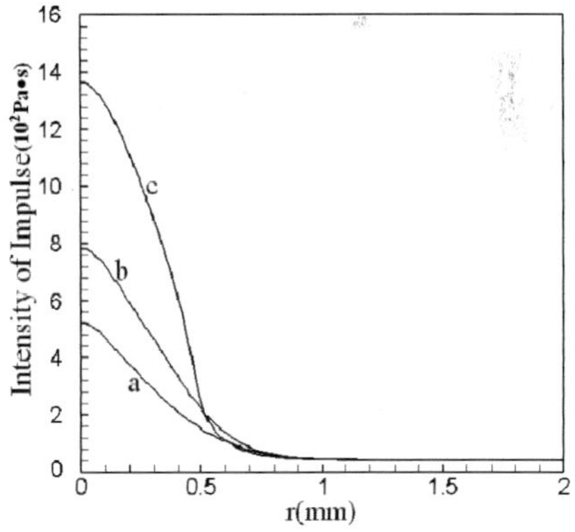


Figure7. Intensity of impulse distribution on the calculus surface for different water distance (a. 0.2mm, b.0.5mm, c. 1.5mm, $E$=2J)

## 4. Conclusions

In summary we have developed a fluid dynamics model to describe the water vapor micro-explosion during Ho:YAG laser lithotripsy. From the calculation results, we can see that the micro-explosion has its special duration and dynamic characters. They are mainly determined by such parameters—laser pulse energy, pulse duration and water distance between calculus and fiber tip. High pulse laser energy gets high force on the calculus surface and mains high efficiency during ablation process. The Intensity of impulse on the calculus surface will increase rapidly with the water distance increasing. This may cause the calculus retropulsion more and has disadvantage to the surgery, so the fiber tip should be put near the calculus during laser lithotripsy. The ideal surgery results need property parameters.

This simulation investigated the water vapor explosion dynamics evolution only. The Ho:YAG laser will act on the calculus directly after the water vapor hole formed. The more detail in laser lithotripsy needs more investigation.

## 5.References


1. Jahja Kokaj, Mustafa Mara., J. Mathew,Optical investigation of dynamics of phenomena of laser-based lithotripsy,Optics and Lasers in Engineering , (46) 535– 540,2008.
2. Hyun Wook Kanga, Ho Leeb, Jason Petersen, Investigation of Stone Retropulsion as a Function of Ho:YAG Laser, Proc. of SPIE, Vol. 6078 607815-1,2006.
3. Yu.E.Geints, A. A. Zemiyanov and R. L. Armstrong, "Explosive boiling of Water droplets irradiated with intense $CO_2$-laser radiation: numerical experiments," *Appl. Opt.* **33**, 5805-5810 ,1994.
4. M. Stanislawki, J. Meister, T. Mitra, M.M. Ivanenko, K. Zanger, P. Hering, "Hard tissue ablation with a free running Er:YAGand a Q-switched CO2 laser: a comparative study," *Appl. Phys. B* **72**, 115–120, 2001.
5. G. Paltauf, H. Schmidt-Kloiber, "Microcavity dynamics during laser-induced spallation of liquids and gels," *Appl. Phys. A* **62**, 303–311 , 1996.
6. S.M.Chitanvis, "Explosion of Water Droplets," *Appl. Opt.* **25**, 1837-1839 , 1986.
7. N. Arnold, J. Gruber, J. Heitz "Spherical expansion of the vapor plume into ambient gas:an analytical model," *Appl. Phys. A* **69** [Suppl.], S87–S93 , 1999.
8. Jun-ya Kohno, Fumitaka Mafune and Tamotsu Kondow, "Ejection of a Large Neutral Cluster from a Liquid Beam Surface Following IR Laser Irradiation," *J. Phys.Chem. A* **108**, 971-977 , 2004.
9. Raimund Hibst and Ulrich Keller "Mechanism of Er:YAG laser induced ablation of dental hard substances," *Proc. SPIE Int. Soc. Opt. Eng.* **1880**, 156-162 , 1993.
10. Jose´ Ma Marti and Ewald Muller, "Extension of the Piecewise Parabolic Method to One-Dimensional Relativistic Hydrodynamics," *J. Comp. Phys.* **123,** 1–14 , 1996.
11. Colella P, Woodward PR, "The piecewise parabolic method (PPM) for gas-dynamical simulations," *J. Comp. Phys.* **54**: 174-201 , 1984.
12. A. Vila Verde, Marta M. D. Ramos, R. Mendes Ribeiro, Marshall Calculusham, "Modeling of the human enamel laser ablation process at the mesoscopic scale," *Proc. SPIE* **4950**, 72-82 , 2003.
13. Quanming Lu, "Thermodynamic evolution of phase explosion during high-power nanosecond laser ablation," *Phys. Rev. E* **67**, 016410 , 2003.
14. NahalAshouria, Ramesh Shorib, Jason M. Cheunga, Daniel Frieda "Infrared Laser Ablation of Dental Enamel: Influence of an Applied Water Layer on Ablation Rate and Peripheral Damage," *Proc. SPIE* **4249**, 72-78 , 2001.
15. A. Polt and G. Maurer, "The Bender equation of state for describing thermodynamic properties of krypton, neon, fluorine, sulfur dioxide and water over a wide range of state," *Fluid Phase Equilibria,* **73**, 27-38, 1992
16. N I Kuskova, S I Tkachenkoy and S V Koval, "Investigation of liquid metallic wire heating dynamics," *J. Phys.: Condens. Matter* **9**, 6175–6184 (1997).
17. Wang Zhongqi "Objected-oriented multi-materials numerical method and application study of explosion mechanics," [M], Beijing University of Science and Technology, 20001216.

# Intravascular photoacoustic detection of vulnerable plaque based on constituent selected imaging

**Jian Zhang and Da Xing***

MOE Key Laboratory of Laser Life Science & Institute of Laser Life Science, College of Biophotonics, South China Normal University, Guangzhou 510631, China

E-mail: xingda@scnu.edu.cn

**Abstract.** Atherosclerosis, a disease of the large arteries, is the primary cause of heart disease and stroke. Over decades, atherosclerosis is characterized by thickening of the walls of the arteries, only advanced atherosclerotic disease could be observed. Photoacoustic imaging is a hybrid imaging technique that combines the advantages of high spatial resolution of ultrasound with contrast of optical absorption. In this paper, we present an intravascular photoacoustic (IVPA) imaging system to characterize vulnerable plaques by using the optical absorption contrast between different constituents. Epidemiological studies have revealed several important plaque constituents associated with early atherosclerosis, such as macrophage, cholesterol, lipid, calcification, and so on. We chose a section of lipid rich atherosclerosis artery and a section of normal artery as the phantom. Two IVPA images of them are given to show the difference between sick and normal. As a new method of detecting vulnerable plaque, IVPA constituents imaging will provide more details for diagnosis that offer an enticing prospect in early detecting of atherosclerosis.

## 1. Introduction

Atherosclerosis is a progressive disease which is the leading death world wide, because of plaque rupture causing heart disease and stroke. All of factors causing to a vulnerable plaque can be divided into morphology factors and functional factors. Traditionally angiography, diagnosis of atherosclerosis was possible only at advanced stages of disease, because they only evaluating the morphological of the large arteries. Vulnerable plaque typically has a thin fibrous cap and a lipid-rich necrotic core. At the same time, plaques can become increasingly complex because of functional changes. For example, calcification, ulceration at the luminal surface, and haemorrhage from small vessels that grow into the lesion from the media of the blood vessel wall, which can not be differentiate by the traditional methods easily [1-3].

In the past decades, a variety of diagnostic imaging techniques have been developed, such as optical coherence tomography (OCT), intravascular ultrasound (IVUS), computed tomography angiography (CAT) and magnetic resonance angiography (MRA) [4-8]. However, OCT must block blood flow, the resolution of IVUS is low, CTA has to use contrast agent containing iodine and patients suffer nuclear radiation, MRA techniques are sensitive to turbulent flow, which can cause proton spins to rapidly

* Correspondence author: Tel: +86-20-85210089, Fax: +86-20-85216052, Email: xingda@scnu.edu.cn

dephased thus causing a significant loss of signal, this can cause mis-diagnosis of stenosis. Further more, all of them are specific to imaging the structural characteristics.

Photoacoustic (PA) imaging is a relatively new kind of technique that has the potential to visualize constituents of the vulnerable plaques. In PA imaging, a short-pulsed laser source is used to illuminate a biological sample. The laser-generated photoacoustic signals that are excited by thermoelastic expansion resulting from a transient temperature rise on the order of 10 mk can be measured by a wide-band ultrasonic transducer; they are used to reconstruct an PA image. Therefore, PA imaging is a hybrid imaging technique that combines the advantages of high spatial resolution of ultrasound with contrast of optical absorption. Further more, PA imaging could provide functional information of tissues based on mapping of the differential optical absorption of tissue constituents, because of the different optical properties of different absorber [9, 10], we could differentiate the constituents from their mixture by using mult-wavelength PA imaging. In this paper, multi-wavelength PA imaging [11-13] technique and an IVPA imaging system based on IVUS imaging catheter were used to detect vulnerable plaque by constituents selected imaging.

## 2. Materials and methods

### *2.1. Animal model of atherosclerosis*

A well-characterized animal model of atherosclerosis is used in the IVPA imaging experiments. Rabbits fed with high-fat diet are classical models for the study of atherosclerosis. In this study, we establish the atherosclerosis models by feeding high-fat diet plus and arterial intimal injury of the ventral aorta with balloon in a 1 year old New Zealand rabbit. After arterial intimal injury of the ventral aorta with balloon, rabbit was fed on a high-fat diet (normal rabbit feed 90%, cholesterol 2% and lard 8%) over a long period of time (6 months) , The high-fat dietary regimen was utilized to induce fibro-cellular lesions comprised of inflammatory macrophage cells and lipids. For comparison, another rabbit was placed on a normal diet for the same period of time, this rabbit served as a control animal.

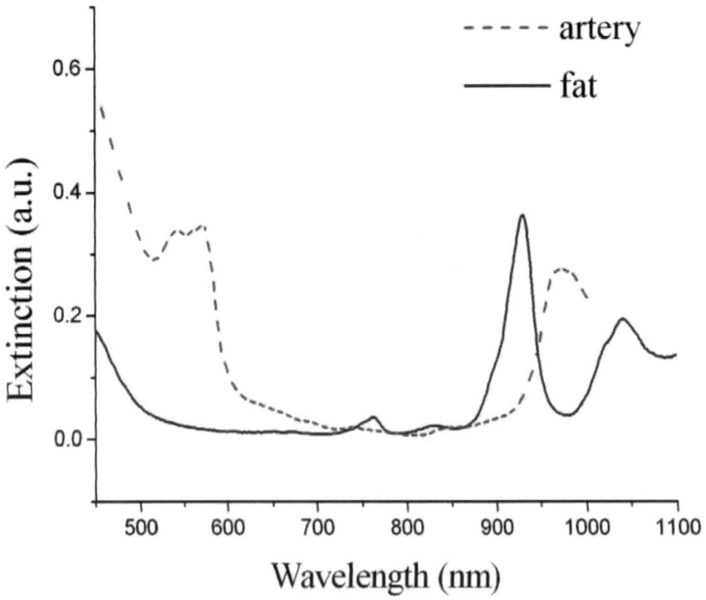


**Figure 1.** The extinction spectrum of artery and fat.

The extinction spectroscopy [14] of artery (hard line) and fat (dash line) is shown in Figure. 1, which is measured by spectrometer (Lambda 35, Perkin Elmer). According to the spectroscopy, the wavelength of laser is adjusted to 930 nm during the experiment, where the absorption of fat is higher than that of artery. A section of a lipid-rich atherosclerosis aorta obtained from a rabbit subjected to a high-fat for 6 months and a section of a normal aorta from the control sample obtained from a rabbit subjected to a normal diet for the same period of time were used in the IVPA experiment. The excised tissue was stored in saline for no more than 4 hours prior to the imaging experiments.

*2.2. Experimental setup of multi-wavelength IVPA imaging*

The setup for multi-wavelength intravascular photoacoustic imaging system is shown in Figure. 2. A tunable pulsed Nd:YAG pumped optical parametric oscillator laser source (Vibrant B 532I, Opotek, USA) operating at the wavelength range of $\lambda = 690 - 960$ nm with a pulse wide of 10 ns and a pulse repetition rate of 10 Hz, was used to provided the optical illumination for photoacoustic imaging. The sample was immersed in a custom-built water tank, and a single element, 2.5 F, 0.83-mm diameter; 40 MHz IVUS imaging catheter (Atlantis SR Plus, Boston Scientific Inc.) was used to detect the photoacoustic waves generated by optical excitation. A revolving detection was driven by a computer-controlled stepper motor to rotate the phantom with a step size of 1.8 deg. The photoacoustic signals detected by the transducer were amplified by a low-noise preamplifier (ZFL-500LN, Min-circuit) and an amplifier (Ha2, Precision Acoustics LTD) before being fed to a digital oscilloscope (TDS 3032, Tektronix), which digitized the photoacoustic signals. Finally, a computer acquired the signals and stored the data for image reconstruction. A modified back-projection algorithm was employed to reconstruct the artery images from the photoacoustic signals.

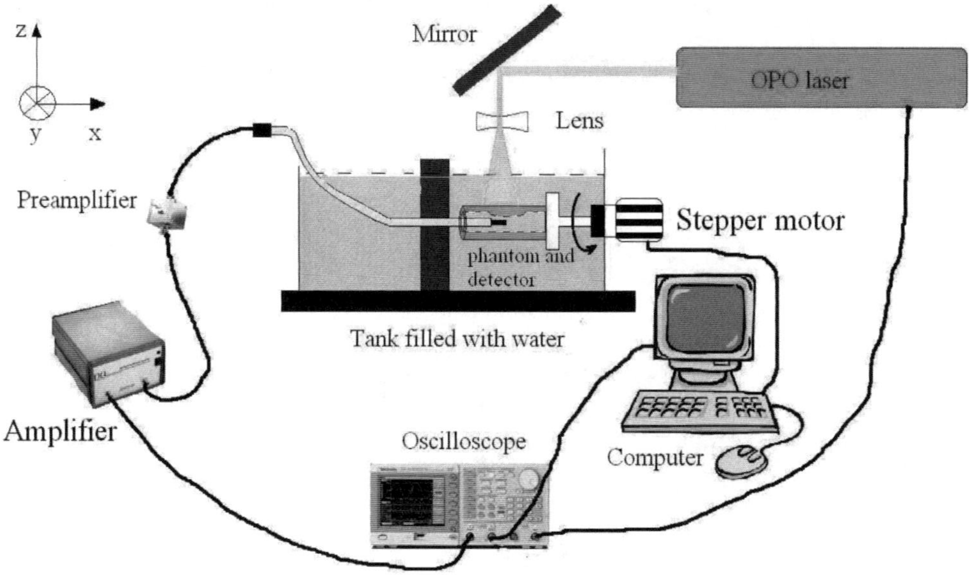


**Figure 2.** Schematic of the multi-wavelength IVPA imaging system.

## 3. Results

Experiment was taken to demonstrate the ability of the IVPA imaging system. A black thermoplastic pipe with the diameter of 6 mm was used as a phantom. During the experiment; the phantom is irradiated by 532nm laser from outside while the detector is placed inside the lumen of it. The IVPA image (a) and cross-section photograph (b) is shown inFig.3. It can be seen that IPVA image visualize the morphology of black tube, as well as, it's unified optical absorption property.

(a) (b)

**Figure 3.**Photoacoustic image (a) and photography (b) of a black tube cross-section using IVPA imaging

The IVPA images and the histological cross-sectional images of the control normal aorta (a, b) and atherosclerotic aorta (c, d) covering a field of view of 8 mm are presented in Figure.4.

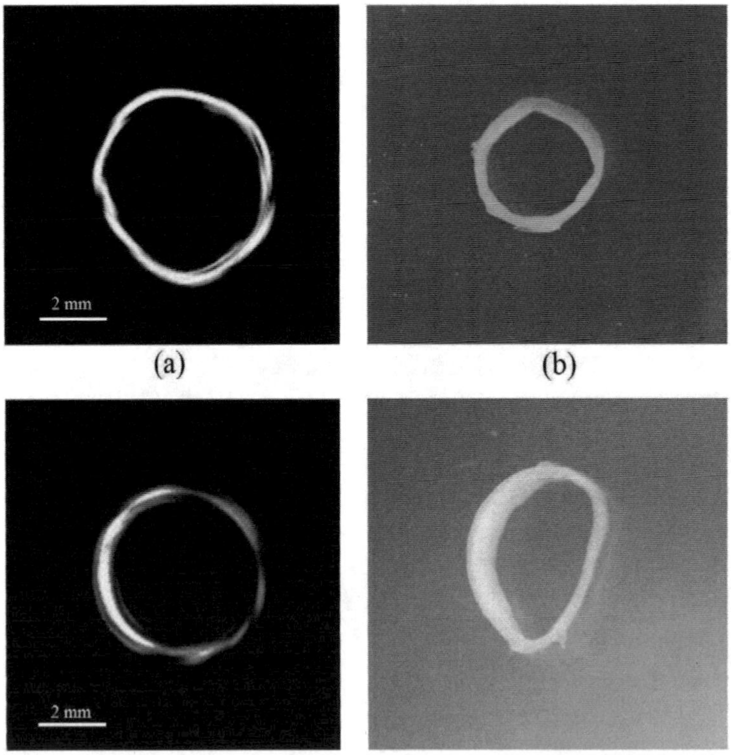


(a) (b)

(c) (d)

**Figure 4.** IVPA images and the histology of the cross-section of the normal control aorta (a, b) and aorta containing plaques(c, d).

The histological cross-sectional images of the atherosclerotic aorta (d), the normal aorta (b) shows a thin wall, at the same time, the difference can be seen clearly from the IVPA images. Figure. 4(a) shows a uniform thin wall while the IVPA image morphology of Figure. 4(b) shows a different character, the left region of it is thicker than the right region, which matched well with its histological cross-sectional image.

## 4. Discussion and Conclusions

IVPA images of atherosclerotic aorta (Figure. 4 (c)) indicate the presence of plaques and are clearly different from the IVPA images of normal aorta (Figure. 4 (a)). There are significant spatial and spectral variations in the energy of the photoacoustic signal within the plaque-rich. Figure. 4(a) shows uniform photoacoustic signal intensity, while the photoacoustic signal intensity of left region is stronger than the right region in the Figure. 4(c). As we know the intensity of photoacoustic signal associate with two factories, the intensity of laser pulse and the extinction coefficient of absorber at the wavelength. In our experiment, the laser beam intensity per pulse is limited to 1 mJ/cm2 and keeps the same in the two experiments. This energy is well below the maximum permissible exposure of 20 $mJ/cm^2$ specified by the American National Standards Institute (ANSI-Z136.1). Therefore, the different absorption ability of the normal aorta and atherosclerotic aorta is the reason of signal intensity uneven. Laser of 930 nm is chosen in the experiment, where the extinction coefficient of fat is bigger than artery, as we can see in the spectroscopy of Figure. 1. The whole aorta wall of the normal aorta is uniform, and its photoacoustic signal is uniform. Yet, the left region of atherosclerotic aorta is lipid rich, where shows a strong absorption and produces strong photoacoustic signals at the wavelength of 930 nm.

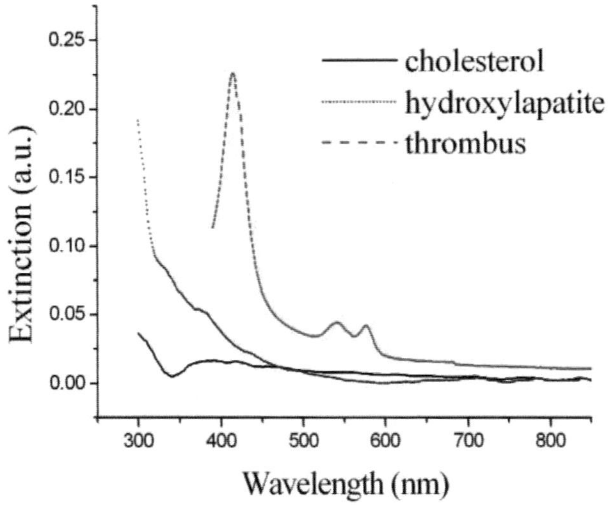


**Figure 5.** The extinction spectrum of cholesterol, hydroxylapatite and thrombus.

Furthermore, we measured the extinction spectrum of cholesterol, hydroxylapatite and thrombus presented in the Figure. 5, all of them are the constituents of the vulnerable atherosclerotic, which show quite different optical absorption. IVPA imaging utilizing the variation in the optical absorption may play a major role in the detection of vulnerable plaque by mult-wavelength.

The results of IVPA photoacoustic imaging of a lipid rich atherosclerotic artery at wavelength 930 nm indicated the potential to detect a vulnerable plaque by using mult-wavelength constituent selected imaging. However, more studies and significant understanding of the imaging technique are required to confirm the ability.


*Acknowledgments*

This research is supported by the National Basic Research Program of China (2010CB732602), the Program for Changjiang Scholars and Innovative Research Team in University (IRT0829), the

National Natural Science Foundation of China (30627003; 30870676), and the Natural Science Foundation of Guangdong Province (7117865).We would like to acknowledge professor Pengfei Zhang (Shandong University, China) for help with the animal experiments.

**References**

[1] R. Ross 1993 *Nature* **362** 801

[2] P. Libby 1999 *J. Intern. Med.* **247** 349

[3] A . H. Gershlick, M. de. Belder, J Chambers, D Hackett, R Keal, A Kelion, S Neubauer, D J Pennell, M Rothman, M Signy, P Wilde 2007 *Heart* **93** 423

[4] M. A. Cordeiro, J. A. Lima 2006 *J. Am. Coll. Cardiol.* **47** C40

[5] T. Saam., T. S. Hatsukami, N. Takaya, et al 2007 *Radiology* **244** 64

[6] Sakuma H. *J.* 2007 *Magn. Reson. Imaging* **26** 3

[7] S. L. Jiao, Z. X. Xie, H. F. Zhang and C. A. Puliafito 2009 *Optics Letters* **34** 2961

[8] G. Rioufol, G. Finet, I. Ginon, X. André-Fouët, R. Rossi, E. Vialle, E. Desjoyaux, G. Convert, J. F. Huret and A.Tabib 2002 *circulation* **106** 804

[9] S. H. Yang, D. Xing, Y. Q. Lao, D. W. Yang, L. M. Zeng, L. Z. Xiang, and W. R. Chen 2007 *Appl. Phys. Lett.* **90**, 243902 1 -3

[10] Y. Q. Lao, D. Xing, S. H. Yang and L. Z. Xiang 2008 *Phys. Med. Biol.* **53** 4203

[11] M. L. Li, J. T. Oh, X. Y. Xie, G. Ku, W. Wang, C. Li, G. Lungu, G. Stoica, and L. V. Wang. 2008 *IEEE* **96**, No. 3

[12] B. Tromberg, N. Shah, R. Lanning, A. Cerussi, J. Espinoza, T. Pham, L. Svaasand, and J. Butler 2000 *Neoplasia* **2**, 26- 40 .

[13] D. Razansky, M. Distel, C. Vinegoni, R. Ma, N. Perrimon, R. W. Köster and V. Ntziachristos. 2009 *nature photonics* **3** 412

[14] http://omlc.ogi.edu/spectra/.

# Yes-Associated Protein ( YAP ) Promotes the Nuclear Import of p73


Heng Zhang, Shengnan Wu[1]

MOE Key Laboratory of Laser Life Science & Institute of Laser Life Science,
College of Biophotonics, South China Normal University,
Guangzhou 510631, China



## ABSTRACT

p73 has been identified as a structural and functional homolog of the tumor suppressor p53. However, mechanisms that regulate the localization of p73 have not been fully clarified. The Yes-associated protein (YAP) is a transcriptional coactivator. As a transcriptional coactivator, YAP needs to bind transcription factors to stimulate gene expression. p73 is a reported YAP target transcription factors and YAP has been shown to positively regulate p73 in promoting apoptosis. Previous studies show that p73 interacts with YAP through its PPPY motif, and increases p73 transactivation of apoptotic genes. In this study, we focused on YAP's regulation of the localization of p73. After transient transfection into Rat pheochromocytoma (PC12) cells and Human embryonic kidney 293T cells with GFP-YAP and/or YFP-p73, and incubated for 24 hours expression. p73 was fused to YFP to allow the examination of its subcellular localization. When expressed alone, YFP-p73 was distributed throughout the cell. When coexpressed with YAP, nuclear accumulation of YFP-p73 became evident. We quantitated the effect of YAP on the redistribution of YFP-p73 by counting cells with nuclear-only YFP signal. We found that YAP can influence the subcellular distribution of p73. Altogether, coexpression with YAP affected the subcellular distribution of the p73 protein. Our studies attribute a central role to YAP in regulating p73 accumulation and YAP, at least in part, might promote the nuclear import of p73.



**Keywords: YAP, p73, nuclear import**


## 1 INTRODUCTION

The p73 protein is a transcription factor related to the tumor suppressor p53 and shares the proapoptotic function of p53 [1]. Like p53, p73 is activated by DNA damage and can stimulate p53-independent apoptosis [2]. Gerry Melino et al.[3] have proved that p73 elicits apoptosis via the mitochondrial pathway using PUMA and Bax as mediators. Most studies report that p73 predominately mediate transcription of Bax in response to DNA damaging agents and the resulting cell death [4,5,6,7,8]. However, mechanisms that regulate the localization of p73 have not been fully clarified.

Studies by Yagi et al. [9] have proved that the Yes-associated protein (YAP) is a transcriptional coactivator. As a transcriptional coactivator, YAP needs to bind transcription factors to stimulate gene expression. Reported YAP target


[1] Corresponding author: Shengnan Wu, Ph.D. Phone: +86-20-85211436-8209; Fax: +86-20-85216052;
E-mail: wushn@scnu.edu.cn

transcription factors include TEAD, p73, Runx2, and the ErbB4 cytoplasmic domain[9,10,11,12]. Previous other studies show that YAP interacts with the p53 family member p73, resulting in an enhancement of p73's transcriptional activity[10], and that YAP promotes p73-mediated transcription of proapoptotic genes in response to DNA damaging agents and the resulting cell death [4,5,13,14,15]. Thus, YAP recruitment is a critical event in turning on the transcriptional activity of p73 by favoring the formation of transcriptionally active competent complexes that play a pivotal role in eliciting apoptosis in response to anticancer treatment [4,5,10,13,14,15].

In this study, we focused on YAP's regulation of the localization of p73. After transient transfection into Rat pheochromocytoma (PC12) cells and 293T cells with GFP-YAP and/or YFP-p73, and incubated for 24 hours expression. p73 was fused to YFP to allow the examination of its subcellular localization. When expressed alone and coexpressed with YAP, we quantitated the effect of YAP on the redistribution of YFP-p73 by counting cells with nuclear-only YFP signal.

## 2 Materials and Methods

**2.1** *Cell Culture*

The Rat pheochromocytoma (PC12) were cultured in a humidified (5% $CO_2$, 37°C) incubator in in Dulbecco's modified Eagle's medium (DMEM, Life Technologies, Inc.), supplemented with 15% fetal bovine serum, 5% horse serum, penicillin (100 U/ml), and streptomycin (100 μg/ml). Human embryonic kidney 293T cells were cultured under the same condition as PC12 cells but not supplemented with horse serum. The medium was refreshed every other day, and the cells were plated at appropriate densities according to each experimental protocol. Before transfection, the cells were cultured in a custom-built dish. When the cells came to 70-80% confluence, plasmids DNA of GFP-YAP and YFP-p73 were transfected with Lipofectamine TM 2000 reagent into the PC12 and 293T cells.

**2.2** *Chemicals and Plasmids*

We used Lipofectin$^{TM}$ 2000 reagent (Invitrogen, USA) to transfect plasmids GFP-YAP and YFP-p73 into PC12/293T cells. The pGFP-YAP was a gift from Prof. Subham Basu, Cell Survival Signalling Laboratory Centre for Molecular Oncology and Imaging Barts and The London School of Medicine and Dentistry, John Vane Science Centre,London and Prof. Julian Downward, Signal Transduction Laboratory, Cancer Research UK London Research Institute,, London UK ( Subham Basu et al., 2003 ). The pYFP-p73 was a gift from Prof. Hedeki Shimodaira, Department of Clincial Oncology Institute of Development, Aging and Cancer Tohoku University, Japan (Hideki Shimodaira et al., 2002 and Ivana Marinovic-Terzic et al.,2008)

***2.3*** *Subcellar Distribution of Yellow Fluorescent Protein (YFP)-Fusion Proteins.*

The 293T cells were transfected with pGFP-YAP and/or pYFP-p73 by using Lipofectamine TM 2000. To quantitate the nuclear-only YFP signal, cells were counted and the fluorescence images were collected via a Zeiss C-Apochromat objective (40x, NA=1.3). Cells in 10 randomly chosen fields were scored per experiment.

## 3 RESULTS

**3.1 p73 locates to cytosol and nucleus under physiological conditions**

To know the subcellular location of p73, we transfected plasmid YFP-p73 into cells and analyzed the subcellular distribution of fluorescently labeled p73 by confocal microscopy. As is shown in Fig. 1, we found that p73 locates to both cytosol and nucleus under physiological conditions .

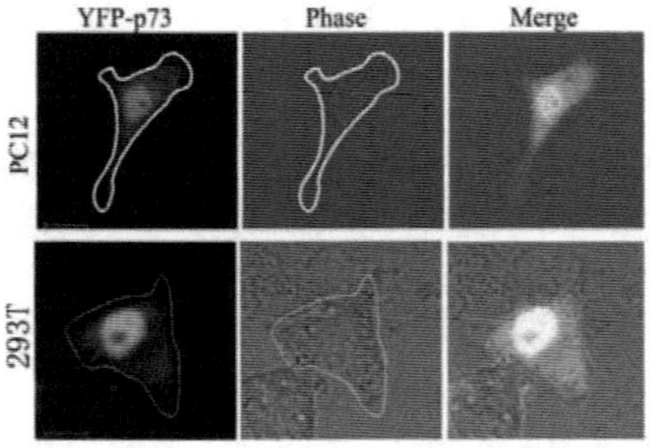


Figure 1. Subcellular localization of YFP-p73 in PC12 and 293T cells. Cells expressing YFP-p73 were imaged by confocal microscopy. The following specific settings were used for light excitation and emissions: $Ar^+$ laser, Ex. 514 nm, Em. BP LP530 nm.

### 3.2 YAP predominantly locates to cytosol under physiological conditions

Next, we study the subcellular location of YAP, we transfected plasmid GFP-p73 into cells and analyzed the subcellular distribution of fluorescently labeled YAP by confocal microscopy. As is shown in Fig. 2, we found that YAP predominately maintained in cytosol.

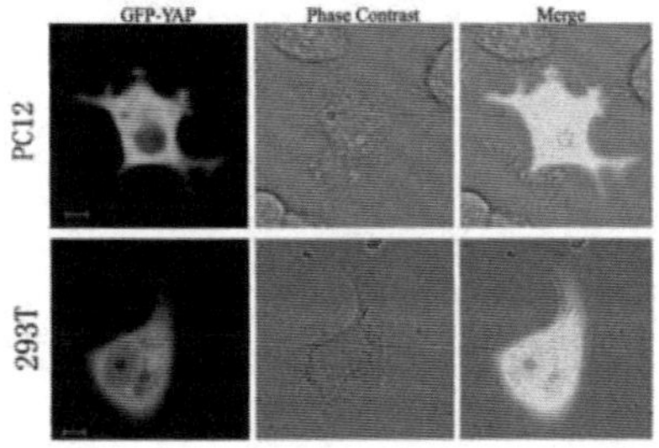


Figure 2. Subcellular localization of GFP-YAP in PC12 and 293T cells. Cells expressing GFP-YAP were imaged by confocal microscopy. The following specific settings were used for light excitation and emissions: $Ar^+$ laser, Ex. 488 nm, Em. BP BP500-550 nm.

### 3.3 YAP promote the nuclear import of p73

Furthermore, we study on YAP's regulation of the localization of p73. After transient transfection into PC12 cells and 293T cells with GFP-YAP and/or YFP-p73, and incubated for 24 hours expression. p73 was fused to YFP to allow the examination of its subcellular localization. When expressed alone, we found that YFP-p73 was distributed throughout the cell. When coexpressed with YAP, we found that nuclear accumulation of YFP-p73 became evident. We quantitated the effect of YAP on the redistribution of YFP-p73 by counting cells with nuclear-only YFP signal. As is show in Figure.3. we found that YAP can influence the subcellular distribution of p73, and at least in part, might promote the nuclear import of p73.

A.

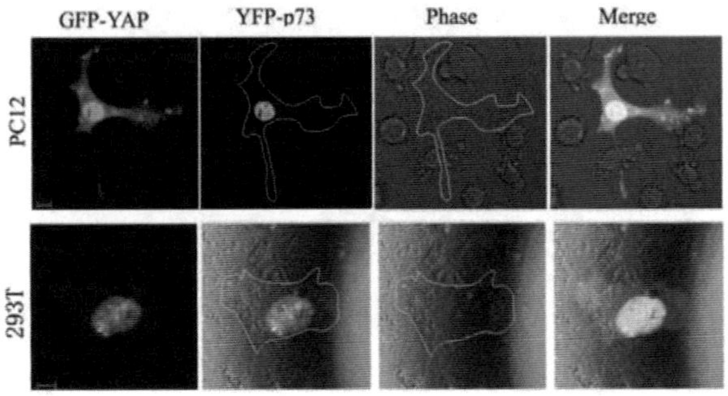


B.

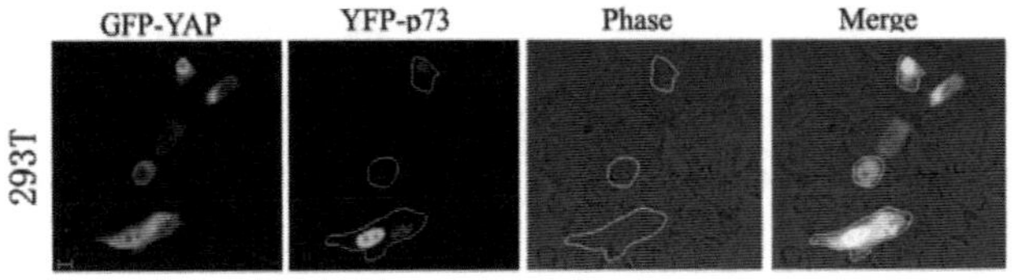


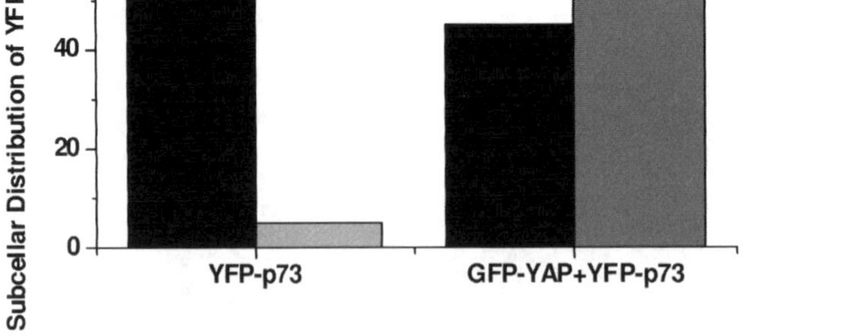

Figure 3.YAP promotes the nuclear import of p73.
A. After transient transfection into PC12 cells and 293T cells with GFP-YAP and/or YFP-p73, Cells were imaged by confocal microscopy. The following specific settings were used for light excitation and emissions: $Ar^+$ laser, Ex: GFP-YAP, 458 nm; YFP-p73,514nm;
Em: GFP-YAP,BP500-530;YFP-p73,LP530.
B. Subcellar Distribution of Yellow Fluorescent Protein (YFP)-Fusion Proteins. To quantitate the nuclear-only YFP signal, cells were counted and the fluorescence images were collected via a Zeiss C-Apochromat objective (40x, NA=1.3). Cells in 10 randomly chosen fields were scored per experiment.

## 4 DISCUSSION

p73 has been identified as a structural and functional homolog of the tumor suppressor p53. However, mechanisms that regulate the localization of p73 have not been fully clarified. Previous studies show that p73 is a reported YAP target transcription factors,and YAP interacts with p73, resulting in an enhancement of p73's transcriptional activity[4,5,10,13,14,15]. Thus, we speculated that YAP could regulate the localization of p73. After transient transfection into Rat pheochromocytoma (PC12) cells and 293T cells with GFP-YAP and/or YFP-p73, and incubated for 24 hours expression. p73 was fused to YFP to allow the examination of its subcellular localization. When expressed alone, we found that YFP-p73 was distributed throughout the cell (Figure 1). When coexpressed with YAP, nuclear accumulation of YFP-p73 became evident (Figure 3). We quantitated the effect of YAP on the redistribution of YFP-p73 by counting cells with nuclear-only YFP signal. We found that YAP can influence the subcellular distribution of p73. Altogether, coexpression with YAP affected the subcellular distribution of the p73 protein. Our studies attribute a central role to YAP in regulating p73 accumulation and YAP, at least in part, might promote the nuclear import of p73.

Some studies have reported that laser irradiation diverse signaling pathways and ultimately affects the cell physiological processes [16]. Consistent with $Ca^{2+}$ increase [17], Gao et al. shown that PKC kinases are activated in human lung adenocarcinoma (ASTC-a-1) cells [18] and rat pheochromocytoma (PC12) cells [19] in response to Low-power laser irradiation (LPLI). It is also reported that LPLI triggers a significant activation of ROS/Src pathway [20] and promotes formation of circular dorsal ruffles via PI3K/Ras pathway [21]. Meanwhile, increase of the level of Akt phosphorylation is observed in African green monkey SV40-transformed kidney fibroblast cells (COS-7)when stimulated by LPLI ($<50J/cm^2$) [22-23]. Subham Basu et al.[4] proved that Akt promotes YAP localization to the cytoplasm, resulting in loss from the nucleus where it functions as a coactivator of transcription factors including p73. Thus, we speculated that YAP maintained in cytoplasm by LPLI via PI3K /Akt pathway, and needs further investigation. This extends the range of mechanisms whereby Akt can promote cellular survival in the face of apoptotic stimuli-mediated by YAP.

## ACKNOWLEDGMENTS

The authors thank Prof. Subham Basu and Prof. Julian Downward for providing the plasmid GFP-YAP, and thank Prof. Hedeki Shimodaira for providing the plasmid YFP-p73. This research is supported by the National Basic Research Program of China (2010CB732602), the Program for Changjiang Scholars and Innovative Research Team in University (IRT0829), and the National Natural Science Foundation of China (30870676; 30870658).

## REFERENCES


[1] Ozaki T, Nakagawara A., p73, a sophisticated p53 family member in the cancer world. *Cancer Sci,* 96:729–737(2005).

[2] .Levrero M, V. De Laurenzi, A. Costanzo1, S. Sabatini, J. Gong, J. Y. J. Wang and G. Melino, The p53/p63/p73 family of transcription factors: Overlapping and distinct functions. *J Cell Sci* 113:1661–1670(2000).

[3] .Gerry Melino, Francesca Bernassola, Marco Ranalli, Karen Yee, Wei Xing Zong, Marco Corazzari, Richard A. Knight, Doug R. Green, Craig Thompson, and Karen H. Vousden., p73 Induces Apoptosis via PUMA Transactivation and Bax Mitochondrial Translocation. *J Biol Chem*, 279(9):8076-8083(2004).

[4] .Subham Basu, Nicholas F. Totty, Meredith S. Irwin, Marius Sudol, and Julian Downward, Akt Phosphorylates the Yes-Associated Protein, YAP, to Induce Interaction with 14-3-3 and Attenuation of p73-Mediated Apoptosis. *Molecular Cell*, 11: 11–23(2003).

[5] Sabrina Strano, Olimpia Monti, Natalia Pediconi, Alessia Baccarini, Giulia Fontemaggi, Eleonora Lapi, Fiamma Mantovani, Alexander Damalas, Gennaro Citro, Ada Sacchi, Giannino Del Sal, Massimo Levrero, and Giovanni Blandino, The Transcriptional Coactivator Yes-Associated Protein Drives p73 Gene-Target Specificity in Response to DNA Damage. *Molecular Cell*, 18: 447–459(2005).

[6] Claudie Hooper, Mahvash Tavassoli, J. Paul Chapple, Dafe Uwanogho, Richard Goodyear,Gerry Melino, Simon Lovestone and Richard Killick, TAp73 isoforms antagonize Notch signalling in SH-SY5Y neuroblastomas and in primary neurons. *Journal of Neurochemistry*, 99: 989–999(2006).

[7] Chee-Onn Leong,1 Nick Vidnovic, Maurice Phillip DeYoung, Dennis Sgroi and Leif W. Ellisen, The p63/p73 network mediates chemosensitivity to cisplatin in a biologically defined subset of primary breast cancers. *The Journal of Clinical Investigation*, 117:1370-1380(2007).

[8] Dan Levy, Yaarit Adamovich, Nina Reuven, and Yosef Shaul,.Yap1 Phosphorylation by c-Abl Is a Critical Step in Selective Activation of Proapoptotic Genes in Response to DNA Damage. *Molecular Cell* ,29: 350–361(2008).

[9] Ryohei Yagi, Lin-Feng Chen, Katsuya Shigesada1, Yota Murakami and Yoshiaki Ito,. A WW domain-containing Yes-associated protein (YAP) is a novel transcriptional co-activator. *The EMBO Journal*, 18(9): 2551–2562(1999).

[10] Sabrina Strano, Eliana Munarriz, Mario Rossii, Luisa Castagnolii, Yosef Shaul, Ada Sacchi, Moshe Oren, Marius Sudol, Gianni Cesarenii, and Giovanni Blandino, 2001. Physical Interaction with Yes-associated Protein Enhances p73 Transcriptional Activity. *J. Biol. Chem.*, 276(18): 15164–15173.

[11] Alex Vassilev, Kotaro J. Kaneko, Hongjun Shu, Yingming Zhao, and Melvin L. DePamphilis,. TEAD/TEF transcription factors utilize the activation domain of YAP65, a Src/Yes-associated protein localized in the cytoplasm. *Genes & Development*, 15:1229–1241(2001).

[12] Akihiko Komuro, Makoto Nagai, Nicholas E. Navin, and Marius Sudol, WW Domain-containing Protein YAP Associates with ErbB-4 and Acts as a Co-transcriptional Activator for the Carboxyl-terminal Fragment of ErbB-4 That Translocates to the Nucleus. *J. Biol. Chem.*, 278(35): 33334–33341(2003).

[13] D Levy, Y Adamovich, N Reuven and Y Shaul, The Yes-associated protein 1 stabilizes p73 by preventing Itch-mediated ubiquitination of p73. *Cell Death and Differentiation*,14: 743–751(2007).

[14] David Matallanas, David Romano, Karen Yee, Katrin Meissl, Lucia Kucerova, Daniela Piazzolla,Manuela Baccarini, J. Keith Vass, Walter Kolch, and Eric O'Neill1, RASSF1A Elicits Apoptosis through an MST2 Pathway Directing Proapoptotic Transcription by the p73 Tumor Suppressor Protein. *Molecular Cell*, 27: 962–975(2007).

[15] Eleonora Lapi, Silvia Di Agostino, Sara Donzelli, Hilah Gal, Eytan Domany, Gideon Rechavi, Pier Paolo Pandolfi, David Givol Sabrina Strano, Xin Lu, and Giovanni Blandino, PML, YAP, and p73 Are Components of a Proapoptotic Autoregulatory Feedback Loop. *Molecular Cell* , 32: 803–814 (2008).

[16] Gao, X., Xing, D., Molecular mechanisms of cell proliferation induced by low power laser irradiation. *J Biomed Sci,* 12, 16-24 (2009).

[17] Lavi, R., Shainberg, A., Friedmann, H., Shneyvays, V., Rickover, O., Eichler, M., Kaplan, D., and Lubart, R., Low energy visible light induces reactive oxygen species generation and stimulates an increase of intracellular calcium concentration in cardiac cells. *J Biol Chem,* 278, 40917–40922 (2003).

[18] Gao, X., Chen, T., and Xing, D., Wang, F., Pei, Y., and Wei, X., Single cell analysis of PKC activation during proliferation and apoptosis induced by laser irradiation. *J Cell Physiol*, 206, 441–448 (2006).

[19] Zhang, L., Xing, D., Zhu, D., and Chen, Q., Low-Power Laser Irradiation Inhibiting Aβ25-35-induced PC12 Cell Apoptosis via PKC Activation. *Cell Physiol Biochem* 22, 215–222 (2008).

[20] Zhang, J., Xing, D., Gao, X., Low-Power Laser Irradiation activates Src tyrosine kinase through reactive oxygen species-mediated signaling pathway. *J Cell Physiol*, 217(2), 518-28 (2008).

[21] XUEJUAN GAO, DA XING,* LEI LIU, AND YONGHONG TANG, H-Ras and PI3K Are Required for the Formation of Circular Dorsal Ruffles Induced by Low-Power Laser Irradiation. *J Cell Physiol*, 219, 535-543 (2009).

[22] Zhang, LL. Xing, D., Gao, X., and Wu, S., Low-Power Laser Irradiation Promotes Cell Proliferation by Activating PI3K/Akt Pathway. *J Cell Physiol*, 219, 553-562 (2009).

[23] Zhang, LL., Zhang YJ., X., and Xing, D., LPLI Inhibits Apoptosis Upstream of Bax Translocation via a GSK-3β-Inactivation Mechanism. *J Cell Physiol*, 224, 218-228 (2010).

# Dynamin-Related Protein 1 Translocates from the Cytosol to Mitochondria during UV-Induced Apoptosis

Zhenzhen Zhang, Shengnan Wu *, Jie Feng

MOE Key Laboratory of Laser Life Science & Institute of Laser Life Science, College of Biophotonics, South China Normal University, Guangzhou 510631, China

## ABSTRACT

Mitochondria are dynamic structures that frequently divide and fuse with one another to form interconnecting network. This network disintegrates into punctiform organelles during apoptosis. However, the mechanisms involved in these processes are still not well characterized. In this study, we investigate the role of dynamin-related protein 1 (Drp1), a large GTPase that mediates outer mitochondrial membrane fission, in mitochondrial dynamics in response to UV irradiation in human lung adenocarcinoma cells (ASTC-α-1) and HeLa cells. Using time-lapse fluorescent imaging, we find that Drp1 primarily distributes in cytosol under physiological conditions. After UV treatment, Drp1 translocates from cytosol to mitochondria, indicating the enhancement of Drp1 mitochondrial accumulation. Our results suggest that Drp1 is involved in the regulation of transition from an interconnecting network to a punctiform mitochondrial phenotype during UV-induced apoptosis.

**Keywords:** Drp1, fission, UV, apoptosis

## 1 INTRODUCTION

Mitochondria are dynamic organelles that frequently divide and fuse with one another to maintain their architecture and functions [1]. Mitochondrial fission and fusion are normal events in healthy cells. While the precise mechanism for the shape changes is not well understood, perturbations of these processes affect cell survival and the evolutionarily conserved proteins mediate these processes [2], indicating that mitochondrial fission and fusion are important for cellular functions. The dynamin-related GTPase Drp1 is an evolutionarily conserved protein that contributes to outer mitochondrial membrane fission in mammalian cells and its functiondrial impairment results in elongated mitochondria [3]. Similar to other dynamins, Drp1 contains an N-terminal GTPase domain, a middle domain involved in protein self-assembly and a C-terminal GTPase effector domain [4-7]. In vitro, DRP1 assembles into spirals at division sites around the outer mitochondrial membrane and tubulates liposomes in a nucleotide-dependent manner, suggesting that it functions directly in membrane constriction and/or scission [8]. In vivo, DRP1 forms cytosolic tetramers and is recruited to defined loci along the mitochondrial surface, which often mark sites of mitochondrial fission [3,5,6,9] .Drp1 exists primarily in the cytoplasm but partially associates into defined foci on the outer surface of mitochondria that coalesce at sites of organelle fission [3].

UV irradiation is a DNA-damaging agent that activates a p53-dependent apoptotic response to induce apoptosis [10]. Our previous studies demonstrate that Bax translocation induced by UV irradiation is a p53 transcription-dependent event and PUMA promotes Bax translocation both by directly interacting with Bax and by competitive binding to Bcl-XL in UV-induced apoptosis [11-12]. Furthermore, BimL is involved in Bax activation during UV irradiation-induced apoptosis [13-14]. These results indicated that UV-induced apoptosis is a highly complex process.

* Corresponding author: Shengnan Wu, Ph.D. Phone: +86-20-85211436-8209; Fax: +86-20-85216052;
E-mail: wushn@scnu.edu.cn

During apoptosis, the mitochondrial network fragmented into smaller and more numerous mitochondria [15]. Drp1 and other proteins participates in regulating the fission of mitochondria. However, the precise role of mitochondrial fission during apoptosis needs to be further clarified.

In this study, we demonstrate that Drp1 translocates from cytosol to mitochondria and accumulates on mitochondria during UV-induced apoptosis. Furthermore, Drp1 regulates the transition from an interconnecting network to a punctiform mitochondrial phenotype, which may be contribute to cell dysfunction and UV-induced apoptosis.

## 2 Materials and Methods

### 2.1 Cell Culture

Human lung adenocarcinoma cells (ASTC-α-1) and HeLa cells were grown on 22-mmculture glasses, in Dulbecco's modified Eagle's medium (DMEM, Life Technologies, Inc.), supplemented with 15% fetal bovine serum, 50 units/mL penicillin, and 50 μg/mL streptomycin in 5% $CO_2$, 95% air at 37°C in a humidified incubator.

### 2.2 Plasmids and Treatments

The pYFP-Drp1 was kindly supplied by Heidi M. McBride (Rodolfo Zunino *et al.*, 2009) and and pDsRed-mit was kindly supplied by Dr. Y. Gotoh (Tsuruta *et al.*, 2002). Transfections were performed with Lipofectamine 2000 reagent (Invitrogen), according to the manufacturer's protocol. The medium was replaced with fresh culture medium after 5 h. Cells were examined at 24–48 h after transfection. For UV treatment, medium was removed and saved. The cells were rinsed with phosphate-buffered saline (PBS) and irradiated, and the medium was restored. Unless otherwise specified, cells were exposed to UV irradiation at a fluence of 120 mJ/cm$^2$ and observed at the times indicated.

### 2.3 Time-lapse confocal fluorescence microscopy

YFP and DsRed fluorescence were monitored confocally using a commercial laser scanning microscope (LSM 510/ConfoCor 2) combination system (Zeiss, Jena, Germany) equipped with a Plan-Neofluar 40×/1.3 NA Oil DIC objective. Excitation wavelength and detection filter settings for each of the fluorescent indicatorswere as follows: YFP fluorescence was excited at 488 nm with an argon ion laser and emission was recorded through a 500–550 nm band pass filter. DsRed fluorescence was excited at 543 nm with a helium-neon laser and emitted light was recorded through a 560 nm long pass filter. For time-lapse imaging, culture dishes were mounted onto the microscope stage that was equipped with a temperature-controlled chamber (Zeiss, Jena, Germany).

### 2.4 FRAP Analysis

For fluorescence recovery after photobleaching (FRAP) analyses, the same 40×objective was used. HeLa cells were plated on 22 mm culture glasses. After 24 hours, cells were transfected with DsRed-Mit to label mitochondria. The RFP signal from mitochondria was bleached by using 100% of the 543- and 514-nm lines. Recovery was monitored by scanning with the 543-nm line every 5s for 5 min. Quantification of recovery kinetics was performed by using the Zeiss LSM software.

### 2.5 Drp1-YFP Translocation Assay

To monitor Drp1-YFP translocation in living cells, ASTC-α-1 cells were cotransfected with Drp1-YFP and DsRed-Mit. Using an LSM 510 confocal microscope (Carl Zeiss), we imaged both the distribution pattern of Drp1-YFP and that of DsRed-Mit simultaneously during UV-induced apoptosis. Drp1 redistribution was assessed by the matching fluorescence of Drp1-YFP and DsRed- Mit emission. The cells exhibiting strong punctate staining of YFP, which overlapped with the distribution of DsRed, were counted as the cells with mitochondrially localized Drp1.

# 3 RESULTS

## 3.1 The Conversion of the Mitochondrial Network into a Punctiform Phenotype during UV-induced apoptosis.

To visualize mitochondria, ASTC-α-1 cells and HeLa cells were transiently transfected with pDsRed-mit for localizing mitochondria. 24-48hours after transfection, cells were treated with UV (120 mJ/cm$^2$) to induce apoptosis and visualized by confocal microscopy.

A.

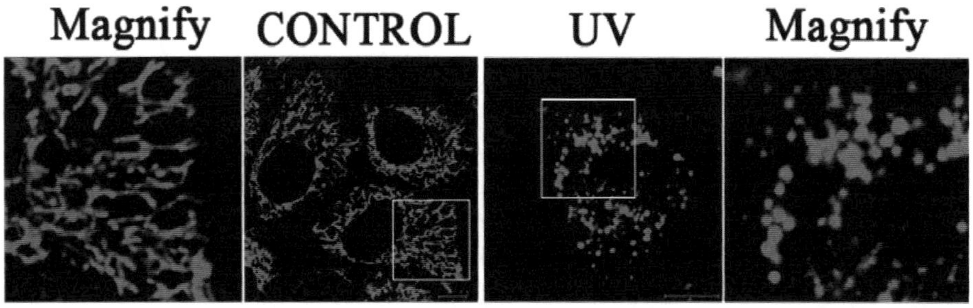


B.

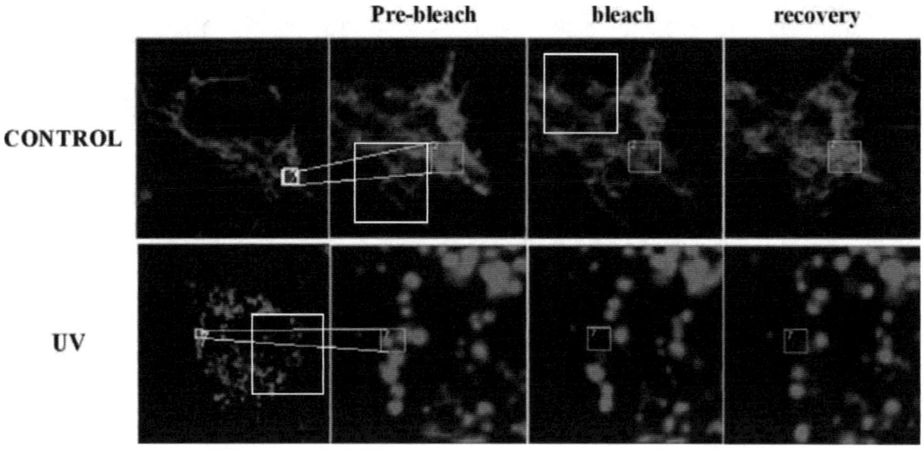

C.

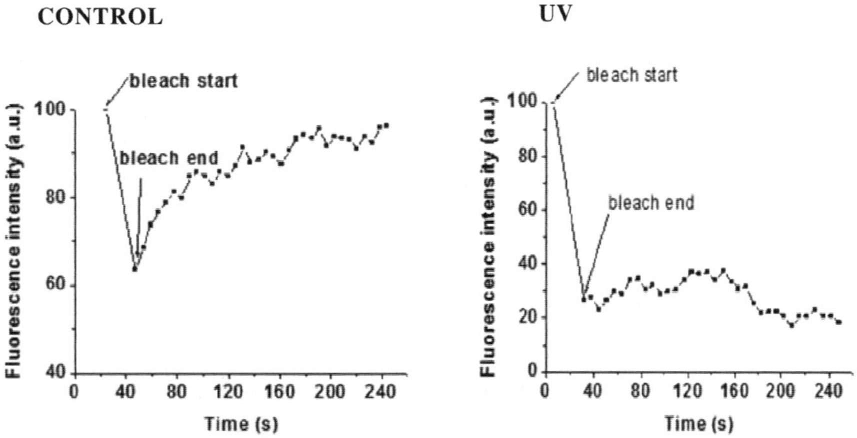


Figure.1. The Conversion of the Mitochondrial Network into a Punctiform Phenotype during UV-induced apoptosis. ASTC-α-1 (A) and HeLa (B, C) cells undergoing profound changes of their mitochondrial phenotype with UV ($120mJ/cm^2$) treatment. In both cells, the characteristic tubular mitochondrial morphology disintegrates into numerous round fragments. Rapid recovery of fluorescence after photobleaching (B, C) is indicative of mitochondrial connectivity, while a failure to recover fluorescence is indicative of mitochondrial discontinuity or fragmentation. Bar=10μm.

### 3.2 Drp1-YFP predominantly locates to cytosol under physiological conditions

To visualize mitochondria, ASTC-α-1and HeLa cells were transiently co-transfected with DsRed-mit and Drp1-YFP. 24-48 hours after transfection, cells were evaluated using laser confocal microscopy.

A.

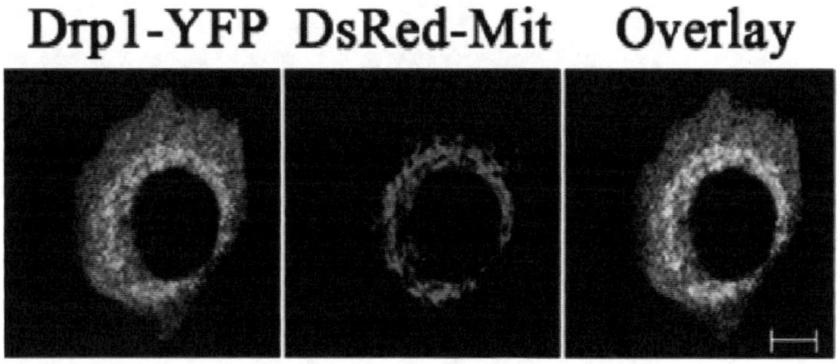


B.

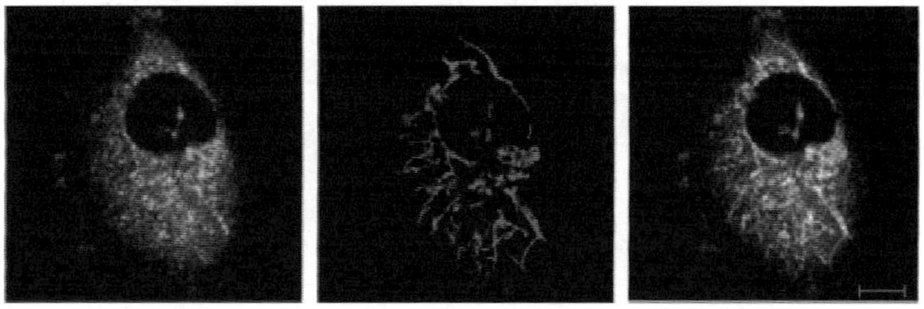

Figure.2. Subcellular localization of Drp1-YFP in ASTC-α-1 and HeLa cells.

ASTC-α-1 cells (A) and HeLa cells (B) were co-transfected with DRP1-YFP and DsRed-mit. In healthy mammalian cells, Drp1 has primarily a cytosolic subcellular distribution. Bar=10μm.

### 3.3 Drp1 Translocates to the Mitochondria during UV-induced Apoptosis

A.

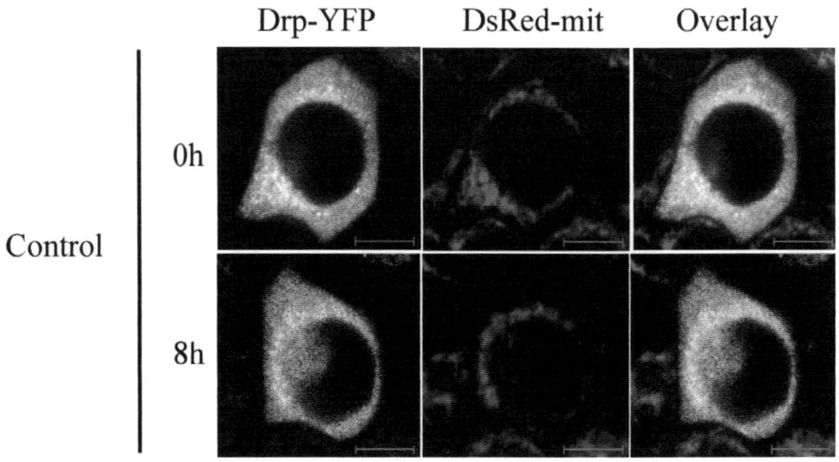


B.

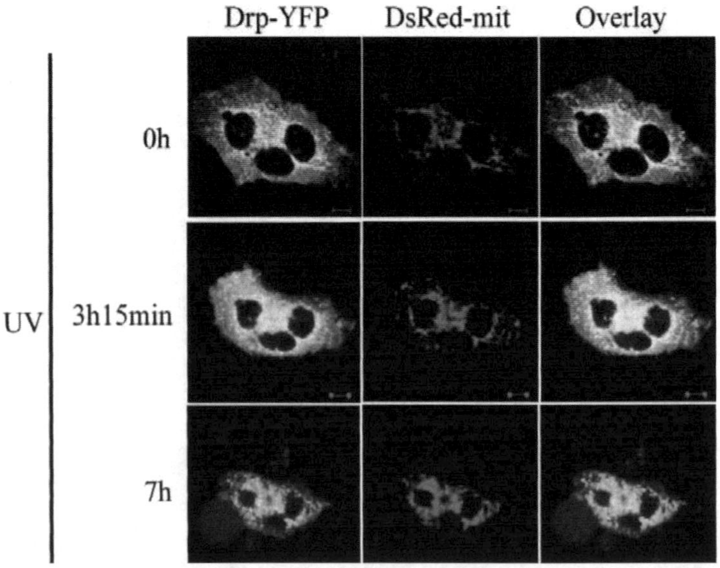


Fig.3. (A) ASTC-α-1 cells were transfected with Drp1-YFP and DsRed-mit. In healthy ASTC-α-1 cells, most of the Drp1-YFP has a diffuse cytosolic distribution. During UV-induced apoptosis (120mJ/cm$^{2)}$, Drp1-YFP translocates from cytosol to mitochondria, indicating the enhancement of Drp1 mitochondrial accumulation. Bar=10μm.

## 4 DISCUSSION

In this study, we demonstrate that during UV-induced apoptosis the mitochondrial reticulum of mammalian cells disintegrates into multiple punctiform mitochondria. We find that Drp1, dynamin-related GTPase, is involved in the profound switch of mitochondrial phenotypes. Furthermore, we document the role of Drp1 in mitochondrial fission at steady state and during UV-induced apoptosis. Under apoptotic stimuli, Drp1 translocates to the mitochondria, and the typical reticular organization of mitochondria disintegrates into multiple small fragments. The apoptosis-associated translocation of Drp1 is consistent with the conclusion that Drp1 is involved in the transition from a tubular to a punctiform mitochondrial phenotype.

Previous reports indicate that Drp1 plays an important role in the progression of apoptosis [16]. However, mechanisms that are responsible for regulation of the mitochondrial fission during cell apoptosis are unknown. Furthermore, the mechanisms for recruitment of Drp1 to the mitochondria for spiral formation and mitochondrial fission remains unclear [17]. Although Drp1 is known to function as a mitochondrial fission enzyme, overexpression of Drp1 within cells does not significant increase mitochondrial fission [18-19], indicating that there are other limiting factors in the mitochondrial fission process. These factors may recruit cytosolic Drp1 to the outer mitochondrial membrane and/or activate the Drp1 GTPase, leading to mitochondrial fission. In future studies, we will focus on the mechanisms for the recruitment of Drp1 to mitochondria surface in response to UV-induced apoptosis.

## ACKNOWLEDGMENTS

The authors thank Prof. Heidi M. McBride for providing the plasmid Drp1-YFP and Dr. Y. Gotoh for providing the plasmid DsRed-mit. This research is supported by the National Basic Research Program of China (2010CB732602), the Program for Changjiang Scholars and Innovative Research Team in University (IRT0829), and the National Natural Science Foundation of China (30870676; 30870658).

## REFERENCES

[1] Xiao-Jian Han, Yun-Fei Lu, Shun-Ai Li, Taku Kaitsuka, Yasufumi Sato, Kazuhito Tomizawa, Angus C. Nairn, Kohji Takei, Hideki Matsui, and Masayuki Matsushita, "CaM kinase Ia–induced phosphorylation of Drp1 regulates mitochondrial morphology," J. Cell Biol, 182, 573–585 (2008).

[2] Karbowski, M. and Youle, R. J., "Dynamics of mitochondrial morphology in healthy cells and during apoptosis," Cell Death Differ, 10, 870-880 (2003).

[3] Smirnova, E., Griparic, L., Shurland, D. L. and van der Bliek, A. M., "Dynamin-related protein Drp1 is required for mitochondrial division in mammalian cells," Mol. Biol. Cell, 12, 2245-2256 (2001).

[4] Fukushima, N.H., E. Brisch, B.R. Keegan, W. Bleazard, and J.M. Shaw, "The GTPase effector domain sequence of the Dnm1p GTPase regulates self-assembly and controls a rate-limiting step in mitochondrial fission," Mol. Biol. Cell, 12, 2756–2766 (2001).

[5] Pitts, K.R., M.A. McNiven, and Y. Yoon, "Mitochondria-specifi c function of the dynamin family protein DLP1 is mediated by its C-terminal domains," J. Biol. Chem., 279, 50286–50294 (2004).

[6] Zhu, P.P., A. Patterson, J. Stadler, D.P. Seeburg, M. Sheng, and C. Blackstone, "Intra- and intermolecular domain interactions of the C-terminal GTPase effector domain of the multimeric dynamin-like GTPase Drp1," J. Biol. Chem, 279, 35967–35974 (2004).

[7] Ingerman, E., E.M. Perkins, M. Marino, J.A. Mears, J.M. McCaffery, J.E. Hinshaw, and J. Nunnari, "Dnm1 forms spirals that are structurally tailored to fi t mitochondria," J. Cell Biol, 170, 1021–1027 (2005).

[8] Yoon, Y., K.R. Pitts, and M.A. McNiven, "Mammalian dynamin-like protein dlp1 tubulates membranes," Mol. Biol. Cell, 12, 2894–2905 (2001).

[9] Shin, H.W., H. Takatsu, H. Mukai, E. Munekata, K. Murakami, and K.Nakayama, "Intermolecular and interdomain interactions of a dynamin-related GTP-binding protein," Dnm1p/Vps1p-like protein, J. Biol. Chem, 274, 2780–2785 (1999).

[10] Vousden, K. H., and Lu, X, "Live or let die: the cell's response to p53," Nat. Rev. Cancer, 2, 594–604 (2002).

[11] Wu, Y., Xing, D., Chen, W. R., and Wang, X, "Bid is not required for Bax translocation during UV-induced apoptosis," Cell Signal, 19, 2468–2478 (2007).

[12] Yingjie Zhang, Da Xing, and Lei Liu, "PUMA Promotes Bax Translocation by Both Directly Interacting with Bax and by Competitive Binding to Bcl-XL during UV-induced Apoptosis," Mol. Biol. Cell, 20, 3077 – 3087 (2009).

[13] Chen, M.J., Xing, D., Chen, T. and Zhang, L., "BimL involvement in Bax activation during UV irradiation-induced apoptosis," Biochem. Biophys. Res. Commun, 358, 559–565 (2007).
[14] Xianwang Wang, Da Xing, Lei Liu, Wei R. Chen, "BimL directly neutralizes Bcl-xL to promote Bax activation during UV-induced apoptosis," FEBS Letters, 583, 1873–1879 (2009).
[15] Karbowski, M., Norris, K.L., Cleland, M.M., Jeong, S.Y., and Youle, R.J., "Role of Bax and Bak in mitochondrial morphogenesis," Nature, 443, 658–662 (2006).
[16] Martinou, J.C., and R.J. Youle, "Which came first, the cytochrome c release or the mitochondrial fission?" Cell Death Differ, 13, 1291–1295 (2006).
[17] Der-Fen Suen, Kristi L. Norris, and Richard J. Youle, "Mitochondrial dynamics and apoptosis," Genes & Dev, 22, 1577-1590 (2008).
[18] Smirnova, E., Shurland, D. L., Ryazantsev, S. N. and van der Bliek, A. M., "A human dynamin-related protein controls the distribution of mitochondria," J. Cell Biol, 143, 351-358 (1998).
[19] Pitts, K. R., Y. Yoon, E. W. Krueger, and M. A. McNiven, "The dynamin-like protein DLP1 is essential for normal distribution and morphology of the endoplasmic reticulum and mitochondria in mammalian cells," Mol. Biol. Cell, 10, 4403–4417 (1999).

# A New Theoretical Approach to Single-Molecule Fluorescence Optical Studies of RNA Dynamics

**Xinghai Zhao** [1], **Guangcun Shan** [2,3], **and Shuying Bao** [3,4]

[1] Institute of Electronic Engineering, China Academy of Engineering Physics, Mianyang 621900, China

[2] Institute of Advanced Materials, Nanjing University of Posts and Telecommunications, Nanjing 210003, China

[3] State Key Laboratory of Functional Materials for Informatics, Shanghai Institute of Microsystem and Information Technology, Shanghai 200050, China

[4] Department of Physics, Fudan University, Shanghai 200433, China

E-mail: spinor@163.com

**Abstract.** Single-molecule fluorescence spectroscopy in condensed phases has many important chemical and biological applications. The single-molecule fluorescence measurements contain information about conformational dynamics on a vast range of time scales. Based on the data analysis protocols methodology proposed by X. Sunney Xie, the theoretical study here mainly focuses on the single-molecule studies of single RNA with interconversions among different conformational states, to with a single FRET pair attached. We obtain analytical expressions for fluorescence lifetime correlation functions that relate changes in fluorescence lifetime to the distance-dependent FRET mechanism within the context of the Smoluchowski diffusion model. The present work establishes useful guideline for the single-molecule studies of biomolecules to reveal the complicated folding dynamics of single RNA molecules at nanometer scale.

## 1. Introduction

In the past few years, single-molecule fluorescence resonant energy transfer (FRET) spectroscopy is undergoing an increasingly popular method of probing the complex dynamics of single biomolecules [1-12]. Site-specific labeling of biomolecules with appropriate dyes is a prerequisite for these experiments. Fluorescently labeled single biomolecules can be detected in two configurations, either immobilized by a surface attachment or freely diffusing in solution. These two detection formats provide different and complementary kinds of information. Single-molecule FRET measurements of surface immobilized RNA have been a powerful approach for studying equilibrium conformational dynamics of DNA and RNA [5-9, 12-14]. The ability to watch one biomolecule at a time helps us to obtain unique information on relevant observables, resolve subpopulations in a heterogeneous sample and record asynchronous time trajectories of observables that would otherwise be hidden during biochemical reactions [1-16]. This ability is crucial in many biological contexts, in which triggering is not possible or ensembles moving stochastically on complex reaction landscape quickly lose their coherence, resulting in averaged behavior.

In a single-molecule FRET experiment, one uses a single pair of dye molecules attached on a biomolecule. If the donor is excited by an incident laser, the excitation energy can be transferred from the donor to the acceptor, provided that the emission spectrum of the donor overlaps with the absorption spectrum of the acceptor. This leads to a reduction in the donor's fluorescence intensity and excited state lifetime, and an increase in the acceptor's emission intensity. The fluorescence excitation energy from the donor to the acceptor then depends on the FRET rate [4-26]

$$\Gamma_{FRET} = \gamma_D \left( \frac{R_0}{r} \right)^6 \tag{1}$$

where $R_0$ is the Förster distance at which 50% of the energy is transferred, $\gamma_D$ is the fluorescence decay rate of the donor in the absence of the acceptor, and $r$ the distance between the donor and the acceptor.

The FRET rate is thus very sensitive to r which could be used as a "nano-ruler" that allows one to infer the distance changes of the probed molecule to monitor the conformational changes. Thus, the energy transfer rate, in turn, depends sensitively on the donor and acceptor distance r. The detection of fluorescence from single pair of dye molecules involves repeated cycling of the molecule between ground and excited states and detection of the series of emitted photons; nonradiative relaxation of the excited state results in a reduction of the maximum photon flux. When diffusing or flowing single molecules in a liquid traverse the laser excitation volume, fluorescence photon bursts are generated. Although short, the bursts can be analyzed for their duration, brightness, spectrum, and fluorescence lifetime, thereby providing molecule information on identity, size, diffusion coefficient, and conformation dynamics. Because a large number of events (photon bursts) can be collected in a relatively short time, statistical analyses of these data are possible, and histograms can be constructed. Most studies of single-molecule fluorescence bursts were to measure distributions in burst size or in fluorescence lifetime. For example, two-color burst analysis and multiple-lifetime approaches for identification and separation were suggested and implemented.

Fluorescence lifetime spectroscopy at the single molecule level allow the measurement of translational, rotational, and conformational dynamics of biomolecules over a wide range of time scales. In combination with fluorescence resonant energy transfer (FRET), intermolecular and intramolecular distances between about 2nm to 10 nm can be probed. Indeed, single-molecule FRET analysis, which is based on fluorescence lifetime, is an alternative to the popular intensity-based methods. The advantages include the less susceptibility to variation in the excitation intensity, photon-detection efficiency, or spectral cross talk. The related techniques, fluorescence correlation spectroscopy and two-color cross-correlation spectroscopy have been used to analyze, sort, and detect conformational states of single biomolecules in the excitation volume.

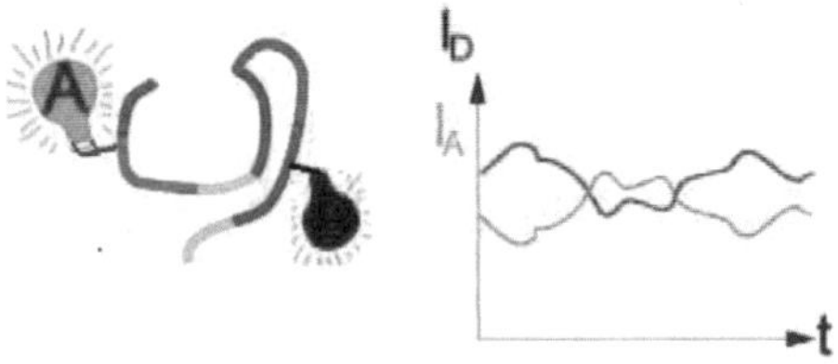


**Figure 1.**. Schematic representation of the single-molecule FRET experiments, in which a single FRET pair attached to a RNA molecule. (*A* stands for acceptor dye molecule and *D* stands for donor dye molecule, respectively).

By using a laser probe as an excitation source, fluorescence of a single molecule has been detected. The emission spectra [17] and excited state lifetime [17–20] of a single molecule have also been measured. Examples for single-molecule fluorescence lifetime measurements include tetramethylrhodamine (TMR) dye molecules linked to $tRNA^{Phe}$ and DNA. The fluctuations of the single-molecule lifetime contain detailed dynamic information. In this light, single-molecule fluorescence lifetime holds the promise of serving as a sensitive measure for conformational dynamics. We note that time resolved fluorescence lifetime measurements provide more information than just the single quantity $E_{obs}$. Consequently, time resolved data—when available—can be used to constrain the underlying distance distribution $p(R)$ more accurately [11]. Moreover, such experiments can be used to obtain information about chain dynamics [12]. To gain insights from the measured single-molecule fluorescence lifetime correlations, Xie *et.al.* have presented the data analysis protocols that afford an efficient extraction of conformational dynamics on a broad range of time scales from single-molecule fluorescence lifetime trajectories photon by photon [1,2], in cases where fluorescence quenching is either due to Förster fluorescence resonance energy transfer or due to the excited-state electron transfer. It is concluded that the method not only extends the fast measurable time scale to a microsecond or less, but also allows the disclosure of information from the limited number of detected photons of a single molecule over the time scales that are important in characterizing the function of a biological macromolecule. A similar detection scheme has been used by Seidel and co-workers for single-molecule identification and conformational dynamics studies [26]. Their data were processed by averaging over every 80 photons, from which the distribution of fluorescence lifetime is constructed and analyzed. Photon statistics, however, demands that a number of photons be obtained, or "binned", in order to reach a statistically reasonable estimate of the lifetime [1, 2, 16-20]. The time evolution of the experimental observable, or trajectory, is analyzed to extract the information on molecular dynamics [1,2,6]. The stochastic delay times of individual photons require that many photons be binned in order to obtain the fluorescence lifetime. Therefore, in combination with single-molecule FRET technique, fluorescence lifetime measurement using time-correlated single-photon counting (TCSPC) is oblivious to events that occur on a time scale faster than or comparable to the bin time [1,2,16] circumventing such a shortcoming. It has been recognized that there is more information available should the single-molecule trajectory be analyzed photon by photon.

However, several challenges need to be overcome to further study the single-molecule FRET related techniques. To improve the time resolution and statistics for single-molecule FRET studies, we have proposed a versatile strategy for studying conformational dynamics of a single biomolecule [27]. In this work, we focus on the lifetime fluctuation-correlation analysis of single RNA molecule using single-molecule FRET experiments, in which the lifetime of single FRET pair of dyes is dependent on donor-acceptor distance. We present a Smoluchowski diffusion model that affords an efficient extraction of conformational dynamics from lifetime trajectories of single-molecule FRET, based on the fluctuation-correlation analysis of observables.

## 2. Model and Results

Consider a RNA molecule that undergoes a Smoluchowski diffusion in the free-energy potential *V*(*r*) with interconversions among several conformational states with different fluorescence lifetimes, and a single FRET pair attached to it undergoes the FRET mechanism. The fluorescence lifetime of the donor fluorophore is $\gamma(t)^{-1} = \left[\gamma_D + \gamma_{FRET}\left(r(t)\right)\right]^{-1}$, where $_D$ is the fluorescence decay rate of the donor in the absence of acceptor and $_{FRET}$ is the energy transfer rate from the donor to acceptor. Conformational dynamics or structural fluctuations of a RNA molecule may cause changes in the donor-acceptor distance *r*, which, in turn, may result in a variation of the energy transfer rate. In this light, fluorescence lifetime can serve as a sensitive and distance-dependent probe for conformational dynamics.

Firstly, briefly review the data analysis methodology proposed by X. Sunney Xie [1, 2]. Given a kinetic scheme of interconversion, all the dynamic information is contained in the time-dependent state propagator $G\left(f,t_f;i,t_i\right)$, which describes the conditional probability of the system arriving at state f at time $t_f$ given the system being at state i at time $t_i$. The structure of $G\left(f,t_f;i,t_i\right)$ is determined by the rate constants that characterize the kinetic scheme. One goal of statistical analyses of single molecule trajectories is therefore to determine the interconversion rates from experimental observables – which, in the present case, are the time-stamped photon sequences from TCSPC. The approach utilizes correlation analyses of the observables. Specifically, we calculate the two-time fluctuation-correlation function, or autocorrelation, of two events separated by period t [1]

$$C_2(t)=\left\langle \delta\gamma(t)^{-1}\,\delta\gamma(0)^{-1}\right\rangle=\left\langle \delta\gamma^{-2}\right\rangle\tilde{C}_2(t) \tag{2}$$

where $\delta\gamma(t)^{-1}=\gamma(t)^{-1}-\left\langle\gamma^{-1}\right\rangle$, and $\tilde{C}_2(t)$ is the normalized autocorrelation function, defined as

$$\tilde{C}_2(t)=\frac{\left\langle \gamma(t)^{-1}\,\gamma(0)^{-1}\right\rangle-\left\langle\gamma^{-1}\right\rangle^2}{\lim\limits_{t\to 0}\left\langle \gamma(t)^{-1}\,\gamma(0)^{-1}\right\rangle-\left\langle\gamma^{-1}\right\rangle^2} \tag{3}$$

Similarly, we define a three-time fluctuation-correlation function of three events separated by periods $t_1$ and $t_2$ [1]

$$C_3(t_1,t_2)=\left\langle \delta\gamma(t_1+t_2)^{-1}\,\delta\gamma(t_1)^{-1}\,\delta\gamma(0)^{-1}\right\rangle=\left\langle \delta\gamma^{-3}\right\rangle\tilde{C}_3(t) \tag{4}$$

where the normalized three-time correlation function is

$$\tilde{C}_3(t_2,t_1)=\frac{\left\langle \delta\gamma(t_1+t_2)^{-1}\,\delta\gamma(t_1)^{-1}\,\delta\gamma(0)^{-1}\right\rangle}{\lim\limits_{t_1\to 0}\lim\limits_{t_2\to 0}\left\langle \delta\gamma(t_1+t_2)^{-1}\,\delta\gamma(t_1)^{-1}\,\delta\gamma(0)^{-1}\right\rangle} \tag{5}$$

where

$$\begin{aligned}\left\langle \delta\gamma(t_1+t_2)^{-1}\,\delta\gamma(t_1)^{-1}\,\delta\gamma(0)^{-1}\right\rangle&=\left\langle \gamma(t_1+t_2)^{-1}\,\gamma(t_1)^{-1}\,\gamma(0)^{-1}\right\rangle\\&-\left(\left\langle \gamma(t_2)^{-1}\,\gamma(0)^{-1}\right\rangle+\left\langle \gamma(t_1)^{-1}\,\gamma(0)^{-1}\right\rangle+\left\langle \gamma(t_1+t_2)^{-1}\,\gamma(0)^{-1}\right\rangle\right)\left\langle\gamma^{-1}\right\rangle+2\left\langle\gamma^{-1}\right\rangle^3\end{aligned} \tag{6}$$

The temporal correlation comes from either the $\left\langle \gamma(t)^{-1}\,\gamma(0)^{-1}\right\rangle$ or the

$\left\langle \gamma(t_1+t_2)^{-1}\,\gamma(t_1)^{-1}\,\gamma(0)^{-1}\right\rangle$ terms. They can be expressed in terms of the propagator $G(f,t_f;i,t_i)$ as

$$\left\langle \gamma(t)^{-1}\,\gamma(0)^{-1}\right\rangle=\sum_{f,i}\gamma_f^{-1}G_{fi}(t)\,\gamma_i^{-1}\rho_i^{eq} \tag{7}$$

And

$$\left\langle \gamma(t_1+t_2)^{-1}\,\gamma(t_1)^{-1}\,\gamma(0)^{-1}\right\rangle=\sum_{f,k,i}\gamma_f^{-1}G_{fk}(t_2)\,\gamma_k^{-1}G_{ki}(t_1)\,\gamma_i^{-1}\rho_i^{eq} \tag{8}$$

where $G_{fi}(t)=G(f,t_f;i,t_i)$, $t=t_f-t_i$ and $\gamma_i^{-1}$ and $\rho_i^{eq}$ are the fluorescence lifetime and equilibrium probability for finding the system at state i, respectively.

Next, we turn to considering a single RNA molecule that undergoes the Smoluchowski diffusion in a free-energy potential $V(r)=\frac{1}{2}m\omega^2r^2=\frac{1}{2}\kappa r^2$, where *m* is the reduced mass of RNA molecule, and $\kappa$ is the potential constant, with interconversions among several conformational states with different fluorescence lifetimes, and a single FRET pair attached to it undergoes the fluorescence resonance energy transfer. The fluorescence lifetime of the donor fluorophore is $\gamma^{-1}=(\gamma_D+\Gamma_{FRET})^{-1}$, where $_D$ is the fluorescence decay rate of the donor in the absence of the acceptor and $_{FRET}$ is the energy transfer rate from the donor to the acceptor. Conformational dynamics or structural fluctuations of a RNA molecule may cause changes in the donor-acceptor distance *r*, which, in turn, may result in a variation of the energy transfer rate. In this light, fluorescence lifetime can serve as a sensitive and distance-dependent probe for conformational dynamics. It is then possible to probe directly the fluctuation of donor-acceptor distance $r$ (t). The rate of widely used FRET mechanism can be expressed as $\Gamma_{FRET}(t) = \gamma_D (R_0 / r(t))^6$, where $r$ is the center-to-center donor-acceptor distance. As a result of the structural fluctuations, the fluorescence lifetime of donor varies in time $\gamma(t)^{-1} = [\gamma_D + \gamma_{FRET}(r(t))]^{-1}$.

To illustrate the trajectories $r$ (t) of this motion, we perform diffusion dynamics simulation for a RNA chain using the velocity Verlet algorithm [6, 28]. The instantaneous value of the distance $r$ (t) in Fig.2 determines, at each moment, the probability of excitation transfer from the donor (D) to the acceptor (A), and thus the fluorescence lifetime according to equation 1.

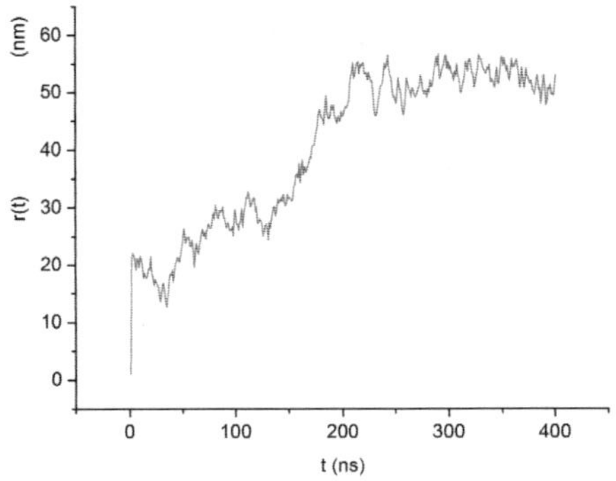


**Figure 2.** The simulated trajectories r (t) of Smoluchowski diffusion motion for a RNA chain using the velocity Verlet algorithm.

As we all know, the corresponding dynamic equation can be described by the fractional Fokker-Planck equation (FFPE) developed by Klafter and co-workers [29]. Now we use the FFPE solutions to calculate the fluorescence lifetime correlation functions for single-molecule FRET measurements. The definitions of and the solutions to FFPE are given in Refs. 29 and 30. The FFPE has the form [1, 29, 30]

$$\frac{\partial}{\partial t} H_\alpha(r,t \mid r_0,t_0) = {}_0D_t^{1-\alpha} \left\{ \frac{\partial}{\partial r} \left[ \lambda_\alpha \frac{dV(r)}{dr} \right] + \lambda_\alpha \theta \frac{\partial^2}{\partial r^2} \right\} H_\alpha(r,t \mid r_0,t_0), \quad (9)$$

where = ²/ is the fractional drift coefficient with the dimension s with being the generalized friction constant, and θ=$k_BT/\kappa$ characterizes the magnitude of the fluctuation at room temperatures (fluctuation–dissipation theorem). The Riemann–Liouville fractional operator is defined by [1, 29, and 30]

$${}_0D_t^{1-\alpha} H_\alpha(r,t) = \frac{1}{\Gamma(\alpha)} \frac{\partial}{\partial t} \int_0^t H_\alpha(r,t') / (t-t')^{1-\alpha} dt' \quad (10)$$

The solution to equation (9) is

$$H_\alpha(r,t \mid r_0,t_0=0) = \frac{1}{\sqrt{2\pi\theta}} \sum_{n=0}^{\infty} \frac{1}{2^n n!} E_\alpha(-n\tilde{t}^\alpha) h_n\left(\frac{\tilde{r}_0}{\sqrt{2}}\right) h_n\left(\frac{\tilde{r}}{\sqrt{2}}\right) e^{-\tilde{r}^2/2} \quad (11)$$

where the reduced parameters are $\tilde{r} = r/\sqrt{\theta}$ and $\tilde{t} = t(\lambda_\alpha)^{1/\alpha}$. The $h_n$'s are Hermite polynomials and their eigenvalues here are $_{n,}$ $=n$ . The Mittag–Leffler function is defined through [29]

$$E_\alpha\left(-n\tilde{t}^\alpha\right)=\sum_{m=0}^{\infty}\frac{\left(-n\tilde{t}^\alpha\right)^m}{\Gamma(1+m\alpha)} \tag{12}$$

where $\Gamma(z)=\int_0^\infty y^{z-1}e^{-y}dy$ is the gamma function. The analytical expressions for the probability density function $H_{,S}$ at short-to-intermediate times and $H_{,L}$ at long times is given by, respectively[1,21]

$$H_{\alpha,S}(r,t\,|\,r_0,t_0=0)=\left[2\pi\theta\left(1-e^{-2\lambda_\alpha t^\alpha/\Gamma(1+\alpha)}\right)\right]^{-1/2}\exp\left[-\frac{\left(r-r_0e^{-\lambda_\alpha t^\alpha/\Gamma(1+\alpha)}\right)^2}{2\theta\left(1-e^{-2\lambda_\alpha t^\alpha/\Gamma(1+\alpha)}\right)}\right] \tag{13}$$

$$H_{\alpha,L}(r,t\,|\,r_0,t_0=0)=\left[2\pi\theta\left(1-\lambda_\alpha^2t^{-2\alpha}/\Gamma(1-\alpha)^2\right)\right]^{-1/2}\exp\left[-\frac{\left(r-r_0\lambda_\alpha t^{-\alpha}/\Gamma(1-\alpha)\right)^2}{2\theta\left(1-\lambda_\alpha^2t^{-2\alpha}/\Gamma(1-\alpha)^2\right)}\right] \tag{14}$$

The donor-acceptor displacement correlation functions in these two limiting cases show a stretched exponential

$$\left\langle r(t)r(0)\right\rangle_{\alpha,S}=\theta e^{-\lambda_\alpha t^\alpha/\Gamma(1+\alpha)} \tag{15}$$

and a power-law behavior

$$\left\langle r(t)r(0)\right\rangle_{\alpha,L}=\theta\lambda_\alpha t^{-\alpha}/\Gamma(1-\alpha) \tag{16}$$

respectively. Using equation (10)-(14), the fluorescence lifetime fluctuation correlation at short-to-intermediate time for FRET dominated lifetime mechanism is given by [1, 29]

$$\left\langle\delta\gamma(t)^{-1}\,\delta\gamma(0)^{-1}\right\rangle_{FRET}=\gamma_D^{-2}\frac{6\theta}{R_0^{12}}\sum_{n=1}^{6}c_n\exp\left[-n\lambda_\alpha t^\alpha/\Gamma(1+\alpha)\right] \tag{17}$$

At long time, the fluorescence lifetime correlation function becomes [1, 29]

$$\left\langle\delta\gamma(t)^{-1}\,\delta\gamma(0)^{-1}\right\rangle_{FRET}=\gamma_D^{-2}\frac{6\theta}{R_0^{12}}\sum_{n=1}^{6}c_n\left[\lambda_\alpha^n t^{-n\alpha}/\Gamma(1-\alpha)^n\right] \tag{18}$$

where $c_n$ (n=1,2,...,6) are given by $c_1=6\left(x_{eq}^5+10x_{eq}^3\theta+15x_{eq}\theta^2\right)^2$, $c_2=75\theta\left(x_{eq}^4+6x_{eq}^2\theta+3\theta^2\right)^2$, $c_3=400x_{eq}^2\theta^2\left(x_{eq}^2+3\theta\right)^2$, $c_4=900\theta^3\left(x_{eq}^2+\theta\right)^2$, $c_5=720x_{eq}^2\theta^4$ and $c_6=120\theta^5$. Equation (17) and (18) clearly show that the time scale of the Smoluchowski diffusion dynamics spans several decades. In this case, the dynamics stretch across a greater span of time scales as the value becomes smaller higher degree of anomaly in diffusion.

The three-time fluorescence lifetime correlation function of equation (4) for this model at short-to-intermediate time for FRET dominated lifetime mechanism is

$$\begin{aligned}C_3(t_1,t_2)=&\left\langle\delta\gamma(t_1+t_2)^{-1}\,\delta\gamma(t_1)^{-1}\,\delta\gamma(0)^{-1}\right\rangle_{FRET}=\gamma_D^{-3}\frac{6\theta}{R_0^{18}}\sum_{n=1}^{6}c_n\\&\exp\left[-n\lambda_\alpha\left(t_1^\alpha+t_2^\alpha\right)/\Gamma(1+\alpha)\right]\times\exp\left[-n\lambda_\alpha(t_1+t_2)^\alpha/\Gamma(1+\alpha)\right]\\&-\sum_{n=1}^{6}c_n\exp\left[-n\lambda_\alpha t_1^\alpha/\Gamma(1+\alpha)\right]-\sum_{n=1}^{6}c_n\exp\left[-n\lambda_\alpha t_2^\alpha/\Gamma(1+\alpha)\right]\\&-\sum_{n=1}^{6}c_n\exp\left[-n\lambda_\alpha(t_1+t_2)^\alpha/\Gamma(1+\alpha)\right]\end{aligned} \tag{19}$$

At long time, the three-time correlation function of equation (4) becomes

$$C_3(t_1,t_2) = \left\langle \delta\gamma(t_1+t_2)^{-1}\,\delta\gamma(t_1)^{-1}\,\delta\gamma(0)^{-1} \right\rangle_{FRET} = \gamma_D^{-3}\frac{6\theta}{R_0^{18}}\sum_{n=1}^{6} c_n$$
$$\left[\lambda_\alpha^n\left(t_1^{-n\alpha}+t_2^{-n\alpha}\right)/\Gamma(1-\alpha)^n\right]\times\left[\lambda_\alpha^n\left(t_1+t_2\right)^{-n\alpha}/\Gamma(1-\alpha)^n\right]$$
$$-\sum_{n=1}^{6} c_n\left[\lambda_\alpha^n t_1^{-n\alpha}/\Gamma(1-\alpha)^n\right]-\sum_{n=1}^{6} c_n\left[\lambda_\alpha^n t_2^{-n\alpha}/\Gamma(1-\alpha)^n\right]$$
$$-\sum_{n=1}^{6} c_n\left[\lambda_\alpha^n\left(t_1+t_2\right)^{-n\alpha}/\Gamma(1-\alpha)^n\right] \tag{20}$$

It is obvious that three-time correlation function analysis is capable of handling models of degenerate state observables and especially handling more complicated conformational dynamics of a single RNA molecule. The analytical expressions that relate changes in fluorescence lifetime to the distance-dependent fluorescence resonance energy transfer within the context of Smoluchowski diffusion model have been obtained. The results corroborate the notion that a methodology that allows characterization of structural dynamics over several decades of time scales is crucial for a full understanding of such a complicated biomolecule system. The dynamics of such a RNA molecule that undergoes a diffusion dynamics would be fully understood with strategies that are capable of characterizing single-molecule dynamics on a wide range of time scales based on single-molecule FRET experiments.

## 3. Conclusions

The single-molecule FRET experiments provide a unique opportunity for exploring the relationship between structure and dynamics of a single biomolecule. The ultimate challenge is to measure conformational dynamics *in vivo*. The theoretical study here mainly focused on the aspects of single-molecule FRET experiments that utilize the data analysis methodology proposed by X. Sunney Xie used in single-molecule spectroscopy. To this end, we've derived analytical expressions of two-time and three-time correlation functions that relate changes in fluorescence lifetime to the distance-dependent FRET mechanism within the context of the Smoluchowski diffusion model. Our study establishes potentially useful guidelines for the experimental study of monitoring the dynamical processes in real time and revealing the mechanism of complex reaction of a single RNA at molecule scale. We hope that this subject will pave its way to the understanding of different biochemical systems at the molecule level.

## References


[1] Haw Yang and X. Sunney Xie, "Probing Single Molecule Dynamics Photon by Photon," *J. Chem. Phys.* **117,**10965 (2002).

[2] Haw Yang and X. Sunney Xie, "Statistical approaches for probing single-molecule dynamics photon-by-photon," *Chem. Phys*, **284**, 423-437 (2002).

[3] Nie, S., Chiu, D. T., Zare, R. N., "Real-time detection of single molecules in solution by confocal fluorescence microscopy", *Anal. Chem.* **67,** 2849–2857 (1995).

[4] Selvin, P.R., "The renaissance of fluorescence resonance energy transfer," *Nature Struct. Biol.* **7**, 730–734 (2000).

[5] Weiss S., "Measuring conformational dynamics of biomolecules by single molecule fluorescence spectroscopy," Nat Struct Biol, 7(9),724-729(2000).

[6] S.Y. Bao, G.C. Shan, and X.H. Zhao, "RNA Dynamics probed by single-molecule FRET studies," J. Comput. Theor. Nanosci., in press.

[7] Schmidt, T., Schutz, G. H., Baumgartner, W., Gruber, H. J., and Schindler, H. "Imaging of single molecule diffusion," *Proc. Natl. Acad. Sci.* **93,** 2926–2929 (1996).

[8] Ha, T., Enderle, T., Ogletree, D. F., Chemla, D. S., Selvin, P. R., and Weiss, S. "Probing the interaction between two single molecules: fluorescence resonance energy transfer between a single donor and a single acceptor," *Proc. Natl. Acad. Sci.* **93,** 6264–6268 (1996).

[9] Guangcun Shan, Wei Huang, "Optical Single Molecule Spectroscopy for Monitoring

Conformational Dynamics of a Single Biomacromolecule," *J. Phys.: Conf. Ser.* 28,57-61,(2006).
[10] Sako Y, Minoghchi S, Yanagida T. Nat. Cell. Biol., 2, 168-172 (2000).
[11] A. Hoffmann, A. Kane, D. Nettels, D. E. Hertzog, P. Baumgartel, J. Lengefeld, G. Reichardt, D. A. Horsley, R. Seckler, O. Bakajin, and B. Schuler, Proc. Natl. Acad. Sci. U.S.A. **104**, 105 (2007).
[12] A. Moglich, K. Joder, and T. Kiefhaber, Proc. Natl. Acad. Sci. U.S.A. **103**, 12394 (2006).
[13] X. Zhuang, L. E. Bartley, H. P. Babcock, R. Russell, T. Ha, D. Herschlag, and S. Chen, Science 288, 2048 (2000).
[14] Taekjip Ha, Alice Y. Ting, Joy Liang, W. Brett Caldwell, Ashok A. Deniz, Daniel S. Chemla, Peter G. Schultz, and Shimon Weiss, "Single-molecule fluorescence spectroscopy of enzyme conformational dynamics and cleavage mechanism", *Proc. Natl. Acad. Sci.* **96,** 893–898 (1999).
[15] Szabo Jr G.,Weaver J.L.,Pine P. S., "Cross-linking of CD4 in a TCR/CD3 juxtaposed inhibitory state: a sp-FRET study, " *Biophys J*, 68, 1170(1995).
[16] Guangcun Shan, Wei Huang, "Theoretical Study of Potential Profile of Single Polypeptides Revealed by Single-pair FRET Technique,"Acta Physico-Chimica Sinica, (submitted).
[17] Mergny JL. "Fluoresence energy transfer as a probe for tetraplex formation: the i-motif," *Biochemistry*, 33,1573(1999).
[18] S. Brasselet, E. J. G. Peterman, A. Miyawaki, W. E. Moerner, "Exploring a Dual GFP Construct at the Single Molecule Level",J. Phys. Chem. B,104, 3676–3682(2000).
[19] Schütz, G.J., Trabesinger, W., Schmidt, T., "Direct Observation of Ligand Colocalization on Individual Receptor Molecules," Biophysical Journal, 74, 2223-2226 (1998).
[20] Ambrose, W. P., Goodwin, P. M., Martin, "Alterations of Single Molecule Fluorescence Lifetimes in Near-Field Optical Microscopy," *Science* **265,** 364–366 (1994).
[21] X. Zhuang. "Single-molecule RNA Science." Annu. Rev. Biophys. Biomol. Struct. 34, 399-414 (2005).
[22] Hoffman A, Kane A, Nettels D. Hertzog D, Baumgartel P, Lengefeld J, Reichardt G, Horsley DA, Seckler R, Bakajin O, Schuler B, "Mapping protein collapse with single-molecule fluorescence and kinetic synchrotron radiation circular dichroism spectroscopy," *Proc. Natl. Acad. Sci.*, 104,105–110 (2007).
[23] Ishii, Y., Yoshida, T., Funatsu, T., Wazawa, T. & Yanagida, T. "Fluorescence resonance energy transfer between single fluorophores attached to a coiled-coil protein in aqueous solution," *Chem. Phys.* **247**, 163–173 (1999).
[24] Ashok A. Deniz, Maxime Dahan, Jocelyn R. Grunwell, Taekjip Ha, Ann E. Faulhaber, Daniel S. Chemla, Shimon Weiss, and Peter G. Schultz, "Single-pair fluorescence resonance energy transfer on freely diffusing molecules: Observation of Förster distance dependence and subpopulations," *Proc. Natl. Acad. Sci. USA* **96**, 3670–3675 (1999).
[25] A. A. Deniz, T. A. Laurence, G. S. Beligere, M. Dahan, A. B. Martin, D. S. Chemla, P. E. Dawson, P. G. Schultz, and S. Weiss. " Single-molecule protein folding: Diffusion fluorescence resonance energy transfer studies of the denaturation of chymotrypsin inhibitor 2," *Proc. Natl. Acad. Sci. USA* 97, 5179-5184 (2000).
[26] C. Eggeling, J. R. Fries, L. Brand, R. Gunter, and C. A. M. Seidel, "Identification of different emitting species in the red fluorescent protein DsRed by means of ensemble and single-molecule spectroscopy", *Proc. Natl. Acad. Sci. U.S.A.* 95, 1556 (1998).
[27] Guangcun Shan, Wei Huang, "Theoretical study on Single-pair fluorescence resonance energy transfer on freely diffusing molecules in microcavity", *Acta Optica Sinica*, 29(4),1049-1053 (2009).
[28] Allen, M. P.; Tildesley, D. J. "Computer Simulation of Liquids," Clarendon Press, Oxford (1987).
[29] R. Metzler, E. Barkai, and J. Klafter, "Anomalous Diffusion and Relaxation Close to Thermal Equilibrium: A Fractional Fokker-Planck Equation," *Phys. Rev. Lett.* **82**, 3563 (1999).
[30] Ralf Metzler, and Joseph Klafter, "Subdiffusive transport close to thermal equilibrium: From the Langevin equation to fractional diffusion", *Phys. Rev. E* 61, 6308 - 6311 (2000).

# Noninvasive monitoring of glucose concentration using differential absorption low-coherence interferometry based on rapid scanning optical delay line

**Yong Zhou, Nan Zeng, Yonghong He***
Laboratory of Optical Imaging and Sensing, Graduate School at Shenzhen, Tsinghua University, Shenzhen, 518055, P. R. China

E-mail address: heyh@sz.tsinghua.edu.cn

**Abstract** A non-invasive method of detecting glucose concentration using differential absorption low-coherence interferometry (DALCI) based on rapid scanning optical delay line is presented. Two light sources, one centered within (1625 nm) a glucose absorption band, while the other outside (1310 nm) the glucose absorption band, are used in the experiment. The low-coherence interferometry (LCI) is employed to obtain the signals back-reflecting from the iris which carries the messages of material concentration in anterior chamber. Using rapid scanning optical delay line (RSOD) as the reference arm, we can detect the signals in a very short time. Therefore the glucose concentration can be monitored in real-time, which is very important for the detection in vivo. In our experiments, the cornea and aqueous humor can be treated as nearly non-scattering substance. The difference in the absorption coefficient is much larger than the difference in the scattering coefficient, so the influence of scattering can be neglected. By subtracting the algorithmic low-coherence interference signals of the two wavelengths, the absorption coefficient can be calculated which is proportional to glucose concentration. To reduce the speckle noise, a 30 variation of signals were used before the final calculation of the glucose concentration. The improvements of our experiment are also discussed in the article. The method has a potential application for noninvasive detection of glucose concentration in vivo and in real-time.

## 1 introduction

Diabetes mellitus is a medical condition in which a person cannot maintain a normal blood glucose level effectively. There are two common types of the diabetes. Type-I diabetes is characterized by the loss of the pancreatic beta cells responsible for the production of insulin [1]. Therefore, Type-I diabetes need therapeutic insulin to compensate for this loss and is known as Insulin-Dependent Diabetes Mellitus. Type-II diabetes is characterized by insulin resistance which causes an individual to secrete more insulin, but often this is still insufficient to regulate the blood glucose levels properly [2]. It has been estimated that over 23.6 million individuals in the US have diabetes mellitus [3]. Frequent monitoring of blood glucose is crucial to facilitate the treatment of this disease.

The current glucose sensing techniques approved by FDA are invasive methods. These methods often require taking a sample of blood from the individual to measure their corresponding glucose levels. This type of testing not only causes pain to individuals, but coupled with possibility of infection at the test site. Therefore, noninvasive methods have been thought to be an attractive alternative. Several teams have made research in this field and several approaches have been under investigation for more than thirty years [4]. These approaches include the glucose absorption measurement with Raman spectroscopy [5], near infrared spectroscopy [6], polarimetric [7], optical coherence tomography (OCT) [8], etc. Because of the complexity of biological media, there are few repeatable and quantifiable results that have reported in vivo. As will be explained below, the method we used was different from these mentioned above.

We present a noninvasive method to monitor the glucose concentration in aqueous humor using a differential absorption low-coherence interferometry. Experiment shows that glucose concentration in the aqueous humor correlate well with it in blood and there is a minimal delay on the order of 5 minutes [9]. And as the aqueous humor is optically more accessible glucose-contain biological fluid, it is suggested that it could serve as a surrogate for blood for noninvasive analysis of glucose concentration. Schmitt et al. were probable the first one to introduce differential absorption technique in the field of OCT [10]. Then Pircher et al developed this technique to measure water concentration in human cornea [11]. Recently, our group utilized the DALCI technique to measure glucose concentration in aqueous humor, and the feasibility of it has been proved [12]. In this article, we measure glucose concentration in real-time, which is very important for the detection in vivo. Two light sources, one centers within (1625 nm) an absorption band of glucose, the other locates outside (1310 nm) the absorption band of glucose, are used to monitor the concentration of glucose. Two OCT systems with rapid scanning optical delay line are used to measure glucose concentration in real-time and eye model experiments are reported in this article to demonstrate the viability of our improvement.

## 2 Methods and Materials

Anterior chamber is located in the front part of eye between the iris and the cornea, with the thickness of 3.13 ± 0.50 mm [13]. The aqueous humor is filled within the anterior chamber and the total volume of it is about 0.25 ml [14].

After passing through a medium, the light is attenuated because of the scattering and absorption. The quantity relationship can be described by Beer's Law

$$I(\lambda, d) = I_0(\lambda)e^{-[\mu_a(\lambda)+\mu_s(\lambda)]d} \tag{2.1}$$

where I(λ, d) denotes the output light after passing through the medium, λ the wavelength of the light, d the depth of the medium, $I_0$ the incident light, $\mu_a$ the absorption coefficient of the medium, $\mu_s$ the scattering coefficient. The OCT signal can be described as:

$$S_{OCT}(\lambda, z) = K(\lambda)\sqrt{I_r(\lambda)I_s(\lambda, z)} \tag{2.2}$$

where K(λ) is a constant factor, which is decided by the system specifications, $I_r(\lambda)$ the light reflected from the reference arm, $I_s(\lambda, z)$ the light reflected from the sample at the depth of z. Subtracting the logarithm of the signals got with two wavelengths, we got

$$\ln(\frac{S(\lambda_1, z)}{S(\lambda_2, z)}) = \ln(\frac{I_0(\lambda_1)}{I_0(\lambda_2)}) - (\mu_a(\lambda_1) + \mu_s(\lambda_1) - \mu_a(\lambda_2) - \mu_s(\lambda_2))z \tag{2.3}$$

where $\lambda_1$ denotes the light of wavelength in 1625 nm, λ2 the light of wavelength in 1310 nm. And $I_0(\lambda_1)$ represents the light 1310 nm reflected from the back of the cornea, and $I(\lambda_1,z)$ the light 1310 nm reflected from the front of the iris, and $I_0(\lambda_2)$, $I(\lambda_2,z)$ have the same meaning but in the wavelength of 1625 nm. Because the absorption coefficient in 1625 nm is much larger than 1310 nm, so the absorption coefficient in 1310 nm can be neglected. In our experiment, as the cornea and the anterior chamber where the light propagate through are transparent and one light source is located in an absorption band of glucose, the scattering coefficients are much smaller than the absorption coefficient; therefore, the influence of the scattering can be neglected. In Eq. (2.2), the double pass configuration have been taken into account, i.e. we have set the medium thickness d=2z, where z denotes the depth of medium in our experiment.

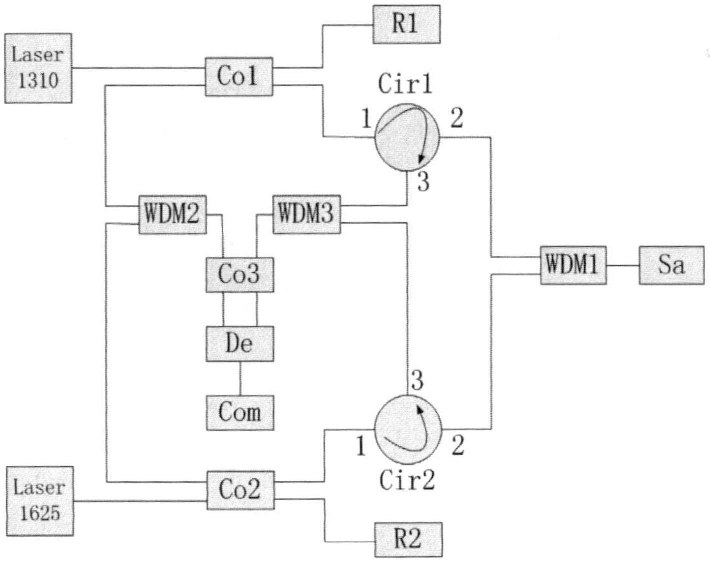


**Figure 1.** Col Co2 Co3, couplers. WDM1 WDM2 WDM3, wavelength division multiplexers. R1 R2, two sets of RSOD with two wavelengths, respectively. Cirl Cir2, circulators. De, detector. Com, computer.

Above all, we can rewrite the Eq. (2.3) as follow,

$$\ln(\frac{S(\lambda_1, z_c)S(\lambda_2, z_i)}{S(\lambda_1, z_i)S(\lambda_2, z_c)}) = \mu_a(\lambda_1)z_{ci} \tag{2.3}$$

where $S(\lambda_1, z_c)$, $S(\lambda_1, z_i)$ denote the interference signal from the back of the cornea and the front of the iris with wavelength 1625 nm, respectively, and the same meaning as $S(\lambda_2, z_c)$, $S(\lambda_2, z_i)$ but in wavelength 1310 nm, And $z_c$, $z_i$ the depth of the back of the cornea and the front of the iris, respectively, $z_{ci}$ the distance from the back of the cornea and the front of the iris. The absorption coefficient can be calculated by the signals detected and the depth $z_{ci}$, which can be measured directly by the images got from OCT. In Eq. (2.3), we denote $R=S(\lambda_1,z_c)\,S(\lambda_2,z_i)/\,S(\lambda_1,z_i)\,S(\lambda_2,z_c)$ just for short. If we know the absorption cross section σa of the pure substance, which can be measured by traditional spectroscopy, the concentration c of the absorbing material can be calculated via

$$C = \frac{\mu_a(\lambda_1)}{\sigma_a(\lambda_1)} \tag{2.4}$$

We prepared an eye model to simulate the human eye. The outside and inside diameters of this glass shell are 16 mm and 10 mm, respectively and the diameter of its cross-section is 20 mm. The bottom of it is adhered to a paper sheet with the center painted black to simulate the iris and the pupil. A set of glucose water solution with the concentrations of 0, 100, 200, 300, 400, 500 and 600 mg/dl were prepared.

## 3 Experiment setup

The experiment schematic is shown in Figure 1. The whole system is based on a fiber based Michelson interferometer. The 1310 nm wavelength light (FWHM bandwidth $\Delta\lambda$=50nm) (Denselight Inc., Singapore) first input into a coupler, which with an optical divide ratio 5:95. The 5 percents light from the output of couple was used as the input of the reference arm of this wavelength. The 95 percents light first passed through a circulator and then was coupled with the other wavelength light 1625 nm (FWHM bandwidth $\Delta\lambda$=28nm) (Denselight Inc., Singapore) into a wavelength division multiplexer (WDM). The output light of this WDM was the input of the sample arm. The backscattering light from the sample propagated back through the WDM and the circulator, and then was coupled with the light 1625 nm which was also the backscattering light from the sample into another WDM. The coupled light used as an input of the coupler before detected by the detector. The other input of this coupler is from another WDM with the coupled the light from the two reference arm of two wavelengths.

We used two rapid scanning optical delay lines (RSOD) as the reference arms of this two wavelengths system. Using this type, at least 400 A-scans can be detected in one minute. First, we coupled the light from fiber to free space using an objective, and then a oscillating mirror was used to achieve B-scanning of the sample. Finally, a normal achromatic lens was used to focus the light on the spot of sample we wanted to detect. Here, we averaged the B-scans to decrease the speckle noise.

## 4 Results and discussions

The results of the eye model experiments are shown in Figure 2. The correlation coefficient of the linear fitting is 0.99. The resolution $\delta C$ can be expressed as

$$\delta \mathrm{C} = \frac{\Delta C}{\Delta(\ln \mathrm{R})}\delta(\ln \mathrm{R}), \tag{4.1}$$

where ΔC is the concentration difference (100mg/dL in this experiment), Δ(lnR) the average difference of lnR, δ(lnR) the mean value of the standard deviation in each measurement.

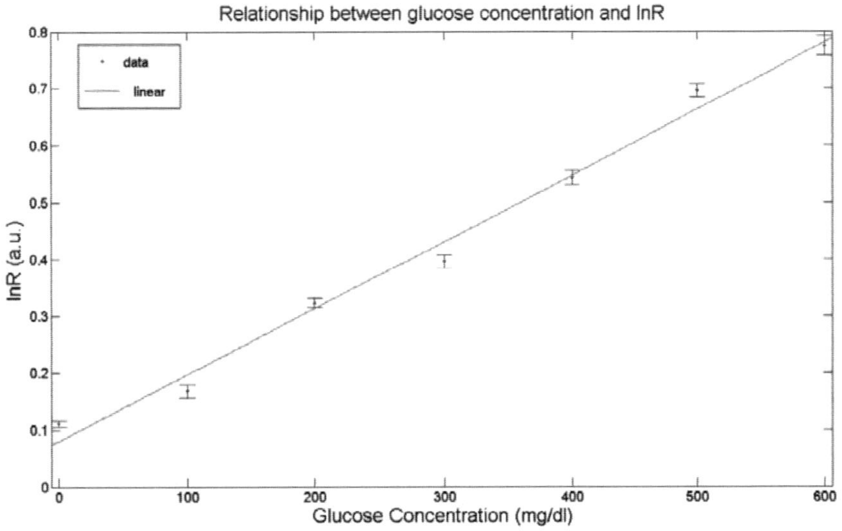


**Figure 2.** The relationship between lnR and concentration C.

After we measured a series of glucose water solution, 7 groups of data were recorded. Taking a weighted average from 30 times experiments over each group, lnR was plotted in Fig.5 (A). As is shown in Fig.3, the resolution calculated by equation (4.1) in this experiment is 25.4mg/dL. Although we can detect and measure the concentration in real time using RSOD, we will have more high frequent noise in the signals. Therefore, in the future, the thing we need to do is finding a balance between fast detection and low noise.

## 5 Conclusions

A developed differential absorption low-coherence interferometry to measure glucose concentration and in vitro experiments were carried out using an eye model. Results show the good linearity between the concentration C and lnR. Using RSOD, we can detect the concentration of glucose in real time. The results show the considerable promise of this method towards application in noninvasive glucose sensing.


## Acknowledgement

This research was supported by the 863 project, China (2006AA6Z402) and the projects 60608019 and 30770592, NSFC, China.


## Reference


[1] Fox, S. Human Physiology (7ed.). McGraw-Hill, New York, 620 (2002).

[2] Widmaier E., Raff H. & Strang K. Vander's Human Physiology: The Mechanisms of Body Function (10 ed.). McGraw-Hill, New York, 628. (2006).

[3] Centers for Disease Control and Prevention. National diabetes fact sheet: general information and national estimates on diabetes in the United States, 2007. Atlanta, GA: U.S. Department of Health and Human Services, Centers for Disease Control and Prevention. (2008).
[4] R. J. McNichols and G. L. Coté, "Optical glucose sensing in biological fluids: an overview," J. Biomed. Opt 5(1), 5–16 (2000).
[5] A.M.K. Enejder, T.G. Scecina, J. Oh, M. Hunter, W.-C. Shih, S. Sasic, G.L.Horowitz, and M. S. Feld, "Raman spectroscopy for noninvasive glucose measurements," J. Biomed. Opt 10 (2005)
[6] H. W. Siesler, Y. Ozaki, S. Kawata, and HM Heise, Near-Infrared Spectroscopy, WILEY-VCH, Weinheim, Germany(2002).
[7] Cameron, B D; Gorde, H W; Satheesan, B, et al. "The use of polarized laser light through the eye for noninvasive glucose monitoring." Diabetes Technol Ther 1(2), 135-43(1999)
[8] R.O. Esenaliev, K.V. Larin, I.V. Larina, and M. Motamedi, "Non-invasive monitoring of glucose concentration with optical coherence tomography," Opt. Let. 26(13). 992–994(2001).
[9] Cameron, B D; Baba, J S; Cote, G L, "Measurement of the glucose transport time delay between the blood and aqueous humor of the eye for the eventual development of a noninvasive glucose sensor." Diabetes Technol Ther 3(2), 201-207(2001)
[10] J. M. Schmitt, S. H. Xiang, K. M. Yung "Differential absorption imaging with optical coherence tomography," J. Opt. Soc. Am. A 15, 2288-2296 (1998)
[11] M. Pircher, E. Götzinger, R. Leitgeb, A. F. Fercher, and C. K. Hitzenberger, "Measurement and imaging of water concentration in human cornea with differential absorption optical coherence tomography," Opt. Express 11, 2190–2197 (2003).
[12] Yao Li; Yonghong He; Xiangsong Dai; Peng Li; Lian Duan; Yong Zhou; Nan Zeng, "Noninvasive detection of glucose concentration by differential absorption OCI with two wavelengths", Proceedings of SPIE, (2009).
[13] F. Y. Sun, Z.J. Sun, The Medical Handbook of Eye Disease, Dalian University of Technology Press, Da Lian (2007).
[14] Mortimer M. Civan "The eye's aqueous humor, from secretion to Glaucoma", Current Topics in Membranes, Volume 45 ,Academic Press, 1998
[15] Rollins, AM; Kulkarni, MD; Yazdanfar, S, et al. "In vivo video rate optical coherence tomography" Opt. Express 3(6), 219-229(1998)

# AUTHOR INDEX

# AUTHOR INDEX